Sustainable Winter Fodder

Sustainable Winter Fodder

Production, Challenges, and Prospects

Edited by
Imran Ul Haq
Siddra Ijaz

CRC Press
Taylor & Francis Group
Boca Raton London New York

CRC Press is an imprint of the
Taylor & Francis Group, an **informa** business

First edition published 2022
by CRC Press
6000 Broken Sound Parkway NW, Suite 300, Boca Raton, FL 33487–2742

and by CRC Press
2 Park Square, Milton Park, Abingdon, Oxon, OX14 4RN

CRC Press is an imprint of Taylor & Francis Group, LLC

ISBN: 978-0-367-51798-4 (hbk)
ISBN: 978-0-367-51836-3 (pbk)
ISBN: 978-1-003-05536-5 (ebk)

DOI: 10.1201/9781003055365

Typeset in Times
by Apex CoVantage, LLC

Contents

Preface

Livestock makes a significant contribution to the livelihoods and food security of the ever-increasing population of the world. It is a vital sub-sector and an integral part of Pakistan's agriculture, contributing to value addition and the national GDP. The livestock sector is expanding, with high rates of approximate standing populations of 1.43 billion large ruminants and 1.87 billion small ruminants worldwide. Milk is the largest commodity obtained from dairy livestock, while small and large ruminants are also a significant source of meat production in the world. Forage is the most crucial component in the diet of these animals. Production of sufficient quality forage is fundamental to the development of an efficient and productive livestock industry.

Plant diseases are the primary threat to crop plants' health and productivity. Among plant diseases, fungal diseases are the core danger. Pakistani farmers primarily use locally produced but substandard and non-certified seed with infestations and infections by fungal pathogens, resulting in reduced germination rate and field diseases. Upon interaction of pathogens with plants, the physiological and biochemical dynamics of plants are changed adversely. Forage crop breeding advancement has begun with the invention of the polycross, the development of synthetic varieties, progeny testing, tetraploidy, and molecular tools.

This book aims to collect current knowledge on winter fodders for academicians, breeders, researchers, and students. We have tried to make the book a scientific knowledge platform equipped with recent scientific information on winter fodders with practical knowledge for biologists dealing with breeding fodder crops' idiosyncrasies. This book focuses on winter fodders, production, bioecology, agronomy, breeding using conventional breeding programs, and biotechnology. It also provides valuable and comprehensive information on the ecological dynamics of winter fodders and their management. Microbial determinants, pests, and toxin-related challenges to winter fodders are also integral components of the book.

Dr. Imran Ul Haq

Dr. Siddra Ijaz

Editor Biographies

Dr. Imran Ul Haq, with a bright career in agriculture, plant pathology, and fungal molecular biology, has a post-doc from the University of California Davis, USA. He is currently serving as Associate Professor in the Department of Plant Pathology, University of Agriculture, Faisalabad, Pakistan. He has supervised more than 40 graduate students and established the Fungal Molecular Biology Laboratory Culture Collection (FMB-CC-UAF) and is an affiliated member of the World Federation for Culture Collections (WFCC). He has published more than 50 articles, four books, four laboratory manuals, and several book chapters. He has made colossal contributions in fungal taxonomy by reporting novel species of a fungal pathogen in plants. His research interests are fungal molecular taxonomy and nanotechnology integration with other control strategies for sustainable plant disease management.

Dr. Siddra Ijaz, with a vibrant career in agriculture and biotechnology, has a post-doc from the Plant Reproductive Biology Laboratory, University of California Davis, USA. She is currently serving as Assistant Professor in the Center of Agricultural Biochemistry and Biotechnology (CABB), University of Agriculture, Faisalabad, Pakistan. She has supervised more than 50 M.Phil and Ph.D. students. She has published more than 50 articles, five books, and several book chapters. Her research focus includes plant genome engineering using transgenic technologies, genome editing through CRISPR/Cas systems, nanobiotechnology, and exploration of genetic pathways in plant-fungus interactions.

Contributors

Ghulam Hassan Abbasi
Department of Soil Science, University College of Agriculture and Environmental Sciences, The Islamia University of Bahawalpur, Bahawalpur, Pakistan

M. Abdullah
Department of Animal Nutrition, University of Veterinary and Animal Sciences, Lahore

Muhammad Adil
Department of Soil Science, University College of Agriculture and Environmental Sciences, The Islamia University of Bahawalpur, Bahawalpur, Pakistan

Jamshaid Ahmad
University of Veterinary and Animal Sciences, Ravi Campus, Pattoki

Maqshoof Ahmad
Department of Soil Science, University College of Agriculture and Environmental Sciences, The Islamia University of Bahawalpur, Bahawalpur, Pakistan

Tanveer Ahmad
Department of Horticulture, MNS University of Agriculture, Multan, Pakistan

Usama Ahmad
Department of Plant Pathology, University of Agriculture Faisalabad, Pakistan

Zahoor Ahmad
Faculty of Sciences, University of the Central Punjab, Bahawalpur campus, Bahawalpur, Pakistan

Manzoom Akhtar
University Institute of Management Sciences, PirMehr Ali Shah Arid Agriculture University, Rawalpindi

Muhammad Waqar Akhtar
Department of Soil Science, University College of Agriculture and Environmental Sciences, The Islamia University of Bahawalpur, Bahawalpur, Pakistan

Sajjad Akhtar
Informatics College of Science, Arifwala, Pakistan

Asghar Ali
Institute of Agricultural Extension, Education and Rural Development, University of Agriculture Faisalabad, Pakistan

Muhammad Ali
Department of Soil Science, University College of Agriculture and Environmental Sciences, The Islamia University of Bahawalpur, Bahawalpur, Pakistan

Muqarab Ali
Department of Agronomy, Muhammad Nawaz Sharif University of Agriculture, Multan, Pakistan

Usman Ali
Department of Animal Nutrition, University of Veterinary and Animal Sciences, Lahore

Tahira Amjad
Department of Agronomy, University of Agriculture, Faisalabad, Pakistan

Muhammad Asaad Bashir
Department of Soil Science, University College of Agriculture and Environmental Sciences, The Islamia University of Bahawalpur, Bahawalpur, Pakistan

Umair Ashraf
Department of Botany, University of Education, Faisalabad, Pakistan

M. Talha Aslam
Department of Agronomy, University of Agriculture, Faisalabad, Pakistan

Asadullah Azam
Plant Protection Department, Agriculture Faculty, Kabul University

Shahan Aziz
Department of Agriculture & Agribusiness Management, University of Karachi, Pakistan

Rabia Tahir Bajwa
Department of Plant Pathology, Faculty of Agriculture and Environment, The Islamia University of Bahawalpur

Farhana Bibi
Department of Agronomy, University of Agriculture, Faisalabad, Pakistan

Bushra
Department of Soil Science, University College of Agriculture and Environmental Sciences, The Islamia University of Bahawalpur, Bahawalpur, Pakistan

Barbaros Cetinel
Bornova Plant Protection Research Institute, Izmir/Turkey

M. Umer Chattha
Department of Agronomy, University of Agriculture, Faisalabad, Pakistan

Anjum Faraz
Department of Plant Pathology, University of Agriculture Faisalabad, Pakistan

Zakia Habib
Centre of Agricultural Biochemistry and Biotechnology (CABB), University of Agriculture Faisalabad, Pakistan

Muhammad Shakeel Hanif
Dairy Technology, Fodder Research Institute Sargodha, Pakistan

Faiza Hassan
Faculty of Veterinary Sciences, University of Agriculture Faisalabad, Pakistan

M. Umair Hassan
Department of Agronomy, University of Agriculture, Faisalabad, Pakistan

Sajid Hussain
Department of Agronomy, University of Agriculture, Faisalabad, Pakistan

Misbah Ijaz
Faculty of Veterinary Sciences, University of Agriculture Faisalabad, Pakistan

Siddra Ijaz
Centre of Agricultural Biochemistry and Biotechnology (CABB), University of Agriculture Faisalabad, Pakistan

Momina Iqbal
Department of Agronomy, University of Agriculture, Faisalabad, Pakistan

Muhammad Irfan
Department of Soil Science, University College of Agriculture and Environmental Sciences, The Islamia University of Bahawalpur, Bahawalpur, Pakistan

Ali Hassan Khan
Ayub Agricultural Research Institute, (AARI), Pakistan

Imran Khan
Department of Agronomy, University of Agriculture, Faisalabad, Pakistan

Nabeeha Aslam Khan
Department of Plant Pathology, University of Agriculture Faisalabad, Pakistan

Muhammad Zunair Latif
Department of Plant Pathology, University of Agriculture Faisalabad, Pakistan

Faisal Mahmood
Department of Environmental Sciences and Engineering, Government College University Faisalabad, Pakistan

Zaffar Malik
Department of Soil Science, University College of Agriculture and Environmental Sciences, The Islamia University of Bahawalpur, Bahawalpur, Pakistan

Asad Manzoor
Faculty of Veterinary Sciences, University of Agriculture Faisalabad, Pakistan

Muhammad Tahir Mohyuddin
Faculty of Veterinary Sciences, University of Agriculture Faisalabad, Pakistan

Zia Mohyuddin
Faculty of Veterinary Sciences, University of Agriculture Faisalabad, Pakistan

Samara Mukhtar
Centre of Agricultural Biochemistry and Biotechnology (CABB), University of Agriculture Faisalabad, Pakistan

Tariq Mukhtar
Department of Plant Pathology, PMAS-Arid Agriculture University, Rawalpindi, Pakistan

Bukhtawer Nasir
Centre of Agricultural Biochemistry and Biotechnology (CABB), University of Agriculture Faisalabad, Pakistan

Saima Naveed
Department of Animal Nutrition, University of Veterinary and Animal Sciences, Lahore

Muhammad Naveed-ul-Haque
Department of Animal Nutrition, University of Veterinary and Animal Sciences, Lahore

Mohsin Nawaz
Key Laboratory of Genetics and Germplasm Innovation of Tropical Special Forest Trees and Ornamental Plants, Ministry of Education, College of Forestry and College of Tropical Crops, Hainan University, Haikou, P. R. China

Muhammad Raheel
Department of Plant Pathology, Faculty of Agriculture and Environment, The Islamia University of Bahawalpur

Ifrah Rashid
Department of Plant Pathology, Muhammad Nawaz Sharif University of Agriculture, Multan

Hafiz Younis Raza
Department of Plant Pathology, Faculty of Agriculture and Environment, The Islamia University of Bahawalpur

Freeha Sabir
Department of Soil Science, University College of Agriculture and Environmental Sciences, The Islamia University of Bahawalpur, Bahawalpur, Pakistan

Sakhidad Saleem
Plant Protection Department, Agriculture Faculty, Kabul University

Syeda Rashida Saleem
Department of Plant Pathology, Faculty of Agriculture and Environment, The Islamia University of Bahawalpur

Muhammad Kaleem Sarwar
Department of Plant Pathology, University of Agriculture Faisalabad, Pakistan

Qaiser Shakeel
Department of Plant Pathology, Faculty of Agriculture and Environment, The Islamia University of Bahawalpur

Rana Muhammad Sabir Tariq
Department of Agriculture & Agribusiness Management, University of Karachi, Pakistan

Imran Ul Haq
Department of Plant Pathology, University of Agriculture Faisalabad, Pakistan

Sajid Usman
Department of Soil and Environmental Sciences, University of Agriculture, Faisalabad, Pakistan

M. Shahid Ibni Zamir
Department of Agronomy, University of Agriculture, Faisalabad, Pakistan

Section I

Production Technologies of Winter Fodders

1 Conventional Breeding Strategies and Programs for Winter Fodders

Siddra Ijaz, Ali Hassan Khan, and Muhammad Kaleem Sarwar

CONTENTS

1.1 IMPORTANCE OF FODDER CROP BREEDING

Achievement of food security, improved nutrition, and sustainable agriculture are important goals and challenges in today's world. The projected upsurge in world population by 2050 requires a significant increase in food production. Developing countries, as compared to the developed world, are expected to have a high population

DOI: 10.1201/9781003055365-1

growth rate. Hitherto inadequate food supply in the developing world will face increasing threats of hunger, malnutrition, and famine for its inhabitants. It will further widen the nutritional and economic gap between the developing and developed world. Hence, improved food production and strengthening the world's food supply to ensure food security without compromising quality are the world's biggest challenges. Additionally, these challenges must be achieved in a sustainable and benign way without impacting the environment (Kingston-Smith et al., 2013). Ruminants play a vital role in the human food supply, as they convert raw plant material into protein products for human consumption. Livestock is an integral component of human food supply and requires integrating conventional and new technologies for sustainable animal production systems that do not compromise human food supply. A high population growth rate is expected to increase global demand for livestock products such as milk and meat. A better understanding of livestock's nutritional requirements has recently improved forage crops' productivity, quality, and soil for upgrading overall animal performance and system sustainability. Forage crops have been improved through conventional breeding strategies that have successfully improved forage consumption by livestock and subsequent livestock production. There is a need to identify livestock's critical nutritional requirements for enhanced productivity and breeding strategies to manipulate forage crops to satisfy livestock needs. This can be achieved by the collaboration of plant and animal sciences. In this chapter, conventional breeding strategies for forage crop improvement will be discussed (Poehlman, 1987; Moose and Mumm, 2008).

Increased forage production is accomplished by improving crop environment and heredity through successful cultural and breeding strategies. Heredity improvement of forage crop varieties results in better productivity and nutrition. Successful breeding results in improved varieties that are vigorous in growth and highly productive under unfavorable environmental conditions, as these efficiently use resources; their altered structure reduces harvesting losses and is more tolerant to drought, temperature, diseases, and pests. However, it is worth mentioning here that the benefits of superior variety are linked with good cultural practices, without which the production potential of variety would be wasted. The combination of high-yielding varieties and successful production practices has significantly overall improved the hectare yield of field and forage crops globally. So it can be concluded that variety improvement and good cultural practices reinforce each other (Rognli, 2013).

Since the beginning of agriculture, farmers intentionally or unintentionally have been changing and improving domesticated plants' genetic make-up through selection. Early farmers found that specific crop plants could be cross-pollinated and desirable features could be transferred and combined from parent material to its offspring for better yields. In earlier days, farmers used to select superior plants with vigorous growth and desirable seed production characteristics to plant the next year. These primitive efforts at artificial selection, though unconsciously, have considerably contributed to changing and improving the heredity of domesticated crop pants through evolutionary development (Barth, 2012; Kingston-Smith et al., 2013).

1.2 BASICS OF PLANT BREEDING STRATEGIES

Development in genetics allowed plant breeders to manipulate specific genes to improve crops' genetic make-up for developing improved varieties. This has significantly improved crop plants' productivity and quality for satisfying human and animal food and feed requirements. Selection in plant breeding to improve specific desirable features has dramatically improved field and forage crop species compared to their wild relatives. Corn cobs grown thousands of years ago on the American continent were as small as the little finger, whereas today, as a result of selective breeding, they are considerably longer, with hundreds of corn varieties being grown globally. Hybridization was being practiced long before Mendel's discovery, but its significance as an essential breeding technique was not fully understood. Hence, Mendel's work helped humans understand the inheritance mechanism and how heredity could be manipulated to develop high-yielding varieties. Conventional plant breeding strategies are based on simple elements.

1. First, morphological traits and the plant's crucial responses for adaptation, productivity, and quality of produce must be recognized.
2. The genetic potential of appropriate plant species' strains is evaluated for traits of interest through specific techniques.
3. Search for gene source with desired features to be utilized in forage crop breeding.
4. Combining desired traits into a new variety through appropriate means of genetic transfer.

Farmers and breeders are familiar with the variety that breeders develop, evaluate, and provide in their seeds to farmers. Genera of plant species are divided into species, which are further subdivided into various agricultural cultivars (cultivated varieties). Hence, a variety contains identical plants and can be distinguished from other cultivars of the same species based on the growth pattern, seed color, plant type, sowing and harvesting time, performance, and other structural features. Strains or lines with the same or different genotypes, when proved superior during scientific experiments, are named and provided commercially to farmers as a distinct variety. A variety must be recognized and differentiated from other cultivars of the plant species, and its prominent features must be reproduced in the progeny (Julier et al., 2018).

Winter fodder crops such as berseem, oat, barley, and mustard are predominately self-pollinated forage crops, except alfalfa. The genetic make-up of predominately self-pollinated crops differs considerably from cross-pollinated crops, as these will have a nearly homozygous plant population. Although 100% homozygosity is challenging to attain, most plants are identical, and slight differences are negligible.

For successful breeding of self-pollinated winter fodder crops, chronological order of introduction (locally grown source varieties), selection (isolation of high-yielding genetic mixtures), and hybridization (a combination of desired features) are followed.

1.2.1 Selection of the Desired Genotype

During selection, single genotypes or a mixture of genotypes from segregated populations are sorted out. Environmental variations could lead to unsuccessful selection, so genetic variability among plant populations must be identified. For a successful selection procedure, two strategies are adopted, namely mass selection and pure-line selection.

1.2.2 Mass Selection and Pure-Line Selection for Successful Breeding

Phenotypically similar plants are chosen and harvested, and seed is composited without testing its reproducibility in progeny. The resulting variety/cultivar will be phenotypically identical but will vary in quantitative inherited features such as yield and produce quality. The purpose of mass selection is to improve the mixed population and propagate phenotypically high-performing plants. However, during mass selection, heterozygous plants could be selected, which will require repeating this whole process in the progeny. Besides, phenotypically superior plants could be the result of favorable environmental conditions rather than heredity. Mass selection is an efficient strategy, as it can purify a mixed population without the need for further testing, plus heterozygous plants may be effective by serving as a buffer against environmental variations (Nakaya and Isobe, 2012).

As compared to mass selection in the pure-line selection strategy, progeny of a homozygous plant are acquired through self-pollination. The resulting plant population is comparatively uniform as, theoretically, all plants have identical genotypes. Many successful cultivars are developed by identifying high-performing plants and increasing their progeny and selection after hybridization of segregating progeny. In pure-line selection, progeny testing is a must, as it evaluates the chosen plant. After confirming the selected plant's superior genotype through strenuous testing, the line or strain is declared a new variety. It will remain pure and serve a purpose depending on the crop type, genotype stability, natural cross-pollination, and vigilance and expertise with which it is developed. Mixing of seed from other varieties during the harvesting or seed-cleaning process and natural cross-pollination with varieties of adjacent fields, breeding nurseries, or the same field that occur as mixture and mutation can comprise the pure-line selection methodology. Despite this, pure-line selection only identifies the already present superior genotype rather than altering or creating hereditary characteristics. Conclusively, it can be stated that pure-line selection is effective in a genetically diverse population, and once a line with superior characteristics is selected, further variations in its progeny are primarily due to environmental variations rather than hereditary factors, and further selection will be ineffective (Poehlman, 2013; Capstaff and Miller, 2018).

In earlier days, farmers desired to plant uniform varieties in large areas to have consistent appearance and performance. However, recent studies have suggested that highly uniform varieties are undesirable and unnecessary, as these will be highly susceptible to environmental variations and potential insect pests and disease organisms. Large areas with the same variety without genetic variations are more likely to develop pandemic situations and extreme losses. Genetically variable plant populations can ensure stable yields under changing climate and potential insect pests or diseases. Nonetheless, genetic variability in variety makes it less attractive, more difficult to identify, and more low yielding than a superior genotype.

1.2.3 Hybridization

In hybridization, two varieties are crossed, and from their progeny, plants with desirable characteristics are selected for further testing and verification of superior features. It involves breeding of self-fertilizing parent crop plants, and from their advanced generation, plants with visible and quantitative traits such as yield and quality of produce are considered. Predominantly self-pollinated fodder crop parents with desirable traits are artificially crossed, which is comparatively unchallenging in case of large floral parts but tedious for forage grasses with small flowers. During this technique, plants are emasculated before pollens are shed, and viable pollens from the male parent are transferred on the emasculated plant's stigma. Emasculation could be achieved using male-sterility genes or cytoplasmic sterility, as in barley fodder. The first generation will be heterozygous, which is reduced by successive self-pollination (Abberton, 2007).

Genetic diversity among self-pollinated crops could be attained by mixing seeds from other varieties. This strategy is based on the assumption that the blend of varieties will perform better under environmental variations, as diverse genotypes will negatively buffer environmental changes, ensuring stable yield over various locations for years. While making a variety blend, it should be ensured that these have uniform harvesting time and produce quality (García-Espinosa and Robinson, 2009).

1.3 SYNTHETIC VARIETY FOR CROSS-POLLINATED FORAGE CROPS

Synthetic variety implies a plant generation artificially produced through cross-pollination of a seed mixture containing various lines, clones, hybrids, and inbreeds. The seed mixture is maintained, and the synthetic variety is reconditioned after regular intervals. In developing synthetic variety, regular reconstitution is a must. It is commonly used for developing forage crop varieties because of less seed production to perform progeny tests, self-incompatibility, and difficulty in controlling cross-pollination. During synthetic variety development, heterosis is exploited, making it popular among forage breeders, as traditional breeding strategies limit the use of heterosis in forage crops. Essential characteristics of synthetic variety development include those constituted from clones in forage species that are reproducible units through random inter-pollination, selection based on the progeny performance test, and reconstitution at regular intervals (Tamaki et al., 2007).

1.4 THE METHODOLOGY OF SYNTHETIC VARIETY DEVELOPMENT IN FODDER CROPS INVOLVES THE FOLLOWING DISTINCT STEPS

1.4.1 Source Population with a Broad Range of Genetic Variability

Different sources are exploited to ensure an extensive range of genetic variability producing strong and high-performing progeny. The selected population must be easy to maintain and perform best when tested for desirable characters' combining abilities. This could be obtained from established pastures and cultivars or a population that has undergone careful and repetitive selections, introductions, or related sources.

1.4.2 Establishment of Phenotypically Selected Clonal Line Nursery

Superior plants, vegetatively propagated from the parent plant, are selected based on phenotype with desirable traits. This clonal line consisting of almost 25 plants undergoes screening to identify vigorous clones with desirable features depending on the breeding program's objective. Progeny testing on the selected superior clones of the polycross nursery will be performed.

1.4.3 Development of Polycross for Progeny Test

Polycross is a distinguished group of clonal lines replicated so that each clone will be pollinated from all other selected clones. Their harvested seed is bulked separately to perform a progeny test.

1.4.4 Polycross Progeny Test to Evaluate Desirable Traits

The harvested seed from each clone is evaluated for quantitative characteristics such as yield and quality by planting during progeny testing. After the progeny test, superior clones are selected to include in synthetics.

1.4.5 Syn Generations

Selected clones are vegetatively propagated and transplanted in the field for cross-pollination. This will foster gene recombination among selected synthetic clone components and is named Syn 0 generation. Harvested seeds from the Syn 0 generation are multiplied in isolation to produce the Syn 1 generation whose harvested seed in isolation will be multiplied, if necessary, to generate Syn 2 generation. The Syn 1 and Syn 2 generations are comparable to the F1 and F2 generations of conventional breeding. Hence, beyond Syn 1, the synthetic variety successively loses its vigor and will require reconstitution through original clones when needed. As compared to hybrids (maize), farmers can produce their seed of synthetic varieties without the necessity of buying new each year.

1.5 BREEDING STRATEGIES FOR BARLEY

Barley is globally distributed and has seven chromosomes. It is a self-fertilized crop plant and can accomplish artificial hybridization. More than 27 male-sterile genes identified in barley eliminate the need for emasculation during artificial crossing or backcrossing. Introduction of a male-sterile gene in the parent allows further progeny to continue segregation of sterile plants, promoting outcrossing. With these strategies used for the self-pollinated crop, breeding is used to develop new varieties and barley hybrids. Barley is used mainly as feed for livestock. Its adaptive disease-free varieties with high yield and heavy grains are satisfactory to fulfill livestock feed requirements. Barley varieties with thinner hulls and awns, strong structure to avoid lodging, no grain sprouting during humid environments before harvesting, and resistance to insect pests and disease, especially head blights, would improve the quality and feeding value of its forage to successful livestock production.

In the last few decades, the breeding of fodder-crop varieties has considerably improved in the developed world, but this is not the case in developing countries, where the seed of many fodder species is still sold without identity and approval from authorities. Compared to cereals and other field crops, forage crops have not been cultivated and domesticated for thousands of years and still are in the domestication process. Some fodder species are highly susceptible to severe diseases, such as root and crown rot in berseem and bacterial wilt in lucerne. This needs special attention from forage breeders to develop resistant forage varieties with high yield and nutritional content (Roy et al., 2019).

1.6 FORAGE CROP CHARACTERISTICS TO BE CONSIDERED BEFORE BREEDING

Even though breeding fodder crops involve the same genetic principles and strategies described previously, fodder breeding poses new breeder challenges. Fodder crops have special features such as pollination diversity, fertilization and seed-setting irregularities, and perennial nature.

- Forages with cross-pollination are challenging to propagate and maintain because of high heterozygosity, and their primarily small flowers make the process of artificial hybridization tedious and challenging to achieve.
- Forage species commonly have self-incompatibility that limits the extent of self-pollination, which is a crucial step in finalizing the new hybrid or synthetic variety.
- Fodder species, primarily grasses, reproduce through apomixes that cause complexities in crossing and gene recombination or produce low-quality seeds with less viability that results in low crop stands. Moreover, the polyploid nature of major fodder crops increases genetic complexity.
- Less availability of suitable land for developing and propagating new strains. In nursery conditions, plants are well spaced compared to actual thick conditions in the field, which may make performance evaluation results misleading. This situation is further aggravated, as in-field fodder crop species are mostly seeded in a combination of other species.
- Different forage production and livestock systems influence individual strains' performance, and due to their perennial nature, it is difficult to determine the persistency and productivity of newly developed strains.
- Many fodder crop species, scarcity of resources, maintenance of germplasm, and challenges, as mentioned earlier, divide the focus of forage breeders, as they have to work on various species simultaneously.

Clover plants have high variability in compatibility, ranging from absolute incompatibility to high self-fertility. According to their self-fertility or sterility, alfalfa differs in seed setting after inbreeding. An increase in temperature could increase self-fertility. Self-pollination in forage legumes is ensured by bagging the floral parts, tripping, or hand manipulation of bagged floral parts. Conversely, forage species, for example, red clover, with high sterility do not need emasculation. It is recommended that forage crops be evaluated for self-sterility or fertility genes before breeding. Forage crops are cross-pollinated using hand pollinators such as camel's hairbrush or through insects such as honeybees.

1.7 INBREEDING AND NATURAL SELECTION IN FODDER CROPS

Inbreeding in cross-pollinated forage crops due to incompatibility mechanisms results in reduced size, vigor, fertility, and seed production. However, in most forage crops, pseudo-incompatibility or self-fertile alleles do not entirely prohibit inbreeding. Forage species such as alfalfa or clovers reduce in plant vigor and seed set following inbreeding. However, lines/strains that can seed set after successive progeny are present but rare.

Forage species domesticated in a particular area over an extended period are genotypically well adapted to local climate conditions and can survive better when compared to other genotypes. This phenomenon of evolution and acclimatization in native or domesticated habitats is termed natural selection. Various biotypes that may evolve from a wild fodder species by natural selection well adapted to local environmental and soil conditions are called ecotypes. These ecotypes represent the management and cultural practices under which the fodder crop was maintained over the years. Natural selection has been observed in various grasses, and forage crops such as alfalfa in Arizona become less winter hardy after two generations of seed production compared to when grown in Montana winters. The same was observed in disease susceptibility; strains of red clover successfully cultivated in the midwest of the United States are susceptible to southern anthracnose disease and cannot be grown successfully in southern parts of the United States due to high disease prevalence. Heterozygous and heterogeneous cross-pollinated forage crops could easily be differentiated into distinguished ecotypes due to their higher gene recombination. Short-lived forage species, red clover, compared to long-lived forage species, have a higher ability to adapt to climatic variations due to the high frequency of genetic recombination. This is contrary to the fact that most forage species are perennials, as described earlier in this chapter, because these were established long before anthropogenic-based fodder breeding. These well-adopted ecotypes could be exploited for breeding high-yielding fodder varieties. A new variety must be evaluated in its recommended area with management system and technologies adopted by local farmers before its commercial availability (Argillier et al., 2000; Singh et al., 2019).

1.8 THE BREEDING PROGRAM FOR FORAGE SPECIES

1.8.1 Source Nursery and Domestication of Germplasm of Fodder Species

The first step in developing a breeding program for fodder crops is an assemblage of appropriate germplasm, which serves as a source nursery. It can be acquired as seed lots from a population or individual plant or a plant clone vegetatively propagated from a living plant part. Clones are preferred, as the forage breeder will be unaffected by pollination control issues. Local ecotypes, indigenous areas of species, developed varieties, hybrids, and recurrent selection procedures serve as the principal source of germplasm. Germplasm populations are selected and developed with repetitive selection for desired traits such as yield; plant vigor; and drought, heat, insect pest, and disease resistance. Superior fodder crop plants from available germplasm are placed in the source nursery. Fodder species are utilized as forage in their native habitats even without approved and improved varieties.

1.8.2 Mass Selection and Single-Plant Selection for Fodder Breeding

Both of these processes are discussed previously in this chapter. Mass selection does not involve progeny evaluation through tests. It is usually applied in the case of domesticated forage species. It is helpful when breeding plants against a specific disease or pest, such as red clover adapted in southern parts of the United States. When artificially inoculated with anthracnose pathogen, most of the plants died, but a few surviving ones were selected, and harvested seed from all strains was bulked to plant next generation. This process was repeated for several generations until a resistant variety was obtained. Contrarily, single-plant selection is considered hazardous because of its low genetic variability, as discussed earlier in detail. Propagation through self seed is undesirable due to rapid inbreeding, but it is not always the case; a variety of alfalfa was developed from a single plant for its resistance against bacterial wilt disease. Its seed was multiplied in successive generations, which contained enough genetic variability to maintain a superior yield in subsequent progeny. Hence, apomixes proved efficient in developing high-yielding forage crop species (Kaiser et al., 2020).

1.8.3 Synthetic and Composite Varieties of Fodder Crops

Synthetic varieties are based on the principle that strains with multiple plants are superior to single-plant strains as they have fewer losses in vigor upon inbreeding. Conversely, multiple plants means the possibility of less uniformity and more significant dangers of divergence from original germplasm characteristics in subsequent progeny. Desirable and superior traits from various clones are exploited and combined without compromising the plant's vigor following inbreeding through reduced generations of seed increasing. Additionally, clones are maintained so that synthetics could be reconstituted upon loss of vigor or any other desirable trait. The process of seed development and distribution has already been discussed. A red clover synthetic variety, Kenstar in the United States, has been developed for more remarkable persistence. More than 1000 plants were selected from the field, and after testing these polycross progeny, only ten clones with the desirable trait (persistence) were selected to constitute the synthetic. One hundred seventy-five clones of alfalfa were inter-pollinated through honey bees to develop a variety. The clones in both cases are vegetatively maintained to ensure reconstitution if and when needed. Composite varieties of fodder crops are developed through blending seeds of different strains.

1.8.4 Hybridization for Fodder Crops

Desirable traits of two or more than two varieties are combined using intraspecific or interspecific hybridization mechanisms. During this process, plants of F2 and F3 generations with desirable traits are selected and intercrossed until plants with the desired features are obtained. This process was discussed earlier (Abberton, 2007).

1.9 CONCLUSION

For practical forage crop breeding, different plant species' reproductive systems must be understood to devise appropriate breeding strategies and techniques. Using appropriate conventional forage breeding strategies, fodder crops could be improved for higher forage and seed yields, seedling vigor, persistent crop stands, climatic and disease factors that reduce forage crop stands, resistance to diseases and insect pests, forage quality, nutritive value, and digestibility for livestock.

REFERENCES

Abberton MT (2007) Interspecific hybridization in the genus *Trifolium*. *Plant Breeding* 126: 337–342.

Argillier O, Méchin V and Barrière Y (2000) Inbred line evaluation and breeding for digestibility-related traits in forage maize. *Crop Science* 40: 1596–1600.

Barth S (2012) Breeding strategies for forage and grass improvement. *Annals of Botany* 110: 1261–1262.

Capstaff NM and Miller A (2018) Improving the yield and nutritional quality of forage crops. *Frontiers in Plant Science* 9: 535.

García-Espinosa R and Robinson RA (2009) Conventional plant breeding for higher yields and pest resistance. *Biotechnology-Volume VIII: Fundamentals in Biotechnology* 8: 242.

Julier B, Annicchiarico P, Barre P, et al. (2018) Alfalfa breeding for intercropping and grazing tolerance. *2. World alfalfa congress* 216 p.

Kaiser N, Douches D, Dhingra A, et al. (2020) The role of conventional plant breeding in ensuring safe levels of naturally occurring toxins in food crops. *Trends in Food Science Technology* 100: 51–66.

Kingston-Smith A, Marshall A and Moorby J (2013) Breeding for genetic improvement of forage plants in relation to increasing animal production with reduced environmental footprint. *Animal* 7: 79–88.

Moose SP and Mumm RH (2008) Molecular plant breeding as the foundation for 21st century crop improvement. *Plant Physiology* 147: 969–977.

Nakaya A and Isobe SN (2012) Will genomic selection be a practical method for plant breeding? *Annals of Botany* 110: 1303–1316.

Poehlman JM (1987) Breeding Forage Crops. In: *Breeding Field Crops*. Dordrecht: Springer, pp. 625–679.

Poehlman JM (2013) *Breeding Field Crops*. Dordrecht: Springer Science & Business Media.

Rognli O (2013) Breeding for improved winter survival in forage grasses. In: Imai R, et al. eds) Plant and Microbe Adaptations to Cold in a Changing World. New York: Springer, New York.

Roy AK, Malaviya D and Kaushal P (2019) Breeding strategies to improve fodder legumes with special emphasis on clover and medics. *International Grassland Congress Proceedings. 23rd International Grassland Congress.*

Singh T, Radhakrishna A, Nayak DS, et al. (2019) Genetic improvement of berseem (*Trifolium alexandrinum*) in India: Current status and prospects. *International Journal of Current Microbiology Applied Sciences* 8: 3028–3036.

Tamaki H, Yoshizawa A, Fujii H, et al. (2007) Modified synthetic varieties: A breeding method for forage crops to exploit specific combining ability. *Plant Breeding* 126: 95–100.

2 Bioecology and Agronomy of Winter Fodder Crops

Imran Khan, Mohsin Nawaz, M. Shahid Ibni Zamir, Farhana Bibi, Faisal Mahmood, Muqarab Ali, M. Umer Chattha, Momina Iqbal, Tahira Amjad, Sajid Hussain, Muhammad Shakeel Hanif, M. Umair Hassan and Sajid Usman

CONTENTS

DOI: 10.1201/9781003055365-2

2.1 INTRODUCTION

Forage crops and grasslands have historically played a key role in agricultural production worldwide. Pastures and grasslands, also known as 'mines of nutrients,' serve livestock animals directly for feed material in the food chain system.

It is imperative to adopt such agronomic practices and the latest techniques to increase fodder production for livestock (Muir et al., 2014). The terms fodder and forage have similar meanings concerning practical purposes because both fodder and forage are used as feeding material for livestock. Fodder includes plant species that are used to feed livestock, either raw (green and cut fodder) or after processing (such as silage) (Vogl et al., 2016). Fodder is available for animals in excess quantities only during monsoon season, and during other seasons, the animals depend on crop residues. Fodder crops are grown as temporary and permanent crops (Barbieri et al., 2017). Temporary fodder crops consist of three major fodder groups, grasses, legume crops, and silage, with different nutritive values (Wicke et al., 2020). Grasses (including cereals) are a rich source of crude fibers and protein (Puhakka et al., 2016). Moreover, legumes also have ample protein and mineral content (Amossé et al., 2013; Sumberg, 2002), while root crops are rich in starch and sugars but poor in fiber (Leidi et al., 2018; Moorthy and Padmaja, 2002). For intensive fodder production on a sustainable basis, frequent irrigation plays an essential role in achieving this task (JAT and Kaushik, 2018). Soil with a field capacity of 75% is considered ideal for the production of fodder crops (Ling et al., 2020). Fodder yield reduction (5–30%) is reported in lucerne, berseem, sorghum, cluster bean, and cowpea due to limited soil moisture availability (ICAR, 1992).

On the other hand, forage consists of plant species that animals directly or indirectly consume. Forage is popular in rainfed, irrigated, range grass ecosystems and can also be grown in stress conditions. Good-quality forage crops are high in digestible nutrients and proteins and low in lignin and fiber. It has been observed that elephants prefer certain forage plants over others (Santra et al., 2008). Forage consists of grasses and legumes grazed directly to pasture animals (Dumont et al., 2007). Several factors can influence the productivity and quality of forage crops (Eskandari et al., 2009), while a forage crop's poor genetic resources are considered a significant constraint in its production (Bennett and Cocks, 1999).

Moreover, marginal lands such as saline soils, hillsides, terrace bunds, and risers are used for forage production (V. Singh, 1995). Heavy rainfall on terraces, risers, and hills causes runoff and, ultimately, nutrient loss and forage production (Otero et al., 2011). Forage grown on risers and terraces provides soil cover and limits nutrient loss by runoff during heavy rainfall conditions.

Agronomic Management Practices of Forage Crops Are:

- Establishment of pasture lands
- Legumes and grasses mixture for better nutrient management
- Management of fertilizer
- Weed and bush control

- Control of insect/pest attack
- Cutting and grazing management

2.2 CHARACTERISTICS OF WINTER FODDERS

The characteristics of winter fodders (Figure 2.1 and Table 2.1) are given in the following:

- Quick and short growing period
- High palatability and ready digestibility
- Capacity to supply fodder throughout the season
- Contain significant amount of all essential minerals and vitamins
- Ability to grow under different climatic regions and in all types of soils
- Ability to resist harsh environmental conditions

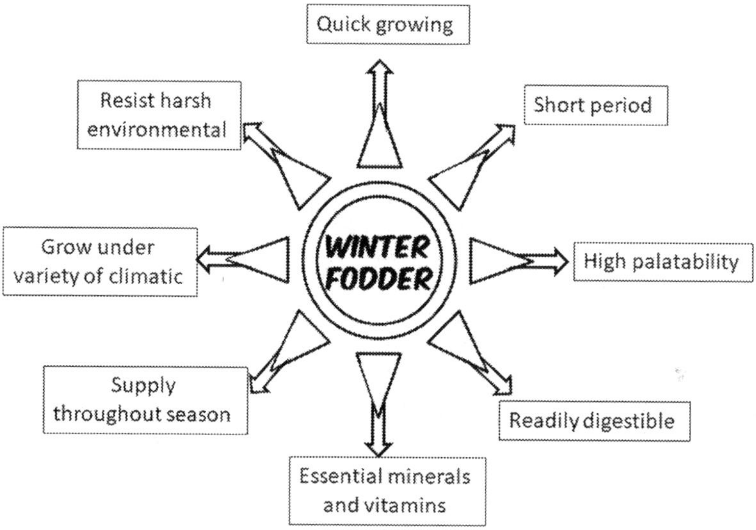

FIGURE 2.1 A pictorial description of the quality of winter fodders.

TABLE 2.1
Classification of Winter Fodder

Crops	Scientific Name
	Non-leguminous fodder
Oat	*Avena sativa* L.
	Leguminous fodder
Berseem Sweet clover, Indian clover	*Trifolium alexandrium* L.
Alfalfa	*Melilotus parviflora* L.
Persian clover	*Medicago sativa* L.
Mustard	*Trifolium resupinatum* L.
	Brassica nigra
	Brassica rapa var dichotoma (Toria)

2.3 BIOECOLOGY AND AGRONOMY OF WINTER FODDER CROPS

2.3.1 Oat (*Avena sativa* L.)

Avena sativa L. belongs to the Poaceae family. It is a winter non-leguminous cereal fodder crop enriched with proteins, vitamin B1, phosphorus, and iron. Compared to berseem, oat can provide fodder for a limited period due to its single-cut capacity. The leaves and grains of oat fodder are rich in carbohydrates and carotenes. Moreover, oats provide nutritious fodder in May's dry month and are relished by all animals (Anwar et al., 2010; Chaudhary and Mukhtar, 1985). Mostly *Avena sativa* L. is cultivated for grain purposes and fodder and forage, hay, silage, straw for bedding, and chaff. Oat grains are still primary used for livestock feed, accounting for an average of around 74% of the world's total usage (Welch, 1995).

2.3.1.1 Bioecology

Initially, it was grown as a minor crop in the Mediterranean zone and then moved to other countries like Russia, the United States, Canada, Poland, Australia, Germany, and France (Sánchez-Martín et al., 2014). In Canada, an area of 1.8 million ha is under *Avena sativa* L. cultivation for human feed and fodder (Reynolds and Suttie, 2004). *Avena sativa* is also cultivated as animal feed used in hay silage, grain, and pasture.

After discovering the continent, the Spanish introduced oat in Latin America as a grain for milling industries, as fodder for animal grazing, and as horse feed (Edwards et al., 2014; Reynolds and Suttie, 2004). Now oat is cultivated in different regions such as South America: the temperate area of Uruguay and Argentina, the temperate area of Chile, tropical and sub-tropical areas of Brazil. The Maghreb is the region of Africa bordering the Mediterranean Sea, except Egypt. It consists of Morocco, Tunisia, Algeria, and Libya. In the Maghreb, *Avena sativa* L. is cultivated on half a million hectares and is known as the most important forage crop of the rainfed area. European varieties were introduced in those regions during the colonial period (Chaouki et al., 2004).

In Pakistan, it was introduced as a fodder crop during the early British era, and 400 cultivars were obtained from different countries like Canada, Australia, Europe, the United States, and New Zealand for Pakistan's oat improvement program (Dost, 1970). Now in Pakistan, it is known as the most important winter fodder. It was introduced in the Himalayas in the early 1970s (Maikhuri et al., 1996). Oat seed was given at the time of the flood in the Paro valley in 1969 when there was nothing left for animal feed.

China is known as the center of oat diversity. Oat (*Avena sativa* L.) is famous as an important fodder crop in 18 regions of China. The predominant form of Chinese oats has been the hexaploid naked type of *Avena sativa*, and wild species of oats can be found in some areas compared to other oat-producing countries (Yang and Sun, 1989). In Japan, its cultivation started in 1872, and the United States provided seed. An area of about 8000 ha of Japan was under *Avena sativa* L. cultivation in 2000, and now its cultivation area has shifted to Kyushu.

In New Zealand and Australia, it is famous as a multi-purpose crop such as for feed, fodder, straw, and grain. Export of its hay has increased for a few years, of which 75% is derived from Australia (Stubbs, 2000).

2.3.1.2 Agronomic Practices

2.3.1.2.1 Climate and Soil

Avena sativa L. is grown in cold and moist conditions. A 20–30°C temperature is suitable for oat fodder, and it can also grow well in areas receiving 80–100 mm rainfall. *A. sativa* L. can be grown in all soil types except alkaline, waterlogged, and sandy soils. Soils with pH 5–6.6 and clay loam soil are considered the best soil for oat growth.

2.3.1.2.2 Land Preparation and Fertilizer

One plowing with a furrow turning plough and planking is required to prepare an adequate and pulverized seedbed. *A. sativa* L. should not grow on residual fertility; instead, an additional fertilizer dose should be applied to get good germination and more fodder yield. The oat crop requires 75 kg of nitrogen and 20 kg phosphorous per hectare. A half dosage of nitrogen (37.5 kg) and phosphorus full dosage (20 kg) should be applied at sowing and reaming nitrogen at first irrigation.

2.3.1.2.3 Seed Rate and Sowing

If an oat crop is grown for fodder purposes, then a 75 kg/ha seed rate is recommended, and for seed crops, 30 kg/ha is recommended. In the case of late sowing, 5% more seed should be added than the recommended seed rate at the optimum time to get more fodder. Drill sowing with rows 20 cm apart is suitable for sowing. Moreover, the broadcast sowing method is also used. However, in this method, it is challenging to spread the seed in the whole field uniformly. The best planting time starts from the end of September, and then it can be grown up to December.

2.3.1.2.4 Inter-Culture and Irrigation

Inter-culture is mostly not practiced with this fodder crop. If weeding is done in the crop's early growth stages, weeds cannot interfere with the crop's growth. If plants become successful in making a good crop stand, then weeding is not necessary. One to two hoeings can be given as inter-culture operations. The first irrigation for oat is needed after 20 days of sowing, while subsequent irrigations should be applied when required. Three to four irrigations are considered sufficient for proper growth. In rains, skip irrigation to prevent water-logging, which causes severe damage to the fodder yield.

2.3.1.2.5 Insect Pests and Disease

Control of insect pests and diseases with the best techniques and skills also comes under crop management and agronomy practices (Moyer et al., 2003). Stem and whitefly mostly attack the crop. If severe insect/pest attack is observed in the seed crop, apply Parathion and Endosulfan @ 1200–1500 ml per hectare. Helminthosporium, leaf spot, loose smut, covered smut, and root rot (caused by root parasites) are the major oat crop diseases. To overcome this problem, pretreatment of seed with Carboxin or Benomyl at the rate of 2.5 g per kg of seed is appropriate as a preventive measure; otherwise, rouge out the affected plants from the field. Additionally, resistant cultivars should be selected to get rid of this problem.

2.3.1.2.6 Harvesting

Oat is single-cut fodder; therefore, cutting should be done before heads appear.

2.3.2 Berseem (*Trifolium alexandrinum*)

Berseem (*Trifolium alexandrinum*) is considered a vital winter fodder crop (Tufail et al., 2019), also known as Egyptian clover. It belongs to the family Leguminosae. Being very palatable and readily digestible, it is known as the king of fodder and fed as green fodder to animals (Tyagi et al., 2018). It is an imperative source of phosphorus, calcium, and vitamin A. Additionally, it also improves soil porosity, bulk density, and soil structure (Nanda et al., 2003).

2.3.2.1 Bioecology

Berseem originated from Egypt and is also known as Egyptian clover (Singh et al., 2019). In the 19th century, it was introduced into India, and it was cultivated at Manipur Khas Farm of Sindh, which is now part of Pakistan. In the 20th century, it was introduced in Pakistan, South Africa, the United States, and Australia. It is mainly grown in some southern European areas, and now there are efforts to grow it in northern areas, such as Brittany. In Egypt, the cultivated area was 1.3 million ha in 2007, and in India, it was 1.9 million ha (Heuzé et al., 2016).

2.3.2.2 Climate and Soil

In the subtropics, it is valued as a winter crop and grows well in mild winter. It completes its vegetative growth at 15–20°C and requires a high temperature at the reproductive stage. It prefers fertile, loamy soils to clay soils with mildly acidic to slightly alkaline pH (6.5–8) but can grow in a wide range of soils. It is also moderately tolerant of salinity (Hackney et al., 2007; Hannaway and Larson, 2004).

2.3.2.3 Agronomic Practices

2.3.2.3.1 Land Preparation

Proper land leveling helps to prevent waterlogging because berseem does not grow well in these conditions. Three plowings with planking are needed to pulverize and soften the soil. The seed size is tiny; therefore, it requires a fine seedbed.

2.3.2.3.2 Sowing Time and Method

Timely sowing provides green fodder during the early season, which is a scarce period of fodder. Late sowing decreased the germination percentage, root length, and protein content of the seed (Patel, 2004). The best sowing time is the last week of September to the first week of October to get more cuttings. In the case of late sowing, late-sown varieties can be sown up to mid-November. After land preparation, divide the field into equal-sized plots and flood the field with almost 4–5 cm water depth. The seed should be broadcast in standing water, and it should be mixed with an equal quantity of soil to ensure uniform distribution. Inoculation of berseem seed with bacterial culture is needed to ensure the best growth in areas where berseem is grown for the first time. Pests and disease-resistant varieties should be used to get more fodder production.

2.3.2.3.3 Seed Rate

Usually, a 20–25 kg/ha seed rate is used for berseem fodder (Singh, 1995). To get higher fodder yield at first cuttings, farmers mix 1.5 kg mustard seed with berseem seed in the sub-continent.

2.3.2.3.4 Fertilizer and Irrigation

This crop fixes atmospheric nitrogen; therefore, it requires less nitrogen. To meet its fertilizer requirement, 50 kg di-ammonium phosphate (DAP) should be applied at the time of sowing. First irrigation should apply after three to ten days of sowing, and subsequent irrigations should apply with 15–20 day intervals during winter and 10–12 day intervals at the time of the last cuttings or summer. Depth of irrigation depends upon the type of soil and the number of cuttings.

2.3.2.3.5 Insect Pest and Disease

Stem rot: Stem rot, also known as white mold, is a fungus infection caused by *Sclerotinia sclerotiorum*. Initial symptoms appear at the pod development stage. First, leaves start wilting and change to a gray-green color. In the end, curling and dying take place.

Root rot: *Fusarium* spp. causes this disease. Light brown spots appear on roots below the soil surface; in a severe attack, the plants die quickly. This problem generally appears after the second cut of berseem. First, to control this disease, remove the affected plants; otherwise, in case of a severe attack, Carboxin should be used.

Armyworm: This pest damages fodder economically at the end of winter and the summer season. Suitable pesticides should be used to avoid damage, particularly during severe attacks. The most care should be taken with seed crops that are usually heavily affected by this pest.

Harvesting: The first cutting is ready after two months and subsequent cuttings with an interval of one month. The total number of cuttings depends upon the variety. However, a maximum of six cuttings can be obtained throughout the season.

2.3.3 Indian Clover (*Melilotus parviflora*)

Indian clover (*Melilotus parviflora*) belongs to the Leguminosae family (Hayder et al., 2012). It contains 16.57% protein and 60% total digestible nutrients. It is a drought-tolerant legume and requires less water than alfalfa and berseem. The stem height is equal to or more than 1 m, and its plant is annual or biennial. It can be fed as green or dry fodder.

2.3.3.1 Bioecology

Initially, it was grown like a weed, but now it is cultivated as a fodder crop because it is useful for animals. Sweet clover originated in Bukhara (Blázovics, 2016). It is considered a crop of Asia and Europe's temperate zone, from where it moved toward Argentina and southern Australia. Its cultivation was started in the United States, giving the highest green fodder return of all clovers.

2.3.3.2 Climatic and Soil Conditions

It can grow well under various types of climatic and soil conditions. It is considered a temperate zone crop and therefore cultivated in cooler, rainfed areas. Moreover, it has vast adaptability and can be grown in moist and dry areas. Well-drained, loamy types of soils are considered appropriate for its growth. On the other hand, it is also a fodder of hilly areas, plains and sandy and average soil types.

2.3.3.3 Agronomic Practices

2.3.3.3.1 Land Preparation and Fertilizer

To prepare a seedbed, one or two plowings followed by planking are enough for proper germination. Basal application of 50 kg DAP (containing 46% phosphorus and 9% nitrogen) is enough to fulfill its nutrient requirements.

2.3.3.3.2 Seed and Sowing of Clover

Clover is generally sown by the broadcasting method. To get maximum yield, the required seed rate is 20–25 kg/ha. September to November is the optimum sowing time to get more fodder. The best time for the planting of clover as a relay crop in standing cotton is the second to the third week of September and as a catch crop in November after harvesting maize and sugarcane.

2.3.3.3.3 Irrigation and Inter-Culture Practices

Three to four irrigations are enough because it requires less water in comparison with berseem and alfalfa. The first irrigation should be given after ten days of sowing and the next ones at intervals of 20–25 days. No intercultural practices are needed.

2.3.3.3.4 Pests and Diseases

Lucerne weevil and aphids damage crops, and to control this problem, spraying the seed crop with Carbaryl at a rate of ½ kg per acre in 300 liters of water is recommended.

2.3.3.3.5 Harvesting

The cutting starts after 2.5–3 months after sowing when the crop reaches the flowering stage.

2.3.4 Alfalfa (*Medicago sativa* L.)

Alfalfa (*Medicago sativa* L.) belongs to the family Leguminosae. It is known as the fodder queen (Ahmad et al., 2016; Halagić, 2005). It can provide green fodder throughout the year, especially during scarce periods in May–June and October–November. It also improves soil fertility by adding nitrogen and reducing salinity levels in irrigated land. It is an extremely drought- and heat-resistant crop. *Medicago sativa* is a multi-cut, perennial legume crop.

2.3.4.1 Bioecology

Alfalfa is one of the oldest fodder crops that originated in Iran and Afghanistan, and from there, it was brought to Arabs and known as alfalfa (Quiros and Bauchan, 1988).

Then it was introduced in Europe through Turkey and known as lucerne. It reached England in 1650 AD, and after that, it spread to America and New Zealand. In the next 100 years, it was a popular fodder crop in many countries, including Indo-Pak.

Alfalfa is a temperate crop and grows best in low-rainfall and high-sunshine areas. It can also grow well under diverse climatic conditions, including extremely hot and cold conditions. It requires well-drained, loamy soil with neutral pH for its proper germination and subsequent growth stages (Table 2.2).

2.3.4.2 Agronomic practices

2.3.4.2.1 Seedbed Preparation and Sowing Time

Alfalfa (*Medicago sativa* L.) can be sown 15th October and 15th November for an excellent fodder return. Late sowing affects germination, plant growth, and fodder yield. Lucerne has a small size seed and therefore requires a fine bed for good germination. Two to three plowings followed by two harrowings are required to prepare suitable tilth. Planking should be used to attain a powdery condition of the field.

2.3.4.2.2 Seed Rate and Sowing Method

The recommended seed rate is 15 kg/ha for line sowing at a 30 cm distance, but it requires 6–12 kg/ha in inter-cropping. For bold seeded varieties, the seed rate must be increased to 20 kg/ha (Relwani, 1979). Line sowing facilitates hoeing and cultural operations in subsequent years as compared to a broadcast-sown crop. In line sowing, the seed is sown in shallow furrows, and in the case of the broadcast method, the seed is sown after preparing a fine seedbed.

2.3.4.2.3 Fertilizer Management

Comparatively, less nitrogen is required because it is a leguminous crop and fixes nitrogen. It is a perennial fodder, so it is better to apply 20–30 tons of farmyard manure (FYM) per hectare about one month before sowing. Otherwise, 70 kg phosphorus and 50 kg nitrogen per hectare are sufficient to meet its nutrient needs. Sulfur has a vital role in legume crops, especially when the sink has sulfur-containing amino acids in large quantities (Hazra and Sinha, 1996).

TABLE 2.2
Botanical Difference between Alfalfa and Berseem Plant

Botany	Alfalfa	Berseem
Color	Dark green	Parrot green
Leaves	Round and serrated	Long and plain
Petiole	Three leaflets, central leaf petiole longer than the other two	Three leaflets
Internode	Solid, thick, and hard	Hollow, visible node, and internode
Stem	Stem reddish from base	Green
Flowers Purple	White	
Seed	Yellow	Yellow

2.3.4.2.4 *Irrigation Management*

A fodder-purpose crop requires first irrigation after three weeks of sowing, and subsequent irrigations should be given at 15–20 days. The number and depth of irrigations depend upon the soil, crop, and weather conditions. After first irrigation, half a urea bag per acre helps in the early maturation of the crop. For seed-purpose crops, the first irrigation during the last cut of fodder and the second one at the time of flowering produce more seed yield than only one irrigation during the last cut (Malik et al., 2004).

2.3.4.2.5 *Hoeing*

After the seasonal rains of summer, hoeing is needed to keep the field free from weeds. After hoeing, add some fertilization; this practice helps to get more fodder. Dodder (*Cuscuta reflexa* Roxb.) weed, which is parasitic, thrives on alfalfa. Frequent cuttings of grassy weeds can control it during the monsoon season to properly grow alfalfa (Langer, 1990).

2.3.4.2.6 *Pests and Diseases*

Three pests, armyworm, gram caterpillar, and aphids, often attack alfalfa crops. If insect pests are observed in crops near the flowering stage, Carbaryl should be dusted; if observed at the flowering stage, 675 ml Nogas in 675 L of water per hectare should be sprayed. The crop should not be harvested immediately after spraying; wait for two weeks to harvest the crop after pesticide application.

2.3.4.2.7 *Time of Harvesting*

The first cutting of lucerne can be taken after three months of sowing and subsequent cuttings after 45 days. The best time to cut the fodder is when 25% of flowers have emerged. On average, six cuts and 40–50 tons of fodder per hectare can be obtained per year.

2.3.4.2.8 *Seed Production*

For seed production, stop the cutting after February and leave the crop for seed. To ensure pollination, provide ten beehives per hectare, which ensure excellent seed set. At the blooming stage, no irrigation should be needed because irrigation at this stage may adversely affect the seed yield.

2.3.5 Persian Clover (*Trifolium resupinatum* L.)

Persian clover (*Trifolium resupinatum* L.) is a Leguminosae family crop known as the winter season's leguminous fodder. *T. resupinatum* L. is multi-cut fodder and is also known as the best soil builder. It is mostly used as a fodder crop because it provides highly palatable and nutritive fodder for animals (Horadagoda et al., 2009).

2.3.5.1 Bioecology

The Persian clover plant originated in South Asia. It was cultivated as an ornamental plant for many years in Europe (Rendeková and Mičieta, 2017); then its cultivation started as a fodder crop (Table 2.3) in Australia, West Germany, South America, and

TABLE 2.3
Nutrient Composition in Persian Clover (Chaudhry, A.R., 1987)

Nutrients	Percentage
Protein	16.3
Crude fibers	30.8
Calcium	1.62
Phosphorus	0.32
Trace minerals	10.3

Central Asia. Now it is known as the best fodder crop in Egypt and Iran. Persian clover is an annual and semi-erect branched legume. It has hollow stems branching from the lower part, and its leaves are trifoliate with oval-oblong leaflets and pinkish flowers.

2.3.5.2 Climate and Soil

Persian clover (*T. resupinatum* L.) can germinate in various soil and climatic conditions, but well-drained and loamy soils are considered best. Light to heavy loam soils with irrigation facilities is best. Cool and moist weather is suitable for better growth.

2.3.5.3 Agronomic Practices

2.3.5.3.1 Seedbed Preparation and Fertilizer

Three to four cultivations are required, each followed by planking to get a fine seedbed for good germination. This crop does not require fertilizers, but two and a 25 kg of DAP per hectare are recommended to increase fodder yield.

2.3.5.3.2 Sowing Time and Seed Rate

In seed broadcasting in standing water, a 12–16 kg/ha seed rate is enough to get maximum fodder yield. The sowing time suggested is the entire month of October to mid-November. However, for seed crops, according to Mukherjee et al. (2000), early sowing in October gives a higher seed yield (2.21 q/ha) than late sowing in November (1.69 q/ha) and December (1.56 q/ha). This crop is usually sown by the broadcasting method, and due to this sowing method, there is no need to have intercultural practices.

2.3.5.3.3 Irrigation

The first irrigation should be applied after 8–10 days of planting, and the remaining irrigations should apply with 15–20 days. In all, 10 to 12 irrigations are necessary from sowing to harvesting (Relwani, 1979).

2.3.5.3.4 Weeds and Pest Control

Chicory/kasni (*Cichorium intybus*) can reduce fodder yield. Therefore, it should rouge out from the field earlier or at planting time. Commonly, weeding is required in this crop when it is sown in the month of September. Avoid the cutting of fodder for two weeks after using weedicides. This fodder crop mainly remains free from pest and disease attacks.

2.3.5.3.5 Harvesting

Persian clover (*T. resupinatum* L.) gives four to five cuttings during the period of December–May. This crop should be harvested before it lodges (Relwani, 1979). Crop seed ripens in April–May, and yield varies from 750–1000 kg/ha.

2.3.6 Mustard (*Brassica nigra*)

Mustard belongs to the genus *Brassica* of the family Cruciferae. *Brassica* is the dominant winter season oilseed and includes different mustards [*Brassica nigra, Brassica rapa* var *dichotoma (Toria), Brassica napus (Gobhi sarson)*].

Among them, *Brassica nigra* is considered the third most important oilseed in the world. It belongs to the family Brassicaceae and is also known as black mustard. It is a broadleaf, cruciferous, and cool seasoned crop. The first true leaves are covered with hairs on both upper and lower leaf surfaces (Raza et al., 2020).

2.3.6.1 Bioecology

It is native to temperate regions of Europe, different parts of Asia, and tropical regions of North Africa. Black mustard (*Brassica nigra*) is also cultivated in Chile, the United States, and Argentina (Zajac et al., 2016).

2.3.6.2 Climatic and Soil Conditions

Mustard is a tropical and temperate-zone crop, requires relatively cool temperature for satisfactory growth and yield, and is grown well in areas receiving 350–550 mm rainfall. Sarson is preferred in low-rainfall areas, whereas these are grown in medium and high rainfall areas. In the tropics, brassicas grow well at a higher elevation around 1000 m and prefer a moderate temperature between 25°C and 28°C with an optimum around 20°C.

The optimum growth of *B. nigra* was observed at 25°C; *B. nigra* is considered sensitive to climatic conditions compared with other crops grown in the same season. Light to heavy soils with 7–8 pH are suitable for mustard growth and development.

2.3.6.3 Agronomic Practices

2.3.6.3.1 Land Preparation and Fertilizer

Brassicas can be grown on a wide range of soils such as alluvial, red loams, and block soils. Brassicas do not stand waterlogging, especially at seedling and vegetative stages. They prefer neutral soil reactions but can withstand alkalinity. However, they cannot come up well on rice fallows due to high residual soil moisture. A high water table is undesirable.

Mustard required a fine seedbed for its good germination. Two to three plowings, each with planking followed by harrowing, are required to prepare a fine seedbed. Fallow lands should be periodically plowed and harrowed for weed control and soil moisture conservation. For *Brassica*, recommended N, P, K is 100, 30, and 15 kg/hectare using urea, single super phosphate (SSP), and murate of potash (MOP).

2.3.6.3.2 *Sowing Time and Seed Rate*

The optimum sowing time for germination of the mustard crop is September to October. However, toria can be sown in the first fortnight of September to October. Row-to-row distance should be 30 cm in rapeseed or mustard. For mustard, keep row-to-row distance to 45 cm and plant-to-plant distance 10 cm. A seed drill is preferred for proper sowing. For the first monoculture of mustard or rapeseed in the field, the recommended seed rate is 3.75 kg/hectare.

For a sole crop, the optimum seed rate is 5 kg ha^{-1} for all brassicas. Profuse branching compensates for missing plants or low plant population. Seed rate, mixed or intercrop, depends on the desired proportion of brassica crop relative to the component crop. It usually ranges from 1 to 2 kg ha^{-1}.

2.3.6.3.3 *Weeds, Their Control, and Irrigation*

To avoid soil-born pests and diseases, seed should be treated with Thiram @ 3 g per kg of seed. Two to three weedings and two hoeings at an interval of two weeks are enough to control weeds. A pre-emergence spray of Isoproturon @ 400 g per 200 L should be applied after two days of sowing and post-emergence spray after 25–30 days of sowing. To get good growth and fodder yield, the crop should be given three irrigations. Added organic manure at the time of sowing in the soil will help conserve the soil's moisture.

2.3.6.3.4 *Diseases and Their Control*

2.3.6.3.4.1 White Rust White pustules are observed on leaves, stems, and flowers. If infestation is noted in the field, then spray with Metalaxyl 8% + Mancozeb 64% @ 2 mg/L of water or with copper oxychloride @ 25 mg/L of water.

2.3.6.3.4.2 Downy Mildew The lower surface of leaves gives a whitish appearance, and then they change to a yellow coloration. Destroy debris (remains of the previous crop) from field before sowing. Spray the crop four times with Indofil M-45 @ 400 gm in 150 L with an interval of 15 days.

2.3.6.3.4.3 Blight The symptoms of this disease appear on branches, stems, pods, and leaflets. They are mostly withering of stem and pods seen in a severe infestation. Using resistant varieties and in-field spray of Indofil M-45 or Captan @ 260 gm/100 L of water per acre are recommended.

2.3.6.3.5 *Pests and Their Control*

2.3.6.3.5.1 Hairy Caterpillar Young larva feeds on leaves and can also damage the leaves completely. Spray Malathion 5% dust @ 15 kg per acre or Dichlorvos @ 200 ml/acre in 100–125 liters of water.

2.3.6.3.5.2 Painted Bug It influences the crop at the germination stage as well as at maturity and leads to drying. It is noted that irrigation application after three to four weeks of sowing reduces the pest population. Malathion @ 400 ml/acre helps to control this in the field.

TABLE 2.4
Harvesting Time of Winter Fodders for Fodder Purposes

Fodder Crops	Harvesting Time/Harvesting Stage
Oat	Early December to end of May
Berseem	Mid December to mid-May
Lucerne	Perennial crop available round the year
Sweet clover	Flowering (three months after sowing)
Persian clover	December–May
Mustard	60–70% of pods turn yellow

2.3.6.3.5.3 Aphid Aphids are sucking pests that suck the sap, and plants become weak, pale, and stunted without bearing pods. Therefore, to control aphids, avoid more nitrogen application, and sow the crop in a timely manner. In-field, to overcome this problem, spray pesticide Oxydemeton-methyl @ 250 ml or Chlorpyrifos @ 200 ml in 100–125 L water per acre.

2.3.6.3.5.4 Harvesting Harvesting should be done when pods turn yellow in color and the seed becomes hard. For fodder purposes, it can be harvested along with berseem if grown in mixed cropping (Table 2.4).

REFERENCES

Ahmad J, Iqbal A, Ayub M and Akhtar J (2016) Forage yield potential and quality attributes of alfalfa (*Medicago sativa* L.) under various agro-management techniques. *Journal of Animal and Plant Science* 26(2): 465–474.

Amossé C, Jeuffroy M-H, Celette F and David C (2013) Relay-intercropped forage legumes help to control weeds in organic grain production. *European Journal of Agronomy* 49: 158–167.

Anwar A, Ansar M, Nadeem M, Ahmad G, Khan S and Hussain (2010) Performance of non-traditional winter legumes with oats for forage yield under rainfed conditions. *Journal of Agricultural Research* 48(2): 171–179.

Barbieri P, Pellerin S and Nesme T (2017) Comparing crop rotations between organic and conventional farming. *Scientific reports* 7(1): 1–10.

Bennett S and Cocks PS (1999) *Genetic resources of Mediterranean pasture and forage legumes.*

Blázovics A (2016) *Avicenna és a modern farmakognózia (Avicenna and the modern pharmacognosy).*

Chaouki AF, Chakroun M, Allagui MB and Sbeita A (2004) Fodder oats in the Maghreb. *Fodder oats: a world overview.* Rome: Food and Agriculture Organization of the United Nations, pp. 53–69.

Chaudhary MH and Mukhtar MA (1985) Performance of three new high fodder yielding varieties of oats. *Pakistan Journal of Agricultural Research* 6(3): 218–222.

Chaudhry AR (1987) *Position Paper on Fodders* (Monograph). Sargodha: Fodder Research Institute.

Dost M (1970) *The Introduction and Use of Oat (Avena sativa) Cultivars in Pakistan.* Gilgit, Pakistan: Aga Khan Rural Support Program (AKRSP).

Dumont B, Garel J-P, Ginane C, Decuq F, Farruggia A, Pradel P, Rigolot C and Petit M (2007) Effect of cattle grazing a species-rich mountain pasture under different stocking rates on the dynamics of diet selection and sward structure. *Animal* 1(7): 1042–1052.

Edwards GR, De Ruiter JM, Dalley DE, Pinxterhuis JB, Cameron KC, Bryant RH, Di HJ, Malcolm BJ and Chapman DF (2014) Urinary nitrogen concentration of cows grazing fodder beet, kale and kale-oat forage systems in winter. *Proceedings of the 5th Australasian Dairy Science Symposium* 144–147.

Eskandari H, Ghanbari A and Javanmard A (2009) Intercropping of cereals and legumes for forage production. *Notulae Scientia Biologicae* 1(1): 7–13.

Hackney B, Dear B and Crocker G (2007) Berseem clover. Primefact 388. *New South Wales Department of Primary Industries*. http://www.Dpi.Nsw.Gov.Au/Primefacts.

Halagić S (2005) Lucerna (*Medicago sativa* L.) kraljica krmnih kultura. *Glasnik Zaštite Bilja* 28(5): 10–15.

Hannaway DB and Larson C (2004) *Berseem clover (Trifolium alexandrinum L.)*. Oregon State University, Species Selection Information System.

Hayder B, Agarwala BK and Kaddou IK (2012) New records of aphids of the subfamily Aphidinae (Homoptera: Aphididae) infested herbaceous plants and shrubs for Iraqi aphid fauna. *Advances in Bioresearch* 3(4): 66–75.

Hazra CR and Sinha NC (1996) *Forage Seed Production*. New Delhi: A technological development published by South Asian Publishers Pvt. Ltd, pp. 118–139.

Heuzé V, Tran G, Boudon A, Bastianelli D and Lebas F (2016) *Berseem (Trifolium alexandrinum)*. Feedipedia (cirad.fr)

Horadagoda A, Fulkerson WJ, Nandra KS and Barchia IM (2009) Grazing preferences by dairy cows for 14 forage species. *Animal Production Science* 49(7): 586–594.

ICAR (1992) *Handbook of Agriculture*. New Delhi: Published by ICAR.

Jat H and Kaushik MK (2018) Quality and economic of fodder oat (*Avena sativa* L.) as influnced by irrigation and nitrogen under Southern Rajasthan. *Forage Resarch* 44(1): 28–31.

Langer RHM (1990) *Forage crop production. Principles and practices: BN Chatterjee and PK Das, 1989. Price Rs. 120*. Oxford and IBH, New Delhi: Elsevier. ISBN 81-204-0398-3, 484 pp.

Leidi EO, Altamirano AM, Mercado G, Rodriguez JP, Ramos A, Alandia G, Sørensen M and Jacobsen S-E (2018) Andean roots and tubers crops as sources of functional foods. *Journal of Functional Foods* 51: 86–93.

Ling Q, Zhao X, Wu P, Gao X and Sun W (2020) Effect of the fodder species canola (*Brassica napus* L.) and daylily (*Hemerocallis fulva* L.) on soil physical properties and soil water content in a rainfed orchard on the semiarid Loess Plateau, China. *Plant and Soil* 453(1): 209–228.

Maikhuri RK, Rao KS and Saxena KG (1996) Traditional crop diversity for sustainable development of Central Himalayan agroecosystems. *The International Journal of Sustainable Development & World Ecology* 3(3): 8–31.

Malik JS, Singh J and Dhankar RS (2004) Effect of cutting management, irrigation and phosphorus levels on seed production of lucerne. *Forage Research* 30: 104–105.

Moorthy SN and Padmaja G (2002) Starch content of cassava tubers. *Journal of Root Crops* 28(1): 30–37.

Moyer JL, Fritz JO and Higgins JJ (2003) Relationships among forage yield and quality factors of hay-type sorghums. *Crop Management* 2(1): 1–9.

Muir JP, Pitman WD, Dubeux Jr JC and Foster JL (2014) The future of warm-season, tropical and subtropical forage legumes in sustainable pastures and rangelands. *African Journal of Range & Forage Science* 31(3): 187–198.

Mukherjee AK, Mandal SR and Patra BC (2000) Effect of dates of sowing, cutting management and levels of phosphate application on seed yield of shaftal. *Environment and Ecology* 18(2): 506–508.

Nanda V, Sarkar BC, Sharma HK and Bawa AS (2003) Physico-chemical properties and estimation of mineral content in honey produced from different plants in Northern India. *Journal of Food Composition and Analysis* 16(5): 613–619.

Otero JD, Figueroa A, Muñoz FA and Peña MR (2011) Loss of soil and nutrients by surface runoff in two agro-ecosystems within an Andean paramo area. *Ecological Engineering* 37(12): 2035–2043.

Patel JR (2004) Effect of sowing date and seed rate on late sown berseem. *Environment and Ecology* 22(2): 328–331.

Puhakka L, Jaakkola S, Simpura I, Kokkonen T and Vanhatalo A (2016) Effects of replacing rapeseed meal with fava bean at 2 concentrate crude protein levels on feed intake, nutrient digestion, and milk production in cows fed grass silage–based diets. *Journal of Dairy Science* 99(10): 7993–8006.

Quiros CF and Bauchan GR (1988) The genus Medicago and the origin of the *Medicago sativa* comp. *Alfalfa and Alfalfa Improvement* 29: 93–124.

Raza A, Hafeez MB, Zahra N, Shaukat K, Umbreen S, Tabassum J, Charagh S, Khan RSA and Hasanuzzaman M (2020) The plant family Brassicaceae: Introduction, biology, and importance. In *The Plant Family Brassicaceae*. Springer, pp. 1–43.

Relwani LL (1979) *Fodder Crops and Grasses.*

Rendeková A and Mičieta K (2017) Interesting and rare plant taxa and community in the ruderal flora and vegetation of Bratislava and Malacky. *Acta Botanica Universitatis Comenianae* 52: 11–27.

Reynolds SG and Suttie JM (2004) *Fodder Oats: A World Overview.* Citeseer.

Sánchez-Martín J, Rubiales D, Flores F, Emeran AA, Shtaya MJY, Sillero JC, Allagui MB and Prats E (2014) Adaptation of oat (*Avena sativa*) cultivars to autumn sowings in Mediterranean environments. *Field Crops Research* 156: 111–122.

Santra AK, Pan S, Samanta AK, Das S and Halder S (2008) Nutritional status of forage plants and their use by wild elephants in South West Bengal, India. *Tropical Ecology* 49(2): 251.

Singh T, Radhakrishna A, Nayak DS and Malaviya DR (2019) Genetic improvement of berseem (*Trifolium alexandrinum*) in India: Current status and prospects. *International Journal of Current Microbiology and Applied Sciences* 8(1): 3028–3036.

Singh V (1995) Technology for forage production in hills of Kumaon. *New Vistas in Forage Production* 197–202.

Stubbs AK (2000) *Atlas of the Australian Fodder Industry: Outline of Production & Trade: A Report for the Rural Industries Research and Development Corporation.* RIRDC.

Sumberg J (2002) The logic of fodder legumes in Africa. *Food Policy* 27(3): 285–300.

Tufail MS, Nielsen S, Southwell A, Krebs GL, Piltz JW, Norton MR and Wynn PC (2019) Constraints to adoption of improved technology for berseem clover (*Trifolium alexandrinum*) cultivation in Punjab, Pakistan. *Experimental Agriculture* 55(1): 38–56.

Tyagi VC, Wasnik VK, Choudhary M, Halli HM and Chander S (2018) Weed management in berseem (*Trifolium alexandrium* L.): A review. *International Journal of Current Microbiology and Applied Sciences* 7: 1929–1938.

Vogl CR, Vogl-Lukasser B and Walkenhorst M (2016) Local knowledge held by farmers in Eastern Tyrol (Austria) about the use of plants to maintain and improve animal health and welfare. *Journal of Ethnobiology and Ethnomedicine* 12(1): 1–17.

Welch RW (1995) *The Oat Crop: Production and Utilization.*

Wicke B, Kluts I and Lesschen JP (2020) Bioenergy potential and greenhouse gas emissions from intensifying european temporary grasslands. *Land* 9(11): 457.

Yang HP and Sun ZM (1989) *China Oat.* Beijing: China Agriculture Press.

Zajac T, Klimek-Kopyra A, Oleksy A, Lorenc-Kozik A and Ratajczak K (2016) Analysis of yield and plant traits of oilseed rape (*Brassica napus* L.) cultivated in temperate region in light of the possibilities of sowing in arid areas. *Acta Agrobotanica* 69(4).

3 Biotechnological Applications for Developing Resistance against Biotic and Abiotic Stresses and Other Quality Traits in Fodder Crops

Siddra Ijaz, Imran Ul Haq, Zakia Habib, Samara Mukhtar, and Bukhtawer Nasir

CONTENTS

DOI: 10.1201/9781003055365-3

3.1 INTRODUCTION

Fodder crops are crops that are primarily cultivated to feed animals such as rabbits, cattle, horses, sheep, goats, and chickens. Fodder crops include a variety of crop plants ranging from *Trifolium* (white clover, red clover, subterranean clover, and alsike clover) to brassica species (kale, rapeseed) and many other crops such as barley, maize, ryegrass, alfalfa, oats, millet, sorghum, soybeans, and wheat that inhabit twice the area of grain crops worldwide because of their adaptability (Jauhar, 1993). Animals utilize these crops in the form of hay, silage, rotation pasture, forage, molasses, seeds, sprouted grains, legumes, bran, and single-cell proteins. Fodder crops also contribute to environmental sustainability, protection, and wildlife conservation irrespective of their use as animal-based diets (Moser and Hoveland, 1996). These crops are produced by many under developing countries like Pakistan, India, Iran, Bangladesh, Sri Lanka, Egypt, Turkey, Afghanistan, and Nepal (Monfreda et al., 2008).

Forage is defined as the feed that animals directly or indirectly eat, while fodders are the plant species that are used as feed for livestock, including cattle, sheep, and horses, either directly or in the form of hay (dry green fodder) and silage (preserved under anaerobic conditions), although both terms are considered similar for all practical purposes (Kumar and Singh, 2012). Globally, about 26% area is represented by forage grasslands (Capstaff and Miller, 2018). In India and Pakistan, about 5% is under forage production (Singh and Kalamdhad, 2011; Haqqani et al., 2000). Forage crops usually belong to the Fabaceae or Leguminaceae family, such as herbaceous legumes (beans and legumes) and the Poaceae family, such as grasses (Singh et al., 2010). The most common cultivated fodder crops throughout the world are medics (*Medicago* spp.), trefoil (*Lotus corniculatus*), cowpea, stylos, vetches (*Vicia* spp.), and clover/berseem (*Trifolium spp.*), and these crops belong to the legume family (Capstaff and Miller, 2018).

Khesari (Lathyrus), senji (*Melilotus alba* Desr.), berseem clover (*Trifolium alexandrinum*), *Sesbania aculeata* Pers., sorghum, moth bean, and cluster bean [*Cyamopsis tetragonoloba* (L.)] are the fodder legumes that are mainly grown in Pakistan (Haqqani et al., 2000). Forage crops have a high nutritional and economic value all over the world. More than 47 forage legumes have been identified in India

(Singh et al., 2010). To enhance environmental and nutritional benefits, forage crops can be grown in mixed-species cultivation. For example, alfalfa is the forage legume that provides the highest yield and gives more protein per unit area than any other forage legume. It can be grown alone or mixed with various grass species (Bélanger et al., 2006).

Although the yield of fodder crops can be affected by various abiotic and biotic stresses, among several constraints, diseases alone are responsible for significant loss in many fodder crops and cause severe damage. In India, diseases can cause up to 75% in cluster beans, 74% in cowpea, 72% in lucerne, 55% in oats, 50% in sorghum, and 30% in bajra. Apart from this, forage quality parameters can also be affected by diseases (Kumar and Singh, 2012).

Several biotic and abiotic stresses have affected fodder crops' productivity and yield (Rai et al., 2011). These stresses impose a considerable threat to fodder crops that lead to worldwide losses in the agricultural economy. This situation is more problematic in developing countries like Pakistan, India, Bangladesh, and Indonesia, where more than half of their economy depends on agriculture. Biotic stresses include many disease-causing agents like bacteria, fungi, parasites, insects, pests, nematodes, weeds, and arachnids (Dangl and Jones, 2001). Abiotic stresses include climatic factors such as temperature (heat, cold, and frost), salinity, drought (water stress), waterlogging (excessive water stress), toxicity, and nutrient deficiency (Boyer, 1982; de Vries, 2000; Athar and Ashraf, 2009; Sanghera et al., 2011). These stresses lead to a series of changes in the crops' molecular, biochemical, physiological, and morphological states, harming their growth and development (Wang et al., 2001). Studies have shown that these factors reduce the quantity and quality of crop production by up to 71%. This decline in productivity was examined in heat stress (40%), salinity (20%), cold (15%), drought (20%), and other factors (8%) (Rehman et al., 2005; Ashraf et al., 2008).

Biotechnological approaches have contributed to understanding the genetics of crop plants for developing stress-resistant varieties. These approaches are necessary for improving the quality, productivity, yield, and sustainability of fodder crops against environmental stresses in the agricultural sector (Dalal and Sharma, 2017). Biotechnological techniques are preferable to conventional breeding techniques because they are efficient and cost-effective and require less time to attain desirable traits. In biotechnology, various molecular and genomic tools such as molecular markers, genetic engineering, genetic transformation, microarrays and expressed sequence tags (ESTs), RNA interference (RNAi), gene editing, and so on are being deployed to address the critical problems of fodder crops (Alam et al., 2010). In marker-assisted selection (MAS), molecular markers are used to study the effect of a single trait of a crop plant at an allelic level. These markers are highly polymorphic and reproducible. Restriction fragment length polymorphisms (RFLPs), amplified fragment length polymorphisms (AFLPs), random amplified polymorphic DNA (RAPD), single nucleotide polymorphisms (SNPs), sequence-tagged sites (STS), sequence characterized amplified regions (SCARs), EST-SSR, quantitative trait loci (QTLs), and simple sequence repeats (SSRs or microsatellites) are commonly used markers in fodder crops against abiotic stresses.

3.2 ABIOTIC STRESS RESISTANCE IN FODDER CROPS

The plant genome consists of an abundance of transcriptional factors (TFs) that play an essential role in activating stress-induced genes against abiotic stresses. C-repeat binding factors (CBFs) and dehydration response element-binding factors (DREBs) are cold-related transcriptional factors that belong to the AP2 domain transcriptional factors subfamily. These transcriptional factors bind to their CCGAC sequence motif present of CRT/DRE cis-regulatory elements present in the promoter region of cold-related genes. CBF1, CBF2, CBF3, and CBF4 are the essential members of the CBF gene family that provide resistance against drought, cold, and salt stresses. These CBF proteins regulate the transcription of many stress-inductive genes such as kin1, kin2, cor6.6, cor15a, rd17, and rd29A (Ingram and Bartels, 1996; Stockinger et al., 1997; Gilmour et al., 1998; Liu et al., 1998; Seki et al., 2001; Thomashow et al., 2001). Transcription factors played predominant roles to confer abiotic stress resistance in fodder crops. Several transcriptional factors identified from Arabidopsis are considered stress-related regulators. One of these transcriptional factors, AtNF-YB1, a subunit of the nuclear factor Y (also known as the CAAT or the HAP family) has been found to regulate transcriptional expression through CCAAT DNA sequences against drought stress. NF-Y is a conserved heterotrimeric protein complex consisting of NF-YA (HAP2), NF-YB (HAP3), and NF-YC (HAP5) subunits. In fungi and other animals, single genes are being used to encode each protein subunit. NF-Y transcription factors, along with other regulatory factors, regulate gene expression against abiotic stresses. Transcriptional factor bZIP is vital to signal transduction and present in many eukaryotes, particularly in crop plants (Cowell, 2002). In plants, the role of some bZIP transcription factors associated with environmental stresses has been identified in *A. thaliana*. Arabidopsis bZIP transcription factors were divided into ten subgroups (Jakoby et al., 2002). AtbZIP39, AtbZIP36, AtbZIP38, AtbZIP66, AtbZIP40, AtbZIP35, and bZIP37 are the stress-related transcriptional factors that belong to subgroup A (Yang et al., 2009).

3.2.1 Molecular Genetics of Abiotic Stress Resistance in *Trifolium alexandrinum*

Trifolium alexandrinum (berseem clover) is not tolerant to abiotic stresses like salt (Abogadallah, 2010) or drought (Iannucci et al., 2000). In *T. alexandrinum*, the overexpression of gene-encoding transcriptional factor proteins regulates the tolerance to abiotic stresses. These stress-related TF proteins belong to the APETAL2/ethylene-responsive factor (AP2/ERF) members (Century et al., 2008). Overexpression of ethylene-responsive factor genes related to drought and salt tolerance have been recently reported in medics (Chen et al., 2010; Jin et al., 2010), rice (Gao et al., 2008), tobacco, and tomato (Zhang et al., 2009; Zhang and Huang, 2010). These stress-related genes improve drought and salt tolerance in transgenic plants. More specifically, overexpression of an ERF gene (HARDY from Arabidopsis) improves the ability of drought resistance in rice by reducing water loss through transpiration and improving photosynthesis (Karaba et al., 2007). This HARDY gene from Arabidopsis was introduced into *T. alexandrinum* through the *Agrobacterium tumefaciens*–mediated transformation method to produce transgenic varieties against drought and salt stress.

Studies showed that salt sensitivity in *T. alexandrinum* is due to excessive accumulation of sodium ions simultaneously with the downregulation of sodium hydrogen exchanger (NHX1), hydrogen adenosine triphosphatase (H^+-ATPase), hydrogen pyrophosphatase (H^+-PPase), and HARDY gene, which is involved in transpiration (Karaba et al., 2007). This study suggested controlling the cytosolic Na^+ concentration and transcription would enable *T. alexandrinum* to tolerate salt and drought stress. So the overexpression of this HARDY gene can complement the expression of stress-induced genes. Nine plants (L1-L9) of *T. alexandrinum* were isolated to study the expression of the HARDY gene. From these plants, transgenic lines L2 and L3 showed 42 and 55% improvement under drought stress and 38 and 95% improvement under salt stress compared with the wild type. Overexpression of this HARDY gene improved the water use efficiency (WSE) by limiting transcription under drought stress and decreased the Na^+ concentration in leaves by enhancing the photosynthesis rate under salt stress (Abogadallah et al., 2011).

3.2.2 Molecular Genetics of Abiotic Stress Resistance in Alfalfa

Medicago sativa (alfalfa) is an essential perennial forage crop sensitive to many abiotic stresses, causing agricultural losses. To overcome these stresses, genetic engineering approaches are being used to produce transgenic alfalfa crops. Plants resist these abiotic stresses by expressing several stress-induced genes. The encoded products of these genes result in the accumulation of many stress proteins such as heat shock proteins (HSPs) and late embryogenesis abundant (LEA) proteins, enzymes such as lipid desaturases, and many other enzymes that catalyze the biosynthesis of transporters, antioxidants, and osmoprotectants (Bartels and Sunker, 2005). One of the most critical osmoprotectants is disaccharide trehalose, which confers resistance to many abiotic stresses. The overexpression of trehalose biosynthetic genes in several plants such as *Nicotiana tabacum*, *Solanum tuberosum*, *Oryza sativa*, *Lycopersicum esculentum*, and *Arabidopsis* resulted in improved tolerance to abiotic stresses (Holmström et al., 1996; Yeo et al., 2000; Garg et al., 2002; Avonce et al., 2004; Cortina and Culiáñez-Macia, 2005).

The multienzyme complex mediates the biosynthesis of trehalose, trehalose-6-phosphate synthase (TPS) encoded by the TPS1 gene trehalose-6-phosphate phosphatase (TPP) encoded by the TPS2 gene in yeast. So alfalfa was transformed with these two genes (TPS1 and TPS2) by *A. tumefaciens*–mediated transformation (LBA4404 strain). The overexpression of these genes led to the accumulation of trehalose in alfalfa crops resulted in increased tolerance to heat, drought, salt, and freezing stress. The results showed that transgenic lines could tolerate salinity stress (up to 15 mM of Nacl), freezing temperature (–8 to –16°C), and heat stress (up to 45°C) without damaging the alfalfa crops (Suárez et al., 2009).

Plant roots play an essential role in accumulating nutrients and water from the soil and contributing to increased crop productivity. The function of roots is influenced by abiotic factors such as salt stress. Studies showed that plants respond to drought/salt stress by increasing root growth and decreasing shoot growth (Creelman et al., 1990; Saab et al., 1990). Root growth can be enhanced by overexpression of root-specific transcriptional factors. Alfn1 is a vital transcription factor primarily expressed in plant roots and acts as a potential candidate for gene expression in rice, Arabidopsis, and many other crops (Bastola et al., 1998; Winicov and Bastola, 1999). This root-specific transcription factor (Alfn1) is a vital root growth regulator that regulates the expression of the root inducible gene known as MSPRP2 under the cauliflower mosaic virus (CaMV35S) in transgenic alfalfa. The binding sites for Alfn1 present in the promoter region of these root genes. The overexpression of Alfin1 resulted in increased expression of the salt-inducible MSPRP2 gene and led to a predominant increase in root growth under saline stress, but it did not adversely affect shoot growth. Ultimately, the increase in root growth led to an increase in crop productivity (Winicov, 2000).

Some leaf wax genes play an essential role in stress tolerance and significantly contribute to fodder crop improvement. These genes are responsible for producing plant cuticular waxes that protect aerial organs from environmental stress like drought. The biosynthesis of this cuticular wax layer and its loading to a plant surface is relatively complex and actively regulated by the number of wax genes (Jenks et al., 2002; Broun et al., 2004). The expression of these genes is regulated by a group of AP2 domain-containing transcriptional factor that belongs to ethylene-responsive element-binding factor (EREBP or ERF)/APETALA 2 gene family (Okamuro et al., 1997; Riechmann et al., 2000). Recently, a dehydration-response element-binding protein/C-repeat binding factor, AP2 domain-containing transcriptional factor has been recognized in *Medicago truncatula*. These factors are responsible for regulating genes induced by abiotic stress (Thomashow, 1999; Shinozaki et al., 2003; Novillo et al., 2004). The overexpression strategy of transcription factors is workable in plants against abiotic stress (Zhang, 2003).

The novel gene designated as WXP1, an AP2 domain-containing transcriptional factor gene from *M. truncatula*, activates wax production in the most important forage species, *Medicago sativa* (alfalfa). The estimated protein of the WXP1 gene has 371 amino acids that confer drought resistance in *M. sativa*. This predicted protein is one of the most extended peptides of all the proteins from AP2 domain-containing genes in *M. truncatula*. The WXP1 gene is significantly distinct from the genes related to abiotic stress in the AP2/ERF transcription factor family that are most studied, such as GL15, AP2s, WIN1/SHN1, CBF/DREB1s, and DREB2s. The expression level of the WXP1 gene is induced by abscisic acid, drought, and high temperature, primarily in shoot tissues of *M. truncatula*. The WXP1 gene from *M. truncatula* is cloned into alfalfa through the *A. tumefaciens* transformation method to generate transgenic lines against drought stress. Overexpression of stress-induced WXP1 gene, controlled by CaMV35S promoter, activated several cuticular wax-related genes that directed a distinct increase in a cuticular wax layer on transgenic alfalfa leaves. Mass spectrometry and gas chromatography indicated that wax deposition per leaf surface increased from 29.6 to 37.7% in transgenic plants. Transgenic alfalfa showed resistance to drought stress by reducing water loss from leaves. These transgenic plants showed a delay in wilting when watering was stopped and quick recovery after rewatering (Zhang et al., 2005).

3.2.3 Molecular Genetics of Abiotic Stress Resistance in Maize

Maize is a widely cultivated crop after rice and wheat, as its annual production is approximately one billion tons worldwide. Conversely, many abiotic factors and, more importantly, drought stress, can affect productivity (Cattivelli et al., 2008). Maize is more susceptible to drought stress during its flowering stage. Genetic approaches can enhance its productivity by overcoming this stress through gene regulation. Studies suggested that a single gene can confer drought resistance to produce genetically modified (GM) crops. More than one stress-induced transgene is required for producing GM crops (Campos et al., 2004). Trehalose is a non-reducing sugar (disaccharide) that responds under drought stress. In the trehalose pathway, a T6P synthase enzyme produces the T6P compound by oxidizing glucose-6-phosphate and UDP-glucose. Another enzyme, trehalose-6-phosphate phosphatase (TPP), produces trehalose sugar from the T6P compound (Nunes et al., 2014). The T6P is a vital growth regulator that acts under drought stress through sucrose accumulation (Schluepmann et al., 2003).

This TPP gene under the promoter of transcriptional regulator gene Mads6 in rice is overexpressed in maize. The TPP gene's overexpression is a sugar signal resulting in a reduced concentration of T6P that regulates growth and development by increased sucrose concentration under drought stress. Results showed that the drought tolerance in transgenic maize improved from 31 to 123% under extreme drought stress (Nuccio et al., 2015).

Ethylene is a phytohormone that modulates plant development and its response to abiotic stress. Recent studies indicated that genetic modification of ethylene biosynthesis and signaling could improve plant tolerance to drought stress. Here novel negative regulators of ethylene signal transduction are reported in Arabidopsis and maize. The ARGOS (auxin-regulated gene involved in organ size) gene family encoded these negative regulators of ethylene. In maize, over-expression of zm-ARGOS1 and zmARGOS8 under the promoter of maize ubiquitin (UBI1) led to an increase in ethylene sensitivity in drought and well-watered conditions. Results showed improved tolerance to abiotic stress in transgenic zm-ARGOS maize crops (Guo et al., 2014; Shi et al., 2015). Other studies indicated that decreasing ethylene biosynthesis by silencing 1-aminocyclopropane-1-carboxylic acid synthase (ACS) in transgenic maize crops led to upgrading grain yield under drought conditions (Habben et al., 2014).

Limited water availability is the principal stress to maize (*Zea mays*) crops and causes adverse crop losses up to 40% only in America (Boyer, 1982). Maize crops are not tolerant to water-limited conditions throughout the vegetative growing season (Claassen and Shaw, 1970; Boyer and Westgate, 2004). RNA chaperones are abundant and ubiquitous proteins and play a role in stress-induced responses in plants such as Arabidopsis and rice. Cold stress proteins (CSPs) containing cold shock domain (CSD) reported from bacteria such as CspA protein from *Escherichia coli* and CspB protein from *Bacillus subtilis* are RNA chaperones that promote stress resistance in many crop plants. These CSPs' function depends on the chaperones/RNA binding activity through the cold shock domain based on the post-transcriptional mechanism. Previously these proteins improved the tolerance ability of wheat, Arabidopsis, and rice against many abiotic stresses, including cold, heat, and drought. However, in maize, productive RNA chaperone activity of CspB protein alone is critical for providing tolerance to drought stress. So these studies suggested that the stress tolerance mediated by CSPs is a novel approach to plant engineering that enhances crop productivity under abiotic stresses (Castiglioni et al., 2008).

Most crop plants accumulate specific organic compounds known as compatible solutes or osmoprotectants to deal with abiotic stresses (Bohnert and Shen, 1998). Glycine betaine $[(CH_3)_3N^+CH_2COO^-]$ is one of the most studied compatible solutes (Le Rudulier et al., 1984; Rhodes and Hanson, 1993; Rontein et al., 2002; Sakamoto and Murata, 2000, 2001). Glycine betaine works as an osmoprotectant and maintains the activities and structures of enzymes and other protein complexes. These organic molecules also stabilize cell membranes' integrity under the stress of excessive salt, heat, cold, freezing, and desiccation (Gorham, 1996; Sakamoto and Murata, 2002).

It has been reported that there is a diverse level of this organic molecule, glycine betaine, in various varieties of *Zea mays* (Brunk et al., 1989). In *Escherichia coli*, a betA gene encodes for a critical enzyme, choline dehydrogenase, which plays a vital role in glycine betaine's biosynthesis from choline. The biosynthesis of glycine betaine can either by a single enzyme or two enzymes. In the first case, the betA gene encodes for choline dehydrogenase that catalyzes choline oxidation into betaine aldehyde (Andresen et al., 1988; Lamark et al., 1991), and then this betaine aldehyde oxidizes into glycine betaine (Landfald and Strom, 1986). In the second case, another enzyme, betaine aldehyde dehydrogenase encoded by the betB gene, catalyzes the betaine aldehyde's oxidation into glycine betaine. In this study, a betA gene from *E. coli* was transferred to the DH4866 inbred line of maize plants. Expression of the betA gene enhanced tolerance against drought/osmotic stress by increasing glycine betaine accumulation. Results showed that transgenic maize lines are more resistant to drought/osmotic stress than the wild-type plants at all developmental stages, specifically seedling and germination. The higher level of glycine betaine in transgenic maize leads to a significant increase in grain yield, cell membrane integrity, and enzyme activity in drought stress (Quan et al., 2004).

3.3 BIOTECHNOLOGICAL APPLICATIONS FOR DEVELOPING RESISTANCE AGAINST ABIOTIC STRESS IN FODDER CROPS

3.3.1 Transgenic Technology

Transgenic technology is an effective strategy to increase resistance against drought stress. Previous studies showed that overexpression of H+-PPase gene AVP1 in transgenic Arabidopsis improved resistance to drought and salt stress (Gaxiola et al., 2001). Park et al. (2005) reported that the upregulation of the AVP1 gene in tomatoes enhanced drought tolerance. Recently studies showed that the heterologous expression of the TsVP gene increased resistance to salt stress in tobacco (Gao et al., 2006). In maize, H^+-PPase and H^+-ATPase pumps play an important role in drought stress. Genetic manipulation of these pump genes proved a valuable strategy under drought stress.

The potassium-dependent TsVP gene from the dicotyledonous *Thellungiella halophila* against drought stress was introduced into maize. The results showed that transgenic maize lines were more resistant to drought and osmotic stress than wild types. The heterologous expression of the TsVP gene resulted in higher activity of H^+-pyrophosphatase (H^+-PPase) and H^+ adenosine triphosphatase (H^+-ATPase) pumps that led to increased seed germination efficiency, enhanced root growth and water uptake, higher accumulation of solutes, more minor damage in the cell membrane, and decreased water loss in transgenic maize crops under drought stress. Most notably, the productivity and grain yields were considerably higher in transgenic maize crops than in wild-type plants under drought treatment. So it was concluded that higher expression of the TsVP gene led to higher proton pump activity in maize and improved its tolerance against drought/osmotic stress. Thus, this heterologous expression of the TvSP gene is a feasible genetic approach to increase other crop productivity under abiotic stresses (Li et al., 2008).

A mitogen-activated protein kinase kinase kinase (MAPKKK) is a stress-induced gene whose expression activates an oxidative signal cascade that leads to resistance against abiotic stresses like freezing, salinity, and heat stress. Previously, studies identified this gene in transgenic tobacco against abiotic stresses. In present studies, a tobacco gene, MAPKKK (NPK1), was expressed in maize to study the role of resulting oxidative signaling against abiotic stresses. NPK1, nicotiana protein kinase 1, induced protective mechanisms in maize from dehydration damage. The expression of NPK1 induced the expression of multiple stress proteins such as GST1, HSPS, and PR1 that confer tolerance to drought and freezing stress in maize. Activation of these stress-related proteins shields maize crops' photosynthesis machinery against drought stress and improves crop yield (Shou et al., 2004).

Genes linked with glycine betaine production in crop plants, trees, and microorganisms have been transferred to those plants that cannot produce this glycine betaine, such as *Brassica nupus* (Huang et al., 2000), *Nicotiana tabacum* (Llius et al., 1996; Holmstrom et al., 2002; Shen et al., 2002), *Arabidopsis thaliana* (Alia et al., 1998a; Sakamoto et al., 2000; Hayashi et al., 2006), *Oryza sativa* (Sakamoto et al., 1998; Mohanty et al., 2002), and *Diospyros kaki* (Gao et al., 2018). The genetic engineering of the glycine betaine gene in these plants enhanced transgenic crop plants' tolerance to drought, extreme temperatures, and salinity stresses (Sakamoto and Murata, 2001; Sulpice et al., 2003).

The NF-YB protein identified from *Arabidopsis thaliana* provides resistance to drought stress. An orthologous NF-YB transcriptional factor, ZmNF-YB2, was found to have the same activity to drought stress from maize. Under a drought stress environment, transgenic maize crops with increased expression of ZmNF-YB2 transcriptional factor showed tolerance to drought response. The maize crops respond in many stress-linked parameters like stomatal conductance, chlorophyll content, reduced wilting, leaf temperature, and photosynthesis continuance. These stress-related adaptations led to increased productivity and grain yield under a water-deficit environment (Nelson et al., 2007). The upregulation of transcriptional factor ZmbZIP17 conferred resistance to drought stress in maize (Jia et al., 2009).

The overexpression of a single cold-related TF in transgenic *Arabidopsis thaliana* improved cold resistance. These CBF transcriptional factors also exist in rice, maize, barley, ryegrass, soybean, and oilseed rape (Jaglo et al., 2001; Choi et al., 2002;

Dubouzet et al., 2003; Qin et al., 2004; Xiong and Fei, 2006; Zhang et al., 2009). In Arabidopsis, cold tolerance is linked with the regulation of cold-related genes (COR) by CRT/DRE elements (Thomashow, 1998). In another study, the CBF1 transcriptional factor's overexpression stimulates COR gene expression (cor78, cor47, cor15a, and cor6.6), resulting in enhanced tolerance to cold temperature in Arabidopsis (Jaglo-Ottosen et al., 1998). Arabidopsis that is engineered with the CBF1 gene either under the strong constitutive promoter from the cauliflower mosaic virus (CaMV35S) or cold-induced promoter in the DRE region from the upstream dehydration-induced gene (rd29A) caused a remarkable increase in capability to tolerate extreme temperature, salinity, and water stress (Kasuga et al., 1999). In Arabidopsis, overexpression of another factor, CBF3, not only enhanced tolerance to freezing temperature but also resulted in several biochemical changes linked with cold tolerance, such as elevated level of soluble sugars such as glucose, fructose, raffinose, and sucrose proline synthetic enzyme, pyrroline-5-carboxylase synthetase (P5CS) that cause increased production of proline (Gilmour et al., 2000).

Transcriptional factors CBF/DREB from Arabidopsis are also found in *Brassica napus* as conserved regions. Constitutive overexpression of this Arabidopsis CBF gene (COR15a, an ortholog to the Bn115 gene in *B. napus*) in transgenic *B. napus* plants resulted in increased expression of these ortholog genes of Arabidopsis to cope with the freezing temperature and many other abiotic stresses. The transgenic lines of *B. napus* are freezing stress tolerance and maintaining and protecting various cellular components' structure and function. These genetically engineered plants result in increased grain productivity and yield (Wang et al., 2003).

3.3.2 Molecular Markers

Sorghum is an important widely produced forage crop worldwide, but its productivity is affected by abiotic stresses, specifically salt stress. RAPD and inter-simple sequence repeat (ISSR) markers had been identified associated with stress tolerance through bulk segregant analysis (BSA). These markers could be used in breeding programs of many other crops through marker-assisted selection. To identify these markers, two inbred sorghum lines were selected as salt sensitive (ICSR89038) and salt tolerant (ATX631) to get F1 generation through the hybridization of these two lines. The F2 generation was obtained by self-crossing of the F1 generation. Then these contrasting genotypes were used to identify DNA-based markers. Thus, bulked segregant analysis was carried out to detect molecular markers in sorghum lines. BSA identifies molecular markers associated with drought tolerance in the F2 generation of contrasting genotypes. As a result, two DNA bulks from this F2 generation were used alongside the F1 generation and their parents to identify ISSR and RAPD markers linked with salt stress tolerance. DNA bulks from F2 generation, F1 generation, and their parents (ATX631, ICSR89038) were tested against ten random primers. Primer O12 showed the lowest band number, four, and the primer C04 showed the highest band number, which was 13. These ten primers developed 78 bands collectively, and from these bands, 49 bands showed 63% polymorphism. As a result, only 14 bands were considered valid for molecular markers related to salt stress.

ISSR markers are newly discovered markers based on short inter-tandem repeats of DNA sequences. The DNA bulks of the F2 generation, F1 generation, and their parents (ATX631, ICSR89038) were tested against ten ISSR primers. Primers 844a and 844 b indicated the lowest number of bands (6), and primer 17899a showed the highest number of bands (12). All ten primers developed 77 bands collectively; 37 bands were polymorphic, showing 48% polymorphism. From these 37 bands, only 15 bands were represented as molecular markers under salinity stress (Younis et al., 2007).

Maize is relatively resistant to drought stress, but identifying genes that improve maize's drought resistance could increase crop production. This increased crop yield can be achieved by allowing for the identification, selection, and production of maize crops with comparatively improved drought resistance. Recently, molecular marker technology became a leading genetic approach against abiotic stresses. Molecular breeding is preferred over conventional breeding by identifying genes and molecular markers from these genes linked with abiotic stresses for crop improvement (Varshney et al., 2005). In recent years, the SNP marker has become a powerful tool for identifying stress-related genes in fodder crops, but in maize, SNP markers based on single nucleotide differences are challenging to apply because of the high cost. So, cleaved amplified polymorphic sequence (CAPS) markers, polymerase chain reaction (PCR)-based markers, are widely utilized markers for detecting SNPs rapidly and reliably that generate unique restriction sites that can differentiate alleles by a difference of even a single nucleotide (Michaels and Amasino, 1998; Neff et al., 1998). In maize, two genes, dhn1 and rsp41, confer resistance against drought stress. So, with CAPS technology, two alleles from these two genes were separated by PAGE to generate single SNPs that act as CAPS markers (dhnC397 and rspC1090). These functional markers can be used in other maize lines (Liu et al., 2015).

Previously, one RAPD marker for salt tolerance and three RAPD markers for drought tolerance were also identified in maize by using bulked segregant analysis (Abdel-Tawab et al., 1997). Many other successful studies to detect RAPD markers for drought or salt tolerance have been reported (Breto et al., 1994; Rahman et al., 1998). RAPD markers are the fastest, most efficient, and simplest marker technique for identifying several molecular markers. In other studies, ISSR markers and microsatellites were detected in soybean against salt stress (Yang et al., 1996).

Molecular markers can be detected by a number of molecular marker techniques such as random amplified length polymorphism, restriction fragment length polymorphisms, DNA amplification fingerprinting, sequence characterized amplified regions, microsatellites, and sequence characterized amplified regions (Lin et al., 1996). This study was used to identify molecular markers linked to salt stress by BSA technique using ISSR, RAPD, and AFLP analysis in salt-tolerant, salt-sensitive genotypes and their F1 and F2 generations.

ISSR primer analysis was conducted to obtain molecular genetic markers linked to salinity tolerance from the two resistant and sensitive genotypes, F1 and F2, using five primers. All five ISSR primers amplified the DNA fragments successfully. Among the 92 amplified DNA fragments across the five primers, 87 fragments were 94.64% polymorphic. The DNA band with a molecular size of 332 bp, amplified by the HB-09 primer, was found in a sensitive genotype, the sensitive bulk of F2, while it was absent in the resistant genotype and resistant bulk of the F2 generation under salt stress. So, this band could be used as a negative genetic marker linked to salinity tolerance in alfalfa crops. Another primer, HB-15, showed one band with a molecular size of 480 bp, which was fully present in the tolerant genotype and the F2 tolerant bulk against salinity stress. This band could be used as a positive molecular marker associated with salinity tolerance in alfalfa plants. The results showed that the ISSR marker is the most acceptable way to evaluate genetic diversity among plant species (Abdel-Tawab et al., 2011).

The RAPD-PCR marker technique can also be used for developing molecular markers associated with salinity stress using 13 random primers (10 mer). All the random primers successfully amplified DNA fragments for all genotypes, and the resultant amplified bands were various in number, ranging from 5 to 19 bands. Only five of these primers showed molecular markers linked to salinity tolerance. Primer OP-G05 amplified two bands with molecular sizes of 390 and 795 bp, present in the sensitive parent and sensitive bulk of F1 and F2 generations under salinity stress, but these molecular bands were not present in the tolerant parent and tolerant bulk of the F2 generation, so these bands can be used as negative molecular markers associated with salinity stress in alfalfa crops.

Another primer, OP-L16, amplified one band with a molecular size of 991 bp, which was found in the sensitive parent and absent in the sensitive bulk of F2 under salinity stress, although it was missing in the tolerant genotype and tolerant bulk of F2. This band can also be used as a negative molecular marker linked to salinity stress in alfalfa crops. Three primers, OP-M17, OP-O18, and OP-O20, amplified one band with molecular sizes of 615, 653, and 658 bp, respectively. These amplifications were found in the tolerant genotype and tolerant bulk of F1 and F2 under salinity stress, but these bands were absent in the sensitive genotype and sensitive bulk of the F2 generation. So, these bands linked to salinity stress can be used as molecular markers for alfalfa crops.

These results agree with those of Fahmy et al. (1997) and carried out this RAPD marker technique to distinguish between tolerant and sensitive genotypes of *Trifolium alexandrinum* (berseem) linked to drought stress and identified two positive molecular markers under drought stress. Wenzel (1992) also emphasized the importance of DNA marker-based analysis of abiotic stress tolerance in crop plants. In other studies, it was reported that RAPD molecular markers seemed to be potentially used for developing genetic information in alfalfa crops (Echt et al., 1992). In other research studies by Yang et al. (2005), BSA, along with the RAPD marker technique, was used to identify molecular markers in the F2 generation of alfalfa, obtained by crossing salt-sensitive and salt-tolerant alfalfa genotypes associated with salinity stress. In the AFLP marker technique, two primer pairs were used for salt stress, and these primers amplified 246 bands. Only five were obtained from these 246 bands to be used as AFLP markers in crops linked to salt stress (Abdel-Tawab et al., 2011).

3.3.3 Quantitative Trait Loci–Based Genetic Approaches

Maize is highly susceptible to drought stress, and the yield loss due to drought stress is quite considerable. Studies showed that drought resistance in maize is visibly a quantitative trait controlled by the number of minor genes known as polygenes that have additively controlled expression (Thi Lang and Chi Buu, 2008). These polygenes' loci on chromosomes are referred to as quantitative trait loci (Ashraf, 2010). In a QTL analysis, quantitative trait evaluation could be carried out in many crop plants from a segregating population with the help of a variety of molecular markers. After that, either the entire population or only a part of this segregating population is genotyped. In the end, a suitable statistical analysis is carried out to identify a specific QTL linked to a particular trait (Asins, 2002).

Now QTL mapping is preferred over phenotypic-based selection by using various molecular markers to encounter abiotic stress conditions (Ashraf, 2010). QTL mapping, along with DNA markers, is valid for exploiting genetic variation within the crop genome under drought stress, followed by marker-assisted selection (Ashraf et al., 2008). QTL mapping allows assessing the pattern of gene function, numbers, magnitude, and locations of phenotypic effects of genomic traits (Vinh and Paterson, 2005). The function of polygenes in controlling a specific trait has been extensively evaluated by traditional means in past years, but now the use of QTL mapping and molecular markers has made it preferable to analyze the complex phenotypic traits (Ashraf, 2010). Several DNA markers such as RAPDs, AFLPs, RFLPs, SNPs, SSRs, PCR indels, and CAPS are now developed and used effectively to identify QTL under stress conditions in many crops (Ashraf et al., 2008). QTL mapping has been carried out in various crops, including maize, barley, wheat, sorghum, rice, and cotton, against drought stress (Sari-Gorla et al., 1999; Ashraf, 2010). In maize, the flowering stage is more susceptible to drought stress that results in the increased anthesis-silking interval (ASI), directly correlated with a decrease in grain yield. The QTLs linked to ASI for drought stress were detected on chromosomes 1, 2, 5, 6, 8, and 10. MAS based on ASI QTLs could be a powerful tool for developing drought resistance in maize crops (Ribaut et al., 1996).

In sorghum, DNA markers associated with QTL, specifically for drought tolerance, are used to increase the effectiveness of breeding techniques for selecting sorghum germplasm with increased drought tolerance. In this study, an RI population was derived by crossing Tx7078, which shows pre-flowering drought tolerance, and B35, which shows post-flowering tolerance. A genetic map for this population was constructed using RFLP and RAPD markers against drought stress. The pre-flowering and post-flowering of these RI lines were evaluated under a drought-stress environment. The result of the single-marker analysis showed that 6 QTLs are linked to pre-flowering drought stress (Tuinstra et al., 1996), and 12 QTLs are linked to post-flowering drought stress (Tuinstra et al., 1997). Stay green is the drought resistance mechanism that confers sorghum resistance to premature senescence associated with the post-flowering stage. A study was conducted for QTLs related to sorghum drought resistance in which B35 lines were used for the stay-green trait. In B35, four stay-green QTLs were identified, including Stg1 and Stg2 on chromosome 3, Stg3 on chromosome 2, and Stg4 on chromosome 5 (Sanchez et al., 2002). Stg1 and Stg2 were also recognized by Tunistra et al. (1997). Another stay-green QTL (Stg3) was identified by Tao et al. (2002) in the RI population resulting from crossing between QL39 and QL41.

In another study, QTL analysis was conducted to identify molecular markers linked to cold seedling tolerance by using the RI population derived from crossing between SRNEP, a cold-sensitive line, and SQR, a cold-tolerant line. These lines were evaluated based on seed germination. QTL mapping analysis showed one QTL linked to germination on chromosome 2 near the txp348 marker. Another QTL was identified on chromosome 1 near the txp43 and txp211 locations. One QTL was identified on chromosome 4 near the position of txp51. After that, these markers could be used to develop new resistant lines through the MAS technique (Ejeta and Knoll, 2007).

3.3.4 Non-Coding RNA-Mediated Applications

Non-coding RNAs such as small interfering RNA (siRNA), microRNA (miRNA), and long non-coding RNAs (lincRNAs) play an essential role in developing the tolerant varieties of crop plants against abiotic stresses by acting on post-transcriptional level (Kumar, 2014; Zhao et al., 2016). In recent studies, miRNA-mediated gene expression regulation at the post-transcriptional level has been carried out to analyze plant responses to abiotic stresses (Phillips et al., 2007; Wen et al., 2008). microRNAs are small non-coding RNA molecules with a molecular size of 20–24 nucleotides that can either downregulate or upregulate specific gene expressions. RNA-mediated interference is carried out by incorporating miRNAs into the RISC-RNA-induced silencing complex. This complex binds to specific stress-related mRNAs based on complementarity and cleaves them out to inhibit the translation (Bartel, 2004; Chen, 2004).

The role of stress-related miRNAs was demonstrated by Rhoades et al. (2002), who identified new miRNAs in *Arabidopsis thaliana* and predicted their target mRNA by using bioinformatics tools. The studies showed that miRNAs play a role in developmental processes and are involved in abiotic (Sunkar and Zhu, 2004; Lu et al., 2005; Zhao et al., 2007), biotic, (Navarro et al., 2006) stress conditions. microRNA, miR398, regulates the expression of the targeted genes that encode antioxidant enzymes, suggesting their role in abiotic stresses (Sunkar et al., 2012). Moreover, heat shock transcription factors (HSFs) regulating the expression of miR398 suggested its role, specifically in heat stress (Guan et al., 2013). Both miR160 and miR169 control phytohormone signaling by regulating genes for auxin response factors (ARFs) in barley, and the expression of these genes is changed under heat stress (Kruszka et al., 2014; Kumar et al., 2015).

In an experiment, Sunkar and Zhu (2004) cloned the small RNAs present in Arabidopsis seedlings. They then exposed these RNAs to many abiotic stresses that resulted in identifying various miRNAs that were either downregulated or upregulated under cold, dehydration, abscisic acid (ABA), and salinity stresses. Previous experiments have shown that *Medicago truncatula* crops adapt to a water deficit environment and then progressively recover after rewatering (Nunes et al., 2008, 2009). This adaptation is achieved by decreasing stomatal conductance (ABA-mediated water deficit avoidance mechanisms) and increasing cell membrane integrity (drought tolerance mechanisms) (Bartels and Sunkar, 2005). These drought tolerance mechanisms can be regulated by miRNAs (Lu and Fedoroff, 2000; Reyes and Chua, 2007). These indications resulted in the finding that miRNAs could be involved in conferring tolerance to crops against drought stresses. In *Medicago truncatula*, the differential expression of miRNAs can confer resistance to drought stress. Upregulation of miR398 and miR408 in both roots and shoots provides tolerance under a water deficit environment. Understanding miRNA-mediated regulation can provide clues for the genetic improvement of crops against abiotic stresses (Trindade et al., 2009).

MicroRNA319 is one of the first characterized and conserved miRNA families in crops that regulate the TCP genes (TEOSINTE BRANCHED, CYCLOIDEA, and PROLIFERATING CELL FACTORS), encoding transcriptional factors related to environmental stresses. Osa-miR319a, a rice miR319 gene, was introduced into creeping bentgrass under the promoter of CaMV35S to produce transgenic lines. This study investigated the role of Osa-miR319a against salinity and drought stress. Results showed that in transgenic bentgrass crops, the overexpression of Osa-miR319a displayed various morphological and physiological changes and conferred enhanced salinity and drought tolerance. These changes include increased leaf and stem area, leaf wax, water retention, cell membrane integrity, less salt accumulation in cytoplasm, and well-maintained photosynthesis. Gene expression analysis showed that the Osa-miR319a gene regulates four putative miR319 target genes: AsPCF5, AsPCF6, AsPCF8, and AsTCP14.

Additionally, Osa-miR319a also affected the transcription factor gene expression; AsNAC60, a homolog of the rice Os12g41680 (NAC-like gene ONAC60), downregulated overexpressing the Osa-miR319a gene. Our results exhibited that miR319 regulates plant responses to salinity and drought stress. The increased tolerance to abiotic stress in transgenic bentgrass crops is significantly correlated to the downregulation of miR319 target genes. This miR319-mediated tolerance implies their potential for developing novel molecular techniques to genetically engineer fodder crop to enhance resistance against environmental stresses (Zhou et al., 2013).

MicroR156 emerged as a useful tool in the biotechnological improvement of crop plants against abiotic stresses. Studies were conducted to characterize the overexpression of miR156 in transgenic alfalfa. Alfalfa was transformed by overexpressing the MsmiR156d (a precursor of miR156) gene to produce miR156OE genotypes (A8, A11, A16b, A17). microRNA156 (miR156) regulates the squamosa promoter-binding protein-like (SPL) genes that subsequently improve plant growth and developmental processes by regulating the expression of downstream genes under abiotic stresses. miR156 targets SPL genes related (SPL6, SPL12, and SPL13) to environmental stress, and these genes undergo silencing by mRNA cleavage. The upregulation of miR156 caused a decline in SPL13 gene expression in miR156OE genotypes compared to wild types under drought and control conditions. Under stress conditions, no significant changes were detected in the expression of SPL6 and SPL12.

Downregulation of the SPL13 gene led to enhanced expression of other stress-related genes such as the P5CS gene (PYRROLINE-5-CARBOXYLATE SYNTHETASE) in both roots and leaves of miR156O3 genotypes (excluding A11) and NCED (9-CIS EPOXYCAROTENOID DIOXYGENASE) in roots under drought stress. The enhanced expression of the NCED and P5CS genes was related to ABA's biosynthesis and proline in miR156OE genotypes under drought stress. These gene expressions led to changes in crop physiological characteristics like decreased transpiration rate, reduced stem water loss, improved root and leaf growth, higher conductance of stomata, increased photosynthesis rate, and increased antioxidant accumulation, ABA, and proline in miR156OE genotypes. Enhanced tolerance to drought stress in miR156OE genotypes led to investigation of the different role of miR156 genes for improving crop quality in its productivity under abiotic stresses (Arshad et al., 2017). In another study, transgenic alfalfa with enhanced expression of miR156 and decreased expression of the SPL13 gene indicated tolerance to heat stress (40°C). Transgenic alfalfa plants (miR156OE) showed high water potential, improved photosynthesis, increased ABA, anthocyanin, chlorophyll, and non-enzymatic antioxidant content than wild types under extreme temperature conditions. The miR156/SPL module plays a significant role in regulating alfalfa's response and many other fodder crops to environmental stresses. (Matthews et al., 2019).

3.3.5 CRISPR-Cas Based Genome-Editing Approach

CRISPR/Cas (clustered regularly interspaced short palindromic repeats–CRISPR-associated protein) is a promising genome editing tool with great versatility, efficiency, and intensive multiplexing that provides opportunities to plant genome engineering (Yu et al., 2017). This latest genome editing technique is preferred over genome targeting nucleases such as transcription activator-like effector nucleases (TALENs) and zinc-finger nucleases (ZFNs), and meganucleases (Gao et al., 2018; Shukla et al., 2009; Li et al., 2012; Li et al., 2013; Čermák et al., 2015). The CRISPR-Cas9-targeted approach forms an intricate system in which guide RNA (gRNA) can direct Cas9, an enzyme, to produce double-stranded DNA breaks (DSBs) at specific DNA sites. These DSBs then activate repair mechanisms non-homologous end joining (NHEJ) and indels (insertions and deletions) that can alter the DNA sequence. In previous years, cas9 technology has confined its editing applications to Arabidopsis, tomato, sorghum, rice, maize, and wheat (Jiang et al., 2013; Liang et al., 2014; Wang et al., 2014; Zhang et al., 2014a; Selosse and Roy, 2009; Shan et al., 2015; Svitashev et al., 2015; Jacobs et al., 2015; Li et al., 2015; Sun et al., 2016; Du et al., 2016). The CRISPR/Cas9 approach has become a predominant molecular technique for genetic engineering linked to the improvement of quantitative and qualitative agronomic traits among various fodder crops against abiotic stresses (Debbarma et al., 2019).

In maize, CRISPR-Cas9 enabled advanced breeding technology to produce new ARGOS8 variants of maize by altering the DNA sequence at the native ARGOS8 locus in maize. The expression pattern of ARGOS 8 (from tissue specific to ubiquitous) was changed by replacing the ARGOS 8 promoter with the maize GOS2 promoter (GOS2 PRO). This change in expression led to a change in transcript level from comparatively low mRNA expression levels to considerably enhanced ARGOS8 expression levels. This well-defined modification in the nucleotide sequence of the ARGOS 8 gene at its native locus was done by cas9 technology and determined by PCR assays for that entire genome region followed by sequencing. Analysis of almost four generations of maize crops indicated that ARGOS 8 variants of maize were stably inherited. Field analysis showed that these novel variants were improved varieties with enhanced grain yield under drought stress. So the genome-editing tool can be used to produce new allelic variations within the genome sequences for increasing crop tolerance to abiotic stresses (Shi et al., 2016).

3.4 BIOTIC STRESS RESISTANCE IN FODDER CROPS

Biotic stress-tolerant cultivars can be developed by identifying the resistance (R) genes and mapping them with closely related molecular markers (Kumar et al., 2018). From many plant species, including monocots (barley, rice, and maize), dicots, pepper, tomato, beet, and potato, about 30 resistance genes have been isolated (Hulbert et al., 2001). It is quite challenging to isolate resistance genes from forage and turf grass species due to their complex and large genomes. However, in grass species, their genomes' conserved regions have been used to isolate resistance genes. According to studies, in Italian ryegrass, the nucleotide-binding site of resistance gene analogs (RGAs) was isolated, and the sequence conserved between resistance genes was used to design primers. From the total 9344 PCR cycles, about 62 open reading frames (ORFs) were identified and considered functional RGAs (Ikeda, 2005).

Several genes for developing resistance against pests and diseases have been introduced and tested in many plants. These are plant defensins, chitinases, phytoalexins, glucanases, viral coat proteins, viral movement proteins, ribosome-inactivating proteins, viral replicase, proteinase inhibitors, Bt toxins, and α-amylase inhibitors. To improve pest and disease resistance, few have been used to develop transgenic forage legumes (Spangenberg et al., 2001). Fungal pathogens usually attack the roots and leaves of many plants. So, the constitutive and organ-specific expression of the genes that usually encode antifungal proteins (AFPs) is collectively or individually responsible for developing resistance against fungal pathogens. It has been reported that the rice class I chitinase gene was introduced in the lucerne (alfalfa) forage crop against two pathogens, *Sclerotium rolfsii* and *Rhizoctonia solani* (Mizukami et al., 2000).

Viral infections caused by numerous viruses, including white clover mosaic potexvirus (WCMV), alfalfa mosaic alfamovirus (AMV), clover yellow vein potyvirus (CYVV), and many others, are causing significant damage to forage crops throughout the world. Many classical methods have been involved in controlling viral infections, but they are labor intensive and expensive economically. However, resistance in *Trifolium* species and alfalfa have been introduced against WCMV, AMV, and CYVV, but no significant results were obtained in resistant forage cultivars. Genetic engineering helps remove species-specific barriers, enhance multigenic resistance, and control the sites and expressions (Spangenberg et al., 2001).

3.4.1 Diseases of *Sorghum*

Several types of diseases, such as other smut (grain smut, loose smut, head smut, and long smut), rust, downy mildew, leaf spot/leaf blight, anthracnose, red rot, zonate leaf spot, and sooty stripe, have been found in sorghum.

Smut: Among all smut types, grain smut is the most destructive one, causing severe loss worldwide. The causing agent of grain smut in sorghum is *Sphacelotheca sorghi.* About 25% of grain yield is affected by grain smut in India (Rangaswami and Mahadevan, 1999). Symptoms of grain smut develop during the grain formation stage by converting into smut sori. Loose smut is caused by *Sphacelotheca cruenta* and is reported in Iran, Africa, China, Italy, and the United States. Loss smut symptoms appear like thinner stalks, premature flowering, and more tillers, and growth remains stunted. Head smut caused by *Sphacelotheca reiliana* is reported in Southern Europe, Africa, America, and Asia. Long smut is reported from many countries, including Iraq, Egypt, West Africa, Pakistan, and India, and the causative agent of this disease is a fungus named *Tolyposporium ehrenbergii.*

Rust: Another severe disease of sorghum is rust, which is caused by *Puccinia purpurea.* The severity of this disease depends on season and variety. This disease occurs at all stages of plant growth, but young plants are more vulnerable to this disease. Symptoms like tiny flecks start to appear on lower leaves.

Downy Mildew: This disease is reported in a mild form in many countries like Africa, Asia, Italy, and the United States. The damage caused by this disease depends on the time of infection and the environment. The cause of downy mildew in sorghum is a fungus, *Peronosclerospora sorghi.* This fungus attacks younger plants' growing tips by forming oospore or conidia and whitish patches on leaves' lower surfaces.

Leaf Spot/Blight: Leaf spot caused by *Exserohilum turcicum* also causes seedling blight and seed rot in sorghum. The symptoms include long elliptical necrotic lesions, along with dark margins (Kumar and Singh, 2012).

Anthracnose and Red Rot: This disease is found throughout the world and is more prevalent in the areas wherever it has grown (Jain, 2001). Anthracnose (leaf spot) and red rot (stalk rot) are both caused by the fungus *Colletotrichum graminicola.*

3.4.2 Diseases of Pearl Millet (Bajra)

Downy Mildew/Green Ear: This disease is mostly found in Iran, India, Israel, Fiji, Japan, the United States, China, and many other Africa countries. This disease's causative agent is *Sclerospora graminicola*, and it causes severe damage to pearl millet by affecting the leaves with chlorosis on the upper surface.

Rust: Another severe disease of pearl millet is rust, which is caused by *Puccinia penniseti.* The disease mainly affects the distal half of the lamina and circular uredosori on leaves' upper and lower surfaces. The plants became unhealthy and stunted during disease severity.

Smut: The disease is prevalent in many countries, including India, Pakistan, Africa, and the United States. Smut is caused by *Tolyposporium penicillariae* and causes damage at the time of grain setting. The typical grains are replaced by top-oval shaped sori, which are 2–3 times bigger and, on maturity, turn into dark black from bright green.

3.4.3 Diseases of Maize

The common diseases that can infect the maize crop are leaf blight, smut, downy mildew, and stalk rot.

Downy Mildew: Several fungal species have been reported that cause downy mildew of maize, and the symptoms of this disease vary and depend on the pathogen which attacks the maize crop.

3.4.4 Diseases of Oats

Smut, rust, and leaf blotch are the diseases found in oats.

Smuts: Two types of smut diseases that can damage oats have been found throughout the world, and these are loose smut and covered smut, caused by fungal species named *Ustilago avenae* and *Ustilago kolleri*, respectively.

Rusts: Two types of rust diseases, stem rust and crown rust, have been found in oats. Stem rust is caused by *Puccinia graminis*, responsible for severe loss to the crop, while crown rust is caused by *Puccinia coronata*, which is prevalent in temperate climatic regions (Kumar and Singh, 2012).

3.4.5 Diseases of Berseem

Root Rot Complex: A few pathogenic fungi, such as *Fusarium semitectum*, *Rhizoctonia solani*, and *Tylenchorhynchus vulgaris*, have been reported to cause root rot complex. Although a fungal pathogen is responsible for the initiation of infection, due to nematodes, the infection rate is enhanced (Hasan and Bhaskar, 2004).

Stem Rot: The disease is caused by a fungus, *Sclerotinia trifoliorum*, present in the soil. In Pakistan (Faisalabad), during February 2014, fluffy mycelium of white color was observed on the plant's aerial parts. Infected plants also showed water-soaked lesions, resulting in wilting and bleaching (Saira et al., 2017).

3.4.6 Diseases of Alfalfa

Significant alfalfa diseases are downy mildew, rust, leaf spot, and bacterial wilt.

Rust: Uromyces striatus causes the disease, and symptoms appear like the presence of red to brown uredia and telia on the stem and leaves.

Bacterial Wilt: Bacterial wilt caused by *Alpanobacter insidiosum* is the most damaging disease of lucerne. Plants with this disease show stunted growth and brown to yellow discoloration of woody tissues.

3.4.7 Diseases of Cluster Bean

Bacterial Blight: This disease of cluster bean is caused by *Xanthomonas campestris*. The symptoms include round and small intra-veinal spots on the leaves' dorsal surface, which may enlarge and become water-soaked lesions.

Alternaria Leaf Spot: The fungal pathogen *Alternaria cyamopsis* causes this severe disease of cluster bean by forming dark brown, round to irregular patches on leaves. In the early stage of growth, the plants fail to make flowers.

3.4.8 Diseases of Cowpea

Dry root rot, anthracnose, rust, and cowpea mosaic are the most common diseases of cowpea.

Rust: The rust bean caused by a pathogen, *Uromyces phaseoli*, typically is found on cowpea. Symptoms like rust pustules appear on the lower surface of leaves, pods, and shoots. First the fungus contains the uredial stage and develops into the telial stage as the disease progresses.

Cowpea Mosaic: This disease is caused by the cowpea mosaic virus transmitted by aphids (*Aphis gossypi*, *Aphis craccivora*, and *Myzus persicae*). The disease mainly affects the leaves by reducing their size and changing their color to pale yellow (Kumar and Singh, 2012).

3.5 BIOTECHNOLOGICAL APPLICATIONS FOR DEVELOPING RESISTANCE AGAINST BIOTIC STRESS IN FODDER CROPS

Several biotic stresses, including insect pest infection, pathogen attack, several diseases, and nematode infections, can cause severe loss to yield quality, nutritional quality, taste, and forage plants' persistence. Forage crops can be improved through biotechnological approaches in various ways by introducing the gene of interest, in-vitro regeneration of plants, and numerous molecular techniques for enhancing the yield and nutritional quality and developing resistance against biotic stresses (Kumar et al., 2013). Molecular breeding is preferable over conventional breeding, as it enhances the nutritional value; saves time; improves resistance against insects, diseases, and several other biotic stresses; increases post-harvest quality like protein and vitamin content, enhances the efficiency of biocontrol agents, and improves metabolic pathways and gene action in crop plants (Sharma et al., 2002).

Various molecular techniques, including molecular markers, microarrays, genetic transformation, and expressed sequence tags, can develop new cultivars with stress tolerance. Isozymes, restriction fragment length polymorphism, random amplified polymorphic DNA, amplified fragment length polymorphism, microsatellites or simple sequence repeats, and single nucleotide polymorphism are the PCR-based molecular markers. These markers can be used to make genetic linkage maps with high density and can ultimately generate QTLs or a set of discrete loci (Zhang et al., 2005).

RFLP markers are codominant and highly reproducible markers, as they have been used to study synteny and DNA conservation among similar plant species, but their implementation in plant improvement is limited due to its slow activity, need for a large amount of template DNA and labor-intensive nature. In 1986, the tomato was the first plant in which the RFLP map was developed, and then alfalfa (*Medicago sativa* L.) was the first forage crop to be mapped. It has also been reported that RFLP probes generated from cereal grasses can be efficiently used in many turf and forage species (Zhang et al., 2005; Brummer et al., 1999).

RAPD markers are made from a single primer of arbitrary sequence to overcome these limitations. They are dominant, highly cost effective, and highly polymorphic and require a minute amount of template DNA. However, due to a short sequence of primers, there are chances of errors like primer mismatch and nonspecific priming during PCR amplification, and hence these markers have low reproducibility

(Mueller and Wolfenbarger, 1999). RFLP and RAPD, both markers, have been the central systems for turf and forage species (Williams et al., 1990).

Another marker that is based on a random sequence of primers is AFLPs. These are fast and easy, have higher reproducibility than RAPD, and generate hundreds of genetic markers at a low cost (Mueller and Wolfenbarger, 1999). Unlike RAPD and AFLP markers, microsatellites or SSRs are more reproducible within or across the species, and they are codominant molecular markers. A considerable amount of time and cost is required to generate PCR-based SSR markers (Akkaya et al., 1992). Single nucleotide polymorphisms are considered an essential DNA polymorphism source (insertion or deletion of a nucleotide). SNPs are the marker alleles developed from candidate genes and are responsible for phenotypic changes.

Expressed sequence tags are small DNA sequences of about 200 to 500 base pairs generated from 3′ or 5′ ends of cDNA clones. They can also be used as molecular markers to study the comparative genomics among closely related species by aligning ESTs and can also be used as a source for SSR and SNP molecular markers. EST data can be used to check the expression of many genes at once in two different ways, like DNA microarrays and protein microarrays (Zhang et al., 2005). Another useful marker is EST-SSR for forage and turf grass species developed from cereal grasses (Warnke et al., 2004).

Quantitative trait loci can control phenotype variation in most turf and forage species. The majority of QTLs involved in the genetic variation of a trait may be determined by genetic mapping, each factor's environment, and the relative magnitude of effect (Foley, 2011; Forster et al., 2001). Cultivars resistant to many complex diseases can be developed by identifying the genetic markers linked with QTLs for disease resistance. QTLs against gray leaf spot were identified in two genomic regions of perennial ryegrass (Curley et al., 2004). Moreover, crown rust resistance was reported in perennial and Italian ryegrass (Zhang et al., 2005).

Resistance derived from pathogens has been used to produce transgenic forage plants with resistance against WCMV, AMV, and CYVV. For example, a lucerne crop expressing the coat protein gene against AMV has been found (Hill et al., 1991). Similarly, in white clover, coat protein-mediated resistance against WCMV indicated the virus's infection was delayed (Kalla et al., 2001; Voisey et al., 2001). White clover transgenic cultivars showed a high level of immunity or resistance when the mutated form of WCMV movement protein (13kDa) or WCMV replicase gene was expressed. Chimeric coat protein-mediated resistance showed against CYVV (Chu et al., 2000; Voisey et al., 2001). Resistance against RMV (ryegrass mosaic virus) has been developed in perennial ryegrass by transferring the RMV coat protein gene through particle bombardment, and the proteins expressed by this gene are responsible for the degradation of targeted RNA. Hence, post-transcriptional gene silencing occurs, and viral RNA replication stops (Zhang et al., 2005).

Insect pests can damage forage crops by either consuming the leaves or roots of the plant directly or indirectly acting as a vector for the transmission of harmful pathogens. A variety of pasture insects may cause adverse effects. Several transgenic approaches can enhance the resistance of forage plants against insects. The expression of these two, proteinase inhibitors (PIs) and *Bt* toxins, is considered effective against the insects of forage crops (Spangenberg et al., 2001). White clover plants expressed a modified chimeric Bt gene (CryIBa), and proteins expressed from this gene accumulated in leaves showed the toxicity against the insect. Proteinase inhibitors like aprotinin or bovine pancreatic trypsin inhibitor have also been shown to increase the resistance of forage crops to insect pests. About 0.07% of aprotinin from total soluble protein provides resistance against *Wiseana* larvae in forage plants (Voisey et al., 2001).

In forage and turf grass species, functional genomics has been used to study gene functions such as the function of chitinase, which can reduce summer patch disease in Kentucky bluegrass (Kobayashi et al., 2002). For expression analysis of disease-resistant genes, DNA microarrays were used. In New Zealand, the first forage gene chip was designed. This gene chip represents uni-genes generated from cDNA libraries of ryegrass and will help discover promoters and new gene expression profiling of ryegrass and other closely related species (Zhang et al., 2005).

3.6 BIOTECHNOLOGICAL APPLICATIONS AGAINST BIOTIC STRESSES IN *SORGHUM*

In the sorghum crop, about 150 insect pest species have been found that can cause high destruction to the forage yield and grain quality. Some of the most destructive pests of sorghum crops are sorghum midge, sorghum shoot fly, head bugs, and spotted stem borer. Several fungicides are used against the pathogens, but they are usually avoided due to high cost and harmful effects on human health and forage crops. Resistant cultivars can be developed or the sorghum genotypes that enhance resistance or tolerance to insect pests can be identified (Maqbool et al., 2001; Mohan et al., 2010). In India, several resistant varieties of sorghum have been developed through conventional breeding (Anahosur, 1992).

From several molecular markers, RFLPs have been significantly used to study the sorghum genome. In past studies, genes resistant against head smut, *Acremonium* wilt, downy mildew, grain mold, zonate leaf spot, leaf blight, and target leaf spot were identified using RAPD RFLP molecular markers (Oh et al., 1996; Mohan et al., 2010). It has also been reported that resistant genes for anthracnose disease have been mapped to the chromosomes (SBI-08 and SBI-05) in the sorghum crop (Singh et al., 2006a, 2006b). Witchweed-resistant sorghum hybrids can be quickly developed by detecting the witchweed-resistant gene's linkage with molecular markers through easy adoption of molecular assisted selection strategies (Mutengwa et al., 2005).

Two sorghum lines named PI 550610 and PI 550607 were considered resistant to greenbug biotype I (Andrews et al., 1993). Then, in 1996, biotype K was identified, which is more virulent than all biotypes of greenbug. Two sorghum resistant varieties, Cargill 607E and PI 550607, possess resistance against biotype K. Through the RFLP marker, the resistance in four sorghum sources was evaluated against Greenberg. About nine loci, named according to the genus of sorghum plant and greenbug, spread on eight linkage groups, were involved in sorghum resistance or tolerance to greenbug. Not even single loci are considered effective against all greenbug biotypes. From three loci, one locus seemed to be responsible for being highly resistant against biotypes (C and E) of greenbug. RFLP analysis showed that resistance in different sorghums might be due to allelic variation at specific loci (Katsar et al., 2002).

Molecular markers also have been used to detect and characterize QTL in sorghum for many traits, including resistance or tolerance to the parasitic weed *Striga*. Single marker analysis identified five QTLs for resistance against *Striga asiatica* and six for resistance to *Striga hermonthica*. The QTLs, which were identified for *S. asiatica* resistance, exhibited a 49% variation in resistance, while QTLs identified against *S. hermonthica* exhibited a 37% variation in resistance (Ejeta and knoll, 2007). According to recent studies, the molecular markers linked to head smut physiological race three are identified through microsatellites or SSR, and it would be easy to select resistant lines in the laboratory if molecular markers were found. Two SSR molecular markers, Xtxp13 and Xtxp145, have been identified and can be used to detect sorghum genes against head smut physiological race 3. Xtxp145 is located in linkage group I, while Xtxp13 is located in linkage group B, and the recombination percentage between the marker and the resistant gene is 10.4% and 9.6%, respectively. It is found that the interaction between two mutual independent nonallelic genes may control resistance to sorghum head smut physiological race 3 (Zou et al., 2010a, 2010b).

Another study was conducted to identify the molecular markers linked with ergot resistance in sorghum and identify the pollen traits, pollen viability (PV), pollen quality (PQ), and ergot-resistant relations in sorghum. Using 303 markers such as two morphological trait loci, 117 AFLP, 36 SSR, and 148 DArt, a genetic linkage map of the recombinant inbred line (RIL) population (R931945–2–2 × IS 8525) of sorghum was designed. Genetic linkage mapping, about four, five, and nine QTLs, was identified linked to molecular markers for PV, PQ, and percentage ergot infection (PCERGOT), respectively. The markers described in this report could be used for marker-assisted selection for ergot disease of sorghum (Parh et al., 2008).

For five-leaf diseases of sorghum, about 12 QTLs for three sorghum linkage groups (SBI-06, SBI-04, and SBI-03) were identified, exhibiting 6.9%-44.9% phenotypic variation (Mohan et al., 2010). It has been reported that four resistant lines of sorghum showed resistance against anthracnose disease (Mehta et al., 2005). Recently, anthracnose-resistant genes linked to SCAR and RAPD markers were mapped on the long arm of chromosome 8 in the G37 sorghum line (Singh et al., 2006a, 2006b). Four AFLP markers have been identified for anthracnose resistance and mapped on at the end of the sorghum (LG-05) linkage group. These markers are linked with the Cg1-resistant gene found in sorghum cultivar (SC748–5). One of the AFLP markers, Xtxa6227, and one SSR marker, Xtxp549, have been mapped between 1.8 and 3.6 cM of the resistance locus. Thirteen breeding lines of sorghum were genotyped to determine the efficacy of Xtxp549 and Xtxa6227 markers for MSA, and 12 breeding lines showed the association of molecular markers with the Cg1 locus. Hence, these two markers could help MSA strategies (Ramasamy et al., 2009).

Crystal proteins (CRYs) from *Bacillus thuringiensis* and several other genes that encode proteins such as proteinase and alpha-amylase inhibitors restrict many insects' digestive systems, including dipteran and lepidopterans, and develop transgenic sorghum plants. Overexpression of pathogen-related plant proteins, including glucanases and chitinases, has been used to inhibit the attacks of many fungi and bacteria (Bennetzen, 1995; Visarada and Kishore, 2007). A mild resistance against spotted stem borer neonate larvae was reported in the sorghum plant by introducing the Btcry1Ac gene under the protease inhibitor gene (mpi) from maize (Girijashankar et al., 2005).

Many plants produce a variety of compounds such as phytoalexins in response to harmful pathogens. In sorghum grains, phenolic compounds are found and help to foil fungal pathogens. Several phytochemicals like sorgoleone, dhurrin, and 3-deoxyanthocyanins are synthesized by *sorghum* (Du et al., 2010). Anthocyanins are synthesized in *sorghum*, called 3-deoxyanthocyanins, due to a hydroxyl group's absence in the 3-position of a carbon ring. These phytoalexins assemble in elementary bodies present inside the cell where fungi penetrate and release their substances. Then these compounds destroy the fungi and the cells that produce these compounds (Chandrashekar and Satyanarayana, 2006). New phytoalexins named flavone, luteolin, and apigenin have been identified against anthracnose in *sorghum* (Du et al., 2010).

Recent studies have reported that two enzymes, phenylalanine ammonia-lyase and chalcone synthase, are essential for phytoalexin production in *sorghum* against pathogens and stress (Nicholson and Wood, 2001). A resistant cultivar (SC748–5) was compared with a susceptible cultivar (BTx623); then phytoalexins and pathogenesis-related genes that encode PR-10 proteins and chalcone synthase were readily accumulated in a resistant cultivar of sorghum. Among all phytoalexin compounds, 3-deoxyanthocyanin phytoalexins account for higher toxicity against anthracnose disease (LO et al., 1999).

The expression analysis of genes involved in defense mechanisms against a fungal pathogen, *Macrophomina phaseolina*, was studied in the sorghum plant. These two antifungal genes were studied under controlled growth conditions using real-time PCR, Chitinase, and stilbene synthase. Two *sorghum* cultivars, resistant (PJ-1430) and susceptible (SU-1080), were used to check the expression of the genes at different times after infection with *M. phaseolina* (MTCC2165). It was observed that the expression of chitinase and stilbene synthase was higher; was produced more rapidly, within 0–24 h, in the resistant cultivar than the susceptible one; and can help increase the resistance against *M. phaseolina* (Sharma et al., 2014).

3.7 BIOTECHNOLOGICAL APPLICATIONS AGAINST BIOTIC STRESSES IN ALFALFA

In alfalfa, also known as lucerne, several insect pests can damage crop quality and yield. Although hundreds of insect species are known, only a few are responsible for crop destruction. These are the spotted alfalfa aphid, pea aphid, alfalfa weevil, blue alfalfa aphid, and potato leafhopper. Of all these insects, the alfalfa weevil is the most serious and can cause 20% to 30% destruction in forage crops' yield. Several fungicides have been used to control these insect pests, but biosafety concerns and environmental pollution are prohibited (Kumar, 2011). Two of the *Medicago* species showed resistance against alfalfa weevil and aphids, but *Medicago sativa*, alfalfa, is sexually incompatible with these two species, and it is quite difficult to transfer these traits in alfalfa through conventional breeding (Chandra, 2009; Mizukami et al., 2006).

Through various biotechnological techniques, somatic hybridization, embryo rescue, and the in-planta gene transfer method could be used to transfer traits and interspecific hybrids (Kumar et al., 2018). The protoplasts from *M. rugosa/M. scutellate* and *M. sativa* were electro-fused and successful genome transfer from *M. rugosa* to *M. sativa* accomplished. Some of the resultant offspring exhibited resistance against alfalfa weevil (Mizukami et al., 2006). Another effective method to develop resistance in alfalfa crops against insects is genetic transformation with the *Bacillus thuringiensis* (*Bt*) gene (McCaslin et al., 2002). A *Bt* gene, Cry1C, which encodes a δ-endotoxin protein, expressed in alfalfa, showed high resistance against beet armyworm and alfalfa weevil (Strizhov et al., 1996). The synthetic Cry3a gene expression in the alfalfa genome showed resistance against weevil pests (Tohidfar et al., 2013).

Another gene transfer–mediated insect resistance was developed through the introduction of the insect proteinase inhibitor gene. A gene from *Manduca sexta*, protease inhibitor anti-elastase, was expressed in alfalfa, and about 0.125% PI protein were produced in leaves, flowers, and roots. This PI anti-elastase indicated insecticidal properties in the alfalfa plant (Thomas et al., 1994). In recent studies, it has been reported that map-based cloning of the resistance to *Colletotrichum trifolii* race 1 (RCT1) gene for R protein exhibiting anthracnose resistance was transferred to the alfalfa plant. The results suggested that potential anthracnose resistance in the alfalfa plant helped to understand translational research progress from *M. trancatula* to *M. sativa* (Yang et al., 2008).

The fungal chitinase gene, endochitinase gene (ech42), was introduced in the alfalfa plant. Results indicated that resistance developed against many fungal pathogens by enhancing chitinase activity by up to 7.5 to 25.7 times in root exudates and vegetative organs compared to untransformed alfalfa plants (Tesfaye et al., 2005). Verticillium wilt is a significant disease of alfalfa caused by *Verticillium alfalfa*. Resistance in alfalfa can be developed against verticillium wilt through the development of molecular markers. Bulk segregant analysis was done in resistant and susceptible cultivars, designed from 13 synthetic alfalfa populations, to identify loci linked with verticillium wilt resistance. Genotyping was done by two markers, single nucleotide polymorphism and simple sequence repeats. About 17 SNP markers were found, located on the 1, 2, 4, 7, and 8 chromosomes; also, SNP markers found on chromosomes 2, 4, and 7 in resistant cultivars shared a location with verticillium-resistant loci in *Medicago trancatula* (Zhang et al., 2014b).

Quantitative trait loci linked with Stagonospora root rot, leaf spot, anthracnose resistance, and many other traits have been reported in alfalfa, but no QTL was identified against verticillium wilt resistance (Musial et al., 2007; Yang et al., 2008; Zhang et al., 2014b). Recently, QTLs associated with *Stagonospora meliloti* resistance or susceptibility were identified in 145 individuals of alfalfa backcross populations. Interval mapping and regression analysis were observed in one region on 2, 6, and 7 linkage groups linked with *S. meliloti* disease reaction in two various experiments. The QTLs involved in *S. meliloti* resistance are located on seven linkage groups and exhibited a 17% variation in phenotype; however, those involved in susceptibility, located on two linkage groups, accounted for 16% phenotypic variation. Markers associated with QTLs for *S. meliloti* resistance will be valid for MAS (Musial et al., 2007).

Using two molecular markers, random amplified polymorphic DNA and amplified fragment length polymorphic, resistance was estimated as quantitatively inherited, and QTLs linked with anthracnose resistance and susceptibility were recognized in a backcross population of alfalfa. A multi-locus region on four linkage groups is considered to exhibit the resistance phenotype (Irwin et al., 2006).

An experiment was conducted in Texas; two alfalfa populations were selected for DNA polymorphism by 14 AFLP primer combinations. A total of 36 resistant cultivars from the UC-143 population and 36 susceptible cultivars from the UC-123 population were selected for downy mildew resistance (isolate I-8). A total of 14 AFLP markers and 4AFLP molecular markers were identified linked with disease resistance susceptibility. By the cross of resistant and susceptible alfa plants (F_1 and S_1 progeny), two AFLP markers ($ACACTC_{486}$ and $ACACTC_{208}$) were linked with downy mildew resistance (Obert et al., 2000).

Cysteine proteinases are two critical digestive enzymes of insect pests such as nematodes, hemipteran, and coleopteran. Cysteine proteinase inhibitors belong to the family phytocystatins, are present in many plant species, and play an essential role in protecting against many insects (Samac and Smigocki, 2003). Two phytocystatins, oryzacystatin I (OC-I) and oryzacystatin II (OC-II), play an inhibitory role against cysteine proteinases in rice seeds. Both OC-I and OC-II have inhibitory activity against papain and cathepsin, respectively (Kondo et al., 1990). OC-I and OC-II from rice were introduced in alfalfa under potato protease inhibitor II (PinII) promoter control. To check gene expression from the PinII promoter, another gene, PinII-β-glucuronidase (GUS), was introduced. GUS expression was observed in root and leaf vascular tissues by inoculating alfalfa plants with root-lesion nematodes. So, oryzacystatinhas can enhance alfalfa's resistance to root-lesion nematodes (Samac and Smigocki, 2003).

The production of flavonoids protects against several diseases, including anthracnose in alfalfa. The activity of three enzymes, cinnamic acid 4-hydroxylase (CA4H), phenylalanine ammonia-lyase (PAL), and isoflavone reductase (IFR), is involved in flavonoid biosynthesis. The induction of phytoalexins like medicarpin and sativan due to host-pathogen interaction as the alfalfa plant is exposed to the avirulent *Collectotrichum trifolli* race. These phytoalexins also protect against virulent *C. trifolli* strains. Rapid accumulation of phytoalexins occurred when alfalfa plants inoculated with both virulent and avirulent types rather than only virulent and avirulent type inoculation (Saunders and O'Neill, 2004). It has also been reported that the highest accumulation of isoflavonoid phytoalexin medicarpin was observed in the roots of two resistant alfalfa plants against root-lesion nematode (Baldridge et al., 1998).

3.8 BIOTECHNOLOGICAL APPLICATIONS AGAINST BIOTIC STRESSES IN BERSEEM

Berseem (*Trifolium alexandrinum*), also known as Egyptian clover, is prone to many insect pests, fungi, and viruses. Hence, the yield and quality of berseem crops are affected due to these pathogens. Different biotechnological approaches have been applied to discover the genes responsible for biotic stress tolerance, disease and insect resistance, salinity, and drought tolerance in berseem. Biotechnological approaches use several molecular breeding methods such as in-vitro regeneration of any part of the plant and deleting or inserting a particular gene into a plant genome for specific traits.

In conventional breeding, molecular techniques like androgenic haploid plant production, embryo rescue, and micro-propagation (somatic embryogenesis and organogenesis) were used in one or many steps. These techniques help generate new genotypes in a short period of about two years with low cost and more genetic purity than conventional breeding, which requires six years to generate new genotypes. The development of genetic variations in berseem through the insertion of genes by physical, chemical, and electrical methods; genetic transformation; somaclonal variation; and somatic hybridization is easy and is impossible through conventional methods (Zayed, 2013). In the middle of the 19th century, biotechnological approaches were initiated in berseem. Plant regeneration from meristematic tissue and developing into regenerable callus culture have been done in several *Trifolium* species, including *Trifolium alexandrinum*, *T. apertum*, *T. glomeratum*, and *T. resupinatum* (Kaushal et al., 2006; Kaushal et al., 2011).

A resistant cultivar of berseem (Bundel berseem-3) was produced through polyploidy, and a mild level of resistance was observed against stem rot and root rot diseases (Singh et al., 2019). Other *Trifolium* species such as *T. apertum*, *T. resupinatum*, *T. constantinopolitanum*, and *T. vesiculosum* showed resistance against profuse basal branching, root rot, stem rot, and several other diseases (Malaviya et al., 2004; Bhaskar et al., 2002; Roy et al., 2004). Limited genomic resources have been found in berseem; hence, molecular markers have been used for genetic improvement in berseem (Chandra, 2011; Verma et al., 2015).

By the development of biochemical markers, isozymes, the genetic variation was studied in *Trifolium alexandrinum* and other related *Trifolium* species (Malaviya et al., 2005). DNA-based markers such as random amplified polymorphic DNA, restriction fragment length polymorphism, inter-simple sequence repeat, amplified fragment length polymorphism, single nucleotide polymorphism, simple sequence repeat, and ribosomal RNA (rRNA) have been developed and utilized for cultivar identification, characterization of germplasm, detection of hybrids, genetic mapping, gene tagging, and quantitative trait loci identification in berseem (Zayed, 2013).

3.9 BIOTECHNOLOGICAL APPLICATIONS AGAINST BIOTIC STRESSES IN MAIZE

Several destructive diseases, including bacterial diseases, viral diseases, leaf blight, ear rot, and stalk and kernel rot, can cause severe damage to the quantity and quality of maize crop grains (Ali and Yan, 2012). To develop resistance against many diseases, chemical and cultural practices have also been used, but chemical control is avoided to treat the seeds, and hence, host-plant resistance has been employed and is considered adequate due to low cost (Yadav et al., 2015). Through modern biotechnological techniques like association mapping and joint linkages, genes have been identified for many traits to improve plant genetic makeup and disease resistance against many diseases (Ali and Yan, 2012).

Four resistance genes have been introduced in maize, and two of them were cloned through a method called transposon tagging. These two genes, Hm1 and Rp1-D, showed resistance in maize against *Cochliobolus carbonum* race-1 and maize common rust, respectively (Johal and Briggs, 1992; Collins et al., 1999). Quantitative trait loci associated with several disease-resistant genes have been identified in maize crops through molecular marker technology. By developing resistant cultivars, the maize crop's significant diseases, including head smut, can be controlled. In a recent study, four QTLs linked with head smut disease resistance were identified by the cross of resistant and susceptible varieties of maize crops and then backcrossed with susceptible ones, although qHSR1 was recognized as the major one and mapped on bin 2.09 (Chen et al., 2008).

A new QTL has been reported in another study, which also provides resistance against head smut disease in maize. The QTL, qHS2.09, was identified in a near-isogenic line (NIL) obtained by the cross of a resistant plant, Mo17, and a susceptible plant, Huangzao4. This QTL is also mapped on 2.09 by using two single nucleotide polymorphism markers. In the maize breeding method, SNPs are positively associated with qHS2.09 QTL and can also help marker-assisted selection (Weng et al., 2012). A minimum of five QTLs were identified in maize using DNA-based markers such as RFLP and RAPD against stalk rot disease caused by *Gibberellazeae*. Moreover, the identification of QTL was dependent on interval mapping, with the help of MAPMAKER-QTL and regression analysis coefficients of family mean value and allelic value in the F_2 population. A linkage map was developed in maize, which showed about four or five regions that may carry genetic factors for developing resistance against *G. zeae* (Pe et al., 1993).

In a recent study, five QTLs were reported from a resistant parent (O61) linked with *Cercospora zeae-maydis* resistance (Clements et al., 2000). Another study was conducted in the United States and South Africa, and several other areas found resistance loci against *C. zeae-maydis*, which causes gray leaf spot disease in maize. From the cross of resistant and susceptible maize varieties, $F_{2:4}$ progeny lines were evaluated. Numerous markers on chromosomes 2 and 4 are linked with resistance in this progeny. On the long arms of chromosomes 2 and 4, QTL determined a 40–47% variation in phenotype. So, the resistant inbred (VO613Y) is considered a resistance source to gray leaf spot disease (Gordon et al., 2004).

Many compounds, such as maysin and other associated compounds, including 3′-methoxymaysin, apimaysin, and chlorogenic acid, exhibited antibiotic properties to corn earworm disease. In maize, a maysin compound (C-glycosyl flavone maysin) exhibits an inhibitory effect in silk tissues towards corn earworms (Rector et al., 2002). The primary purpose was to identify the molecular markers linked with maysin and other associated compounds in maize crops. In the F_2 maize population, which is derived from the cross of two inbred maize lines, QTL for marker-assisted selection strategy can be identified against corn earworm resistance. Two significant QTLs were found, one on the short arm of chromosome 1 and the other on the csu1066-umc176 interval on the 2C-2L genomic region, involved in the production of maysin and other associated compounds (Butrón et al., 2001).

Through segregation analysis, about four RAPD markers were identified associated with mosaic dwarf mosaic virus (MDMV)–resistant genes in maize. These

RAPD markers are located in the 25cM region of the associated gene. The fragment of about 650 base pairs elongated by the UBC376 primer was closely linked with the MDMV gene (Agrama and Moussa, 1996). In previous studies, it has been shown that Mdm1, a gene with the ability to show resistance against MDMV, was found in the Pa405 inbred maize line. Using the RFLP molecular marker, umc85, the gene associated with MDMV resistance was found on the short arm of chromosome 6 (McMullen and Louie, 1989; McMullen and Louie, 1991). So, it was concluded that Mdm1 and other related genes found on the short arm of chromosome 6 are linked with MDMV resistance in maize germplasm (Jones et al., 2007).

Chromosome regions linked with resistance to *Aspergillus flavus* and inhibition of aflatoxin synthesis have been found through RFLP analysis in three resistant inbred lines (R001, LB31, and Tex6) of maize (White et al., 1998). In Central and West Africa, maize inbred lines have been screened out through the kernel screening assay (KSA) method for aflatoxin and ear rot resistance (Brown et al., 2001). In inbred, resistant maize, two kernel proteins with 28 kDa and more than 100 kDa size were identified, and these proteins were involved in inhibition of *A. flavus* growth and toxin formation, respectively (Huang et al., 1997). In maize, numerous QTLs linked with aflatoxin contamination and *A. flavus* inhibition have been reported, as one QTL was found in a resistant inbred Mp313E of maize (Paul et al., 2003; Brooks et al., 2005).

Through log-transformed percent kernel uninfected (PKU) and number of pollen grains germinated (NPG), three and five QTLs associated with markers were confirmed in maize. One QTL associated with the TRAP marker was detected. Similarly, three QTLs were detected through discriminant analysis (DA). Hence, it is proved that several genetic mechanisms are involved in controlling the resistance of kernels and maize to *A. flavus* (Alwala et al., 2008). Maize crops are also involved in producing peptide compounds such as ZmPep3, which plays a defensive role against fungal pathogens and herbivores (Huffaker et al., 2011; Huffaker et al., 2013). Other defensive chemicals of maize include protease inhibitors and phytoalexins that protect against many biotic stresses. A cysteine protease, mirl-CP, a defensive compound in maize, provides resistance against aphids, *Diabrotica* spp., and *Spodoptera frugiperda* (Lopez et al., 2007). Kauralexins and zealexins, two families of phytoalexins, are important against pathogen activity (Schmelz et al., 2011).

3.10 BIOTECHNOLOGICAL APPLICATIONS AGAINST BIOTIC STRESSES IN PEARL MILLET

Pearl millet, also known as bajra, is susceptible to many biotic stresses, including bacterial, fungal, and viral pathogens, resulting in severe loss in pearl millet yield and production. Among several diseases of pearl millet, downy mildew, or green ear, are the most damaging diseases in Africa and India (Shivhare and Lata, 2017). To improve pearl millet against these biotic stresses, several techniques, including evaluation of germplasm, hybrid parental lines, development, and implementation of several screening techniques and breeding material, have been introduced so far (Yadav and Rai, 2013). Host plant resistance for developing new resistant varieties by identifying and introducing foreign gene into plant genome is also a practical biotechnology approach. Pearl millet's germplasm has been evaluated to check resistance towards several fungal pathogens, and many other diseases were detected (Shivhare and Lata, 2017).

The biolistic transformation method, an antifungal protein (afp) gene from immature zygotic embryos of *Aspergillus giganteus*, was introduced in pearl millet, and transformed pearl millet showed about 90% resistance to downy mildew compared to non-transgenic plants (Girgi et al., 2006). Similarly, another pearl millet transgenic plant showed resistance against downy mildew disease. This transgenic plant was grown by introducing the prawn antifungal protein-encoding gene (pin) into the pearl millet embryonic calli, and the transformation was confirmed by PCR, and northern and southern blotting (Latha et al., 2006). In the future, genes like glucanase and chitinase may be inserted into pearl millet to enhance its resistance against numerous fungal pathogens and improve its overall yield (Ceasar and Ignacimuthu, 2009).

Several molecular markers such as RFLP, AFLP, RAPD, SSRs/microsatellites, SNP, EST, CISP, STSs, and DArTs have been developed that assist in studying population structure and detection of QTLs against biotic and abiotic stress tolerance (Shivhare and Lata, 2017). Many QTLs associated with downy mildew resistance have been identified in pearl millet using several molecular markers (Hash et al., 2001; Jones et al., 2002). The genetically complicated forms of disease-resistant genes can be studied easily through QTL mapping. A genetic linkage map was designed to locate QTLs linked with rust disease resistance using diversity array technology (DArT) and SSR molecular markers in pearl millet. A significant QTL for rust resistance was identified on linkage group 1 by developing a genetic map and accounted for a 58% phenotypic difference (Ambawat et al., 2016). Segregant analysis of the pearl millet F_2 population reported that the $SCAR_{ISSR\ 863}$ marker was linked with downy mildew resistance (Jogaiah et al., 2014).

According to research conducted in Mali, Burkina Faso, Niger, India, and Nigeria between 1995 and 1999, two resistant inbred lines (700651 and P 310–17) of pearl millet showed high resistance against downy mildew disease and remained stable across locations (Thakur et al., 2004). Among several pathogenesis-related (PR) proteins, β-1,3-glucanase, synthesized by pearl millet seeds, protects against fungal pathogens, including downy mildew disease. Β-1,3-glucanase acts as a biochemical marker to screen out pearl millet resistant varieties. This enzyme's high level is found in seeds highly infested with downy mildew pathogen, *Sclerospora graminicola* (Kini et al., 2000).

Hydroxy proline-rich glycoproteins (HRGPs) are proteins synthesized in the plant cell wall in response to pathogen attack. High accumulation of HRGPs against *S. graminicola* was observed in the resistant cultivar (IP_18_{292}) of pearl millet by comparing resistant and susceptible cultivars. Hence, it was observed that high content of hydroxyproline in HRGPs, H_2O_2, and a combination of polypeptide cross-linking of isodityrosine were involved in cell wall strengthening of pearl millet resistant cultivar against downy mildew pathogen (Deepak et al., 2007). In a recent study, glycosphingolipids such as cerebrosides obtained from various plant pathogens induced resistance to downy mildew in pearl millet (Deepak et al., 2003). Similarly, polyphenol oxidase (PPO) activity was observed in a resistant cultivar of pearl millet, and this enzyme was also involved in the plant defense mechanism. Higher PPO activity was observed in seedlings of pearl millet inoculated with a downy mildew pathogen. In pearl millet, PPO can also be used as a marker to identify downy mildew resistance (Raj et al., 2006).

3.11 BIOTECHNOLOGICAL APPLICATION FOR QUALITY TRAITS IN FODDER CROPS

Pakistan's economy largely depends on agriculture. The agricultural investment returns are relatively low compared to services and industrial sectors, but this sector provides a cushion to sustain the masses' economy and living standards over time. The livestock sector, an essential pillar in the agro-based economy, ensures national food and nutrition security. The livestock sector shows great potential in developing countries in terms of growth and productivity. It shows a growth of 2.58 this year and the second-highest growth after the crops sector in 2019–2020 (GOP, 2020).

All the attention is given to significant cash crops due to their share in agriculture. Oilseed and pulses gain attention due to high import bills. Livestock was not a priority area for policymakers, but livestock became the largest subsector of agriculture over the last decade. It employs millions of people, as more than 8 million farm families are linked to livestock for their earnings, and it contributes 3.1% to the foreign exchange earnings of Pakistan. The livestock sector is an essential source of non-farm income. Being an agro-based economy, Pakistan has to concentrate on agriculture to sustain its industrial growth, as all the industries have forward and backward linkages with agriculture. Likewise, livestock also provides inputs for industries like dairy, meat, and so on. In the last few years, the livestock sector has become the most significant contributor in value addition in agriculture and contributed more than 60% in agricultural value addition and approximately 12% to the national GDP in 2019–2020 (GOP, 2020). The gross value addition of livestock has great potential due to the adoption of innovative farming practices and high-yielding breeds.

The livestock sector faces many challenges in developing countries due to traditional farm practices and high demands due to population growth. Farmlands are declining over time, and fodder crops are shifting towards marginal lands due to high sustainable cash crop prices. Due to this reason, the feed needs of the livestock sector are being neglected to some extent. Various studies highlighted the importance of livestock feed in terms of higher livestock productivity. The livestock sector must make enhancing farm incomes a priority. Livestock includes milk, meat, manure, hides, and fibers. Infect livestock is one of the significant growth sectors in most develop and developing countries. It became a priority area due to its growth potential and efforts to increase productivity to harness that growth potential.

Fodder crops are the most comprehensive and affordable source of livestock feed. Unfortunately, in underdeveloped and developing countries, the shortage of fodder crops is a significant factor in restrictive livestock production. Fodder crops are cultivated and harvested to feed livestock animals, such as forage, hay, and silage. Fodder crop production is vital for animal production and plays an essential role in soil maintenance through preventing soil erosion, increasing soil fertility, and improving water-holding capacity. Leguminous fodder crops (*Medicago sativa, Trifolium alexandrinum*, etc.) contain 7–11% digestible crude protein, 40–60% total digestible nutrients, and low phosphorus content, so farmers use wheat or rice bran mixed with fodder crop for animal grazing. Based on dry matter, non-leguminous fodder crops (hay, jawar, bajra, etc.) contain 2.5% digestible crude protein and 45–55% total digestible nutrients. Vitamin A is the primary source of green fodder crops. Green fodder contains 100 mg carotenes/kg when compared with about 20 mg/kg in silage. Carotene's requirement in milch livestock animals is 11 mg for growth and 60 mg for production, and the pregnancy requirement is 30 mg per 100 kg live weight.

Alfalfa (*Medicago sativa*) is an essential perennial herbaceous legume with a deep root that reaches down to 4.5 m, but in well-drained soils, it reaches down to 7–10 m. (Frame, J., 2005) Alfalfa is not available throughout the year with high nutritious values and protein content. It is essential for organic farming because of its nitrogen fixation ability, nutritional quality, and adaptive capacity (Torricelli, 2006). Berseem (*Trifolium alexandrinum*) winter fodder provides three-fourths of the winter forage crop known as king of forages. It is used for fodder, increases soil fertility, and has good nutritional value (Hannaway and Larson, 2004). In developing countries like Pakistan, supply and demand for fodder crops had substantially increased because the livestock sector was associated with fodder production (Shaug SuePea et al., 2000). *Trifolium alexandrinum* is also known as a milk multiplier.

White clover (*Trifolium repens*) is a perennial forage crop that is self-sufficient in nitrogen content and provides nitrogen to other plants through nitrogen fixation. The white clover yield is low compared to alfalfa (*Medicago sativa*) and berseem (*Trifolium alexandrinum*). Several investors focus on setting up to increase dairy farm and livestock production in Pakistan. This sector is directly linked with fodder crop production and thus will increase the demand for berseem, barley, alfalfa, and other fodder crops in Pakistan. Higher demand for fodder will ultimately force scientists and farmers to enhance the quantity and quality of fodder crops by increasing their production and enhancing the nutritional value.

Biotic and abiotic stresses are major limiting factors of low yields and cause severe damage to fodder crops. Downy mildew disease alone can cause severe yield losses, up to 70% in alfalfa (Ahmad et al., 1977) and 50% in sorghum (Sundaram et al., 1970). Stem and crown rot and root rot disease reduce the yield and the quality of *Trifolium*. In *Trifolium* diseases, 17% damage the plants in the field before harvest (Heffer and Johnson, 2007).

Moreover, diseases not only reduce the yield of fodder crops but also reduce the quality of fodder. To improve the genetic potential of forage, most fodder crops and genetic data are not available but could be identified from related species; hence for quantity and quality enhancement, biotechnological approaches are implemented in fodder crops. This allows transformation of genes from related and unrelated species to upregulate and downregulate gene expression and introduce novel genes. Agriculture biotechnology has powerful potential to increase production by eliminating severe threats from insects, pests, and pathogens. First transgenic forage-type tall fescue plant production has had tremendous success in bioenergy and fodder crops (Wang et al., 1992).

Digestibility is essential for fodder crops. The major constraint on ruminant digestion of forage cell walls is lignin. For increased digestibility, caffeic acid *O*-methyltransferase (COMT) encoding lignin biosynthetic enzyme was cloned from monocot fodder species *Festuca arundinacea* (tall fescue). The enzymatic properties of antisense downregulation of lignin biosynthesis gene caffeic acid *O*-methyltransferase protein expressed in *Escherichia coli* were determined using six substrates. To improve digestibility, 5-hydroxyferulic acid and caffeoyl aldehyde substrate were more critical for transgenic fodder crops. Transgenic *Festuca arundinacea* showed reduced expression level, reduced lignin content, and significantly increased digestibility (9.8–10.8%) (Lei Chen et al., 2004).

Alfalfa (*Medicago sativa*) is a crucial forage crop to which biotechnological approaches have been extensively applied. In 2006, the first transgenic variety, Roundup Ready alfalfa, was released (Smith et al., 2005). Scientist focus on essential traits was introduced by genetic engineering in fodder crops. Overexpression of cytosolic glutamine synthetase showed a significant increase in photosynthetic and protein content (Fuentes et al., 2001).

Phosphorus is an essential component of quality and increases the biomass of forages. Phosphorus nutrition components could be provided by expressing hydrolise phosphate enzyme. In *Medicago truncatula*, purple acid phosphatase and fungal phytase genes were expressed, and results revealed that significantly increase phosphorus assimilation (Nunes et al., 2009). Polyphenol oxidase and o-diphenol PPO substrates are associated with the detrimental effect of browning. PPO-containing fodder crops were associated with the reduction of nitrogen losses in silo and rumen. Red clover has up to 85% less proteolysis due to the presence of the PPO gene. By expressing the red clover PPO gene in alfalfa (*Medicago sativa*), forage crops showed a fivefold decrease in proteolysis (Sullivan and Hatfield, 2006) and reduced nitrogen losses. To improve the digestibility of the *Medicago sativa* fodder crop, antisense downregulation of lignin biosynthesis gene COMT and caffeoyl CoA 3-O-methyltransferase (CCOMT) increased the protein content and dry matter digestibility (Guo et al., 2001).

Sulfur is an essential amino acid for crops and animal diets that boosts the true protein and soluble sugar content in fodder. Animals need the balance of nitrogen and sulfur nutrients in their feed. The shortage of sulfur nutrient content may reduce the quality and digestibility of feed. δ-zein encodes a sulfur maize seed storage protein that accumulates a higher concentration in the endoplasmic reticulum (Medina-Holguín et al., 2008; Bellucci et al., 2007). These genes were introduced to improve the protein content and forage quality in white clover by agrobacterium-mediated gene transformation under a constitutive promoter (Sharma, 1998). Maize seed storage proteins are an essential source of livestock feed consumption, although these proteins have low nutritional values due to lower amounts of lysine and tryptophan contents. Lysine-rich protein genes (sb401 and SBgLR) from potatoes were transformed into maize under a 35S promoter. Total protein content was significantly improved in transgenic maize (Yue et al., 2014).

The most crucial trait of fodder crops is high biomass production. Nutritional contents also depend on fodder biomass. The ability to shoot meristems to respond with increased growth after cutting is essential for biomass production. The *Arabidopsis* enhanced drought tolerance1 gene was transformed in alfalfa (*Medicago sativa*) to increase root length, height of the shoot, and biomass production. Moreover, it enhances soluble sugar, proline, and chlorophyll content to improve the fodder plant (Zheng et al., 2017).

Scientists help improve the production, quality, and quantity traits of fodder crops that bioinformatics and genomics tools have revolutionized. Moreover, sequence information also helps identify and solve biotic and abiotic stresses (Thorogood et al., 2017). Scientists also focus on identifying different candidate genes, which transform fodder crops to enhance the quality of fodder for livestock and increase the soil's water-holding capacity and fertility.

REFERENCES

Abdel-Tawab FM, Fahmy EM, Bahieldin A and Eissa HF (1997) Molecular markers for salt tolerance in some inbreds of maize, *Zea mays L. Centro Internacional de Mejoramiento de Maíz y Trigo* 1998: 113221.

Abdel-Tawab F, Fahmy E, El-Nahrawy M, Sharawy W and Sayed M (2011) Detection of molecular markers associated with salt tolerance in alfalfa (*Medicago sativa* L.). *Egyptian Journal of Genetics and Cytology* 40(1): 113–127.

Abogadallah GM (2010) Sensitivity of *Trifolium alexandrinum* L. to salt stress is related to the lack of long-term stress-induced gene expression. *Plant Science* 178(6): 491–500.

Abogadallah GM, Nada RM, Malinowski R and Quick P (2011) Overexpression of HARDY, an AP2/ERF gene from Arabidopsis, improves drought and salt tolerance by reducing transpiration and sodium uptake in transgenic *Trifolium alexandrinum* L. *Planta* 233(6): 1265–1276.

Agrama HAS and Moussa ME (1996) Identification of RAPD markers tightly linked to the dwarf mosaic virus resistance gene in maize. *Maydica* 41(3): 205–210.

Ahmad ST, Srinath PR and Gupta MP (1977) Estimation of losses due to downy mildew in lucerne. *Indian Phytopathology* 30(1): 466–468.

Akkaya MS, Bhagwat AA and Cregan PB (1992) Length polymorphism of short tandem repeat DNA sequences in soybean. *Genetics* 132(4): 1131–1139.

Alam I, Sharmin SA and Lee BH (2010) Reviews; Advances in the molecular breeding of forage crops for abiotic stress tolerance. *Journal of Plant Biotechnology* 37(4): 425.

Ali F and Yan J (2012) Disease resistance in maize and the role of molecular breeding in defending against global threat. *Journal of Integrative Plant Biology* 54(3): 134–151.

Alia, Chen THH and Murata N (1998a) Transformation with a gene for choline oxidase enhances the cold tolerance of Arabidopsis during germination and early growth. *Plant, Cell and Environment* 21(2): 232–239.

Alwala S, Kimbeng CA, Williams WP and Kang MS (2008) Molecular markers associated with resistance to *Aspergillus flavus* in maize grain: QTL and discriminant analyses. *Journal of New Seeds* 9(1): 1–18.

Ambawat S, Senthilvel S, Hash CT, Nepolean T, Rajaram V, Eshwar K and Srivastava RK (2016) QTL mapping of pearl millet rust resistance using an integrated DArT-and SSR-based linkage map. *Euphytica* 209(2): 461–476.

Anahosur KH (1992) Sorghum diseases in India: Knowledge and research needs. In de Milliano WA, Frederiksen RA, Bengston GD (eds) *Sorghum and Millets Diseases A Second World Review*. India, Patancheru: International Crops Research Institute for the Semi-Arid Tropics, pp. 45–56.

Andresen PA, Kaasen I, Styrvold OB and Boulnois G (1988) Molecular cloning, physical mapping and expression of the bet genes governing the osmoregulatory choline-glycine betaine pathway of *Escherichia coli*. *Microbiology* 134(6): 1737–1746.

Andrews DJ, Bramel-Cox PJ and Wilde GE (1993) New sources of resistance to greenbug, biotype I, in sorghum. *Crop Science* 33(1): 198–199.

Arshad M, Feyissa BA, Amyot L, Aung B and Hannoufa A (2017) MicroRNA156 improves drought stress tolerance in alfalfa (*Medicago sativa*) by silencing SPL13. *Plant Science* 258(2017): 122–136.

Ashraf M (2010) Inducing drought tolerance in plants: Recent advances. *Biotechnology Advances* 28(1): 169–183.

Ashraf M, Athar HR, Harris PJC and Kwon TR (2008) Some prospective strategies for improving crop salt tolerance. *Advances in Agronomy* 97(2008): 45–110.

Asins MJ (2002) Present and future of quantitative trait locus analysis in plant breeding. *Plant Breeding* 121(4): 281–291.

Athar HR and Ashraf M (2009) Strategies for crop improvement against salinity and drought stress: An overview. In: Ashraf M, Ozturk M and Athar HR (eds) Salinity and Water Stress. Netherlands, Dordrecht: Springer, pp. 1–16.

Avonce N, Leyman B, Mascorro-Gallardo JO, Van Dijck P, Thevelein JM and Iturriaga G (2004) The Arabidopsis trehalose-6-P synthase AtTPS1 gene is a regulator of glucose, abscisic acid, and stress signaling. *Plant Physiology* 136(3): 3649–3659.

Baldridge GD, O'Neill NR and Samac DA (1998) Alfalfa (*Medicago sativa* L.) resistance to the root-lesion nematode, *Pratylenchus penetrans*: Defense-response gene mRNA and isoflavonoid phytoalexin levels in roots. *Plant molecular biology* 38(6): 999–1010.

Bartel DP (2004) MicroRNAs: Genomics, biogenesis, mechanism, and function. *Cell* 116(2): 281–297.

Bartels D and Sunkar R (2005) Drought and salt tolerance in plants. *Critical Reviews in Plant Sciences* 24(1): 23–58.

Bastola DR, Pethe VV and Winicov I (1998) Alfin1, a novel zinc-finger protein in alfalfa roots that binds to promoter elements in the salt-inducible MsPRP2 gene. *Plant Molecular Biology* 38(6): 1123–1135.

Bélanger G, Castonguay Y, Bertrand A, Dhont C, Rochette P, Couture L and Michaud R (2006) Winter damage to perennial forage crops in eastern Canada: Causes, mitigation, and prediction. *Canadian Journal of Plant Science* 86(1): 33–47.

Bellucci M, De Marchis F and Arcioni S (2007) Zeolin is a recombinant storage protein that can be used to produce value-added proteins in alfalfa (*Medicago sativa* L.). *Plant Cell, Tissue and Organ Culture* 90(1): 85–91.

Bennetzen JL (1995) Biotechnology for sorghum improvement. *African Crop-Science Journal (Uganda)* 3(2): 161–170.

Bhaskar RB, Malaviya DR, Roy AK and Kaushal P (2002) Evaluation of exotic *Trifolium* accessions for disease incidence and resistance. In Abstracts of National Symposium on Grassland and Fodder Research in the New Millennium, UAS, Dharwad, 3–14 December, pp. 31–33.

Bohnert HJ and Shen BO (1998) Transformation and compatible solutes. *Scientia Horticulturae* 78(1–4): 237–260.

Boyer JS (1982) Plant productivity and environment. *Science* 218(4571): 443–448.

Boyer JS and Westgate ME (2004) Grain yields with limited water. *Journal of Experimental Botany* 55(407): 2385–2394.

Breto MP, Aśins MJ and Carbonell EA (1994) Salt tolerance in Lycopersicon species. III. Detection of quantitative trait loci by means of molecular markers. *Theoretical and Applied Genetics* 88(3–4): 395–401.

Brooks TD, Williams WP, Windham GL, Willcox MC and Abbas HK (2005) Quantitative trait loci contributing resistance to aflatoxin accumulation in the maize inbred Mp313E. *Crop Science* 45(1): 171–174.

Broun P, Poindexter P, Osborne E, Jiang CZ and Riechmann JL (2004) WIN1, a transcriptional activator of epidermal wax accumulation in Arabidopsis. *Proceedings of the National Academy of Sciences* 101(13): 4706–4711.

Brown RL, Chen ZY, Menkir A, Cleveland TE, Cardwell K, Kling J and White DG (2001) Resistance to aflatoxin accumulation in kernels of maize inbreds selected for ear rot resistance in West and Central Africa. *Journal of Food Protection* 64(3): 396–400.

Brummer EC, Cazcarro PM and Luth D (1999) Ploidy determination of alfalfa germplasm accessions using flow cytometry. *Crop Science* 39(4): 1202–1207.

Brunk DG, Rich PJ and Rhodes D (1989) Genotypic variation for glycinebetaine among public inbreds of maize. *Plant Physiology* 91(3): 1122–1125.

Butrón A, Li RG, Guo BZ, Widstrom NW, Snook ME, Cleveland TE and Lynch RE (2001) Molecular markers to increase corn earworm resistance in a maize population. *Maydica* 46(2001): 117–124.

Campos H, Cooper M, Habben JE, Edmeades GO and Schussler JR (2004) Improving drought tolerance in maize: A view from industry. *Field Crops Research* 90(1): 19–34.

Capstaff NM and Miller AJ (2018) Improving the yield and nutritional quality of forage crops. *Frontiers in Plant Science* 9(1): 1–18.

Castiglioni P, Warner D, Bensen RJ, Anstrom DC, Harrison J, Stoecker M, Abad M, Kumar G, Salvador S, D'Ordine R and Navarro S (2008) Bacterial RNA chaperones confer abiotic stress tolerance in plants and improved grain yield in maize under water-limited conditions. *Plant Physiology* 147(2): 446–455.

Cattivelli L, Rizza F, Badeck FW, Mazzucotelli E, Mastrangelo AM, Francia E and Stanca AM (2008) Drought tolerance improvement in crop plants: An integrated view from breeding to genomics. *Field Crops Research* 105(1–2): 1–14.

Ceasar SA and Ignacimuthu S (2009) Genetic engineering of millets: Current status and future prospects. *Biotechnology Letters* 31(6): 779–788.

Century K, Reuber TL and Ratcliffe OJ (2008) Regulating the regulators: The future prospects for transcription-factor-based agricultural biotechnology products. Plant physiology 147(1): 20–29.

Čermák T, Baltes NJ, Čegan R, Zhang Y and Voytas DF (2015) High-frequency, precise modification of the tomato genome. *Genome Biology* 16(1): 1–5.

Chandra A (2009) Screening global Medicago germplasm for weevil (*Hypera postica* Gyll.) tolerance and estimation of genetic variability using molecular markers. *Euphytica* 169(3): 363–374.

Chandra A (2011) Use of EST database markers from *M. truncatula* in the transferability to other forage legumes. *Journal of Environmental Biology* 32(3): 347–354.

Chandrashekar A and Satyanarayana KV (2006) Disease and pest resistance in grains of sorghum and millets. *Journal of Cereal Science* 44(3): 287–304.

Chen JR, Lü JJ, Liu R, Xiong XY, Wang TX, Chen SY and Wang HF (2010) DREB1C from *Medicago truncatula* enhances freezing tolerance in transgenic *M. truncatula* and China Rose (*Rosa chinensis* Jacq.). *Plant Growth Regulation* 60(3): 199–211.

Chen L, Auh CK, Dowling P, Bell J, Lehmann D and Wang ZY (2004) Transgenic down-regulation of caffeic acid O-methyltransferase (COMT) led to improved digestibility in tall fescue (*Festuca arundinacea*). *Functional Plant Biology* 31(3): 235–245.

Chen X (2004) A microRNA as a translational repressor of APETALA2 in Arabidopsis flower development. *Science* 303(5666): 2022–2025.

Chen Y, Chao Q, Tan G, Zhao J, Zhang M, Ji Q and Xu M (2008) Identification and fine-mapping of a major QTL conferring resistance against head smut in maize. *Theoretical and Applied Genetics* 117(8): 1241.

Choi DW, Rodriguez EM and Close TJ (2002) Barley Cbf3 gene identification, expression pattern, and map location. *Plant Physiology* 129(4): 1781–1787.

Chu P, Holloway B, Venables I, Lane L, Gibson J, Larkin P and Higgins TJ (2000) Development of transgenic white clover (*Trifolium repens*) with resistance to clover yellow vein virus. In: *Abstracts 2nd International Symposium Molecular Breeding of Forage Crops*. Victoria, Australia: Lorne and Hamilton.

Claassen MM and Shaw RH (1970) Water deficit effects on Corn. I. Vegetative components. *Agronomy Journal* 62(5): 649–652.

Clements MJ, Dudley JW and White DG (2000) Quantitative trait loci associated with resistance to gray leaf spot of corn. *Phytopathology* 90(9): 1018–1025.

Collins N, Drake J, AyliVe M, Sun Q, Ellis J, Hulbert S and Pryor T (1999) Molecular characterization of the maize Rpl-D rust resistance haplotype and its mutants. *Plant Cell* 11: 1365–1376.

Cortina C and Culiáñez-Macià FA (2005) Tomato abiotic stress enhanced tolerance by trehalose biosynthesis. *Plant Science* 169(1): 75–82.

Cowell IG (2002) E4BP4/NFIL3, a PAR-related bZIP factor with many roles. *Bioessays* 24(11): 1023–1029.

Creelman RA, Mason HS, Bensen RJ, Boyer JS and Mullet JE (1990) Water deficit and abscisic acid cause differential inhibition of shoot versus root growth in soybean seedlings: Analysis of growth, sugar accumulation, and gene expression. Plant Physiology 92(1): 205–214.

Curley J, Sim SC, Jung G, Leong S, Warnke S and Barker RE (2004) QTL mapping of gray leaf spot resistance in ryegrass, and synteny-based comparison with rice blast resistance genes in rice. In: Hopkins A, Wang ZY, Mian R, Sledge M and Barker RE (ed) *Molecular Breeding of Forage and Turf*. Dordrecht, the Netherlands: Kluwer Acad. Publ., pp. 37–46.

Dalal M and Sharma TR (2017) Biotechnological applications for improvement of drought tolerance. In *Abiotic Stress Management for Resilient Agriculture*. Singapore: Springer, pp. 299–312.

Dangl JL and Jones JD (2001) Plant pathogens and integrated defence responses to infection. *Nature* 411(6839): 826–833.

de Vries GE (2000) Climate changes leads to unstable agriculture. *Trends in Plant Science* 5(9): 367.

Debbarma J, Sarki YN, Saikia B, Boruah HPD, Singha DL and Chikkaputtaiah C (2019) Ethylene response factor (ERF) family proteins in abiotic stresses and CRISPR—Cas9 genome editing of ERFs for multiple abiotic stress tolerance in crop plants: A review. *Molecular Biotechnology* 61(2): 153–172.

Deepak S, Shailasree S, Kini RK, Hause B, Shetty SH and Mithöfer A (2007) Role of hydroxyproline-rich glycoproteins in resistance of pearl millet against downy mildew pathogen *Sclerospora graminicola*. *Planta* 226(2): 323–333.

Deepak SA, Raj SN, Umemura K, Kono T and Shetty HS (2003) Cerebroside as an elicitor for induced resistance against the downy mildew pathogen in pearl millet. Annals of Applied Biology 143(2): 169–173.

Du H, Zeng X, Zhao M, Cui X, Wang Q, Yang H and Yu D (2016) Efficient targeted mutagenesis in soybean by TALENs and CRISPR/Cas9. *Journal of Biotechnology* 217(1): 90–97.

Du Y, Chu H, Wang M, Chu I K and Lo C (2010) Identification of flavone phytoalexins and a pathogen-inducible flavone synthase II gene (SbFNSII) in sorghum. *Journal of Experimental Botany* 61(4): 983–994.

Dubouzet JG, Sakuma Y, Ito Y, Kasuga M, Dubouzet EG, Miura S and Yamaguchi-Shinozaki K (2003) OsDREB genes in rice, *Oryza sativa* L., encode transcription activators that function in drought-, high-salt-and cold-responsive gene expression. *The Plant Journal* 33(4): 751–763.

Echt CS, Erdahl LA and McCoy TJ (1992) Genetic segregation of random amplified polymorphic DNA in diploid cultivated alfalfa. *Genome* 35(1): 84–87.

Ejeta G and Knoll JE (2007) Marker-assisted selection in sorghum. In *Genomics-assisted crop improvement*. Dordrecht: *Springer*, pp. 187–205.

Fahmy EM, Abdel-Tawab FM, Belal AFH and Sharawy WM (1997) Marker-assisted selection for drought-tolerance in berseem clover (*Trifolium alexandrinum* L.). *Journal of Union of Arab Biologists* 4(1): 303–328.

Foley JA (2011) Can we feed the world sustain the planet? *Scientific American* 305(5): 60–65.

Forster JW, Jones ES, Kölliker R, Drayton MC, Dumsday JL, Dupal MP and Smith KF (2001) Development and implementation of molecular markers for forage crop improvement. In: Spangenberg G (ed) *Molecular Breeding of Forage Crops*. Dordrecht: Springer, pp. 101–133.

Frame J (2005) *Forage Legumes for Temperate Grasslands*. Amsterdam: FAO and Science Publishers.

Fuentes SI, Allen DJ, Ortiz-Lopez A and Hernández G (2001) Over-expression of cytosolic glutamine synthetase increases photosynthesis and growth at low nitrogen concentrations. *Journal of Experimental Botany* 52(358): 1071–1081.

Gao F, Gao Q, Duan X, Yue G, Yang A and Zhang J (2006) Cloning of an H+-PPase gene from *Thellungiella halophila* and its heterologous expression to improve tobacco salt tolerance. *Journal of Experimental Botany* 57(12): 3259–3270.

Gao R, Feyissa BA, Croft M and Hannoufa A (2018) Gene editing by CRISPR/Cas9 in the obligatory outcrossing *Medicago sativa. Planta* 247(4): 1043–1050.

Gao S, Zhang H, Tian Y, Li F, Zhang Z, Lu X and Huang R (2008) Expression of TERF1 in rice regulates expression of stress-responsive genes and enhances tolerance to drought and high-salinity. *Plant Cell Reports* 27(11): 1787–1795.

Garg AK, Kim JK, Owens TG, Ranwala AP, Do Choi Y, Kochian LV and Wu RJ (2002) Trehalose accumulation in rice plants confers high tolerance levels to different abiotic stresses. *Proceedings of the National Academy of Sciences* 99(25): 15898–15903.

Gaxiola RA, Li J, Undurraga S, Dang LM, Allen GJ, Alper SL and Fink GR (2001) Drought- and salt-tolerant plants result from overexpression of the AVP1 H+-pump. *Proceedings of the National Academy of Sciences* 98(20): 11444–11449.

Gilmour SJ, Sebolt AM, Salazar MP, Everard JD and Thomashow MF (2000) Overexpression of the Arabidopsis CBF3transcriptional activator mimics multiple biochemical changes associated with cold acclimation. *Plant Physiology* 124(4): 1854–1865.

Gilmour SJ, Zarka DG, Stockinger EJ, Salazar MP, Houghton JM and Thomashow MF (1998) Low temperature regulation of the Arabidopsis CBF family of AP2 transcriptional activators as an early step in cold-induced COR gene expression. *The Plant Journal* 16(4): 433–442.

Girgi M, Breese WA, Lörz H and Oldach KH (2006) Rust and downy mildew resistance in pearl millet (*Pennisetum glaucum*) mediated by heterologous expression of the afp gene from *Aspergillus giganteus. Transgenic Research* 15(3): 313–324.

Girijashankar V, Sharma HC, Sharma KK, Swathisree V, Prasad LS, Bhat BV and Seetharama N (2005) Development of transgenic sorghum for insect resistance against the spotted stem borer (*Chilo partellus*). *Plant Cell Reports* 24(9): 513–522.

GOP, Pakistan economic survey (2020) www.finance.gov.pk/survey_1920.html.

Gordon SGa, Bartsch M, Matthies I, Gevers HO, Lipps PE and Pratt RC (2004) Linkage of molecular markers to *Cercospora zeae-maydis* resistance in maize. *Crop Science* 44(2): 628–636.

Gorham J (1996) Mechanisms of salt tolerance of halophytes. *Halophytes and Biosaline Agriculture* 1996: 31–53.

Guan Q, Lu X, Zeng H, Zhang Y and Zhu J (2013) Heat stress induction of miR398 triggers a regulatory loop that is critical for thermotolerance in Arabidopsis. *The Plant Journal* 74(5): 840–851.

Guo D, Chen F, Inoue K, Blount JW and Dixon RA (2001) Downregulation of caffeic acid 3-O-methyltransferase and caffeoyl CoA 3-O-methyltransferase in transgenic alfalfa: Impacts on lignin structure and implications for the biosynthesis of G and S lignin. *The Plant Cell* 13(1): 73–88.

Guo M, Rupe MA, Wei J, Winkler C, Goncalves-Butruille M, Weers BP and Hou Z (2014) Maize ARGOS1 (ZAR1) transgenic alleles increase hybrid maize yield. *Journal of Experimental Botany* 65(1): 249–260.

Habben JE, Bao X, Bate NJ, DeBruin JL, Dolan D, Hasegawa D and Reimann K (2014) Transgenic alteration of ethylene biosynthesis increases grain yield in maize under field drought-stress conditions. *Plant Biotechnology Journal* 12(6): 685–693.

Hannaway DB and Larson C (2004) *Berseem clover (Trifolium alexandrinum L.)*. Corvallis, OR: Oregon State University, Species Selection Information System.

Haqqani AM, Zahid MA and Malik MR (2000) Legumes in Pakistan. Legumes in rice cropping system of the Indo-genetic planes-constraints and opportunities. *ICRISAT. India* 230: 98–128.

Hasan N and Bhaskar RB (2004) Disease complex of berseem involving nematode, and two soil inhabiting fungi. *Annals of Plant Protection Sciences* 12(1): 159–161.

Hash CT, Rahman MA, Raj AB and Zerbini E (2001) Molecular markers for improving nutritional quality of crop residues for ruminants. In: Spangenberg G (ed) *Molecular Breeding of Forage Crops*. Dordrecht: Springer, pp. 203–217.

Hayashi Y, Maharjan KL and Kumagai H (2006) Feeding traits, nutritional status and milk production of dairy cattle and buffalo in small-scale farms in Terai, Nepal. *ASIAN Australasian Journal of Animal Sciences* 19(2): 189–197.

Heffer LV and Johnson KB (2007) The Plant Health Instructor. Available at: www.apsnet.org/edcenter/disandpath/fungalasco/pdlessons/Pages/WhiteMoldPort.aspx (accessed 12–02–2021).

Hill KK, Jarvis-Eagan N, Halk EL, Krahn KJ, Liao LW, Mathewson RS and Loesch-Fries LS (1991) The development of virus-resistant alfalfa, *Medicago sativa* L. *Biotechnology* 9(4): 373–377.

Holmström C, Egan S, Franks A, McCloy S and Kjelleberg S (2002) Antifouling activities expressed by marine surface associated Pseudoalteromonas species. *FEMS Microbiology Ecology* 41(1): 47–58.

Holmström KO, Mäntylä E, Welin B, Mandal A, Palva ET, Tunnela OE and Londesborough J (1996) Drought tolerance in tobacco. *Nature* 379(6567): 683–684.

Huang J, Hirji R, Adam L, Rozwadowski KL, Hammerlindl JK, Keller WA and Selvaraj G (2000) Genetic engineering of glycinebetaine production toward enhancing stress tolerance in plants: Metabolic limitations. *Plant Physiology* 122(3): 747–756.

Huang Z, White DG and Payne GA (1997) Corn seed proteins inhibitory to *Aspergillus flavus* and aflatoxin biosynthesis. *Phytopathology* 87(6): 622–627.

Huffaker A, Kaplan F, Vaughan MM, Dafoe NJ, Ni X, Rocca JR and Schmelz EA (2011) Novel acidic sesquiterpenoids constitute a dominant class of pathogen-induced phytoalexins in maize. *Plant Physiology* 156(4): 2082–2097.

Huffaker A, Pearce G, Veyrat N, Erb M, Turlings TC, Sartor R and Teal PE (2013) Plant elicitor peptides are conserved signals regulating direct and indirect antiherbivore defense. *Proceedings of the National Academy of Sciences* 110(14): 5707–5712.

Hulbert S, Webb C, Smith S and Sun Q (2001) Resistance gene complexes: Evolution and utilization. *Annual Review of Phytopathology* 39: 285–312.

Iannucci A, Rascio A, Russo M, Di Fonzo N and Martiniello P (2000) Physiological responses to water stress following a conditioning period in berseem clover. *Plant and Soil* 223(1–2): 219–229.

Ikeda S (2005) Isolation of disease resistance gene analogs from Italian ryegrass (*Lolium multiflorum* Lam.). *Grassland Science* 51: 63–70.

Ingram J and Bartels D (1996) The molecular basis of dehydration tolerance in plants. *Annual Review of Plant Biology* 47(1): 377–403.

Irwin JAG, Aitken KS, Mackie JM and Musial JM (2006) Genetic improvement of lucerne for anthracnose (*Colletotrichum trifolii*) resistance. *Australasian Plant Pathology* 35(6): 573–579.

Jacobs TB, LaFayette PR, Schmitz RJ and Parrott WA (2015) Targeted genome modifications in soybean with CRISPR/Cas9. *BMC Biotechnology* 15(1): 1–10.

Jaglo KR, Kleff S, Amundsen KL, Zhang X, Haake V, Zhang JZ and Thomashow MF (2001) Components of the Arabidopsis C-repeat/dehydration-responsive element binding factor cold-response pathway are conserved in *Brassica napus* and other plant species. *Plant Physiology* 127(3): 910–917.

Jaglo-Ottosen KR, Gilmour SJ, Zarka DG, Schabenberger O and Thomashow MF (1998) Arabidopsis CBF1 overexpression induces COR genes and enhances freezing tolerance. *Science* 280(5360): 104–106.

Jain RK (2001) Pests and diseases of fodder crops and their management. In: Trivedi PC (ed) *Plant Pathology*. Jaipur: Pointer Publishers, pp. 422.

Jakoby M, Weisshaar B, Dröge-Laser W, Vicente-Carbajosa J, Tiedemann J, Kroj T and Parcy F (2002) bZIP transcription factors in Arabidopsis. *Trends in Plant Science* 7(3): 106–111.

Jauhar PP (1993) Alien gene transfer and genetic enrichment of bread wheat. In: *Biodiversity and Wheat Improvement*. Chichester, UK: John Wiley & Sons, pp. 103–119.

Jenks MA, Eigenbrode SD and Lemieux B (2002) Cuticular waxes of Arabidopsis. *The Arabidopsis Book/American Society of Plant Biologists* 1.

Jia Z, Lian Y, Zhu Y, He J, Cao Z and Wang G (2009) Cloning and characterization of a putative transcription factor induced by abiotic stress in *Zea mays*. *African Journal of Biotechnology* 8: 1.

Jiang W, Zhou H, Bi H, Fromm M, Yang B and Weeks DP (2013) Demonstration of CRISPR/Cas9/sgRNA-mediated targeted gene modification in Arabidopsis, tobacco, sorghum and rice. *Nucleic acids research* 41(20): e188-e188.

Jin T, Chang Q, Li W, Yin D, Li Z, Wang D and Liu L (2010) Stress-inducible expression of GmDREB1 conferred salt tolerance in transgenic alfalfa. *Plant Cell, Tissue and Organ Culture (PCTOC)* 100(2): 219–227.

Jogaiah S, Sharathchandra RG, Raj N, Vedamurthy AB and Shetty HS (2014) Development of SCAR marker associated with downy mildew disease resistance in pearl millet (*Pennisetum glaucum* L.). *Molecular Biology Reports* 41(12): 7815–7824.

Johal GS and Briggs SP (1992) Reductase activity encoded by the HM1 disease resistance gene in maize. *Science* 258(5084): 985–987.

Jones ES, Breese WA, Liu CJ, Singh SD, Shaw DS and Witcombe JR (2002) Mapping quantitative trait loci for resistance to downy mildew in pearl millet: Field and glasshouse screens detect the same QTL. *Crop Science* 42(4): 1316–1323.

Jones MW, Redinbaugh MG and Louie R (2007) The Mdm1 locus and maize resistance to maize dwarf mosaic virus. *Plant Disease* 91(2): 185–190.

Kalla R, Chu P and Spangenberg G (2001) Molecular breeding of forage legumes for virus resistance. In: Spangenberg G (eds) *Molecular Breeding of Forage Crops*. Dordrecht: Springer, pp. 219–237.

Karaba A, Dixit S, Greco R, Aharoni A, Trijatmiko KR, Marsch-Martinez N and Pereira A (2007) Improvement of water use efficiency in rice by expression of HARDY, an Arabidopsis drought and salt tolerance gene. *Proceedings of the National Academy of Sciences* 104(39): 15270–15275.

Kasuga M, Liu Q, Miura S, Yamaguchi-Shinozaki K and Shinozaki K (1999) Improving plant drought, salt, and freezing tolerance by gene transfer of a single stress-inducible transcription factor. *Nature Biotechnology* 17(3): 287–291.

Katsar CS, Paterson AH, Teetes GL and Peterson GC (2002) Molecular analysis of sorghum resistance to the greenbug (Homoptera: Aphididae). *Journal of Economic Entomology* 95(2): 448–457.

Kaushal N, Gupta K, Bhandhari K, Kumar S, Thakur P and Nayyar H (2011) Proline induces heat tolerance in chickpea (*Cicer arietinum* L.) plants by protecting vital enzymes of carbon and antioxidative metabolism. *Physiology and Molecular Biology of Plants* 17(3): 203–213.

Kaushal P, Tiwari A, Roy AK, Malaviya DR and Kumar B (2006) *In vitro* regeneration of *Trifolium glomeratum. Biologia Plantarum* 50(4): 693–696.

Kini KR, Vasanthi NS and Shetty HS (2000) Induction of β-1, 3-glucanase in seedlings of pearl millet in response to infection by *Sclerospora graminicola. European Journal of Plant Pathology* 106(3): 267–274.

Kobayashi DY, Reedy RM, Bick J and Oudemans PV (2002) Characterization of a chitinase gene from *Stenotrophomonas maltophilia* strain 34S1 and its involvement in biological control. *Applied and Environmental Microbiology* 68(3): 1047–1054.

Kondo H, Abe K, Nishimura I, Watanabe H, Emori Y and Arai S (1990) Two distinct cystatin species in rice seeds with different specificities against cysteine proteinases. Molecular cloning, expression, and biochemical studies on oryzacystatin-II. *Journal of Biological Chemistry* 265(26): 15832–15837.

Kruszka K, Pacak A, Swida-Barteczka A, Nuc P, Alaba S, Wroblewska Z and Szweykowska-Kulinska Z (2014) Transcriptionally and post-transcriptionally regulated microRNAs in heat stress response in barley. *Journal of Experimental Botany* 65(20): 6123–6135.

Kumar B and Singh KP (2012) *Major Diseases of Forage & Fodder Crops and Their Ecofriendly Management.* Ranichauri, Uttarakhand, India, pp. 573–590.

Kumar D, Pannu R, Kumar S and Arya RK (2013) A review: Recent advances in forage crop improvement through biotechnology. *Biomirror* 4(12): 1–5.

Kumar R (2014) Role of microRNAs in biotic and abiotic stress responses in crop plants. *Applied Biochemistry and Biotechnology* 174(1): 93–115.

Kumar RR, Pathak H, Sharma SK, Kala YK, Nirjal MK, Singh GP and Rai RD (2015) Novel and conserved heat-responsive microRNAs in wheat (*Triticum aestivum* L.). *Functional & Integrative Genomics* 15(3): 323–348.

Kumar S (2011) Biotechnological advancements in alfalfa improvement. *Journal of Applied Genetics* 52(2): 111–124.

Kumar T, Bao AK, Bao Z, Wang F, Gao L and Wang SM (2018) The progress of genetic improvement in alfalfa (*Medicago sativa* L.). *Czech Journal of Genetics and Plant Breeding* 54(2): 41–51.

Lamark T, Kaasen I, Eshoo MW, Falkenberg P, McDougall J and Strøm AR (1991) DNA sequence and analysis of the bet genes encoding the osmoregulatory choline–glycine betaine pathway of *Escherichia coli. Molecular Microbiology* 5(5): 1049–1064.

Landfald B and Strøm AR (1986) Choline-glycine betaine pathway confers a high level of osmotic tolerance in *Escherichia coli. Journal of Bacteriology* 165(3): 849–855.

Lang NT and Buu BC (2008) Fine mapping for drought tolerance in rice (*Oryza sativa* L.). *Omonrice* 16(2008): 9–15.

Latha AM, Rao KV, Reddy TP and Reddy VD (2006) Development of transgenic pearl millet (*Pennisetum glaucum* (L.) R. Br.) plants resistant to downy mildew. *Plant Cell Reports* 25(9): 927–935.

Le Rudulier D, Strom AR, Dandekar AM and Smith LT (1984) Valentine RC. Molecular biology of osmoregulation. *Science* 224(4653): 1064–1068.

Li B, Wei A, Song C, Li N and Zhang J (2008) Heterologous expression of the TsVP gene improves the drought resistance of maize. *Plant Biotechnology Journal* 6(2): 146–159.

Li F, Pignatta D, Bendix C, Brunkard JO, Cohn MM, Tung J, Sun H, Kumar P and Baker B (2012) MicroRNA regulation of plant innate immune receptors. *Proceedings of the National Academy of Sciences* 109(5): 1790–1795.

Li S, Li SK, Gan RY, Song FL, Kuang L and Li HB (2013) Antioxidant capacities and total phenolic contents of infusions from 223 medicinal plants. *Industrial Crops and Products* 51: 289–298.

Li Z, Liu ZB, Xing A, Moon BP, Koellhoffer JP, Huang L, Ward RT, Clifton E, Falco SC and Cigan AM (2015) Cas9-guide RNA directed genome editing in soybean. *Plant Physiology* 169(2): 960–970.

Liang Z, Zhang K, Chen K and Gao C (2014) Targeted mutagenesis in *Zea mays* using TALENs and the CRISPR/Cas system. *Journal of Genetics and Genomics* 41(2): 63–68.

Lilius G, Holmberg N and Bülow L (1996) Enhanced NaCl stress tolerance in transgenic tobacco expressing bacterial choline dehydrogenase. *Bio/technology* 14(2): 177–180.

Lin JJ, Kuo J, Ma J, Saunders JA, Beard HS, MacDonald MH and Matthews BF (1996) Identification of molecular markers in soybean comparing RFLP, RAPD and AFLP DNA mapping techniques. *Plant Molecular Biology Reporter* 14(2): 156–169.

Liu Q, Kasuga M, Sakuma Y, Abe H, Miura S, Yamaguchi-Shinozaki K and Shinozaki K (1998) Two transcription factors, DREB1 and DREB2, with an EREBP/AP2 DNA binding domain separate two cellular signal transduction pathways in drought-and low-temperature-responsive gene expression, respectively, in Arabidopsis. *The Plant Cell* 10(8): 1391–1406.

Liu S, Hao Z, Weng J, Li M, Zhang D, Pan G and Li X (2015) Identification of two functional markers associated with drought resistance in maize. *Molecular Breeding* 35(1): 1–10.

Lo SCC, De Verdier K and Nicholson RL (1999) Accumulation of 3-deoxyanthocyanidin phytoalexins and resistance to *Colletotrichum sublineolum* in sorghum. *Physiological and Molecular Plant Pathology* 55(5): 263–273.

Lopez L, Camas A, Shivaji R, Ankala A, Williams P and Luthe D (2007) Mirl-CP, a novel defense cysteine protease accumulates in maize vascular tissues in response to herbivory. *Planta* 226(2): 517–527.

Lu C and Fedoroff N (2000) A mutation in the Arabidopsis HYL1 gene encoding a dsRNA binding protein affects responses to abscisic acid, auxin, and cytokinin. *The Plant Cell* 12(12): 2351–2365.

Lu S, Sun YH, Shi R, Clark C, Li L and Chiang VL (2005) Novel and mechanical stress-responsive microRNAs in *Populus trichocarpa* that are absent from Arabidopsis. *The Plant Cell* 17(8): 2186–2203.

Malaviya DR, Kumar B, Roy AK, Kaushal P and Tiwari A (2005) Estimation of variability of five enzyme systems among wild and cultivated species of *Trifolium*. *Genetic Resources and Crop Evolution* 52(7): 967–976.

Malaviya DR, Roy AK, Kaushal P, Kumar B, Tiwari A and Lorenzoni C (2004) Development and characterization of interspecific hybrids of *Trifolium alexandrinum* X T-apertum using embryo rescue. *Plant Breeding* 123(6): 536–542.

Maqbool SB, Devi P and Sticklen MB (2001) Biotechnology: Genetic improvement of sorghum (*Sorghum bicolor* (L.) Moench). *In Vitro Cellular & Developmental Biology-Plant* 37(5): 504–515.

Matthews C, Arshad M and Hannoufa A (2019) Alfalfa response to heat stress is modulated by microRNA156. *Physiologia Plantarum* 165(4): 830–842.

McCaslin M, Temple S and Tofte J (2002) inventors; Forage Genetics Inc, assignee. Methods for maximizing expression of transgenic traits in autopolyploid plants. *United States patent application* US 09/900,314.

McMullen MD and Louie R (1989) The linkage of molecular markers to a gene controlling the symptom response in maize to maize dwarf mosaic virus. *Molecular Plant Microbe Interaction* 2(30): 314.

McMullen MD and Louie R (1991) Identification of a gene for resistance to wheat streak mosaic virus in maize. *Phytopathology (USA)* 81(6): 624–627.

Medina-Holguín AL, Holguín FO, Micheletto S, Goehle S, Simon JA and O'Connell MA (2008) Chemotypic variation of essential oils in the medicinal plant, *Anemopsis californica*. *Phytochemistry* 69(4): 919–927.

Mehta PJ, Wiltse CC, Rooney WL, Collins SD, Frederiksen RA, Hess DE and TeBeest DO (2005) Classification and inheritance of genetic resistance to anthracnose in sorghum. *Field Crops Research* 93(1): 1–9.

Michaels SD and Amasino RM (1998) A robust method for detecting single-nucleotide changes as polymorphic markers by PCR. *The Plant Journal* 14(3): 381–385.

Mizukami Y, Houmura I, Takamizo T and Nishizawa Y (2000) November). Production of transgenic alfalfa with chitinase gene (RCC2). In *Abstracts 2nd International Symposium Molecular Breeding of Forage Crops*. Vol. 105. Victoria: Lorne and Hamilton.

Mizukami Y, Takamizo T, Kanbe M, Inami S and Hattori K (2006) Interspecific hybrids between *Medicago sativa* L. and annual Medicago containing alfafa weevil resistance. *Plant Cell, Tissue and Organ Culture* 84(1): 80–89.

Mohan SM, Madhusudhana R, Mathur K, Chakravarthi DVN, Rathore S, Reddy RN and Seetharama N (2010) Identification of quantitative trait loci associated with resistance to foliar diseases in sorghum [*Sorghum bicolor* (L.) Moench]. *Euphytica* 176(2): 199–211.

Mohanty A, Kathuria H, Ferjani A, Sakamoto A, Mohanty P, Murata N and Tyagi A (2002) Transgenics of an elite indica rice variety Pusa Basmati 1 harbouring the codA gene are highly tolerant to salt stress. *Theoretical and Applied Genetics* 106(1): 51–57.

Monfreda C, Ramankutty N and Foley JA (2008) Farming the planet: 2. Geographic distribution of crop areas, yields, physiological types, and net primary production in the year 2000. *Global Biogeochemical Cycles* 22(1): 1–19.

Moser LE, Hoveland CS (1996) *Cool-Season Grass Overview*. Hoboken, NJ: Wiley Publisher.

Mueller UG and Wolfenbarger LL (1999) AFLP genotyping and fingerprinting. *Trends in Ecology & Evolution* 14(10): 389–394.

Musial JM, Mackie JM, Armour DJ, Phan HTT, Ellwood SE, Aitken KS and Irwin JAG (2007) Identification of QTL for resistance and susceptibility to *Stagonospora meliloti* in autotetraploid lucerne. *Theoretical and Applied Genetics* 114(8): 1427–1435.

Mutengwa CS, Tongoona PB and Sithole-Niang I (2005) Genetic studies and a search for molecular markers that are linked to *Striga asiatica* resistance in sorghum. *African Journal of Biotechnology* 4(12): 1–7.

Navarro L, Dunoyer P, Jay F, Arnold B, Dharmasiri N, Estelle M and Jones JD (2006) A plant miRNA contributes to antibacterial resistance by repressing auxin signaling. *Science* 31 2(5772): 436–439.

Neff MM, Neff JD, Chory J and Pepper AE (1998) dCAPS, a simple technique for the genetic analysis of single nucleotide polymorphisms: Experimental applications in Arabidopsis thaliana genetics. *The Plant Journal* 14(3): 387–392.

Nelson DE, Repetti PP, Adams TR, Creelman RA, Wu J, Warner DC and Hinchey BS (2007) Plant nuclear factor Y (NF-Y) B subunits confer drought tolerance and lead to improved corn yields on water-limited acres. *Proceedings of the National Academy of Sciences* 104(42): 16450–16455.

Nicholson RL and Wood KV (2001) Phytoalexins and secondary products, where are they and how can we measure them? *Physiological and Molecular Plant Pathology* 59(2): 63–69.

Novillo F, Alonso JM, Ecker JR and Salinas J (2004) CBF2/DREB1C is a negative regulator of CBF1/DREB1B and CBF3/DREB1A expression and plays a central role in stress tolerance in Arabidopsis. *Proceedings of the National Academy of Sciences* 101(11): 3985–3990.

Nuccio ML, Wu J, Mowers R, Zhou HP, Meghji M, Primavesi LF and Basu SS (2015) Expression of trehalose-6-phosphate phosphatase in maize ears improves yield in well-watered and drought conditions. *Nature Biotechnology* 33(8): 862–869.

Nunes C, Araújo SS, Silva JM, Fevereiro P and Silva AB (2009) Photosynthesis light curves: A method for screening water deficit resistance in the model legume *Medicago truncatula*. *Annals of Applied Biology* 155(3): 321–332.

Nunes C, de Sousa Araújo S, da Silva JM, Fevereiro MPS and da Silva AB (2008) Physiological responses of the legume model *Medicago truncatula* cv. Jemalong to water deficit. *Environmental and Experimental Botany* 63(1–3): 289–296.

Nunes-Nesi A, Brito DS, Inostroza-Blancheteau C, Fernie AR and Araújo WL (2014) The complex role of mitochondrial metabolism in plant aluminum resistance. *Trends in Plant Science* 19(6): 399–407.

Obert DE, Skinner DZ and Stuteville DL (2000) Association of AFLP markers with downy mildew resistance in autotetraploid alfalfa. *Molecular Breeding* 6(3): 287–294.

Oh BJ, Frederiksen RA and Magill CW (1996) Identification of RFLP markers linked to a gene for downy mildew resistance (Sdm) in sorghum. *Canadian Journal of Botany* 74(2): 315–317.

Okamuro JK, Caster B, Villarroel R, Van Montagu M and Jofuku KD (1997) The AP2 domain of APETALA2 defines a large new family of DNA binding proteins in Arabidopsis. *Proceedings of the National Academy of Sciences* 94(13): 7076–7081.

Parh DK, Jordan DR, Aitken EAB, Mace ES, Jun-Ai P, McIntyre CL and Godwin ID (2008) QTL analysis of ergot resistance in sorghum. *Theoretical and Applied Genetics* 117(3): 369–382.

Park S, Li J, Pittman JK, Berkowitz GA, Yang H, Undurraga S and Gaxiola RA (2005) Up-regulation of a H+-pyrophosphatase (H+-PPase) as a strategy to engineer drought-resistant crop plants. *Proceedings of the National Academy of Sciences* 102(52): 18830–18835.

Paul C, Naidoo G, Forbes A, Mikkilineni V, White D and Rocheford T (2003) Quantitative trait loci for low aflatoxin production in two related maize populations. *Theoretical and Applied Genetics* 107(2): 263–270.

Pe ME, Gianfranceschi L, Taramino G, Tarchini R, Angelini P, Dani M and Binelli G (1993) Mapping quantitative trait loci (QTLs) for resistance to *Gibberella zeae* infection in maize. *Molecular and General Genetics MGG* 241(1–2): 11–16.

Phillips JR, Dalmay T and Bartels D (2007) The role of small RNAs in abiotic stress. *FEBS letters* 581(19): 3592–2597.

Qin F, Sakuma Y, Li J, Liu Q, Li YQ, Shinozaki K and Yamaguchi-Shinozaki K (2004) Cloning and functional analysis of a novel DREB1/CBF transcription factor involved in cold-responsive gene expression in *Zea mays* L. *Plant and Cell Physiology* 45(8): 1042–1052.

Quan R, Shang M, Zhang H, Zhao Y and Zhang J (2004) Engineering of enhanced glycine betaine synthesis improves drought tolerance in maize. *Plant Biotechnology Journal* 2(6): 477–486.

Rahman H, Sabreen S, Alam S and Kawai S (2005) Effects of nickel on growth and composition of metal micronutrients in barley plants grown in nutrient solution. *Journal of Plant Nutrition* 28(3): 393–404.

Rahman M, Malik TA, Iqbal MJ, Zafar Y, Alam K, Perveen Z and Rehman S (1998) DNA marker for salinity resistance. *Rice Biotechnology Quarterly* 35(1): 12–13.

Rai MK, Kalia RK, Singh R, Gangola MP and Dhawan AK (2011) Developing stress tolerant plants through *in vitro* selection—an overview of the recent progress. *Environmental and Experimental Botany* 71(1): 89–98.

Raj SN, Sarosh BR and Shetty HS (2006) Induction and accumulation of polyphenol oxidase activities as implicated in development of resistance against pearl millet downy mildew disease. *Functional Plant Biology* 33(6): 563–571.

Ramasamy P, Menz MA, Mehta PJ, Katilé S, Gutierrez-Rojas LA, Klein RR and Magill CW (2009) Molecular mapping of Cg1, a gene for resistance to anthracnose (*Colletotrichum sublineolum*) in sorghum. *Euphytica* 165(3): 597–606.

Rangaswami G and Mahadevan A (1999) Diseases of cereals. In: *Diseases of Crop Plants in India*. 4th ed. New Delhi: Prentic Hall of India, Pvt. Ltd., pp. 160–264.
Rector BG, Snook ME and Widstrom NW (2002) Effect of husk characters on resistance to corn earworm (Lepidoptera: Noctuidae) in high-maysin maize populations. *Journal of Economic Entomology* 95(6): 1303–1307.
Reyes JL and Chua NH (2007) ABA induction of miR159 controls transcript levels of two MYB factors during Arabidopsis seed germination. *The Plant Journal* 49(4): 592–606.
Rhoades MW, Reinhart BJ, Lim LP, Burge CB, Bartel B and Bartel DP (2002) Prediction of plant microRNA targets. *Cell* 110(4): 513–520.
Rhodes D and Hanson AD (1993) Quaternary ammonium and tertiary sulfonium compounds in higher plants. *Annual Review of Plant Biology* 44(1): 357–384.
Ribaut JM, Hoisington DA, Deutsch JA, Jiang C and Gonzalez-de-Leon D (1996) Identification of quantitative trait loci under drought conditions in tropical maize. 1. Flowering parameters and the anthesis-silking interval. *Theoretical and Applied Genetics* 92(7): 905–914.
Riechmann JL, Heard J, Martin G, Reuber L, Jiang CZ, Keddie J and Creelman R (2000) Arabidopsis transcription factors: Genome-wide comparative analysis among eukaryotes. *Science* 290(5499): 2105–2110.
Rontein D, Basset G and Hanson AD (2002) Metabolic engineering of osmoprotectant accumulation in plants. *Metabolic Engineering* 4(1): 49–56.
Roy AK, Malaviya DR, Kaushal P, Kumar B and Tiwari A (2004) Interspecific hybridization of *Trifolium alexandrinum* with *T. constantinopolitanum* using embryo rescue. *Plant Cell Reports* 22(9): 705–710.
Saab IN, Sharp RE, Pritchard J and Voetberg GS (1990) Increased endogenous abscisic acid maintains primary root growth and inhibits shoot growth of maize seedlings at low water potentials. *Plant Physiology* 93(4): 1329–1336.
Saira M, Rehman A, Gleason ML, Alam MW, Abbas MF, Ali S and Idrees M (2017) First report of *Sclerotinia sclerotiorum* causing stem and crown rot of berseem (*Trifolium alexandrinum*) in Pakistan. *Plant Disease* 101(5): 835.
Sakamoto A and Murata N (2000) Genetic engineering of glycinebetaine synthesis in plants: Current status and implications for enhancement of stress tolerance. *Journal of Experimental Botany* 51(342): 81–88.
Sakamoto A and Murata N (2001) The use of bacterial choline oxidase, a glycinebetaine-synthesizing enzyme, to create stress-resistant transgenic plants. *Plant Physiology* 125(1): 180–188.
Sakamoto A, Valverde R, Chen TH and Murata N (2000) Transformation of Arabidopsis with the codA gene for choline oxidase enhances freezing tolerance of plants. *The Plant Journal* 22(5): 449–453.
Sakamoto K, Akiyama Y, Fukui K, Kamada H and Satoh S (1998) Characterization; genome sizes and morphology of sex chromosomes in hemp (*Cannabis sativa* L.). *Cytologia* 63(4): 459–464.
Samac DA and Smigocki AC (2003) Expression of oryzacystatin I and II in alfalfa increases resistance to the root-lesion nematode. *Phytopathology* 93(7): 799–804.
Sanchez AC, Subudhi PK, Rosenow DT and Nguyen HT (2002) Mapping QTLs associated with drought resistance in sorghum (*Sorghum bicolor* L. Moench). *Plant Molecular Biology* 48(5–6): 713–726.
Sanghera GS, Wani SH, Hussain W and Singh NB (2011) Engineering cold stress tolerance in crop plants. *Current Genomics* 12(1): 30.
Sari-Gorla M, Krajewski P, Di Fonzo N, Villa M and Frova C (1999) Genetic analysis of drought tolerance in maize by molecular markers. II. Plant height and flowering. *Theoretical and Applied Genetics* 99(1–2): 289–295.
Saunders J and O'Neill N (2004) The characterization of defense responses to fungal infection in alfalfa. *BioControl 49*(6): 715–728.

Schluepmann H, Pellny T, van Dijken A, Smeekens S and Paul M (2003) Trehalose 6-phosphate is indispensable for carbohydrate utilization and growth in Arabidopsis thaliana. *Proceedings of the National Academy of Sciences* 100(11): 6849–6854.

Schmelz EA, Kaplan F, Huffaker A, Dafoe NJ, Vaughan MM, Ni X and Teal PE (2011) Identity, regulation, and activity of inducible diterpenoid phytoalexins in maize. *Proceedings of the National Academy of Sciences* 108(13): 5455–5460.

Seki M, Narusaka M, Abe H, Kasuga M, Yamaguchi-Shinozaki K, Carninci P and Shinozaki K (2001) Monitoring the expression pattern of 1300 Arabidopsis genes under drought and cold stresses by using a full-length cDNA microarray. *The Plant Cell* 13(1): 61–72.

Selosse MA and Roy M (2009) Green plants that feed on fungi: Facts and questions about mixotrophy. *Trends in Plant Science* 14(2): 64–70.

Shan Q, Zhang Y, Chen K, Zhang K and Gao C (2015) Creation of fragrant rice by targeted knockout of the Os BADH 2 gene using TALEN technology. *Plant Biotechnology Journal* 13(6): 791–800.

Sharma HC (1998) Bionomics, host plant resistance, and management of the legume pod borer, *Maruca vitrata*—a review. *Crop Protection* 17(5): 373–386.

Sharma HC, Crouch JH, Sharma KK, Seetharama N and Hash CT (2002) Applications of biotechnology for crop improvement: Prospects and constraints. *Plant Science* 163(3): 381–395.

Sharma I, Kumari N and Sharma V (2014) Defense gene expression in *Sorghum bicolor* against *Macrophomina phaseolina* in leaves and roots of susceptible and resistant cultivars. *Journal of Plant Interactions* 9(1): 315–323.

Shaug S, Lu C, King W, Buu R and Lin J (2000) Forage production and silage making for berseem clover. *Journal of Taiwan Livestock Research* 33(1): 105–110.

Shen ZG, Li XD, Wang CC, Chen HM and Chua H (2002) Lead phytoextraction from contaminated soil with high-biomass plant species. *Journal of Environmental Quality* 31(6): 1893–1900.

Shi H, Chen K, Wei Y and He C (2016) Fundamental issues of melatonin-mediated stress signaling in plants. *Frontiers in Plant Science* 7(1): 1124.

Shi J, Habben JE, Archibald RL, Drummond BJ, Chamberlin MA, Williams RW and Weers BP (2015) Overexpression of ARGOS genes modifies plant sensitivity to ethylene, leading to improved drought tolerance in both Arabidopsis and maize. *Plant Physiology* 169(1): 266–282.

Shinozaki K, Yamaguchi-Shinozaki K and Seki M (2003) Regulatory network of gene expression in the drought and cold stress responses. *Current Opinion in Plant Biology* 6(5): 410–417.

Shivhare R and Lata C (2017) Exploration of genetic and genomic resources for abiotic and biotic stress tolerance in pearl millet. *Frontiers in Plant Science* 7(1): 2069.

Shou H, Bordallo P and Wang K (2004) Expression of the Nicotiana protein kinase (NPK1) enhanced drought tolerance in transgenic maize. *Journal of Experimental Botany* 55(399): 1013–1019.

Shukla VK, Doyon Y, Miller JC, DeKelver RC, Moehle EA, Worden SE, Mitchell JC, Arnold NL, Gopalan S, Meng X and Choi VM (2009) Precise genome modification in the crop species *Zea mays* using zinc-finger nucleases. *Nature* 459(7245): 437–441.

Singh J and Kalamdhad AS (2011) Effects of heavy metals on soil, plants, human health and aquatic life. *International Journal of Research in Chemistry and Environment* 1(2): 15–21.

Singh JV, Chillar BS, Yadav BD and Joshi UN (2010) *Forage Legumes*. Lewis Way, CA: Scientific publishers.

Singh M, Chaudhary K and Boora KS (2006a) RAPD-based SCAR marker SCA 12 linked to recessive gene conferring resistance to anthracnose in sorghum [*Sorghum bicolor* (L.) Moench]. *Theoretical and Applied Genetics* 114(1): 187–192.

Singh M, Chaudhary K, Singal HR, Magill CW and Boora KS (2006b) Identification and characterization of RAPD and SCAR markers linked to anthracnose resistance gene in sorghum [*Sorghum bicolor* (L.) Moench]. *Euphytica* 149(1–2): 179–187.

Singh T, Radhakrishna A, Nayak DS and Malaviya DR (2019) Genetic improvement of berseem (*Trifolium alexandrinum*) in India: Current status and prospects. *International Journal of Current Microbiology and Applied Sciences* 8(1): 3028–3036.

Smith T, Mlambo V, Sikosana JLN, Maphosa V, Mueller-Harvey I and Owen E (2005) *Dichrostachys cinerea* and *Acacia nilotica* fruits as dry season feed supplements for goats in a semi-arid environment. *Animal Feed Science and Technology* 122(1–2): 149–157.

Spangenberg G, Kalla R, Lidgett A, Sawbridge T, Ong EK and John U (2001) Transgenesis and genomics in molecular breeding of forage plants. *Molecular Breeding of Forage Crops (ed. G. Spangenberg)* 219–237.

Stockinger EJ, Gilmour SJ and Thomashow MF (1997) *Arabidopsis thaliana* CBF1 encodes an AP2 domain-containing transcriptional activator that binds to the C-repeat/DRE, a cis-acting DNA regulatory element that stimulates transcription in response to low temperature and water deficit. *Proceedings of the National Academy of Sciences* 94(3): 1035–1040.

Strizhov N, Keller M, Mathur J, Koncz-Kálmán Z, Bosch D, Prudovsky E and Zilberstein A (1996) A synthetic cryIC gene, encoding a *Bacillus thuringiensis* δ-endotoxin, confers Spodoptera resistance in alfalfa and tobacco. *Proceedings of the National Academy of Sciences* 93(26): 15012–15017.

Suárez R, Calderón C and Iturriaga G (2009) Enhanced tolerance to multiple abiotic stresses in transgenic alfalfa accumulating trehalose. *Crop Science* 49(5): 1791–1799.

Sullivan ML and Hatfield RD (2006) Polyphenol oxidase and o-diphenols inhibit postharvest proteolysis in red clover and alfalfa. *Crop Science* 46(2): 662–670.

Sulpice R, Tsukaya H, Nonaka H, Mustardy L, Chen TH and Murata N (2003) Enhanced formation of flowers in salt-stressed Arabidopsis after genetic engineering of the synthesis of glycine betaine. *The Plant Journal* 36(2): 165–176.

Sun Y, Zhang X, Wu C, He Y, Ma Y, Hou H and Xia L (2016) Engineering herbicide-resistant rice plants through CRISPR/Cas9-mediated homologous recombination of acetolactate synthase. *Molecular Plant* 9(4): 628–631.

Sundaram NV, Bhowmik TP and Khan ID (1970) Water-soluble alkaloid content of ergot (*Claviceps microcephala* (Walk.) Tul.) sclerotia of pearl millet (*Pennisetum typhoides* (Burm. F.) Stapf & CE Hubb.). *Indian Journal of Agricultural Science* 40(6): 569–572.

Sunkar R and Zhu JK (2004) Novel and stress-regulated microRNAs and other small RNAs from Arabidopsis. *Plant Cell* 16(8): 2001–2019.

Sunkar R, Li YF and Jagadeeswaran G (2012) Functions of microRNAs in plant stress responses. *Trends in Plant Science* 17(4): 196–203.

Svitashev S, Young JK, Schwartz C, Gao H, Falco SC and Cigan AM (2015) Targeted mutagenesis, precise gene editing, and site-specific gene insertion in maize using Cas9 and guide RNA. *Plant Physiology* 169(2): 931–945.

Tao LZ, Cheung AY and Wu HM (2002) Plant Rac-like GTPases are activated by auxin and mediate auxin-responsive gene expression. *The Plant Cell* 14(11): 2745–2760.

Tesfaye M, Denton MD, Samac DA and Vance CP (2005) Transgenic alfalfa secretes a fungal endochitinase protein to the rhizosphere. *Plant and Soil* 269(1–2): 233–243.

Thakur RP, Rao, VP, Wu BM, Subbarao KV, Shetty HS, Singh G and Gupta SC (2004) Host resistance stability to downy mildew in pearl millet and pathogenic variability in *Sclerospora graminicola*. *Crop Protection* 23(10): 901–908.

Thomas JC, Wasmann CC, Echt C, Dunn RL, Bohnert HJ and McCoy TJ (1994) Introduction and expression of an insect proteinase inhibitor in alfalfa *Medicago sativa* L. *Plant Cell Reports* 14(1): 31–36.

Thomashow MF (1998) Role of cold-responsive genes in plant freezing tolerance. *Plant physiology* 118(1): 1–8.
Thomashow MF (1999) Plant cold acclimation: Freezing tolerance genes and regulatory mechanisms. *Annual Review of Plant Biology* 50(1): 571–599.
Thomashow MF, Gilmour SJ, Stockinger EJ, Jaglo-Ottosen KR and Zarka DG (2001) Role of the Arabidopsis CBF transcriptional activators in cold acclimation. *Physiologia Plantarum* 112(2): 171–175.
Thorogood D, Yates S, Manzanares C, Skot L, Hegarty M, Blackmore T, Barth S and Studer B (2017) A novel multivariate approach to phenotyping and association mapping of multi-locus gametophytic self-incompatibility reveals S, Z, and other loci in a perennial ryegrass (Poaceae) population. *Frontiers in Plant Science* 8(1): 1–13.
Tohidfar M, Zare N, Jouzani GS and Eftekhari SM (2013) Agrobacterium-mediated transformation of alfalfa (*Medicago sativa*) using a synthetic cry3a gene to enhance resistance against alfalfa weevil. *Plant Cell, Tissue and Organ Culture (PCTOC)* 113(2): 227–235.
Torricelli R (2006) *Evaluation of Lucerne Varieties for Organic Agriculture*. Netherland: European Association for Research on Plant Breeding (EUCARPIA) Publisher.
Trindade H, Costa MM, Lima SB, Pedro LG, Figueiredo AC and Barroso JG (2009) A combined approach using RAPD, ISSR and volatile analysis for the characterization of *Thymus caespititius* from Flores, Corvo and Graciosa islands (Azores, Portugal). *Biochemical Systematics and Ecology* 37(5): 670–677.
Tuinstra MR, Grote EM, Goldsbrough PB and Ejeta G (1996) Identification of quantitative trait loci associated with pre-flowering drought tolerance in sorghum. *Crop Science* 36(5): 1337–1344.
Tuinstra MR, Grote EM, Goldsbrough PB and Ejeta G (1997) Genetic analysis of post-flowering drought tolerance and components of grain development in *Sorghum bicolor* (L.) Moench. *Molecular Breeding* 3(6): 439–448.
Varshney RK, Graner A and Sorrells ME (2005) Genomics-assisted breeding for crop improvement. *Trends in Plant Science* 10(12): 621–630.
Verma P, Chandra A, Roy AK, Malaviya DR, Kaushal P, Pandey D and Bhatia S (2015) Development, characterization and crossspecies transferability of genomic SSR markers in Berseem (*Trifolium alexandrinum* L.), an important multi-cut annual forage legume. *Molecular Breeding* 35(1): 23.
Vinh NT and Paterson AH (2005) Genome mapping and its implication for stress resistance in plants. *Abiotic Stresses: Plant Resistance through Breeding and Molecular Approaches* 725(1): 119–124.
Visarada KBRS and Kishore NS (2007) Improvement of sorghum through transgenic technology. Report, CiteSeer^x March.
Voisey CR, Dudas B, Biggs R, Burgess EPJ, Wigley PJ, McGregor PG and White DWR (2001) Transgenic pest and disease resistant white clover plants. In *Molecular Breeding of Forage Crops*. Dordrecht: Springer, pp. 239–250.
Wang W, Vinocur B and Altman A (2003) Plant responses to drought, salinity and extreme temperatures: Towards genetic engineering for stress tolerance. *Planta* 218(1): 1–14.
Wang Y, Cheng X, Shan Q, Zhang Y, Liu J, Gao C and Qiu JL (2014) Simultaneous editing of three homoeoalleles in hexaploid bread wheat confers heritable resistance to powdery mildew. *Nature Biotechnology* 32(9): 947–951.
Wang ZY, Seto H, Fujioka S, Yoshida S and Chory J (2001) BRI1 is a critical component of a plasma-membrane receptor for plant steroids. *Nature* 410(6826): 380–383.
Wang ZY, Takamizo T, Iglesias VA, Osusky M, Nagel J, Potrykus I and Spangenberg G (1992) Transgenic plants of tall fescue (*Festuca arundinacea* Schreb.) obtained by direct gene transfer to protoplasts. *Bio/technology* 10(6): 691–696.
Warnke SE, Barker RE, Jung G, Sim SC, Mian MAR, Saha MC, Brilman LA, Dupal MP and Forster JW (2004) Genetic linkage mapping of an annual 3 perennial ryegrass population. *Theoretical and Applied Genetics* 109(2): 294–304.

Wen PF, Chen JY, Wan SB, Kong WF, Zhang P, Wang W, Zhan JC, Pan QH and Huang WD (2008) Salicylic acid activates phenylalanine ammonia-lyase in grape berry in response to high temperature stress. *Plant Growth Regulation* 55(1): 1–10.

Weng J, Liu X, Wang Z, Wang J, Zhang L, Hao Z and Liu C (2012) Molecular mapping of the major resistance quantitative trait locus qHS2.09 with simple sequence repeat and single nucleotide polymorphism markers in maize. *Phytopathology* 102(7): 692–699.

Wenzel GERHARD (1992) *Application of unconventional techniques in classical plant production*. Oxford: Pergamon Press.

White DG, Rocheford TR, Naidoo G, Paul C, Hamblin AM and Forbes AM (1998) Inheritance of molecular markers associated with, and breeding for resistance to aspergillus ear rot and aflatoxin production in corn using Tex6. In *Proceedings of the USDA-ARS Aflatoxin Elimination Workshop held at St. Louis, MO*, pp. 4–6.

Williams JGK, Kubelik AR, Livak KJ, Rafalski JA and Tingey SV (1990) DNA polymorphisms amplified by arbitrary primers are useful as genetic markers. *Nucleic Acids Research* 18(22): 6531–6535.

Winicov I (2000) Alfin1 transcription factor overexpression enhances plant root growth under normal and saline conditions and improves salt tolerance in alfalfa. *Planta* 210(3): 416–422.

Winicov I and Bastola DR (1999) Transgenic overexpression of the transcription FactorAlfin1 enhances expression of the endogenous MsPRP2Gene in alfalfa and improves salinity tolerance of the plants. *Plant Physiology* 120(2): 473–480.

Xiong Y and Fei SZ (2006) Functional and phylogenetic analysis of a DREB/CBF-like gene in perennial ryegrass (*Lolium perenne* L.). *Planta* 224(4): 878–888.

Yadav OP and Rai KN (2013) Genetic improvement of pearl millet in India. *Agricultural Research* 2(4): 275–292.

Yadav OP, Hossain F, Karjagi CG, Kumar B, Zaidi PH, Jat SL and Yadava P (2015) Genetic improvement of maize in India: Retrospect and prospects. *Agricultural Research* 4(4): 325–338.

Yang O, Popova OV, Süthoff U, Lüking I, Dietz KJ and Golldack D (2009) The Arabidopsis basic leucine zipper transcription factor AtbZIP24 regulates complex transcriptional networks involved in abiotic stress resistance. *Gene* 436(1–2): 45–55.

Yang QC, Han JG, Sun Y and Wu MS (2005) Identification of molecular marker linked to salt tolerance gene in alfalfa. *Agricultural Sciences in China* 4(10): 781–787.

Yang S, Gao M, Xu C, Gao J, Deshpande S, Lin S and Zhu H (2008) Alfalfa benefits from *Medicago truncatula*: The RCT1 gene from *M. truncatula* confers broad-spectrum resistance to anthracnose in alfalfa. *Proceedings of the National Academy of Sciences* 105(34): 12164–12169.

Yang W, de Oliveira AC, Godwin I, Schertz K and Bennetzen JL (1996) Comparison of DNA marker technologies in characterizing plant genome diversity: Variability in Chinese sorghums. *Crop Science* 36(6): 1669–1676.

Yeo ET, Kwon HB, Han SE, Lee JT, Ryu JC and Byu MO (2000) Genetic engineering of drought resistant potato plants by introduction of the trehalose-6-phosphate synthase (TPS1) gene from *Saccharomyces cerevisiae*. *Molecules and Cells* 10(3): 263–268.

Younis RA, Ahmed MF and El-Menshawy MM (2007) Molecular genetic markers associated with salt tolerance in grain sorghum. *Arab Journal of Biotechnology* 10(1): 249–258.

Yu Y, Jia T and Chen X (2017) The 'how' and 'where' of plant micro RNAs. *New Phytologist* 216(4): 1002–1017.

Yue J, Hu X and Huang J (2014) Origin of plant auxin biosynthesis. *Trends in Plant Science* 19(12): 764–770.

Zayed EM (2013) Applications of biotechnology on Egyptian clover (berseem) (*Trifolium alexandrinum* L.). *International Journal of Agricultural Science and Research* 3(1): 99–120.

Zhang G, Chen M, Li L, Xu Z, Chen X, Guo J and Ma Y (2009) Overexpression of the soybean GmERF3 gene, an AP2/ERF type transcription factor for increased tolerances to salt, drought, and diseases in transgenic tobacco. *Journal of Experimental Botany* 60(13): 3781–3796.

Zhang H, Zhang J, Wei P, Zhang B, Gou F, Feng Z and Zhu JK (2014a) The CRISPR/C as9 system produces specific and homozygous targeted gene editing in rice in one generation. *Plant Biotechnology Journal* 12(6): 797–807.

Zhang JY, Broeckling CD, Blancaflor EB, Sledge MK, Sumner LW and Wang ZY (2005) Overexpression of WXP1, a putative *Medicago truncatula* AP2 domain-containing transcription factor gene, increases cuticular wax accumulation and enhances drought tolerance in transgenic alfalfa (*Medicago sativa*). *The Plant Journal* 42(5): 689–707.

Zhang JZ (2003) Overexpression analysis of plant transcription factors. *Current Opinion in Plant Biology* 6(5): 430–440.

Zhang T, Yu LX, McCord P, Miller D, Bhamidimarri S, Johnson D and Samac DA (2014b) Identification of molecular markers associated with Verticillium wilt resistance in alfalfa (*Medicago sativa* L.) using high-resolution melting. *PLoS One* 9(12): e115953.

Zhang Z and Huang R (2010) Enhanced tolerance to freezing in tobacco and tomato overexpressing transcription factor TERF2/LeERF2 is modulated by ethylene biosynthesis. *Plant Molecular Biology* 73(3): 241–249.

Zhao B, Liang R, Ge L, Li W, Xiao H, Lin H and Jin Y (2007) Identification of drought-induced microRNAs in rice. *Biochemical and Biophysical Research Communications* 354(2): 585–590.

Zhao J, He Q, Chen G, Wang L and Jin B (2016) Regulation of non-coding RNAs in heat stress responses of plants. *Frontiers in Plant Science* 7(1): 1213.

Zheng X, Tan DX, Allan AC, Zuo B, Zhao Y, Reiter RJ, Wang L, Wang Z, Guo Y, Zhou J and Shan D (2017) Chloroplastic biosynthesis of melatonin and its involvement in protection of plants from salt stress. *Scientific Reports* 7(1): 1–2.

Zhou M, Li D, Li Z, Hu Q, Yang C, Zhu L and Luo H (2013) Constitutive expression of a miR319 gene alters plant development and enhances salt and drought tolerance in transgenic creeping bentgrass. *Plant Physiology* 161(3): 1375–1391.

Zou C, Li Z and Yu D (2010a) Bacillus megaterium strain XTBG34 promotes plant growth by producing 2-pentylfuran. *The Journal of Microbiology* 48(4): 460–466.

Zou J, Li Y, Zhu K and Wang Y (2010b) Study on inheritance and molecular markers of sorghum resistance to head smut physiological race 3. *Scientia Agricultura Sinica* 43(4): 713–720.

Section II

Ecological Dynamics and Winter Fodders

4 Medicago sativa
Diseases, Etiology, and Management

Nabeeha Aslam Khan, Imran Ul Haq, Siddra Ijaz, and Barbaros Cetinel

CONTENTS

DOI: 10.1201/9781003055365-4

4.1 INTRODUCTION

The Arabic word *al-faṣfaṣa* is modified into the Spanish language as alfalfa (Dozy and Englemann, 1869). Alfalfa (*Medicago sativa*) is leguminous; the herbaceous plant belongs to the Fabaceae family, considered the most important, oldest, and most widely grown fodder crop in the world due to its adaptive nature to a wide range of soils, climates with high nutritional values (17–20% crude protein 60–65% total digestible nutrients), and remarkable productivity used as hay and silage for grazing livestock (Chauhan et al., 2009). Alfalfa is also called the forage queen. The cultivation of alfalfa was started 8000 years ago in south-central Asia (Ivanov, 1980).

According to archaeological evidence, the origin of alfalfa was Iran (Yuegao et al., 2009). It is cultivated in about more than 80 countries on approximately 45 million hectares (Mielmann, 2013) (Radović et al., 2009). It is the crop of temperate regions of the globe (Annicchiarico, 2015). Alfalfa is also known as lucerne in many countries, such as Germany, Britain, France, Australia, and so on. It has a remarkable resemblance to clover. This forage crop is a perennial legume with a five- to seven-year lifespan, while favorable conditions (climate and variety) can extend its lifespan. The plant grows up to 1 meter high with 2–3 meters root depth; sometimes, roots grow to more than 15 meters. This crop has biological nitrogen-fixing properties due to the deep root system and *Sinorhizobium meliloti* bacteria present in root nodules that improve soil fertility and avoid soil erosion (Galibert et al., 2001). Its cultivation also improves the soil structure and penetration of water and breaks pathogen and pest cycles. The plant shows autotoxicity (it is difficult for seeds to grow with existing strands of alfalfa crop) (Nelson et al., 1997). Its cultivation's primary purpose is to produce fodder and seed (Dragovoz et al., 2002); seed production is not a common practice among domestic farmers due to a lack of knowledge about its economic importance. It is used as hay; silage; and feed for dairy cows, beef cattle, sheep, horses, and goats (as green chop) (Jennings and Pennington, 2003). Leaf concentrates and dehydrated alfalfa are used in poultry diets for meat and egg yolk pigmentation due to high carotenoid content (Heuze et al., 2013). Fresh alfalfa is highly nutritious and palatable to livestock, with highly digestible fiber, about 16–20% raw proteins (proteins have high biological value and amino acid content similar to animal proteins), and 8% minerals, and it is rich in vitamins K, E, D, and A. High biomass production is a promising feature of the alfalfa crop. As it has a remarkable ability to regenerate leaves and stems, it can be harvested three to four times in a single growing season, but it can be cut up to 12 times (Smith et al., 1999). Total yield mainly depends upon the climate, region, and growth stage of the crop at harvesting time (Smith et al., 1999; Lloveras et al., 1998). Nitrogen-fixing ability and nutritional profile (animal feed) make it more useful to improve agriculture efficiency (Emerich and Krishnan, 2009).

4.1.1 Problems

As alfalfa is the most important (economically and nutritionally) crop among all fodder crops, production technology (practices are interlinked) and several management practices (factors) affect the profitability (cost and returns) such as the selection of site, seed variety, a-biotic, weed control, method and timing of harvesting, and so on (Ward and Hutson, 2004). Numbers of biotic constraints have been recorded to affect alfalfa crop production, including fungi, bacteria, viruses, nematodes, and insects. Diseases have a drastic impact on crop potential and cause a significant reduction in the forage's feeding quality.

4.2 FUNGAL DISEASES

Fungi include many plant pathogens and are responsible for a variety of severe diseases of crops. More than 19,000 fungi are well known to cause plant diseases globally. They can live on both living and dead plants and remain dormant until conditions become conducive to their growth and dissemination. Fungi may develop their mycelium or spores inside host plants. The primary infections are infected or contaminated seeds, crop debris, soil, weeds, and alternate hosts. Fungal diseases are spread by infected seeds, water, wind, soil, insects, invertebrates, farm machinery, animals, and workers (Lazarovits et al., 2014). Plant pathogenic fungi cause harmful diseases such as wilt, damping-off, root rot, anthracnose, mildew, leaf spot, rust, blight, and dieback. Systemic infections can cause significant yield losses and diminished quality of the crops by killing cells (Iqbal et al., 2018). Fungal plant pathogens may pose damaging threats to food security. The rapid and timely identification of disease leads to inhibiting the progress and wide spread of the pathogen. Various recognition techniques are required to quickly detect diseases at early phases of fungal infection and plant growth (Singh and Misra, 2017).

4.2.1 Wilt

The wilt disease or wilt-causing pathogen of plants badly affects the vascular system. Fungi, bacteria, and nematodes are involved and cause rapid death of the plant. Wilt-causing pathogens invade the vascular bundles and cause xylem clogging to disrupt water transport to the foliage, ultimately causing stem and leaf wilting.

4.2.1.1 *Fusarium* Wilt

Fusarium wilt is caused by *Fusarium oxysporum f.sp. medicaginis* (Weimer, 1928). This fungus does not grow very luxuriantly on oat, corn-meal, or lima-bean hard agars made according to Sherbakoff's directions (1915).

4.2.1.1.1 Symptoms

Fusarium is a vascular bundle–inhibiting pathogen; it causes several diseases, as a typical symptom of fusarium wilt is infection occurring on one side of a plant; first, dark brown–colored cankerous depressions or streaks appear on the roots due to browning of vascular bundles (Cottam, 1921), and red perithecia present on roots (Arnaud, 1910). Discoloration also extends into the stem portion; the color of upper leaves becomes bright yellow to bleached straw, lower leaves show a light pink shade, temporary wilting and drooping of tips of stems occur, and the color of the infected vascular system is usually cinnamon to mummy brown (Ridgway, 1912). The disease progresses slowly in the alfalfa stand. The entire plant gradually dies because the pathogen plugs the vascular vassals (Frosheiser and Barnes, 1978). Wilt becomes severe in association with nematode disease (Griffin and Thyr, 1978). *Fusarium* wilt symptoms may overlap with bacterial wilt (Jones, 1925; Jones and McCulloch, 1926).

4.2.1.1.2 Biology and Ecology

This fungus is soil borne and survives as conidia or resting spores (chlamydospores) in soil. Spores contact plant roots, forming a germ tube, then penetrating the roots through wounds and natural openings. After entry, hyphae grow inter- and intracellularly (Ramírez-Suero et al., 2010). Micro-conidia are found in the vascular bundles. Discoloration of host tissue occurs due to toxins produced by the fungus. Dark brown streaks appear in the taproot. Olive brown discoloration occurs on the stem. Sporulation occurs on dead plant parts, and spores spread by wind, water, equipment, debris, and infested soil (Leath et al., 1988).

4.2.1.2 *Verticillium* Wilt

Verticillium albo-atrum is a causal agent of verticillium wilt. This disease was first reported in 1918 (Hedlund, 1923) and had great importance in Europe (Raynal and Guy, 1977; Kreitlow, 1962). This alfalfa disease is confined to cooler regions such as North America, Europe, and north Japan (Peaden et al., 1985). Approximately 25°C is the optimum temperature required for the growth of *Verticillium albo-atrum* on artificial media. Some strains also spread into warmer regions (Heale, 1985; Christen and French, 1982).

4.2.1.2.1 Infection

Infected plants exhibit the symptoms of stunting, wilting, V-shaped lesions on leaves; stems of diseased plants remain green. In case of severe infection, plants die. Humid conditions are more favorable for heavy sporulation of *Verticillium albo-atrum* (dense verticillate conidiophores, single-celled conidia) on old stems underneath the canopy of the plant. Verticillium wilt can cause significant yield reduction, resulting in substantial economic losses (Page et al., 1992; Viands et al., 1992; Smith et al., 1995). This pathogen poses systemic infection risk because the fungus can be extracted from all plant parts (leaves, roots, and stems) (Christen and Peaden, 1981; Isaac, 1957). However, the sporadic pattern of distribution of *Verticillium albo-atrum in* infected leaves (Huang, 1989), stems (Pennypacker and Leath, 1983), and seeds (Huang et al., 1985) is also observed; it may be due to ineffective colonization of fungi in xylem tissues of diseased stems and leaves (Huang et al., 1985; Pennypacker and Leath, 1983).

Verticillium albo-atrum is a seed-borne pathogen (Isaac and Heale, 1961). Different studies have demonstrated how seeds become contaminated; pathogen carrier seeds are produced from diseased plants (Christen, 1983; Gilbert and Peaden, 1988; Huang et al., 1985) or externally contaminated seeds (Christen and Peaden, 1981). Seed lots are often contaminated with inoculum (Christen, 1983; Sheppard and Needham, 1980). Seed trade is an essential source of the pathogen's international spread. This pathogen can also be spread by water, wind, farm equipment, and direct contact of roots (Lindemann et al., 1982; Davies and Isaac, 1958; Isaac, 1957; Christen and Peaden, 1982; Heale and Isaac, 1963; Howard, 1985).

4.2.1.2.2 Management

Quarantine measures and management practices include crop rotation, sanitation of farm equipment, and the cultivation of resistant varieties to manage the disease. There is no effective chemical to control this disease; that is, methyl bromide fumigant proved ineffective (Isaac and Heale, 1961; Peaden et al., 1985). Seed treatment with thiram (50% WP) thiophanate-methyl (70% WP), Captan (50% WP), and benomyl (50% WP) also failed to control the disease (Peaden et al., 1985). The highly effective method for the management of this disease is the use of resistant varieties.

4.2.2 *Sclerotinia* Stem and Crown Rot

Stem and crown rot of alfalfa was observed in cool and humid regions of North America and Europe, caused by *Sclerotinia trifoliorum* (Adams and Ayers, 1979; Graham et al., 1979). *Sclerotinia sclerotiorum* and *Sclerotinia trifoliorum* are both reported as causal organisms of stem and crown rot in alfalfa (Frate and Long, 2005).

4.2.2.1 Symptoms

This disease's most apparent symptom is flagging or wilting stems, with a bleached area on the base and the stem's tip. Stems become soft and mushy. In moist conditions (fog, dew, or rain), the fungus can be observed as a white mass growing on the plant parts and spreading onto the soil. Gradually, infected plants die. In case of prolonged favorable conditions, the fungus can also grow into the crown. Crowns can survive for longer. In moist conditions, the diagnostic evidence is the presence of sclerotia (a hard resting structure made up of mycelium, round or elongated in shape and 1/8 inch in diameter). Sclerotia form both internally and externally. Internally formed sclerotia are cylindrical, while the external ones are irregular in shape. Sclerotia remain dormant during hot and dry weather, as these are unfavorable conditions for disease.

4.2.2.2 Disease Cycle

Sclerotia play a vital role in crown and stem rot of alfalfa. Apothecia (orange-brown-colored fruiting body) is produced from buried sclerotia in soil under favorable conditions (temperature and moisture) during the fall. Several apothecia produce from single sclerotium, white-colored mycelia grow on the host plant, and disease symptoms appear. Infections initiate due to two types of inoculum as mycelium or germinated ascospores (Lumsden, 1979). Sclerotia may germinate directly to form a mycelium; germination in which ascospores are formed is critical. Commonly ascospores are considered a primary inoculum. Penetration of mycelium into the host plant requires organic matter to fulfill the need for inoculum nutrition. After penetration, mycelium excessively colonizes the host plant's tissues, and death of the plant occurs (wilting of stems). Mycelia become compact and form into clumps, ultimately forming mature sclerotia. Sclerotia remain in the soil during summer. After the formation of sclerotia, the disease cycle is completed (Gilbert, 1987).

4.2.2.3 Dissemination

S. trifoliorum and *S. sclerotiorum*, both pathogens of alfalfa, have a broad host range. Both pathogens can survive and reproduce on other plants. Apothecia form on those plants after germination of sclerotia whose ascospores may infect the alfalfa crop. *S. sclerotiorum* has been reported to infect 52 species of the legume family.

Sclerotinia spp. spreads from one area/field to another in several ways. Ascospores and sclerotia can move from one field to another utilizing air and irrigation water, respectively. Ascospores and sclerotia are also disseminated by humans, farm equipment, infected seedlings, or animals (Starr et al., 1953; Weston et al., 1946). If infected plants use animal feed, they can also cause pathogen dissemination in uncontaminated fields due to the spreading of manure from those animals. Sclerotia were observed on irrigation paths and other waterways, where they remained alive for 10 to 21 days (Steadman et al., 1975). Infected seed (infection of seed with mycelium and contamination with sclerotia) is also a source of long-distance pathogen dissemination (Adams and Ayers, 1979).

4.2.2.4 Epidemiology

Sclerotia of both *Sclerotinia* species can stay alive in the soil for about six to ten years (Tribe, 1957; Brown and Butler, 1936; Weston et al., 1946) under the resting or dormant stage; this varies from species to species of *Sclerotinia*. Moisture contents play a critical role in developing this disease (initiation of infection and growth). The pH has less importance for sclerotia of *Sclerotinia* (Adams and Ayers, 1979). The optimum temperature range is 50–68°F for *Sclerotinia trifoliorum*. If temperature and humidity are not conducive, the disease will completely disappear. It is challenging to find diseased alfalfa plants in the field under unfavorable conditions (change in temperature, absence of rain, dew, and fog). The disease seems prevalent during December to mid-February. Survival can be reduced due to a constant 35°C soil temperature for three weeks (Adams, 1975); this prolonged constant duration would not be expected naturally (Frate and Long, 2005).

Soil microflora affect the survival of sclerotia of *Sclerotinia* species (Moore, 1949); destruction of sclerotia of *Sclerotinia trifoliorum* was observed due to fungi, snails, and bacteria. *Trichoderma viride* can destroy the sclerotia of *Sclerotinia trifoliorum* (Makkonen and Pohjakallio, 1960), and 58% of destruction can be achieved by soil infested with *T. viride*. Under natural environment conditions, the mycoparasite of *Sclerotinia* spp., *Coniothyrium minitans*, is well known (Campbell, 1947; Tribe, 1957).

4.2.2.5 Management

Cultivation of resistant varieties is an ideal solution to manage this disease, but there is no commercial resistant variety. Planting the same crop into old stubbles should be avoided; otherwise, pathogens move to new seedlings. With poorly drained soil or a record of bad stand establishment (allelopathy), seed treatment (metalaxyl) should be adopted. Apply proper irrigation in order to avoid moisture stress. Avoid dense planting because **a** dense plant canopy also triggers this disease, because dense planting can increase relative humidity (humidity is a critical factor in disease development). Delay planting can minimize the risk of disease.

Biological control is helpful because bacteria, fungi, and some insects are reported to antagonize or parasitize the sclerotia of *Sclerotinia*, but to date, bioproducts are not commercially available (Frate and Long, 2005).

Herbicide toxicity to *Sclerotinia* spp. can decrease disease severity and alter the alfalfa plant (Altman and CL, 1977; Katan and Eshel, 1973). Some herbicides can suppress the disease by reducing the weight of sclerotia, retarding the growth of mycelium and sclerotium germination. Herbicides used on alfalfa stands in the autumn season can affect the incidence and severity of crown and stem rot (Reichard et al., 1997).

4.2.3 Damping-Off

Damping-off is a complex disease, and many pathogens (soil-borne), such as *Phytophthora medicaginis*, *Rhizoctonia solani*, *Pythium* spp., and *Fusarium* spp., cause damping-off in alfalfa (Leath et al., 1988; Hancock, 1983; Schmitthenner, 1964).

Fifteen *Pythium* species are involved in seed rot and damping-off (Berg et al., 2017). *Pythium* species were reported as a causal agent of alfalfa damping off, such as *P. ultimum, P. irregular, P. paroecandrum, P. sylvaticum, P. acanthicum, P. dissotocum, P. rostratum, and P. torulosum* from various countries (Hancock, 1983; Hancock, 1985; Larkin et al., 1995).

Pythium and *Phytophthora* are highly destructive pathogens causing huge losses throughout the world (Hansen and Maxwell, 1991; Erwin and Ribeiro, 1996). *Phytophthora medicaginis* can cause pre-and post-emergence damping-off disease (Xiao et al., 2002). Some scientists identified *Pythium* spp. They cause disease on crops sown in rotation with alfalfa crops, such as soybean and corn (Matthiesen et al., 2016; Broders et al., 2007; Alejandro Rojas et al., 2017; Radmer et al., 2017; Zhang and Yang, 2000).

4.2.3.1 Symptoms

Pythium-infected alfalfa crops show brown lesions, soft radicles, and fewer root hairs. Water-soaked lesions appear on hypocotyls and roots; seedlings collapse and ultimately die. The infection causes disruption in nutrients and water absorption and nitrogen fixation due to the destruction of feeder roots. Root injury may cause forking of roots (Leath et al., 1988). Wet, cool weather is favorable for pythium's attack to mature roots (Berg et al., 2017). A prevalence of wet conditions during the plant's early developmental stages is a yield-suppressing element (Erwin and Ribeiro, 1996; Teutsch and Sulc, 1997).

4.2.3.2 Management

Management of damping-off requires a variety of control methods such as cultural practices and biological and chemical treatments. Cultivation of resistant cultivars is an effective control method for *Phytophthora* spp. crops (Erwin and Ribeiro, 1996), but unfortunately, resistant cultivars become immune to this disease under highly available inoculum densities and conducive environmental conditions (Faris and Sabo, 1981; Havey and Grau, 1985; Erwin and Ribeiro, 1996).

Cultural practices such as improving soil drainage, minimizing the compaction of soil, and crop rotation with non-alternative host crops for a longer duration (crop rotation is comparatively less effective) contribute to disease management (Pulli and Tesar, 1975; Schmitthenner and Van Doren, 1985; Erwin and Ribeiro, 1996).

Actinomycetes (Filonow and Lockwood, 1985) and *Sinorhizobium meliloti* play a role as biocontrol agents against *Phytophthora* spp. (Tu, 1979; Tu, 1980; Tu, 1978; Handelsman et al., 1990).

Antibiotic-producing *Streptomyces* spp. has been used as biocontrol for *Phytophthora medicaginis* (Jones and Samac, 1996). The inhibitory potential of *Streptomyces* spp. can be used against many plant pathogens (El-Abyad et al., 1993; Gyenis et al., 1999; Merriman et al., 1974; Liu, 1992; Hiltunen et al., 1995; Jones and Samac, 1996; Paulsrud, 1996).

Apron XL (active ingredient mefenoxam) is commonly used for seed treatment, and various levels of its sensitivity have been observed to different *Pythium* spp. of many crops, such as potato (Taylor et al., 2002), soybean, corn (Matthiesen et al., 2016; Broders et al., 2007; Radmer et al., 2017), pea (Parke et al., 1991), wheat (Cook and Zhang, 1985), and sugar beet (Brantner and Windels, 1998). The QoI fungicide stamina (active ingredient pyraclostrobin) has also been used to treat seed diseases caused by *Phytophthora* and *Pythium*. These fungicides interrupt the respiration process by blocking the electron chain (Venancio et al., 2009). The insensitivity of pathogens to these fungicides is due to a mutation at the binding sites (Bartlett et al., 2002).

Seed-only treatment is not sufficient for the management of damping-off. Phenylamide fungicides (including mefenoxam and metalaxyl) are widely used against oomycetes. These fungicides inhibit the ribosomal RNA polymerases of the pathogen, which develop resistance in pathogens. The sensitivity of metalaxyl to *Pythium* varies with strains of the *Pythium* spp. (Larkin et al., 1995; Hwang, 1988).

4.2.4 Leaf and Stem Diseases

The severity of leaf and stem diseases depends upon the duration of wet weather. Mostly these diseases are favored by cool temperatures and high rainfall. Leaf and stem diseases directly affect the nutritional status of alfalfa and can cause heavy leaf drop. The leaves carry more protein as compared to stems.

4.2.4.1 *Leptosphaerulina* Leaf Spot

Leptosphaerulina briosiana causes leaf spot. Gray-black spots with a tan center, dark margins, and round to oval small pinpoints to 1/8" diameter in size appear on leaves. Severe symptoms can cause death or premature dropping of leaves. Affected petioles and leaflets remain attached to stems.

4.2.4.1.1 Epidemiology

These spots are prevalent during wet and cool weather. *Leptosphaerulina briosiana* overwinters in old leaves present on the surface of the soil. The optimum temperature range is 60°F to 80°F for this disease.

4.2.4.2 Common Leaf Spot

Pseudopeziza medicaginis causes a common leaf spot in alfalfa. These spots occur when the crop is at the initial growth stages. This disease has been reported in many countries of the world, such as Canada, the USSR, Europe, the United States, and Africa. It causes significant losses regarding quality and quantity.

4.2.4.2.1 Symptoms

Brown-colored small, circular spots with dendritic or smooth margins appear on the leaflets. These spots cause discoloration of tissues. In the case of a severe attack, the leaves turn yellow and drop off. This disease's diagnostic character is black raised apothecia in the center of the mature spot. These raised areas usually appear on the leaves' upper side, rarely on the lower side or on both sides. These areas may look like jelly under wet weather. First, symptoms develop on lower leaves. Sometimes disease occurs on succulent stems as small spots.

4.2.4.2.2 Host Range

Pseudopeziza medicaginis has wide host range, such as *Medicago sativa, M. arabica, M. falcata, M. globosa, M. scutellata, M. ciliaris, M. hispida, M. orbicularis, M. truneatula, M. tuberculata, M. lupulina, M. varia* (Jones, 1919; Schmiedeknecht, 1959; Schüepp, 1959), several species of Trigonella, and *Vicia villosa Onobryehis sativa* (Jones, 1919).

4.2.4.2.3 Epidemiology

This disease is favored by 60–75°F temperatures, cool and wet weather, and acidic and less fertile soil. The disease starts from the lower leaves of the plant and gradually progresses, causes premature defoliation, and reduces the plant's vigor; apparently, no permanent damage occurs, but it may reduce hay quality and quantity and damage nutritional values (leaves contain protein).

4.2.4.2.4 Management

Using resistant varieties is the best managing practice (Jones, 1953; Jones et al., 1941).

4.2.4.3 Spring Black Stem and Leaf Spot

The *Phoma medicaginis* var. *medicaginis* causes this disease.

4.2.4.3.1 Symptoms

The spots appear on the stems, petioles, and lower leaves. Irregular small-sized, black spots develop. Disease on young shoots is more severe, as it may kill the shoots. Leaf lesions become enlarged and collapse, and then leaves turn yellow and drop off. The black lesions on the stem and petiole may enlarge near the plant's base. Affected petioles and stems show large lesions and black-colored areas near the plant's base, so the plant becomes brittle and blackens in the case of a severe attack.

4.2.4.3.2 Epidemiology

The mycelium overwinters in crop debris and stubbles, where small black pycnidia are produced. Wet and cool weather favors the infection. This fungus may be a seed-borne pathogen. In wet and cool weather conditions, the infection becomes more severe, as shoots and stems become blackened, brittle, and easily breakable. Yield losses are severe when the cutting of the crop is late.

4.2.4.4 Summer (*Cercospora*) Black Stem

Cercospora medicaginis causes summer black stem.

4.2.4.4.1 Symptoms

As the name indicates, a black stem is its obvious symptom. Small spots are brown while enlarged (1/8 to 1/4" in diameter) and become gray to brown with irregular margins, appearing on both sides of leaves near the midrib and tip of leaflets; tissues present around the spot margins become yellow. Yellow leaflets may drop off; in severe infection, heavy defoliation occurs. Gradually lesions enlarge and cover the surface of the petioles and stem. Stems become brittle to some extent.

4.2.4.4.2 Epidemiology

The temperature range for the development of *Cercospora* in berseem is 80–90°F, and 100% humidity is required at the time of first cutting. The severity of infection is less during cool and dry weather, while warm and humid weather conditions favor the infection. This disease causes a reduction in the quantity and quality of forage.

4.2.4.5 Yellow Leaf Blotch

Yellow leaf blotch is extensively spread in the United States (Jones, 1918), Canada, Argentina, Europe, Austria, France, Germany, and Italy. In some regions, this disease has secondary importance, but under some weather conditions, it causes severe defoliation. Foliage losses have been recorded as 40% (Melchers, 1916). More than 50% of losses are occasional.

4.2.4.5.1 Symptoms

Young lesions convert into yellow blotches that appear parallel to the veins of the leaves. The yellow lesions on the upper surface enlarge and become more profound and yellow to orange in color. On the leaves' upper surface, brown- to black-colored pycnidia are developed, while few pycnidia are formed on the leaves' underside. Under favorable conditions, apothecia may develop as small black dots on lower leaf surfaces. Yellow to chocolate brown lesions on the stem also appear. Stem lesions have less importance compared to leaf blotches.

4.2.4.5.1 Epidemiology

The fungus overwinters in infected stems and leaflets. It is a seed and air-borne disease. Cool, wet spring weather conditions and succulent growth are favorable for yellow leaf blotch disease. Overwintered apothecia contribute to the development of primary inoculum. Cold weather during early spring and hot weather in midsummer pose an inhibiting effect on the production of ascospores. The optimum temperature range is 12–26°C for germination of ascospores. Ascospores are usually disseminated by the wind over short distances (Jones, 1918).

4.2.4.5.3 Management

Removing or burning leaves and stubble in early spring is helpful to reduce the inoculum. Fallen plant material usually becomes a source of inoculum. Growing resistant varieties is a management strategy (Jones, 1949). When disease is severe, early cutting is recommended.

4.2.4.6 *Stemphylium* Leaf Spot

It is caused by *Stemphlium botryosum*, composed of more than one race (Smith, 1940).

4.2.4.6.1 Symptoms

Small dark brown, oval-shaped sunken spots develop on the stems, petioles, and leaves. A yellow halo surrounds the sunken spots. Leaves become yellow and prematurely drop off. Blackened areas develop on the petioles and stems. Spots are irregular in size and shape; almost all above-ground plant parts may show symptoms. Olive brown to dark brown conidia is produced during early summer. Chlorotic and necrotic areas develop on the plant's leaves, and blackened, water-soaked lesions appear on stem and leaves. Older lesions show concentric rings. In wet seasons, severe infection causes the death of the plant. Infection also occurs in seeds and flowers as the color becomes dark (Nelson, 1955).

4.2.4.6.2 Epidemiology

Stemphylium requires periods of warm and wet weather during the summer and fall seasons. It overwinters in plant debris. The disease disseminates by air, water-borne conidia, and ascospores and also by infected seeds.

Stemphylium leaf spot is mostly considered a minor disease of alfalfa, but in some cases, it can cause noticeable defoliation (Nelson, 1955; Benedict, 1954).

4.2.4.6.3 Disease Cycle

Stemphylium overwinters and is dormant in plant debris and dead plant tissues, respectively. Ascospores usually serve as primary inoculum. Optimum weather conditions are 18–22°C temperature, and 100% relative humidity is required for initiation of infection. A humid climate is highly favorable for this disease throughout the world. Penetration of conidia and ascospores is through stomatal openings or wounds (Smith, 1940).

4.2.4.6.4 Management

Inoculum can be reduced by pulling out diseased plants in the fall and burning stubble in the spring. In case of severe infection, the crop should be cut for hay.

4.2.5 Downy Mildew

Peronospora trifoliorum causes downy mildew.

4.2.5.1 Symptoms

Infection appears first on the leaves present at the top portion of the plant (younger leaves are highly susceptible), while other fungal foliar diseases first appear on older leaves. The most apparent symptom is the light green appearance due to yellow-green colored blotchy areas developing on leaves; sometimes leaves become curled. Symptoms of downy mildew may be confused with nutrient deficiency. Due to systemic infection, stems become stunted and thickened underneath. In high humidity, fruiting bodies are generously produced by the fungus on the lower side of leaves.

4.2.5.2 Epidemiology

Optimum weather conditions for infection are 50–65°F temperature and 99–100% relative humidity. That is why this disease is more severe in cool weather and occurs during spring and fall (warmer and dryer areas). This fungus is also considered seed borne.

Young plants are more susceptible. The fungus disappears under unfavorable conditions, then reappears during a suitable environment because it survives systemically in the infected crown and shoots. Oospores remain dormant, overwinter in dead leaves, then germinate in the spring.

4.2.6 Rust

Caused by the *Uromyces straitus* fungus, it can cause considerable damage in seed fields. Rain during summer and fall causes enormous losses. Warm and humid weather conditions favor the disease. In alfalfa, rust is very rare. It is general in humid temperature zones of the world.

4.2.6.1 Symptoms

On leaves, brown spores within the tissues rupture the epidermis surface and appear as scattered pustules, the lesion surrounded by a yellow halo. Pustules may be arranged in a circle. Pustules also establish on the surface of petioles and stems.

4.2.6.2 Epidemiology

Rust disease is more severe during the late summer and fall. Favorable conditions are 70–85°F and high humidity for the development of this fungal disease.

4.2.7 Anthracnose

Anthracnose is caused by *Colletotrichum trifolii*.

4.2.7.1 Symptoms

Infected plants are a straw color; leaves become yellow and turn tan after the death of the plant. Gray-brown spots with purple-colored borders appear near the base of stems, spots enlarge and gradually become lesions, the central portion of the lesions are black due to fruiting bodies of the fungus, and the tips of stems become bend. Lesions create necrotic areas on plant parts.

4.2.7.2 Epidemiology

The fungus survives in infected crowns. Warm and wet weather conditions favor the infection. This disease appears in the fall. Anthracnose causes a reduction in yield by reducing the crown size and plant numbers.

4.2.7.3 Control

Resistant or moderately resistant varieties are the only excellent practical control.

4.3 BACTERIAL DISEASES

Bacterial diseases are challenging to manage (biologically and chemically) and are controlled by adopting preventive measures. The main risk factors for the spread or introduction of bacterial pathogens are insect vectors and infected planting material. Acute bacterial diseases are discussed in the following.

4.3.1 Bacterial Leaf Spot

Xanthomonas alfalfae subsp. alfalfae is reported as a causal agent of bacterial leaf spots (BLSs) in alfalfa (Schaad et al., 2007). It can cause sporadic epidemics on alfalfa crops throughout the world. The first time it was reported was in 1930 in the United States (Riker et al., 1935). It was also reported in India in 1954 (Bull et al., 2010). In Japan, 1964, *X. phaseoli* subsp. alfalfae was reported as the causal agent of this disease (Tominaga, 1964).

4.3.1.1 Symptoms

The water-soaked yellow spots are tiny at first, then size increases (2–3 mm in diameter), and they may coalesce to form lesions. Gradually spots turn to dark brown surrounded by a halo (Riker et al., 1935). Chlorotic and necrotic areas develop on plant parts. The pathogen can cause defoliation of leaves, stunting, and post-emergence damping-off (Riker et al., 1935). Diseased stems show dark brown to black shiny and greasy necrotic spots and lesions. A considerable height difference exists between non-infected and infected seedlings (Stuteville and Sorensen, 1966). Bacteria can penetrate the plant through stomata and wounds. Intercellular invasion of the pathogen causes devastation of mesophyll and palisade cells (Thiers and Blank, 1951).

4.3.1.2 Epidemiology

Bacterium requires a temperature range of 27–32°C for optimum growth. This bacterial disease is favored by rainy, hot, and windy weather. This bacterium overwinters in seed and plant debris. Farm equipment, wind, insects, and rain play an essential role in spreading this pathogenic bacterium (Malvick, 1988).

4.3.1.3 Management

BLS is a seed-transmitted disease. Therefore, it is highly recommended to use only certified seed. High-quality seed will be able to produce vigorous stands. It is essential to start with a high-quality stand of seedlings. *X. alfalfae* subsp. alfalfae are sensitive to copper-based chemicals. This bacterial pathogen can be effectively managed using copper biocides. Copper-based biocides are also used to avoid resistance development by the bacterial pathogen. The selection of resistant cultivars for cultivation is considered a useful and effective management strategy. Proper sanitation practices should be adopted, as the pathogen overwinters in plant debris. Strict quarantine measures should be adopted to avoid the spread of bacteria from one region to another due to the seed-borne nature of *X. alfalfae* subsp. alfalfa (Bradbury, 1981).

4.3.2 Bacterial Wilt

Clavibacter michiganensis subsp. *Insidiosus* causes bacterial wilt. Cold climate (22°C temperature for optimum growth temperature) favors the disease. Bacterial cells are Gram-positive aerobic, short, and non-motile (Jones and McCulloch, 1926; McCulloch, 1925).

4.3.2.1 Symptoms

Diseased plants show stunted growth, yellowing of leaves, and mottling. Leaflets may slightly curve or be severely distorted. Yellow-brown rings can be seen on the cross-section of bark. Yellow-brown streaks appear on wood outer tissues. The infected stems often become spindly and show signs of proliferation (Jones and McCulloch, 1926). The bacterium survives in alfalfa plant tissues (living or dead) present in soil and seed. The infected taproots show brown discoloration in xylem bundles, and yellow-brown discoloration appears under the taproots' bark (Ribaldi and Panella, 1958). Infection may occur through injuries in the roots or stems after mowing.

4.3.2.2 Dissemination

Clavibacter michiganensis subsp. *insidiosus* is a seed-borne pathogen and can survive in dead plant material and seed for 8–10 years (Jones and McCulloch, 1926; Cormack, 1961; Koehler et al., 1932). Under unfavorable conditions (absence of host plant, moist and warm conditions), bacterial cells become inactive (Nelson and Neal, 1974). Seed transmission often occurs at a low frequency (Jones and McCulloch, 1926; Cormack, 1961; Samac et al., 1998). The bacterium may contaminant seed systemically or on the surface seed (Cormack and Moffatt, 1956).

4.3.2.3 Management

Cultivation of resistant varieties is the best way of controlling the disease. Good stand health is vital as a preventive measure. Improving soil fertility (potassium level) can enhance the strength of the stand. Remove and dispose of unhealthy plants from the field. Adopt proper sanitation practices (Stuteville and Erwin, 1990).

4.3.3 Crown Gall

Crown gall disease in plants is caused by *Agrobacterium tumefaciens* (Smith and Townsend, 1907). It is considered an intensively studied disease of plants due to its economic importance (Barrett, 1929; Dowson, 1957; Hoerner, 1945; Kerr, 1969; Lehoczky, 1971; Malenin, 1972), as well as resemblance to the animal tumor (Stapp, 1927; Smith, 1916b; Smith, 1916a; Riker and Berge, 1935; Kupila, 1963; Knopf, 1974; Guillé and Quetier, 1970). The bacterium *Agrobacterium tumefaciens* has a vast host range (Wheeler, 1969; Walker 3rd, 1969; Schilberszky, 1935; Bene et al., 1964; Braun, 1962; Milani and Heberlein, 1972; Manocha, 1970).

4.3.3.1 Host Range

A wide range of hosts belong to 57 dicotyledonous families, 2 gymnosperm families, and 2 monocotyledonous families. Some economically important crops such as apple, cherry, pear, walnut, cotton, tomatoes, almond, beans, grapevine, and alfalfa are highly subject to *Agrobacterium tumefaciens*.

4.3.3.2 Disease Cycle

Agrobacterium tumefaciens (soil, Gram-negative bacterium) causes crown gall disease. Crown galls are hard and robust at the initial stage of infection. The growth of bacteria can be seen approximately 12–14 days after the initial infection. After one year of infection, galls become highly complex cavities. These cavities make the plants more susceptible to many insects/pests. The old galls can easily be detached from plants. *Agrobacterium tumefaciens* overwinters in the soil and weeds close to host plants. Injured plants are more susceptible to infection (wounds caused by cultivating farm practices and frost). The bacteria can survive in different types of soil for several years. This bacteria does not have specific nutritional requirements and is more prevalent in the winter season. The infection starts with the cultivation of infected planting sources; irrigation water aids dissemination. Then secondary infection occurs through contaminated farm equipment (Kado, 2002).

4.3.3.3 Symptoms

The galls are white and spherical at first, but old galls darken, and the outer cells begin to die. In severe infection, less plant vigor, prominent foliage reduction, stunted growth, yield loss, and overall plant stress can be seen. The disease's severity mainly depends upon the number of *Agrobacterium tumefaciens* that have entered the host plant (Kado, 2002) (www.ipm.ucdavis.edu/PMG/r5100411.html).

4.3.3.4 Management

Crop rotation with cereal crops and green manuring can be helpful to lessen the presence of the bacterium. Avoid injury during handling; good drainage and air circulation can also reduce the survival of bacteria.

Copper-based chemicals and creosote-based and strong oxidants (e.g., sodium hypochlorite) are effective chemical treatments.

A. tumefaciens can be managed by adopting a biological control strategy. The bacterium is sensitive to the agrocin (antibiotic) produced by *A. radiobacter* (non-plant pathogenic bacterium). An antibiotic Agrocin-84 (analog of the opine agrocinopine A), agrocinopine A, is typically produced in the gall tumors induced by *A. tumefaciens*. The antibiotic Agrocin-84 can inhibit DNA replication and cellular growth of *A. tumefaciens* (Kado, 2002).

4.3.4 Stem and Leaf Blight

Pseudomonas syringae pv. *Syringae* causes bacterial blight in alfalfa worldwide (Harighi, 2007; Nemchinov et al., 2017), but some scientists confirmed *Pseudomonas viridiflava* as a casual pathogen after a pathogenicity test and molecular analysis (Lipps et al., 2019).

P. syringae is a Gram-negative bacterium, divided into 50 pathovars according to the host range. *Pseudomonas syringae* is a heterogeneous group and can affect more than 200 species of plants. This bacterium can move for long distances (Morris et al., 2013). Bacterial cells can move from plant surface to troposphere by air currents and re-deposit on plants and water systems by rainwater and snow. Interaction between the bacterium and host plant has been studied (Alfano and Collmer, 1996).

4.3.4.1 Symptoms

This disease has two phases, systemic wilt and foliar necrosis (blight). Common symptoms of foliar infection are water soaking, chlorosis, and necrosis. Necrosis appears on the leaves, stems, and petioles due to penetration of bacteria to the host stem at injury sites. The appearance of bacterial exudes on stem cells that become black with age, and ultimately, the infected plant shows stunted growth and becomes wilted (Lipps et al., 2019).

4.3.5 Bacterial Root and Crown Rot

Pseudomonas viridiflava causes this disease on alfalfa crops globally (alfalfa-growing areas) (Lukezic et al., 1983). Root and crown rot causes huge crop loss in terms of quality and quantity. Burkholder recognized *P. viridiflava* as the causal agent for the first time in 1930; Dowson gave the name *Phythomonas viridiflava* in 1938 but changed it to *Pseudomonas viridiflava* (Graham et al., 1979; Stuteville and Erwin, 1990; Anzai et al., 2000). *P. viridiflava* is a highly damaging bacterium because it affects the crop at all stages of growth. It can cause several diseases on different hosts, that is, leaf blight on tomato melon, cauliflower, pumpkin, cabbage, lettuce, grapevine, carrot, poppy peas, turnip, and eggplant; nude and canker blight on kiwi; and onion wilt (Alippi et al., 2003; Gitaitis et al., 1997; Goumans and Chatzaki, 1998). Previously this bacterium was considered *Pseudomonas syringae* but was separated into different groups after 16 sRNA (Anzai et al., 2000). This bacterium can also cause necrosis of pepper and tomato, root rot, and dwarfing of red clover (Lukezic et al., 1983) and can also assist other pathogens (*Pseudomonas corrugate, Meloidogyne hapla, Ditylenchus dipsaci,* and *Pratylenchus penetrans)* in causing disease (Anzai et al., 2000). *P. viridiflava* possesses genes (hrp) that substantially affect plant destruction, cytoplasm, periplasm, and apoplasm. These genes play an essential role in hypersensitive response (Alfano and Collmer, 1997).

4.3.5.1 Dissemination and Symptoms

Primary infection starts after entry of bacterial cells into roots through wounds, mechanical injury, winter injury, nematode feeding sites, or injury at harvest time. Disease progression is maximum under warm, humid conditions and abundant moisture in the soil. Bacterial cells survive in soil, hay, and planting material. Rain and irrigation water and contaminated equipment are the primary sources of its dissemination. In severe infection, bacterial cells present in the intercellular spaces of the root and crown (Stuteville and Erwin, 1990). Wilting of the plant mainly occurs due to bacterial proliferation in xylem bundles, which disrupts water transport by vessel clogging and production of extracellular polysaccharides. Symptoms shown by infected seedlings are chlorosis, stunting, and malformation of the plant. Dwarfing of plants, root rotting, and discoloration are also common symptoms of infected plants. Disease symptoms are more evident under warm, humid, and rainy conditions. The bacterium can spread through rainwater (Jones et al., 1984). The ability of seed production of the infected plant is greatly affected (Sesma and Murillo, 2001). The bacterium grows on infected plants as pale yellow fluidal colonies.

4.4 VIRAL DISEASES

Viral diseases of alfalfa can negatively affect the quality, production, and durability of the crop. If viral diseases are not adequately controlled/managed, viruses will probably limit alfalfa production in that region (Al-Shahwan et al., 1998). Commonly, viruses are detected by serological assays; some frequently detected viruses are alfalfa mosaic virus and alfalfa leaf curl virus.

4.4.1 Alfalfa Mosaic Virus

The alfalfa mosaic virus is a causal agent. *Alfalfa mosaic* virus, also known as *Potato calico* virus, belongs to *Alfamovirus* and the family Bromoviridae. It was reported in alfalfa for the first time in the United States (Weimer, 1931). Several alfalfa mosaic virus strains with slight differences exist (strain S, strain Q, strain 425, strain AlMV-S, strain AlMV-B). The distinction is mainly based on different physico-chemical properties and symptoms in chosen hosts (Choi, 2000).

4.4.1.1 Host Range

It has a wide host range, including forage and food crops. It reduces the winter survival of crops and badly affects the plant's vigor, ultimately causing substantial yield losses (McDonald and Suzuki, 1983). Alfalfa mosaic virus can cause infection in almost 697 species of 167 genera (Edwardson and Christie, 1997). Some commercially essential crops affected by this virus are *Medicago sativa*, *Lactuca sativa*, *Solanum tuberosum*, tomato *Lycopersicon* esculentum, *T. repens* L., and *Trifolium pratense* L. (Fletcher, 2001; Hull, 1969).

4.4.1.2 Symptoms

The disease's severity depends on host variety, the host plant's growth stage, viral strain, and environmental conditions. It can cause yellow streaks, necrosis, and mottling between the veins of leaflets; wilting; mosaics; white flecks; and dwarfing. Leaflets become crinkled and stunted, and often stunting of the whole plant appears. Alfalfa mosaic virus can be detected in all parts of the infected host plant. The inclusion bodies are mainly found in the infected host plant (Kudo and Misawa, 1971). Disease symptoms are mild during spring (cool weather), and summer is highly unfavorable weather for this disease. Alfalfa mosaic virus is most severe in humid weather.

4.4.1.3 Epidemiology

Environmental factors (temperature and light) directly influence the movement and multiplication of viruses and indirectly in plants. Dark conditions usually slow down the multiplication as a reduction in ATP production and inactivation temperature range of the virus is 60–65°C. Infection also disturbs the quantity of minerals (Cu, P, Fe, Zn, and Mn) in plants. Consumption of infected crops by domestic animals is also very harmful to animals' health (Kudo and Misawa, 1971; Alblas and Bol, 1977; Yardimci et al., 2007; Jaspars and Bos, 1980).

4.4.1.4 Transmission

Fifteen aphid species are responsible for virus transmission in a nonpersistent manner. *Myzus persicae* plays a vital role in virus transmission (Kennedy et al., 1962). It is also frequently transmitted by pollen, seed, and parasitic plants and through the sap's mechanical inoculation.

4.4.1.5 Management

Efficient field control has not been developed yet. There is no effective insecticide available to control the aphid population. Some cultural practices are recommended (use of healthy seeds, weed control) to manage this disease (Jenkins, 2011). Research has been done on transgenic-resistant plants, reducing the plant's susceptibility to the virus (Kalla et al., 2001).

4.4.2 Alfalfa Leaf Curl

The alfalfa leaf curl virus belongs to genus Capulavirus of the Geminiviridae family, widely distributed across the world, that is, South Africa, the India subcontinent, southern and northern Europe, the Mediterranean basin, Syria, Jordan, Lebanon, Tunisia, Argentina, and Iran (Bernardo et al., 2013; Roumagnac et al., 2015; Susi et al., 2017; Bernardo et al., 2016; Kumari et al., 2018; Davoodi et al., 2018; Bejerman et al., 2018). Geminiviridae contains diverse and novel viruses (Varsani et al., 2017); full genome sequences of this virus obtained from different countries have also been studied.

4.4.2.1 Symptoms and Distribution

Alfalfa leaf curl disease symptoms are leaf curling, shriveling, crumpling, and stunting of the plant (Alliot et al., 1972; Blattný, 1959; Leclant et al., 1973; Cook and Wilton, 1984; Rodriguez Sardiña and Novales Lafarga, 1973). The alfalfa leaf curl virus is found in diverse climatic and geographical zones worldwide, including hot and cold semi-desert, temperate oceanic, mountainous, and subtropical climatic zones. Some natural barriers limit the spread or distribution of viruses in the world.

4.4.2.2 Transmission

Insect vectors transmit most plant pathogenic viruses. Transmission of 55% of viruses occurs by insects of the Hemiptera order (Hogenhout et al., 2008). Many viruses belonging to a different genus of family Geminiviridae are transmitted by hemipterans, whiteflies, leafhoppers, and treehoppers (Razavinejad et al., 2013; Heydarnejad et al., 2013; Bahder et al., 2016). Another group of insects, Aphididae, is responsible for the transmission of a large number of viruses. The aphid *Aphis craccivora* Koch species is reported as a vector of genus *Capulavirus* (Varsani et al., 2017; Roumagnac et al., 2015; Zerbini et al., 2017). Most geminiviruses are transmitted by insect vectors in a circulative persistent manner (Whitfield et al., 2015; Nault, 1997).

4.4.2.3 Management

Unproductive crops should be plowed down. Cultivate virus-free certified seeds of alfalfa. Do not grow alfalfa close to the other legumes (specifically beans and garden peas). Manage the weeds in the field along with the crop. A highly efficient solution is to develop varieties of alfalfa resistant to viral diseases.

REFERENCES

Adams P (1975) Factors affecting survival of *Sclerotinia sclerotiorum* in soil. *The Plant Disease Reporter* 59: 599.

Adams P and Ayers W (1979) Ecology of Sclerotinia species. *Phytopathology* 69: 896–899.

Al-Shahwan I, Abdalla O and Al-Saleh M (1998) Potato viruses in central Saudi Arabia. *Journal of King Saud University, Agricultural Sciences* 10: 45–53.

Alblas F and Bol J (1977) Factors influencing the infection of cowpea mesophyll protoplasts by alfalfa mosaic virus. *Journal of General Virology* 36: 175–185.

Alejandro Rojas J, Jacobs JL, Napieralski S, et al. (2017) Oomycete species associated with soybean seedlings in North America—Part I: Identification and pathogenicity characterization. *Phytopathology* 107: 280–292.

Alfano JR and Collmer A (1996) Bacterial pathogens in plants: Life up against the wall. *The Plant Cell* 8: 1683.

Alfano JR and Collmer A (1997) The type III (Hrp) secretion pathway of plant pathogenic bacteria: Trafficking harpins, Avr proteins, and death. *Journal of Bacteriology* 179: 5655.

Alippi AM, Dal Bo E, Ronco L, et al. (2003) Pseudomonas populations causing pith necrosis of tomato and pepper in Argentina are highly diverse. *Plant Pathology* 52: 287–302.

Alliot B, Giannotti J and Signoret P (1972) Mise en évidence de particules bacilliformes de virus associés à la maladie à énations de la Luzerne (*Medicago sativa* L.). *Acad Sci Paris CR Ser D.*

Altman J and CL C (1977) Effect of herbicides on plant diseases.

Annicchiarico P (2015) Alfalfa forage yield and leaf/stem ratio: Narrow-sense heritability, genetic correlation, and parent selection procedures. *Euphytica* 205: 409–420.

Anzai Y, Kim H, Park J-Y, et al. (2000) Phylogenetic affiliation of the pseudomonads based on 16S rRNA sequence. *International Journal of Systematic Evolutionary Microbiology* 50: 1563–1589.

Arnaud G (1910) Une nouvelle maladie de la Luzerne (Maladie rouge).

Bahder BW, Zalom FG, Jayanth M, et al. (2016) Phylogeny of geminivirus coat protein sequences and digital PCR aid in identifying *Spissistilus festinus* as a vector of grapevine red blotch-associated virus. *Phytopathology* 106: 1223–1230.

Barrett J (1929) A severe case of aerial "crown gall" on hothouse roses. *Phytopathology* 19: 1145.

Bartlett DW, Clough JM, Godwin JR, et al. (2002) The strobilurin fungicides. *Pest Management Science: Formerly Pesticide Science* 58: 649–662.

Bejerman N, Trucco V, De Breuil S, et al. (2018) Genome characterization of an Argentinean isolate of alfalfa leaf curl virus. *Archives of Virology* 163: 799–803.

Bene R, Girard T and Baldo S (1964) Exigences nutritionnelles des germes de la rhizosphere des plantes tumorisees par *Agrobacterium tumefaciens. Ann. Inst. Pasteur* 107: 86–99.

Benedict W (1954) Stemphylium leaf spot of alfalfa in Ontario. *Plant Dis Rep* 38: 27–29.

Berg LE, Miller SS, Dornbusch MR, et al. (2017) Seed rot and damping-off of alfalfa in Minnesota caused by *Pythium* and *Fusarium* species. *Plant Disease* 101: 1860–1867.

Bernardo P, Golden M, Akram M, et al. (2013) Identification and characterisation of a highly divergent geminivirus: Evolutionary and taxonomic implications. *Virus Research* 177: 35–45.

Bernardo P, Muhire B, François S, et al. (2016) Molecular characterization and prevalence of two capulaviruses: Alfalfa leaf curl virus from France and Euphorbia caput-medusae latent virus from South Africa. *Virology* 493: 142–153.

Blattný C (1959) Virus papillosity of the leaves of lucerne. *Folia Microbiologica* 4: 212–215.

Bradbury J (1981) *Xanthomonas campestris* pv. alfalfae. [Descriptions of fungi and bacteria]. *IMI Descriptions of Fungi* IMI Descriptions of Fungi Bacteria.

Brantner JR and Windels CE (1998) Variability in sensitivity to metalaxyl *in vitro*, pathogenicity, and control of *Pythium* spp. on sugar beet. *Plant Disease* 82: 896–899.

Braun AC (1962) Tumor inception and development in the crown gall disease. *Annual Review of Plant Physiology* 13: 533–558.

Broders K, Lipps P, Paul P, et al. (2007) Characterization of *Pythium* spp. associated with corn and soybean seed and seedling disease in Ohio. *Plant Disease* 91: 727–735.

Brown JG and Butler KD (1936) *Sclerotiniose of Lettuce in Arizona*. Tucson, AZ: College of Agriculture, University of Arizona.

Bull CT, De Boer SH, Denny TP, et al. (2010) Comprehensive list of names of plant pathogenic bacteria, 1980–2007. *Journal of Plant Pathology* 92: 551–592.

Campbell W (1947) A new species of Coniothyrium parasitic on sclerotia. *Mycologia* 39: 190–195.

Chauhan V, Yadav S and Sankar G (2009) Nutritive value of commonly used feed resources in Telangana region of Andhra Pradesh. *J Indian Journal of Animal Nutrition* 26: 23–28.

Choi JK (2000) Characterization and partial nucleotide sequence analysis of Alfalfa mosaic alfamoviruses isolated from potato and azuki bean in Korea.

Christen A (1983) Incidence of external seedborne *Verticillium albo-atrum* in commercial seed lots of alfalfa. *Plant Disease* 67: 17–18.

Christen A and French R (1982) Growth and pathogenicity of alfalfa strain of *Verticillium albo-atrum*. *Plant Diseases*.

Christen A and Peaden R (1981) Verticillium wilt in alfalfa. *Plant Disease* 65: 319–321.

Christen A and Peaden R (1982) Relative importance of sources of verticillium wilt infestation in alfalfa. *Phytopathology*. Amer Phytopathological Soc 3340 Pilot Knob Road, St Paul, MN 55121, 960–960.

Cook A and Wilton A (1984) Alfalfa enation virus in the Kingdom of Saudi Arabia.

Cook R and Zhang B-X (1985) Degrees of sensitivity to metalaxyl within the *Pythium* spp. pathogenic to wheat in the Pacific Northwest. *Plant Disease* 69: 686–688.

Cormack M (1961) Longevity of the bacterial wilt organism in alfalfa hay, pod debris, and seed. *Phytopathology* 51.

Cormack M and Moffatt J (1956) Occurrence of the bacterial wilt organism in alfalfa seed. *Phytopathology* 46.

Cottam WP (1921) A "dry rot" disease of alfalfa roots caused by a fusarium. *Phytopathology* 11: 383.

Davies R and Isaac I (1958) Dissemination of *Verticillium albo-atrum* through the atmosphere. *Nature* 181: 649–649.

Davoodi Z, Heydarnejad J, Massumi H, et al. (2018) First report of alfalfa leaf curl virus from alfalfa in Iran.

Dowson WJ (1957) Plant diseases due to bacteria. *Plant Diseases Due to Bacteria*.

Dozy RPA and Englemann WH (1869) *Glossaire des Mots Espagnols et Portugais dérivés de L'Arabe*: EJ Brill.

Dragovoz I, Kots SY, Chekhun T, et al. (2002) Complex growth regulator increases alfalfa seed production. *Russian Journal of Plant Physiology* 49: 823–827.

Edwardson J and Christie R (1997) *Viruses Infesting Solanaceous Crops*. Gainesville, FL: University of Florida Extension Station, IFAS.

El-Abyad M, El-Sayed M, El-Shanshoury A, et al. (1993) Towards the biological control of fungal and bacterial diseases of tomato using antagonistic *Streptomyces* spp. *Plant* 149: 185–195.

Emerich DW and Krishnan HB (2009) *Nitrogen fixation in crop production*. ASA-CSSA-SSSA.

Erwin DC and Ribeiro OK (1996) *Phytophthora diseases worldwide*: St Paul, MN: American Phytopathological Society (APS Press).

Faris M and Sabo F (1981) Effect of *Phytophthora megasperma* on yield and survival of resistant and susceptible alfalfa cultivars. *Canadian Journal of Plant Science* 61: 955–960.

Filonow A and Lockwood J (1985) Evaluation of several actinomycetes and the fungus *Hyphochytrium catenoides* as biocontrol agents for *Phytophthora* root rot of soybean. *Plant Disease* 69: 1033–1036.

Fletcher J (2001) New hosts of alfalfa mosaic virus, cucumber mosaic virus, potato virus Y, soybean dwarf virus, and tomato spotted wilt virus in New Zealand. *New Zealand Journal of Crop Horticultural Science* 29: 213–217.

Frate C and Long RF (2005) Sclerotinia in alfalfa: Biology and control in the central valley. *Proceedings, California Alfalfa and Forage Symposium*. Citeseer, 12–14.

Frosheiser F and Barnes D (1978) Field reaction of artificially inoculated alfalfa populations to the *Fusarium* and bacterial wilt pathogens alone and in combination. *Phytopathology* 68: 943–946.

Galibert F, Finan TM, Long SR, et al. (2001) The composite genome of the legume symbiont Sinorhizobium meliloti. *Science* 293: 668–672.

Gilbert R (1987) Crown and stem rot of alfalfa caused by *Sclerotinia sclerotiorum. Plant Disease* 71: 739–742.

Gilbert RG, Peaden RN (1988) Dissemination of Verticillium albo-atrum in alfalfa by internal seed inoculum. *Canadian Journal of Plant Pathology* 10(1): 73–77.

Gitaitis R, Sumner D, Gay D, et al. (1997) Bacterial streak and bulb rot of onion: I. A diagnostic medium for the semiselective isolation and enumeration of *Pseudomonas viridiflava. Plant Disease* 81: 897–900.

Goumans D and Chatzaki A (1998) Characterization and host range evaluation of *Pseudomonas viridiflava* from melon, blite, tomato, chrysanthemum and eggplant. *European Journal of Plant Pathology* 104: 181–188.

Graham JH, Stuteville D, Frosheiser F, et al. (1979) *A Compendium of Alfalfa Diseases*. St Paul, MN: American Phytopathological Society.

Griffin G and Thyr BJJoN (1978) Interaction of Meloidogyne hapla and *Fusarium* oxysporum on alfalfa. 10.

Guillé E and Quetier F (1970) Le "crown gall": Modèle expérimental pour l'étude du mécanisme de la transformation tumorale. *CR Acad. Sci. Paris, Sér. D* 270: 3307–3310.

Gyenis L, Anderson N and Ostry M (1999) Biological control of Septoria leaf spot and canker of hybrid poplar under field conditions. *Phytopathology* 89: S31.

Hancock J (1983) Seedling diseases of alfalfa in California. *Plant Disease* 67: 1203–1208.

Hancock J (1985) Fungal infection of feeder rootlets of alfalfa. *Phytopathology* 75: 1112–1120.

Handelsman J, Raffel S, Mester EH, et al. (1990) Biological control of damping-off of alfalfa seedlings with *Bacillus cereus* UW85. *Applied Environmental Microbiology* 56: 713–718.

Hansen E and Maxwell D (1991) Species of the *Phytophthora megasperma* complex. *Mycologia* 83: 376–381.

Harighi B (2007) Occurrence of alfalfa bacterial stem blight disease in Kurdistan Province, Iran. *Journal of Phytopathology* 155: 593–595.

Havey M and Grau C (1985) Decline of established alfalfa in soils naturally infested with *Phytophthora megasperma* f. sp. medicaginis and level of correlation by seedling assay. *Plant Disease* 69: 221–224.

Heale J and Isaac I (1963) Wilt of lucerne caused by species of Verticillium: IV. Pathogenicity of *V. albo-atrum* and *V. dahliae* to lucerne and other crops; spread and survival of *V. albo-atrum* in soil and in weeds; effect upon lucerne production. *Annals of Applied Biology* 52: 439–451.

Heale JB (1985) Verticillium wilt of alfalfa, background and current research. *Canadian Journal of Plant Pathology* 7: 191–198.

Hedlund T (1923) Om Nagrasjukdomar och skador pa vara lantbruksvaxter. *Sver. Allm. Jordbrukstidskr* 5: 166–168.

Heuze V, Tran G, Boval M, et al. (2013) *Alfalfa (Medicago sativa)*, Feedipedia. org. a programme by INRA, CIRAD, AFZ and FAO.

Heydarnejad J, Keyvani N, Razavinejad S, et al. (2013) Fulfilling Koch's postulates for beet curly top Iran virus and proposal for consideration of new genus in the family Geminiviridae. *Archives of Virology* 158: 435–443.

Hiltunen LH, Linfield CA and White J (1995) The potential for the biological control of basal rot of Narcissus by *Streptomyces* sp. *Crop Protection* 14: 539–542.

Hoerner G (1945) Crown gall of hops. *Plant Dis. Reptr* 29: 98–110.

Hogenhout SA, Ammar E-D, Whitfield AE, et al. (2008) Insect vector interactions with persistently transmitted viruses. 46: 327–359.

Howard RJ (1985) Local and long-distance spread of Verticillium species causing wilt of alfalfa. *Canadian Journal of Plant Pathology* 7: 199–202.

Huang H (1989) Distribution of *Verticillium albo-atrum* in symptomed and symptomless leaflets of alfalfa. *Canadian Journal of Plant Pathology* 11: 235–241.

Huang H, Hanna M and Kokko E (1985) Mechanisms of seed contamination by *Verticillium albo-atrum* in alfalfa. *Phytopathology* 75: 482–488.

Hull R (1969) Alfalfa Mosaic Virus. In: Smith KM, Lauffer MA and Bang FB (eds) *Advances in Virus Research*. New York: Academic Press, pp. 365–433.

Hwang S (1988) Effects of VA mycorrhizae and metalaxyl on growth of alfalfa seedlings in sopils from fields with alfalfa sickness in Alberta. *Plant Disease* 72: 448–452.

Iqbal Z, Khan MA, Sharif M, et al. (2018) An automated detection and classification of citrus plant diseases using image processing techniques: A review. *Computers Electronics in Agriculture* 153: 12–32.

Isaac I (1957) Wilt of lucerne caused by species of Verticillium. *Annals of Applied Biology* 45: 550–558.

Isaac I and Heale J (1961) Wilt of lucerne caused by species of Verticillium: III. Viability of V. Albo-atrum carried with lucerne seed; effects of seed dressings and fumigants. *Annals of Applied Biology* 49: 675–691.

Ivanov AI (1980) Lucerne. *Lucerne.*

Jaspars E and Bos L (1980) Alfalfa mosaic virus. *CMI/AAB Descriptions of Plant Viruses* 229.

Jenkins T (2011) *Alfalfa Mosaic Virus Infection Downregulates a Vital Photosynthesis Gene.* West Lafayette, IN: Purdue University.

Jennings JA and Pennington JA (2003) *Alfalfa for Dairy Cattle*: Cooperative Extension Service, University of Arkansas, US Department of Agriculture, and County Governments Cooperating.

Jones CR and Samac DA (1996) Biological control of fungi causing alfalfa seedling damping-off with a disease-suppressive strain of *Streptomyces*. *Biological Control* 7: 196–204.

Jones F (1925) A new bacterial disease of Alfalfa. *Phytopathology* 15.

Jones F (1949) Resistance in alfalfa to yellow leaf blotch. Amer Phytopathological Soc 3340 Pilot Knob Road, St Paul, MN 55121, 1064–1065.

Jones F (1953) Measurement of resistance in alfalfa to common leaf spot. *Phytopathology* 43: 651–654.

Jones F, Allison J and Smith W (1941) Evidence of resistance in alfalfa, red clover, and sweet clover to certain fungus parasites. *Phytopathology* 31: 765–766.

Jones FR (1918) Yellow-leafblotch of alfalfa caused by the fungus *Pyrenopeziza medicaginis*. *Jour. Agr. Res* 13: 307–329.

Jones FR (1919) *The Leaf-Spot Diseases of Alfalfa and Red Clover Caused by the Fungi Pseudopeziza medicaginis and Psuedopeziza [sic] trifolii, Respectively.* Washington, DC: US Department of Agriculture.

Jones FR and McCulloch L (1926) A bacterial wilt and root rot of alfalfa caused by *Aplanobacter insidiosum* L. McC. *Journal of Agricultural Research* 33.

Jones J, Jones J, McCarter S, et al. (1984) *Pseudomonas viridiflava*: Causal agent of bacterial leaf blight of tomato. *Plant Disease* 68: 341–342.

Kado C (2002) Crown gall. *Crown gall.*

Kalla R, Chu P and Spangenberg G (2001) Molecular breeding of forage legumes for virus resistance. *Molecular Breeding of Forage Crops*. Springer, pp. 219–237.

Katan J and Eshel Y (1973) Interactions between herbicides and plant pathogens. *Residue Reviews*. Springer, pp. 145–177.

Kennedy JS, Day MF and Eastop VF (1962) A conspectus of aphids as vectors of plant viruses. *A Conspectus of Aphids as Vectors of Plant Viruses.*

Kerr A (1969) Crown gall of stone fruit. *Australian Journal of Biological Sciences* 22: 111–116.

Knopf U (1974) On a eukaryotic cell-transformation by a bacterium. *Sub-Cellular Biochemistry* 3: 39–48.

Koehler B, Jones FRJB and 378 n. (1932) *Alfalfa Wilt as Influenced by Soil Temperature and Soil Moisture*. Bulletin.

Kreitlow K (1962) Verticillium wilt of alfalfa, a destructive disease in Britain and Europe not yet observed in the United States. US Dep. Agric., Crops Res. Div. *ARS*: 34–20.

Kudo A and Misawa T (1971) Some phenomena observed in systemic infection of alfalfa mosaic virus. II. The influence of air temperature and shading on symptom appearance. *Tohoku J Agr Res.*

Kumari SG, Moukahel AR, Richet C, et al. (2018) *First report of alfalfa leaf curl Virus affecting alfalfa (Medicago sativa)* in Jordan, Lebanon, Syria, and Tunisia.

Kupila S (1963) Crown gall as an anatomical and cytological problem: A review. *Cancer Research* 23: 497–509.

Larkin R, English J and Mihail J (1995) Effects of infection by *Pythium* spp. on root system morphology of alfalfa seedlings. *Phytopathology* 85: 430–435.

Lazarovits G, Turnbull A and Johnston-Monje D (2014) Plant health management: Biological control of plant pathogens. In: Van Alfen NK (ed) *Encyclopedia of Agriculture and Food Systems*. Oxford: Academic Press, pp. 388–399.

Leath KT, Erwin DC and Griffin GD (1988) Diseases and nematodes. *Alfalfa Alfalfa Improvement* 29: 621–670.

Leclant F, Alliot B and Signoret P (1973) Transmission et épidémiologie de la maladie à énations de la luzerne (lev). Premiers résultats. *Ann. Phytopathol* 5: 441–445.

Lehoczky J (1971) Further evidences concerning the systemic spreading of *Agrobacterium tumefaciens* in the vascular system of grapevines. *Vitis* 10: 215–221.

Lindemann J, Arny D and Delwiche P (1982) Detection of Verticillium-albo-atrum in the air over infected alfalfa fields in Wisconsin. *Phytopathology*. Amer Phytopathological Soc 3340 Pilot Knob Road, St Paul, MN 55121, 1382–1382.

Lipps S, Lenz P and Samac D (2019) First report of bacterial stem blight of Alfalfa caused by *Pseudomonas viridiflava* in California and Utah. *Plant Disease* 103: 3274–3274.

Liu D (1992) Biological control of *Streptomyces* scabies and other plant pathogens.

Lloveras J, Ferran J, Alvarez A, et al. (1998) Harvest management effects on alfalfa (*Medicago sativa* L.) production and quality in Mediterranean areas. *Grass Forage Science* 53: 88–92.

Lukezic F, Leath K and Levine R (1983) *Pseudomonas viridiflava* associated with root and crown rot of alfalfa and wilt of birdsfoot trefoil. *Plant Disease* 67: 808–811.

Lumsden R (1979) Histology and physiology of pathogenesis in plant diseases caused by Sclerotinia species. *Phytopathology* 69: 890–895.

Makkonen R and Pohjakallio O (1960) On the parasites attacking the sclerotia of some fungi pathogenic to higher plants and on the resistance of these sclerotia to their parasites. *Acta Agriculturae Scandinavica* 10: 105–126.

Malenin I (1972) L'influence des facteurs écologiques sur l'infection de la vigne par le chancre bactérien dans les conditions naturelles (in Bulgarian). Gradin. lozar. *Nauka, Balg* 9: 129–136.

Malvick D (1988) Leaf and stem diseases of alfalfa. *Report on Plant Disease.*

Manocha M (1970) Fine structure of sunflower crown gall tissue. *Canadian Journal of Botany* 48: 1455–1458.

Matthiesen R, Ahmad A and Robertson A (2016) Temperature affects aggressiveness and fungicide sensitivity of four *Pythium* spp. that cause soybean and corn damping off in Iowa. *Plant Disease* 100: 583–591.

McCulloch L (1925) *Aplanobacter insidiosum* n. sp., the cause of an Alfalfa disease. *Phytopathology* 15.

McDonald JG and Suzuki M (1983) Occurrence of alfalfa mosaic virus in Prince Edward Island. *Canadian Plant Disease Survey* 63: 47–50.

Melchers L (1916) Plant diseases affecting Alfalfa. *Rpt. Kansas State Bd. Agr* 35: 339–353.

Merriman P, Price R, Kollmorgen J, et al. (1974) Effect of seed inoculation with Bacillus subtilis and *Streptomyces* griseus on the growth of cereals and carrots. *Australian Journal of Agricultural Research* 25: 219–226.

Mielmann A (2013) The utilisation of lucerne (*Medicago sativa*): A review. *British Food Journal.*

Milani VJ and Heberlein GT (1972) Transfection in *Agrobacterium tumefaciens. Journal of Virology* 10: 17–22.

Moore W (1949) Flooding as a means of destroying the sclerotia of *Sclerotinia sclerotiorum. Phytopathology* 39: 920–927.

Morris CE, Monteil CL and Berge O (2013) The life history of Pseudomonas syringae: Linking agriculture to earth system processes. *Annual Review of Phytopathology* 51: 85–104.

Nault L (1997) Arthropod transmission of plant viruses: A new synthesis. *Annals of the Entomological Society of America* 90: 521–541.

Nelson CJ, Jennings J, Chon S, et al. (1997) Dealing with alfalfa autotoxicity. *The 25th Central Alfalfa Improvement Conference.* 13.

Nelson G and Neal J (1974) Persistence of a streptomycin-resistant variant of Corynebacterium insidiosum in soil and alfalfa roots in soil. *Plant Soil* 40: 581–588.

Nelson RR (1955) Studies on Stemphylium leafspot of alfalfa. *Phytopathology* 45: 352–356.

Nemchinov LG, Shao J, Lee MN, et al. (2017) Resistant and susceptible responses in alfalfa (*Medicago sativa*) to bacterial stem blight caused by Pseudomonas syringae pv. syringae. *PLoS One* 12: e0189781.

Page M, Gray F, Legg D, et al. (1992) Economic impact and management of Verticillium wilt on irrigated alfalfa hay production in Wyoming. *Plant Disease* 76: 504–508.

Parke J, Rand R, Joy A, et al. (1991) Biological control of *Pythium* damping-off and Aphanomyces root rot of peas by application of Pseudomonas cepacia or P. Fluorescens to seed. *Plant Disease* 75: 987–992.

Paulsrud BE (1996) *Characterization of antagonistic Streptomyces spp. from potato scab-suppressive soils and evaluation of their biocontrol potential against potato and non-potato pathogens.*

Peaden R, Gilbert R and Christen A (1985) Control of *Verticillium albo-atrum* on alfalfa. *Canadian Journal of Plant Pathology* 7: 511–514.

Pennypacker B and Leath K (1983) Dispersal of *Verticillium albo-atrum* in the xylem of alfalfa. *Plant Disease* 67: 1226–1229.

Pulli SK and Tesar MB (1975) *Phytophthora* root rot in seeding-year Alfalfa as affected by management practices inducing stress 1. *Crop Science* 15: 861–864.

Radmer L, Anderson G, Malvick D, et al. (2017) *Pythium*, *Phytophthora*, and *Phytopythium* spp. associated with soybean in Minnesota, their relative aggressiveness on soybean and corn, and their sensitivity to seed treatment fungicides. *Plant Disease* 101: 62–72.
Radović J, Sokolović D and Marković J (2009) Alfalfa—Most important perennial forage legume in animal husbandry. *Biotechnology in Animal Husbandry* 25: 465–475.
Ramírez-Suero M, Khanshour A, Martinez Y, et al. (2010) A study on the susceptibility of the model legume plant *Medicago truncatula* to the soil-borne pathogen *Fusarium oxysporum*. *European Journal of Plant Pathology* 126: 517–530.
Raynal G and Guy P (1977) Répartition et importance des maladies de la luzerne en France et en Europe. *Fourrages* 71: 5–14.
Razavinejad S, Heydarnejad J, Kamali M, et al. (2013) Genetic diversity and host range studies of turnip curly top virus. *Virus Genes* 46: 345–353.
Reichard S, Sulc R, Rhodes L, et al. (1997) Effects of herbicides on Sclerotinia crown and stem rot of alfalfa. *Plant Disease* 81: 787–790.
Ribaldi M and Panella A (1958) On bacterial wilt of alfalfa (*Medicago sativa* L.) caused by *Corynebacterium insidiosum* (McCull.) Jensen, in Italy. *Euphytica* 7: 179–182.
Ridgway R (1912) *Color Standards and Color Nomenclature*. The Author.
Riker A and Berge T (1935) Atypical and pathological multiplication of cells approached through studies on crown gall. *The American Journal of Cancer* 25: 310–357.
Riker AJ, Jones F and Davis MC (1935) Bacterial leaf spot of alfalfa. *J. Agric. Res* 51: 177–182.
Rodriguez Sardiña J and Novales Lafarga J (1973) Una virosis de la alfalfa con producción de "enations". *An. INIA/Ser. Prt. Veg* 3: 131–146.
Roumagnac P, Granier M, Bernardo P, et al. (2015) Alfalfa leaf curl virus: An aphid-transmitted geminivirus. *Journal of Virology* 89: 9683–9688.
Samac DA, Nix RJ and Oleson AE (1998) Transmission frequency of Clavibacter michiganensis subsp. insidiosus to alfalfa seed and identification of the bacterium by PCR. *Plant Disease* 82: 1362–1367.
Schaad N, Postnikova E, Lacy G, et al. (2007) *Xanthomonas alfalfae* sp. nov., *Xanthomonas citri* sp. nov. and *Xanthomonas fuscans* sp. nov. *List of New Names New Combinations Previously Effectively, but not Validly, Published, Validation List* 893–897.
Schilberszky K (1935) Beiträge zur Biologie von *Pseudomonas tumefaciens*. *Zeitschrift für Pflanzenkrankheiten und Pflanzenschutz* 45: 146–159.
Schmiedeknecht MJDZ (1959) Beitrag zur Eigenschaftsanalyse der Resistenz verschiedener Medicago-Arten gegen *Pseudopeziza medicaginis* (Lib.) Sacc. 29: 65–72.
Schmitthenner A (1964) Prevalence and virulence of *Phytophthora*, Aphanomyces, *Pythium*, Rhizoctonia, and *Fusarium* isolated from diseased alfalfa seedlings. *Phytopathology* 54: 1012–1018.
Schmitthenner A and Van Doren D (1985) Integrated control of root rot of soybean caused by *Phytophthora megasperma* f. sp. glycinea. *Ecology and Management of Soilborne Plant Pathogens*: 263–266.
Schüepp H (1959) Untersuchungen über Pseudopezizoideae sensu NANNFELDT. ETH Zurich.
Sesma A and Murillo J (2001) The Biochemistry and Molecular genetics of host range definition in "*Pseudomonas syringae*". *The Biochemistry Molecular Genetics of Host Range Definition in "Pseudomonas syringae"*: 1000–1024.
Sheppard J and Needham S (1980) Verticillium wilt of alfalfa in Canada: Occurrence of seed-borne inoculum. *Canadian Journal of Plant Pathology* 2: 159–162.
Sherbakoff C (1915) Fusaria of potatoes. New York (Ithaca) Agric. *Exp. Sta. Mem*: 87–270.
Singh V and Misra AK (2017) Detection of plant leaf diseases using image segmentation and soft computing techniques. *Information Processing in Agriculture* 4: 41–49.

Smith DH, Beck KG, Peairs F, et al. (1999) *Alfalfa: Production and Management*. Colorado State University. Libraries.

Smith EF (1916a) Further evidence that crown gall of plants is cancer. *Science* 43: 871–889.

Smith EF (1916b) Studies on the crown gall of plants: Its relation to human cancer. *The Journal of Cancer Research* 1: 231–309.

Smith EF and Townsend CO (1907) A plant-tumor of bacterial origin. *Science* 25: 671–673.

Smith EG, Acharya SN and Huang HC (1995) Economics of growing Verticillium wilt-resistant and adapted Alfalfa cultivars in Western Canada. *Agronomy Journal* 87: 1206–1210.

Smith OF (1940) Stemphylium leaf spot of Red Clover and Alfalfa. *Journal of Agricultural Research.*

Stapp K (1927) *Der bakterielle Pflanzenkrebs und seine Beziehungen zum tierischen und menschlichen Krebs.*

Starr GH, Walters H and Bridgmon GH (1953) White Mold (Sclerotinia) of Beans.

Steadman JR, Maier C, Schwartz H, et al. (1975) Pollution of surface irrigation waters by Plant Pathogenic Organisms 1. *JAWRA Journal of the American Water Resources Association* 11: 796–804.

Stuteville D and Sorensen E (1966) Distribution of leaf spot and damping-off (*Xanthomonas alfalfae*) of alfalfa in Kansas, and new hosts. *Plant Disease Reporter* 50: 1.

Stuteville DL and Erwin DC (1990) *Compendium of Alfalfa Diseases*: Aps Press.

Susi H, Laine A-L, Filloux D, et al. (2017) Genome sequences of a capulavirus infecting *Plantago lanceolata* in the Åland archipelago of Finland. *Archives of Virology* 162: 2041–2045.

Taylor R, Salas B, Secor G, et al. (2002) Sensitivity of North American isolates of *Phytophthora* erythroseptica and *Pythium* ultimum to mefenoxam (metalaxyl). *Plant Disease* 86: 797–802.

Teutsch CD and Sulc RM (1997) Influence of seedling growth stage on flooding injury in alfalfa. *Agronomy Journal* 89: 970–975.

Thiers H and Blank L (1951) A histological study of bacterial blight of cotton. *Phytopathology* 41: 499–510.

Tominaga T (1964) On the bacterial leaf spot of alfalfa. *Japanese Journal of Phytopathology* 29: 162–166.

Tribe H (1957) On the parasitism of Sclerotinia trifoliorum by Coniothyrium minitans. *Transactions of the British Mycological Society* 40: 489–499.

Tu J (1979) Evidence of differential tolerance among some root rot fungi to rhizobial parasitism *in vitro*. *Physiological Plant Pathology* 14: 171–177.

Tu J (1980) Incidence of root rot and overwintering of alfalfa as influenced by rhizobia. *Phytopathologische Zeitschrift* 97: 97–108.

Tu JC (1978) Protection of soybean from severe *Phytophthora* root rot by Rhizobium. *Physiological Plant Pathology* 12: 233–240.

Varsani A, Roumagnac P, Fuchs M, et al. (2017) Capulavirus and Grablovirus: Two new genera in the family Geminiviridae. *Archives of Virology* 162: 1819–1831.

Venancio WS, Rodrigues MAT, Begliomini E, et al. (2009) Physiological effects of strobilurin fungicides on plants. *Ciências Exatas e da Terra, Ciências Agrárias e Engenharias* 9.

Viands D, Lowe C, Bergstrom G, et al. (1992) Association of level of resistance to Verticillium wilt with alfalfa forage yield and stand. *Journal of Production Agriculture* 5: 504–509.

Walker J (1969) *Plant Pathology*. 3rd. New York: McGraw-Hill, pp. 345.

Ward CE and Hutson AL (2004) *Economics of Producing Alfalfa*.

Weimer J (1928) A wilt disease of alfalfa caused by *Fusarium oxysporum* var. medicaginis, n. var. *Journal of Agricultural Research* 37.

Weimer J (1931) Alfalfa mosaic. *Phytopathology* 21.

Weston WARD, Loveless AR and Taylor RE (1946) Clover rot. *The Journal of Agricultural Science* 36: 18–28.
Wheeler BEJ (1969) An introduction to plant diseases. *An Introduction to Plant Diseases.*
Whitfield AE, Falk BW and Rotenberg D (2015) Insect vector-mediated transmission of plant viruses. *Virology* 479: 278–289.
Xiao K, Kinkel LL and Samac DA (2002) Biological control of *Phytophthora* root rots on alfalfa and soybean with *Streptomyces. Biological Control* 23: 285–295.
Yardimci N, Eryigit H and Erda I (2007) Effect of alfalfa mosaic virus (AMV) on the content of some macro-and micronutrients in alfalfa. *Journal of Culture Collections* 5: 90–93.
Yuegao H, Cash D, Kechang L, et al. (2009) Global status and development trends of alfalfa. *Alfalfa Management Guide for Ningxia*: 1–14.
Zerbini FM, Briddon RW, Idris A, et al. (2017) ICTV virus taxonomy profile: Geminiviridae. *The Journal of General Virology* 98: 131.
Zhang B and Yang X (2000) Pathogenicity of *Pythium* populations from corn-soybean rotation fields. *Plant Disease* 84: 94–99.

5 *Trifolium* Species
Diseases, Etiology, and Management

Anjum Faraz, Imran Ul Haq, Siddra Ijaz, and Muhammad Zunair Latif

CONTENTS

5.1 INTRODUCTION

Trifolium alexandrinum, common name berseem clover, Egyptian clover, is an important leguminous winter fodder crop cultivated in irrigated tracts in areas like Egypt, Israel, Syria, Pakistan, India, and the United States. It is also known as the king of fodder crops due to its high nutrition, digestibility, palatability, and succulence. It has 18–28 crude percent that is better than or equal to crimson clover and alfalfa; it also contains 70% dry matter digestible nutrient, a strong milk multiplier (Karsli et al., 1999; Sharma and Murdia 1974). Iqbal and Iqbal (2014) and Manjunatha et al., (2017). Berseem clover faces fungal, viral, and nematode diseases that reduce the yield (Mishra et al., 1980; Windham and Pederson, 1988; Maghazy et al., 2008;

DOI: 10.1201/9781003055365-5

Singh et al., 2020). Red clover *(Trifolium pratense)* is one of the most important perennial forage crops. It is widely cultivated for its high nutritional value either in mixture with other legumes or as a pure stand. Several diseases attack red clover; some cause extensive damage to its growth, persistence, and overwintering capacity, while others are economically less important. The material described in this chapter will provide an opportunity to understand the nexus of plant–microbe interaction and help initiate a focus on clover diseases.

5.2 DISEASES OF *T. ALEXANDRINUM*

5.2.1 Fungal Diseases of *T. alexandrinum*

5.2.1.1 Stem and Crown Rot

Stem rot of clover is a significant fungal disease of berseem clover caused by *Sclerotinia trifoliorum* (Faruqui et al., 2002; Mattila et al., 2010; Manjunatha et al., 2019). This disease is spread all over India, but the most severe is in the northwest regions (Manjunatha et al., 2017). For the last few years, this disease has been a threat to fodder and seed production due to favorable environmental conditions during its growing season (Rathi et al., 2010). The pathogen *Sclerotinia* spp. is a necrotrophic fungus that causes different diseases on various crop plants, including winter fodder crops like *Medicago* sp. and *Trifolium* sp. (Gargouri et al., 2017; Ficker, 2019). It infects the stem at the soil surface, and symptoms appear during the cool season in January and February (Rathi et al., 2010). In India, many scientists determined *S. trifoliorum* is an actual pathogen of stem rot of berseem in India (Kumar et al., 2003; Rathi et al., 2007; Pande et al., 2008). The same disease was reported in 2016 in Pakistan but with a different etiological agent, *S. sclerotiorum*, identified based on morphological and molecular characterization (Saira et al., 2016; Singh et al., 2020).

Pathogen Morphology: Whitish fluffy growth on potato dextrose agar medium; mycelium is hyaline and septate. Black hard, cylindrical, or irregularly shaped sclerotia appeared on culture medium and are difficult to remove (Kim and Cho, 2003; Vleugels et al., 2013; Manjunatha et al., 2019). The pathogen grows at 5–30°C temperature and 5–8 pH, but optimum temperature and pH for mycelial growth and sclerotial formations are 20 ± 3°C, 6.5 ± 0.5 and 20 ± 5°C, 6 ± 0.5, respectively (Rathi et al., 2010; Manjunatha et al., 2017, 2019; Singh et al., 2020). Faraz et al., 2021 identified *S. trifoliorum* using morph-cultural and molecular characteristics (ITS, LSU, G3PDH, CAL, HSP60 genetic loci) and claimed that it caused stem and crown rot of berseem in Pakistan.

Symptoms: Water-soaked lesions appeared around the stem, which later turned to patches in severe conditions under favorable conditions, and ultimately, the crop fails to grow. Many sclerotia produces on infected tissues and plant debris, which is the disease's source (Vleugels et al., 2013) (Figure 5.1).

Management: Use healthy plants collected from disease-free fields or certified government or private sectors. Cultural practices: Fields affected by this disease flood in summer seasons and avoid over-irrigation (Singh and Singh, 1995). Chemical control: spray 0.1% of bavistan at 15 days or 0.4% of brasicol drenching after each cutting (Singh and Singh, 1995). Thiophenate methyl (Topsin-M) at 500 g/

FIGURE 5.1 Symptoms of stem and crown rot of *T. alexandrinum*.

ha provides significant control of berseem rotting (Iqbal and Iqbal, 2014). Biological control: *Trichoderma harzianum* and *Bacillus thuringiensis* are biocontrol agents and have a significant effect against stem and crown rot of Egyptian clover (Zaher et al., 2013). *Coniothyrium minitans* is a mycoparasite that reduces the sclerotia present in the soil by feeding on them (Johnson and Atallah, 2014).

The Fungal Molecular Biology Laboratory (FMB), Department of Plant Pathology, University of Agriculture Faisalabad (UAF), Pakistan, was awarded a project by research funding agency Punjab Agriculture research board (PARB), Project PARB Project No. 951, entitled 'Improvement of berseem clover (*Trifolium alexandrinum*) seed quality by managing fungal seed infections and infestations with special emphasis on stem and crown rot.' The findings of research project: From all seed batches of *Fusarium delphinoides*, *F. fujikuroi*, *F. falciforme*, *F. equiseti*, *Didymella pinodella*, and *Stemphylium trifolii* were isolated and morphogenomics bases identified. A seed germination test was done after 8, 15, and 30 days of inoculation with fungal pathogens frequently associated with collected seed batches. After 8, 15, and 30 days of inoculation, 67.32%, 62.66%, and 49.32% seeds were germinated, respectively. The results showed that, as the time period of inoculation increased, the germination percentage of seeds decreased. Furthermore, pathogenicity was confirmed with selected fungal isolates to determine the etiology of stem and crown rot disease in berseem. Koch's postulates confirmed *Sclerotinia trifoliorum* as a pathogen of stem and crown rot of berseem clover. We have formulated the following management plan for better management of stem and crown rot of berseem clover. Healthy, clean, graded, good-quality seed of cultivar Lyallpur Faisalabad should be purchased or obtained from authentic, certified sources. Seed batches must be harvested and collected from healthy crops, because they may be infested or infected with seed mycoflora or contain sclerotia as seed contaminant mixed with seed batches. Sowing of such a seed-borne batch may introduce stem and crown rot in the field. Select a field with no previous history of stem and crown rot disease, as fungus survives in the soil in the form of sclerotia (hard structures which can survive in soil for years even under adverse conditions). Pre-sowing seed treatment with thiophanate methyl at the rate of 2–3 gram/kg of berseem clover seed should be used. There is no resistant cultivar against stem and crown rot disease in Pakistan, but Lyallpur late is tolerant to the disease. The disease appears in patches. As disease patches appear in the field, thiophenate methyl should be applied as spray/flood in the field @ 800–1000 grams per acre, right after getting the crop cut. Don't cut the fodder until at least 25–30 days from the application of fungicide. The same fungicide with the previously mentioned dose may also be used as a preventative measure.

5.2.1.2 Root Rot

Causative agent and disease development: Three fungal pathogens cause this disease, *Fusarium moniliforme, Rhizoctonia solani,* and *Sclerotinia bataticola,* known as root rot complex. Association of these fungal pathogens with nematode *Tylenchorhynchus vulgaris* produced severe injury to the crop (Jobshy et al., 1981; Faruqui et al., 2002; Kumar and Singh, 2012; Hasan and Bhaskar, 2004). Resting spores of fungi survived in the soil for a long time, and when favorable conditions (temperature, host) were available, they attacked the crop and caused the disease (Rathi et al., 2010; Singh et al., 2020).

Symptoms: Initial symptoms were wilting of tillers, which appeared in patches under favorable conditions. White hyphae turned to gray with time by *R. solani. Fusarium* species enter the healthy roots of plants and move upward through xylem tissues, and vascular bundles become blocked. The nematode also enters through the root system and destroys it (Faruqui et al., 2002; Iqbal and Iqbal, 2014). Due to dense cropping, high humidity favors sclerotia present in the soil; however, all of these pathogens infect plants and produce root rot complex (Hasan and Bhaskar, 2004; Rathi et al., 2010; Singh et al., 2020).

Management: Cultural control showed significant control of the disease, such as fallowing the field or crop rotation two to three years, with deep plowing in summer and 2 gm/kg seed of carbendazim as seed treatment (Chaudhry et al., 1992). Metalaxyl fungicide showed the best results as drenching in western Australia against root rot in subterranean clover. Thiophenate methyl at 500 grams/ha showed significant results against this (Iqbal and Iqbal 2014). Seed treatment with bromothalnil @ 2.5 g/kg seed showed significant results against disease (Asghar et al., 2019; Singh et al., 2020).

5.2.1.3 Damping-Off

Causative Agent and Disease Development: This disease is caused by *Pythium spinosum Sawada* var. *spinosum*, a soil-borne fungal pathogen that affects Egyptian clover seeds and seedlings (Barbetti, 1983).

Pathogen Biology: Hyphae 7.7 wide and septate, lemon-shaped swelling at the tip; globose shape; and zoospores produced at above 38°C. Oogonia are usually globose; intercalary; 23–28 μm (25 μm average) in diameter, with digitate, blunt-shaped projections 2–12 × 2–4 μm in size. Plerotic, sometimes aplerotic, oospores are produced 19–25 μm (21 μm average) in diameter, and walls are 2 μm thick. It is frequently produced in wet soils with inadequate soil drainage systems (Maghazy et al., 2008; Singh et al., 2020).

Symptoms: It usually refers to stem and root tissues rotting below the soil surface. In pre-emergence conditions, seeds decay or sometimes germinate water-soaked blighted seedlings. In post-emergence conditions, seedlings collapse and wilt, and roots dry up and die below the soil line (Maghazy et al., 2008).

Management: Fungicides such as metalaxyl control the diseases of roots caused by *P. irregular* (Greenhalgh, 1983). Drenching infested soil with metalaxyl and benomyl separately and in combination reduces root rot caused by *Pythium* spp. and *F. avenaceum* on subterranean clover (Greenhalgh and Clarke, 1985). Seed treatment of subterranean clover with metalaxyl or benomyl could reduce root rot (Smiley et al., 1986). Root diseases caused by *P. clandestine* could be controlled using metalaxyl (Hochman et al., 1990; Burnett et al., 1994). Potassium phosphate application, crop rotation, deep plowing in summer, and certified seeds are effective methods to manage root rot caused by *P. clandestine* (Greenhalgh et al., 1994). Seed coatings of *Trifolium alexandrinum* L. with biocontrol agents *Penicillium funiculosum, Aspergillus carneus, A. cervinus, A. sulphureus, Phoma pomorum, P. islandicum, P. nigricans, Paecilomyces lilacinus*, and *Chaetomium globosum* had significant results against pre-emergence damping-off caused by *Pythium spinosum* var. *spinosum*. The use of these biocontrol agents in infested soils also significantly protected berseem clover in the field (Maghazy et al., 2008; Singh et al., 2020).

5.2.1.4 Stemphylium Leaf Spot

Causative Agent and Disease Development: This disease is caused by *Stemphylium globuliferum* (Vestergren). *Stemphylium* species, including *botryosum, alfalfa, trifolii, globuliferum*, and *vesicarium,* are involved in leaf spot development on *Trifolium* sp. (Bradley et al., 2003). Reactive oxygen species such as hydrogen peroxide (H_2O_2) and superoxide (O^-_2) have a pivotal role in developing necrotic symptoms of leaf spots in clover caused by *S. globuliferum.*

Pathogen Morphology: Colonies of fungi on potato dextrose agar culture medium are grayish with amber pigmentation. Conidiophores are unbranched to rarely branched, three to five septations, 6-8 × 45-65 μm in size, cylindrical, smooth, 3–5 μm apical, and dark brown or olive-brown and slightly curved with 7–8-μm swollen apex. Conidia are oblong 4-μm horizontal and longitudinal septations, medium brown with smooth external walls, 27-29 × 29-31 μm in size. Numerous stomata are present in artificial culture medium in 28–31 days with dimensions of

0.3–0.4 μm broad and tall ascocarps and dark pigmentation and are normally sub-globose tapered. Mature asci are cylindrical to elliptical, 45–53 μm broad and 24–31 μm tall. Ascospores are oval shaped, 19-22 × 42-46 μm in size with 7–1 μm, 2–3 μm horizontal and longitudinal septa, respectively, usually olive-brown mid-septum. Based on the previously mentioned perfect and imperfect stages, Omar et al., 2015 confirmed this pathogen for the first time as *S. globuliferum* causing leaf spot on *T. alexandrinum* in Egypt.

Symptoms: Small brown multiple oval symptoms are produced initially, which later enlarge to brown to black lesions. Sometimes these lesions are surrounded by yellow margins and interact with lesions of leaves. As the disease progresses, lesions are enlarged and expanded, and finally leaves wither from tips. Dark brown lesions are also produced on the stem (Omar et al., 2015).

Management: Among systemic and contact fungicides, triazoles, tebuconazole, difenoconazole, hexaconazole, and Kavach, Antracol, and Indofil M-45, respectively, had a significant effect against stemphylium blight of onion in laboratory and field conditions (Mohan et al., 2003). Systemics and fungicides azoxystrobin and propiconazole and Indofil M-45 (mancozeb) and Antracol (propineb) had significant inhibition of *Stemphylium* species (Mishra and Gupta, 2012; Ihsanul and Nowsher, 2007). Foliar spray of propiconazole at 0.1% and mancozeb at 0.25% four times reduced *Stemphylium* blight intensity in the field (Gupta and Gupta, 2013). *Pseudomonas fluorescens* and hexaconazole used in combination or alone as seed treatment, foliar spray, or root dip had a good effect on controlling *S. botryosum* leaf blight severity (Barnwal et al., 2003). Different biocontrol agents such as *Trichoderma harzianum, Gliocladium* spp., and *Saccharomyces cerevisiae* had significant inhibition of *S. vesicarium* mycelium in *in-vitro* conditions, but the highest inhibition was observed with *T. harzianum.* In *in-vivo* conditions among biocontrol agents, *B. subtilis, P. fluorescens,* and *T. harzianum* had significant inhibition of disease severity, but *T. harzianum* gave the lowest inhibition (Hussein et al., 2007). Extracts of different plants; *Lantana camara, Azadirachta indica, Argimone mexicana Datura metel, ciumum* spp. gave sigfinificant results against disease. Furthermore, *Parthenium hystorophorus* and biocontrol agents *Trichoderma harzianum, T. viride, Aspergillus niger, Penicillium citrinum,* and *G. virens* significantly inhibited the growth of *S. botryosum.* Among plant extracts and biocontrol agents, *A. indica* and *T. harizianum* gave maximum inhibition under field conditions, and two foliar sprays of *T. harzianum* and *A. indica* at 0.2% 15 days interval suppressed disease severity (Kumar et al., 2012; Mishra and Gupta, 2012). For integrated management of disease, chemical (mancozeb) and biocontrol agents (*Trichoderma viride* Tv-1) at spore concentration of 1 × 109 and plant extract of *C. indica* at 10% significantly reduced the severity of *Stemphylium* leaf spot.

5.2.1.5 *Curvularia lunata* Leaf Spot

Causative Agent and Disease Development: Causal organism: *Curvularia lunata.*

Pathogen Morphology: Grayish-black from front side and light to dark black from margin to center on potato dextrose agar culture medium. Hyphae is septate, branched, and brownish with a smooth wall. Conidiophores are dark brown, with no septations, and branched. Conidia are boat shaped, with septations and round tips and are 26±7 ×12±3 μm in size. Based on three genetic loci ITS, RPB2, and GAPDH, this pathogen is identified as *Curvularia lunata* (Ellis, 1971; Manamgoda et al., 2015; Haq et al., 2020).

Symptoms: V-shaped, brown color, and 0.8–2-cm diameter, which merged and produced black burning on leaflets. This disease was first reported in 2020 on *Trifolium alexandrinum* in Pakistan (Haq et al., 2020).

Management: *Chaetomium cupreum* is a biological control agent and used in both *in-vitro* and *in-vivo* conditions against *Curvularia lunata*. Sporulation of *C. cupreum* had significant inhibition of *C. lunata* using the dual culture technique. Under field experiment use of spore suspensions, *Chaetomium* bioproduct had significant suppression of disease; however, bioproduct showed maximum disease severity reduction compared to spore suspensions (Kanokmedhakul et al., 2006; Soytong, 2014; Tann and Soytong, 2016). Good agriculture practices (GAP), organic methods, and chemicals are different methods to manage curvularia leaf spot. Among these methods, the GAP method showed a maximum reduction of disease compared to other methods (Stanhill, 1990; Maeder et al., 2002; Tann and Soytong, 2016). The organic method showed better results than GAP but required many factors to evaluate for successful management of disease (Paull, 2011). Cymbopogon citratus essential oil is a phytotoxic chemical against *Curularia lunata* leaf spot. Conidial germination is completely inhibited at 5 and 7.5 μL mL-1 concentrations (Mourão et al., 2017; Sarmento-Brum et al., 2013, 2014; Souza Júnior et al., 2009). The concentration of oil at 7.5 to 50 μL mL-1 is phytotoxic against pathogen and suppresses the disease. A 7.5-μL mL-1 concentration is used as preventive control, not a curative control (Santos et al., 2013). Geranial and Neral are *C. citratus* components that suppress the disease 41.46% and 32.43%, respectively (Pinto et al., 2014; Gonçalves et al., 2015; Aquino et al., 2014).

5.2.2 Viral Diseases of *T. alexandrinum*

5.2.2.1 Mosaic Disease in Berseem Clover

Causative Agent and Disease Development: The mosaic virus that causes this disease belongs to the genus *Alfamovirus* and the family Bromoviridae. This virus is a single standard, with positive sense RNAs and proteins P1 and P2. These proteins encode RNAs1 and two, which are involved in replicating the virus (Nassuth and Bol, 1983; Herranz et al., 2012). This is sap inoculated, transmitted through aphids (*Aphis gossypii*), but other species of aphids (*A. rumicis or A. craccivora*) are not involved in transmission. This virus resembles the alfalfa mosaic virus that is confirmed by serological assay (Babadoost, 1990). This virus's new strain is transmitted 60–70% through Egyptian clover seeds (Mishra et al., 1980; Pineyro et al., 2002; Singh et al., 2020).

Symptoms: Early symptoms, including vein clearing, appear during January and February. Symptoms are prominent after each cutting, such as light green mottling, which appears along veins and veinlets. Chlorotic patches appear when the season warms (Mishra et al., 1980; Norton and Johnstone, 1998; Singh et al., 2020).

Management: Use genetically and physically certified pure seed. Systemic insecticides like Dimethioate 20% EC at 1-1.5 L/ha can be used to control the aphids involved in the transmission of disease (Martin et al., 1997; Singh, 2001; Singh et al., 2020).

5.2.3 Nematode Diseases of *T. alexandrinum*

5.2.3.1 Root-Knot Nematode

Causal Organism/Pathogen: This disease is a major limiting factor involved in the reduction of legume production. This is caused by two species of *Meloidogyne* (*incognita* and *arenaria*) (Baltensperger et al., 1985; Singh et al., 2010; Singh et al., 2020).

Symptoms: Although there are no above-ground symptoms by the root-knot nematode, affected plants with yellowing, stunted growth, and wilting in severe conditions. Knots are produced on plant roots (Khan, 2015; Singh et al., 2020).

Management: Once the nematode develops in the field, it is difficult to manage this disease; crop rotation with non-host crops is a suitable method to reduce nematode populations in soil. In the summer seasons, two to three times deep plowing at 15-day intervals could help manage nematodes. Microorganisms like *Paecilomyces lilacinus* and *Pochonia chlamydosporia* parasitize nematode eggs (Khan, 2015). Nematicide Carbofuran at 25-30 kg/ha had a significant effect on the control nematode (Kaushal et al., 2001; Kanwar and Bajaj, 2010). Organic amendments like the application of neem cake in diseased fields (*Trifolium alexandrinum*) significantly suppressed the nematodes (Hasan and Jain, 1984; Siddiqui, 1997; Azmi et al., 2000). As mentioned earlier, the integration of management could be economical and effective against root-knot nematode (Khan et al., 2009; Singh et al., 2020).

5.3 DISEASES OF *TRIFOLIUM PRETENSE*

5.3.1 Fungal Diseases of *T. pratense*

5.3.1.1. *Sclerotinia* Crown and Stem Rot: Sclerotinia crown and stem rot and root rot are important red-clover diseases in Canada, Europe, Finland, Germany, the Soviet Union, and the United States. Sclerotinia stem rot is attributed to *S. trifoliorum* or *S. sclerotiorum,* although *S. trifoliorum* is most common. It causes significant losses of red clover foliage and seeds. The disease spreads and develops most rapidly during cool, wet weather conditions, but the casual fungus can grow and infect plants below freezing temperature. The stem rot fungus attacks all plant parts. It may destroy large patches in red clover fields during winter. *S. trifoliorum* has a broad host range. This disease usually occurs in patches (Ficker, 2019). Wilting leaves and stems are the most common symptoms of the disease. Disease symptoms initiate as small brown spots on the leaves and petioles. Heavily infected leaves turn grayish brown, wither, and fall to the soil surface. A white mycelium growth develops from them, which spreads to the crowns and roots. Afterward, a brown, soft rot appears on crowns and basal parts of the young stems, which extend down into the roots. As a consequence, part or all of the new growth of infected plants wilts and dies. A mass of mycelium may develop on dead stems and petioles. Some of the masses then change to sclerotia, small, hard, black, cartilaginous bodies embedded in the surface of diseased tissue or soil (Clements, 1996). The sclerotia remain in the soil for several years and are the primary means of infection in the following years. Sclerotia develop apothecia in which numerous ascospores are produced, which are wind-carried to healthy leaves to cause infection of leaves. According to several reports, the pathogen advances from plant to plant via mycelium, but the primary inoculum is ascospores according to several reports. Clover rot management is wholly based on cultural practices and cultivar resistance (Hanson and Kreitlow, 1953; Purdy, 1979; Yli-Mattila et al., 2010; Mikaliūnienė et al., 2015).

5.3.1.2. *Aphanomyces* Root Rot is caused by oomycete fungus *Aphanomyces* spp. Infected seedlings have yellow cotyledons, stunted growth, and root decay, leading to symptoms resembling nitrogen deficiency. The disease is easily spread in wet soils. Balanced use of potassium and phosphorous and three-year crop rotation with non-legumes can help protect plants from disease (Hanson and Kreitlow, 1953; Nyvall, 2013).

5.3.1.3. Anther Mold: Spores of the fungus *Botrytis anthophila* partially replace pollen grains in the anther. Anthers of diseased plants look like a characteristic ash-gray instead of normal bright yellow. The disease has been reported in Germany, Russia, and Wales. Bees disseminate the spores. The disease has an adverse influence on fertility in red clover (Silow, 1933).

5.3.1.4. Northern Anthracnose or clover scorch is a severe disease of red clover in cool, wet weather. *Kabatiella cauliuora* overwinters as mycelium and acervuli in diseased clover tissues. It may also be seed borne. Spores are produced in cool, wet weather and are splashed. It is prevalent in cooler areas of Asia, Europe, and North America. It appears just before flowering and attacks both stems and foliage. Lesions on leaf petioles and stems are elongated, dark brown to black. An infected field gives a scorched appearance from a distance. Lesions may also girdle and kill stems, causing drying and browning of the foliage and wilting of leaves or drooping flower heads. Always grow resistant cultivars to avoid the disease. Crop rotation will reduce fungal inoculum (Heusinkveld, 1948; Hanson and Kreitlow, 1953; Undersander, 1990; Nyvall, 2013).

5.3.1.5. Southern Anthracnose is mainly confined to the southern clover-belt. *Colletotrichum destructivum* overwinters as ascervuli on diseased tissues. Conidia are produced during cool, moist weather, which is rain-splashed or wind-blown to healthy plants. The disease occurs in Africa, southern Europe, and the southern United States. It is frequently found on the upper part of the taproot and crown and more on the second crop's new growth. Dark brown spots of irregular shape appear on leaves that may vary in size from a barely visible dot to entire leaf general necrosis. Water-soaked spots that become elongated, dark brown to black, with a gray or light brown center, develop on petioles. Crown infections cause wilting and death of the plants. Disease management grows adapted, resistant cultivars and uses clean, disease-free seed (Heusinkveld, 1948; Hanson and Kreitlow, 1953; Nyvall, 2013).

5.3.1.6. Powdery Mildew fungi *Erysiphe polygoni* overwinters as black specks, cleistothecia. Ascospores are the primary source of infection. These wind-blown ascospores can infect at any time during the growing season but preferably when nights are cool and days are warm. Prolonged dry weather conditions facilitate disease development. The disease has widely distributed in red clover–growing areas in the temperate zones of the world. At first, a white to light-gray mycelium develops on the leaves. The patches become enlarged and coalesce, covering a larger area of the leaf surface. Infected leaves may turn yellow and wither prematurely, but most remain green. In the case of a severe attack, the entire field may appear white (Heusinkveld, 1948; Hanson and Kreitlow, 1953; Undersander, 1990; Nyvall, 2013).

5.3.1.7. Downy Mildew fungus, *Peronospora trifoliorum*, overwinters in crown buds and dead leaves. Wet and humid weather conditions favor disease development. The disease has widespread distribution among red clover–growing regions. Infected leaves are light green with gray fuzz consisting of mycelium on the underside. Severely infected plants become twisted, and stem growth is retarded. Early harvesting and crop rotation with non-leguminous crops may provide effective disease control (Hanson and Kreitlow, 1953; Nyvall, 2013).

5.3.1.8. *Cercospora* Leaf Spots have several other names like summer black stem, angular leaf spot, and stem spot. This disease has widespread in Europe and central and eastern North America. *Cercospora zebrine* is a seed-borne fungus. Spots on leaves are dark brown, angular, and more or less encircled by veins. Older spots may appear ash-gray. Lesions on stem and petioles are somewhat sunken and dark brown and may coalesce to form extensive dark areas on the stem's lower parts. Infected seeds become shriveled and discolored. Seed treatment with protectant fungicide and early harvesting is crucial in managing this disease and reducing inoculum (Hanson and Kreitlow, 1953; Nyvall, 2013).

5.3.1.9. Rust is caused by *Uromyces trifolii*. It has several stages and overwinters mainly as teliospores on clovers. Urediospores are blown north each summer. The disease is favored by cool, wet weather with an optimum temperature of 16–22°C. It is common in humid and semi-humid areas. Disease severity is higher in areas with delayed harvesting. Small reddish-brown pustules can be seen on the petioles, stems, and leaves. Each pustule is composed of thousands of wind-blown spores. Telia occur in old uredia or independently. The aecial stage appears as swollen, light yellow to orange-yellow pustules on stems, petioles, and leaves. Aecia appear as clusters of tiny cuplike structures under magnification (Hanson and Kreitlow, 1953; Diachun and Henson, 1974; Nyvall, 2013).

5.3.1.10. Brown Root Rot: *Phoma sclerotioides* is the primary cause of this disease. It overwinters in residues in soil. In the early spring, affected plants show chlorosis and stunted growth followed by death. Circular, brown, necrotic areas can be found on roots, eventually spreading into the crown. The fungus produces numerous pycnidia in diseased areas. Crop rotation with non-leguminous crops is useful to reduce inoculum pressure (Nyvall, 2013).

5.3.1.11. Pythium Blight *Pythium* spp., particularly *P. debaryanum*, overwinters in soil and plant debris as oospores. The oospores germinate during more soil moisture to produce sporangium. The zoospores are released and swim through free water to root tips or seeds, where they germinate. The disease has a wide distribution. The fungus usually infects young plants but rarely older plants. The fungus causes pre-emergence damping-off of seedlings. Small seedlings show discoloration of the roots. The use of protectant fungicide as a seed treatment is sufficient to escape the disease (Nyvall, 2013).

5.3.1.12. Black Patch is a foliage and stem disease which is caused by *Rhizoctonia leguminicola*. The fungus survives as mycelium in residues. The disease occurs in the southern United States. It typically appears in scattered patches and infects leaves, stems, flowers, and seeds. Brown to black lesions with concentric rings affect large areas of a leaf. The disease usually kills lower leaves. Stems and petioles develop black mycelial growth. On the flowers and seeds, it produces dark lesions and dark coarse mycelium. Seedlings are often blighted and overgrown with black mycelium. The fungus does not produce spores and spreads further via mycelial growth or mechanically at some distances. Seeds also transmit it. Disease management options include seed treatment with protectant fungicides, crop rotation, and early harvesting to reduce foliage loss (Nyvall, 2013).

5.3.1.13. Curvularia Leaf Blight has worldwide distribution. The leaf blight fungus, *Curvularia trifolii*, is most severe during warm, wet weather. It sometimes causes wilting and premature death of leaves. Water-soaked, angular, yellow to brown lesions develop on leaflets. Lesions often develop on petioles, resulting in necrotic leaves remaining attached to the plant (Nyvall, 2013).

5.3.1.14. Cylindrocladium Root and Crown Rot is caused by *Cylindrocladium crotalariae* and prevails in Florida. Black lesions appear on the crown and roots and cause infected leaves' mortality (Nyvall, 2013).

5.3.1.15. Pseudoplea Leaf Spot or pepper spot is due to *Pseudoplea trifolii* that survives as perithecia or mycelium in the residue. During cool, wet weather, ascospores are produced and carried by the wind. It is commonly found in temperate, humid regions across Asia, Canada, Europe, and the United States. Several small, sunken black spots of only a few millimeters develop on leaves and petioles. These spots ultimately turn gray with a red-brown margin. Heavy infections cause yellowing, withering, and ultimate killing of leaves. In some cases, part of the flowers and flower stalks may also become infected and die. Early harvesting is a good option (Nyvall, 2013).

5.3.1.16. Common Leaf Spot is also called *Pseudopeziza* leaf spot and is caused by *Pseudopeziza trifolii*. The fungus overwinters either as apothecia or mycelium in diseased debris. Windborne ascospores are produced during cool, wet weather. Angular to round small, dark spots of various colors develop on either leaf surface (Nyvall, 2013).

5.3.1.17. Crown Wart is caused by *Physoderma trifolii*, which overwinters in debris and soil. The sporangia germinate during wet soil conditions and liberate zoospores. Zoospores germinate and penetrate the plant via infective hyphae. The disease has spread primarily in warmer red clover–growing areas of Europe and North and South America. Irregularly shaped galls are produced around the crowns at or below ground level (Nyvall, 2013).

5.3.1.18. Gray Stem Canker is caused by *Ascochyta caulicola*. It occurs in Canada. Silver-white cankers develop on stems, leaf stalks, and midribs. Severely diseased stems may be stunted, swollen, and twisted, with a few small leaves (Nyvall, 2013).

5.3.1.19. Myrothecium Leaf Spot prevails in Pennsylvania. Two fungal species, *Myrothecium roridum* and *M. verrucaria*, are considered the cause of the disease.

Wounds and hot, humid weather favor disease development. At first, small, circular, dark, water-soaked lesions occur on both leaf surfaces. Sporodochia develop in all parts of the lesion. As lesions coalesce, the leaf becomes flaccid, and necrosis may extend to the petioles. Prolonged humid conditions of about five or more days cause curling and the plants' ultimate death (Nyvall, 2013).

5.3.1.20. Winter Crown Rot, distributed in Canada, is also called snow mold. It is caused by *Coprinus psychromorbidus*, a fungus previously known as low-temperature basidiomycete. The disease occurs at temperatures near 0°C at the soil surface under snow cover. Young stands have irregular areas of dead plants, while older stands have scattered plants with rotted crowns. White fungal mycelia are visible at the edge of melting snow (Nyvall, 2013)

5.3.1.21. Sooty Blotch or black blotch is caused by *Cymadothea trifolii*. This fungus overwinters as a stroma in which the formation of perithecia takes place. Ascospores are liberated from these perithecia and carried by wind to healthy leaves. The disease is found in temperate zones of the world. Disease initiates as tiny olive-green spots, usually on the lower leaf surface. The dots enlarge and become thicker and darker, giving a velvety, black, elevated cushion look. Afterward, these cushions are replaced by other black areas that have a shiny surface. Severe infection causes the leaf to turn brown, die, and drop off (Hanson and Kreitlow, 1953; Nyvall, 2013).

5.3.1.22. Spring Black Stem or Ascochyta leaf spot caused by *Phoma trifolii* is present in Europe, North America, and South America. The fungus may be seed borne or overwinter as pycnidia. The most apparent symptoms are small, dark, brown to black lesions on the stems. Several spots coalesce, covering a large surface area, especially on the lower part of the stem. Leaves may develop irregular dark brown to black spots, some with gray centers. Leaf and stem infections, when they appear in combination, may cause severe defoliation. Managing the disease always uses certified seed and follows cultural practices (Hanson and Kreitlow, 1953; Nyvall, 2013).

5.3.1.23. Stagnospora Leaf Spot This disease is also called gray leaf spot. *Stagnospora meliloti* overwinters as mycelium, pycnidia, and perithecia. Splashed or wind-blown conidia and ascospores are produced during moist weather. Spots are usually 3–6 mm in diameter and circular to irregular in shape. The spot's center is a pale buff to almost white with a light to dark brown margin. In some instances, the spots tend to show faint concentric zones the same color as the margin. Older spots have small dark spots, which are the pycnidia scattered throughout the affected area (Nyvall, 2013).

5.3.1.24. Stemphylium Leaf Spot is caused by *Stemphylium sarcinaeforme* and *S. botryosum*. This disease is also called target spot, zonate leaf spot, and ring spot. Conidia are produced from overwintering residue and are blown to healthy leaves. Fallen infected leaves provide inoculum for next season's infection. Optimum disease development has been observed under wet weather and cool temperatures ranging from 20–24°C. Small, dark brown spots appear that become enlarged. The entire leaf becomes wrinkled and dark brown, with a sooty appearance, but usually remains attached to the plant. Sunken brown lesions infrequently appear on stems, petioles, and pods. Early harvesting may reduce crop loss (Nyvall, 2013).

5.3.1.25. *Fusarium* Root Rot or common root rot mainly caused by *Fusarium* spp., including *F. oxysporum*, *F. avenaceum*, *F. acuminatum*, and *F. solani*, causes severe infections in older red clover crops. Some other soil-borne fungi like *Rhizoctonia* and *Phoma* can also cause root rot and slow decline. The fungus survives as chlamydospores and saprophytic mycelium in the residue. *Fusarium* root rot is widely distributed among red clover–growing areas. Red clovers can be infected at any growth stage but are most devastated during the second year. Initial symptoms are yellowing and wilting of plants during hot, dry weather. Roots and crowns exhibit light- to dark-brown discoloration. The inner core of the roots becomes rotted. Management practices include using certified disease-free seeds, avoiding overgrazing, proper use of fertilizers, and maintaining soil pH between 6.2 and 7.0 (Nyvall, 2013; Yli-Mattila et al., 2010).

5.3.2 Viral Diseases of *Trifolium pretense*

5.3.2.1. Red Clover Mottle Virus has been distributed in England. The virus is transmitted by mechanical inoculation (Hanson and Hagedorn, 1961). The first visible symptoms are the youngest leaves' yellowing along the veins, followed by chlorosis, mottling, chlorotic rings and spots, and leaf crinkling (Sinha, 1960; EPPO, 2013; Nyvall, 2013).

5.3.2.2. Clover Yellow Mosaic Virus is seed borne and occurs in Canada. Plants show stunted growth with a bushy leaf appearance. Leaves may also have a mosaic appearance with yellowing of leaf veins (Nyvall, 2013).

5.3.2.3. Red Clover Vein Mosaic Virus is a critical virus that can cause devastating losses. Symptoms of the disease are vein yellowing and little or no apparent stunting. This virus is transmitted by aphids and by seed and mechanical means and overwinters in legumes. Infected clover plants serve as the primary inoculation source. The disease prevails in the United States, particularly the eastern and north-central states. Symptoms are faint yellowing of the leaf veins; gradually, it becomes more chlorotic until veins and adjacent tissues turn white, causing reduced yields (Nyvall, 2013; Fletcher et al., 2016).

5.3.3 Nematode Diseases of *T. pratense*

5.3.3.1. Clover Cyst Nematode, *Heterodera trifolii*, is mainly present in Europe. Plants show stunting (Nyvall, 2013).

5.3.3.2. Root-Knot Nematode, *Meloidogyne hapla*, is probably widely distributed. During moisture stress, infected plants wilt. Leaves frequently have a purple underside. Roots have galls, causing them to be excessively branched and brittle (Nyvall, 2013).

5.3.3.3. Clover Stem Nematode is caused by *Ditylenchus dipsaci*. The disease has limited distribution (Smith, 1919; Skipp and Christensen, 1990; Nyvall, 2013).

5.3.4 Bacterial Diseases of *T. pratense*

5.3.4.1. Bacterial Leaf Spot, also known as bacterial blight or bacterial leaf blight, caused by *Pseudomonas syringae*, is widespread throughout Australia, Europe, Great Britain, Italy, Japan, and North America. *P. syringae* survives in infected soils and becomes active during cool, wet weather. It is disseminated through rain splashes and penetrates the host either through natural openings or wounds. The disease may affect different plant parts, mainly leaflets. Small, dark brown, water-soaked spots develop on the lower leaf surface, enlarge filling areas between leaf veins, and turn black, leaving water-soaked margins. During wet weather, milky white bacterial exudates may develop in the spot, which on drying forms a thin, crusty film. Dark, elongated, slightly sunken lesions can be seen on petioles and stems (Nyvall, 2013).

5.3.5 Mycoplasma of *T. pratense*

5.3.5.1. Phyllody Leaves turn into leaf-like organs, reduced in size with mild chlorosis, while old leaves may have a bronze color (Nyvall, 2013).

5.3.5.2. Proliferation Infected plants give a witches' broom appearance, and flowers transform into leaf-like appendages (Nyvall, 2013).

5.3.5.3. Witches' Broom occurs in Western Canada. Symptoms include dwarfing, chlorosis of leaves, and a bunchy appearance (Nyvall, 2013).

5.3.5.4. Yellow Edge Symptoms are mild to severe chlorosis along leaf margins, red-brown discoloration, reduced flower size and production, stunted plant growth, and a witches' broom appearance (Nyvall, 2013).

REFERENCES

Aquino CF, Sales NLP, Soares EPS, Martins ER and Costa CA (2014) Chemical composition and *in vitro* activity of three essential oils on *Colletotrichum gloeosporioides* from passion fruit. *Rev Bras Planta Med* 16: 329–336.

Asghar M, Baig MMQ, Raza AM, Arshad M, Hussian M and Afzal MS (2019) Evaluation of the effect of chemicals used as seed dressing for the control of berseem root rot. *Fuuast Journal of Biology* 9(1): 47-50.

Azmi MI, Pandey KC and Bhaskar RB (2000) Effect of some treatments on the nematode fauna in an IPM experiment. *Indian Journal of Nematology* 30: 105-105.

Babadoost M (1990) Virus Diseases of Alfalfa and Clovers in Illinois. Reports on Plant Diseases RPD.

Baltensperger DD, Quesenberry KH, Dunn RA, Abd-Elgawad (1985) Root-Knot nematode interaction with berseem clover and other temperate forage legumes. *Crop Science* 25: 848-851.

Barbetti MJ (1983) Fungicidal control of damping-off and seedling root rot in subterranean clover. Fungicide Nematicide Tests 38: 47.

Barnwal MK, Prasad SM and Maiti D (2003) Efficacy of fungicides and bioagents against Stemphylium blight of onion. *Indian Phytopathology* 56: 291–292.

Bradley DJ, Gregory SG and Parker IM (2003) Susceptibility of clover species to fungal infection: The interaction of leaf surface traits and environment. *Amer J Bot* 90: 857–864.

Burnett VF, Coventry DR, Hirth JR and Greenhalgh FC (1994) Subterranean clover decline in permanent pastures in North-Eastern Victoria. *Plant and Soil* 164: 231-241.

Chaudhry AR, Haq I and Reman N (1992) A cultural approach towards control of berseem root rot. *Pakistan Journal of Agricultural Sciences* 29(1): 65-68.

Clements RO (1996) Pest and disease damage to white clover (*Trifolium repens*) in Europe. REU technical series.

Diachun S and Henson L (1974) Dominant resistance to rust in red clover. Phytopathology 64: 758–759.

Ellis MB (1971) *Dematiaceous hyphomycetes*. Kew Surrey: CMI.

EPPO (2013) PQR–EPPO database on quarantine pests. Available at www.eppo.int (accessed 11 March 2014).

Faraz A, Haq IU and Ijaz S (2021) First report of *Sclerotinia trifoliorum* stem and crown rot on *Trifolium alexandrinum* in Pakistan. *Journal of Plant Pathology* 1–2.

Faruqui SA, Pandey KC and Singh JB (2002) Forage Plant Protection. Jhansi, India: Indian Grassland and Fodder Research Institute.

Ficker AL (2019) *Sclerotinia sclerotiorum* impacts on host crops. *Creative Components*. 309.

Fletcher J, Tang J, Blouin A, Ward L, Macdiarmid R and Ziebell H (2016) Red clover vein mosaic virus—a novel virus to New Zealand that is widespread in legumes. *Plant Disease* 100(5): 890–895.

Gargouri S, Berraies S, Gharbi MS, Paulitz T, Murray TD, Burgess LW (2017) Occurence of sclerotinia stem rot of fenugreek caused by *Sclerotinia trifoliorum* and *S. sclerotiorum* in Tunisia. *European Journal of Plant Pathology* 149(3): 587–597.

Gonçalves AH, Pereira AS, Santos GR and Guimarães LGL (2015) Fungitoxicity *in vitro* of essential oils from *Lippia sidoides* Cham., Cymbopogon citratus (DC.) Stapf. and their major constituents in the control of *Rhizoctonia solani* and *Sclerotium rolfsii*. *Rev Bras Planta Med* 17: 1007–1015.

Greenhalgh FC (1983) Growth cabinet evaluation of fungicides for control of *Pythium* damping-off and root rot of subterranean clover. *Fungicide and Nematicide Tests* 38-47.

Greenhalgh FC, de Boer RF, Merriman PR, Hepworth G and Keane PJ (1994) Control of *Phytophthora* root rot of irrigated subterranean clover with potassium phosphonate in Victoria Australia. *Plant Pathology* 43: 1009-1019.

Greenhalgh FC and Clarke RG (1985) The use of fungicides to study the significance and etiology of root rot of subterranean clover in dryland pastures of Victoria. In: Parker CA, Moore KJ, Wong PTW, Rovira AD and Kollmorgen JF (eds) *Ecology and Management of Soil-borne Plant Pathogens*. St Paul, USA: American Phytopathological Society, pp. 234–236.

Gupta RC and Gupta RP (2013) Effect of integrated disease management packages on disease incidence and bulb yield of onion (*Allium cepa* L.). *SAARC Journal of Agriculture* 11: 49–59.

Hanson EW and Kreitlow KW (1953) The many ailments of clover. Plant Diseases, US. Dept. of Agric. Yearbook. United States Government Printing Office, Washington, DC. 217–28.

Hanson EW and Hagedorn DJ (1961) Viruses of red clover in Wisconsin 1. *Agronomy Journal* 53(2): 63–67.

Haq IU, Ijaz S, Faraz A and Khan NA (2020) First report of *Curvularia lunata* leaf spot on *Trifolium alexandrinum* in Pakistan. *Journal of Plant Pathology* 1-1.

Hasan N and Bhaskar RB (2004) Disease complex of berseem involving nematode and two soil inhabiting fungi. *Annals of Plant Protection Sciences* 12: 159–161.

Hasan N and Jain RK (1984) Effect of soil amendments on fodder production, photosynthetic pigments and nematodes associated with berseem (*Trifolium alexandrinum* L.) followed by Bajra (*Pennisetum typhoides* L.). *Agriculture Science Digest* 4: 12–14.

Herranz MC, Pallas V and Aparicio F (2012) Multifunctional roles for the N-terminal basic motif of alfalfa mosaic virus coat protein: Nucleolar/cytoplasmic shutting, modulation of RNA-binding activity and virion formation. *Molecular Plant Microbe Interaction* 25: 1083-1103.

Heusinkveld D (1948) Red clover for Illinois. Circular (University of Illinois (Urbana-Champaign campus). *Extension Service in Agriculture and Home Economics* 627.

Hochman Z, Osborne GJ, Taylor PA and Cullis B (1990) Factors contributing to reduced productivity of subterranean clover pastures on acid soils. *Australian Journal of Agricultural Research* 41: 669-682.

Hussein MAM, Hassan MHA, Allam ADA and Abo-Elyousr KAM (2007) Management of stemphylium blight of onion by using biological agents and resistance inducers. *Egyptian Journal of Phytopathology* 35: 49–60.

Ihsanul HM and Nowsher AK (2007) Efficacy *in-vivo* of different fungicides in controlling stemphylium blight of lentil during 1998–2001. *Bangladesh Journal of Scientific Industrial Research* 42: 89–96.

Iqbal MF and Iqbal Z (2014) Efficacy of fungicides sprayed against rottening of berseem. *International Journal of Advanced Multidisciplinary Research* 1(2): 22-24.

Jobshy ZM, Syed EI, Rammah A and Satter MA (1981) Pathogenicity and control of three fungi associated with damping-off and root rot of Egyptian clover *Trifolium alexandrinum. Research Bulletin* 1674, 14.

Johnson DA and Atallah ZK (2014) Disease cycle, development and management of Sclerotinia stem rot of potato. *American Journal of Plant Sciences* 5: 317-3726.

Kanokmedhakul S, Kanokmedhakul K, Nasomjai P, Louangsysouphanh S, Soytong K, Isobe M, Kongsaeree P, Prabpai S and Suksamrarn A (2006) Antifungal azaphilones from the fungus *Chaetomium c upreum* CC3003. *Journal of Natural Products* 69(6): 891–895.

Kanwar RS and Bajaj HK (2010) Cereal cyst nematode infestation in wheat. In: Khan MR and Jairajpuri MS (eds) Nematode Infestation, Part-I: Food crops. Allahabad, India: The National Academy of Sciences, 325p.

Karsli MA, Russell JR and Hersom MJ (1999) Evaluation of berseem clover in diets of ruminants consuming corn crop residues. *Journal of Animal Science* 77(11): 2873-2882.

Kaushal KK, Sharma GL and Paruthi IJ (2001) Nematode diseases of wheat and barley and their management. In: National Congress on Centenary of Nematology in India—Appraisal and Future Plans, 5–7 December, 2001, Division of Nematology, Indian Agricultural Research Institute, New Delhi, India.

Khan MR (2015) Nematode Disease of Crops in India. Published at https://www.researchgate.net/publication/294285952.

Khan MR, Bhattacharya I, Chattopadhyay SB and Ghosh S (2009) Integrated approach for managing root knot nematode (*Meloidogyne incognita*) in pointed gourd (*Trichosanthes dioica* Roxb.). *Indian Journal of Nematology* 39(1): 25-28.

Kim WG, Cho WD (2003) Occurrence of Sclerotinia rot in solanaceous crops caused by *Sclerotinia* spp. *Mycobiology* 31(2): 113–118.

Kumar A, Singh RAM, Jalali SL (2003) Evaluation of resistance to stem rot and yield losses caused by the disease in rice. *Indian Phytopathology* 56: 403–407.

Kumar B and Singh KP (2012) Major Diseases of Forage and Fodder Crops and Their Ecofriendly Management. Published at https://www.researchgate.net/publication/327011585.

Kumar U, Naresh P and Bishwas SK (2012) Ecofriendly management of Stemphylium blight of garlic by plant extracts and bioagents. *Hort Flora Research Spectrum* 22: 231–245.

Maeder P, Fliessbach A, Dubois D, Gunst L, Fried P and Niggli U (2002) Soil fertility and biodiversity in organic farming. *Science* 296(5573): 1694–1697.

Maghazy SMN, Abdelzaher HMA, Haridy MS and Moustafa MN (2008) Biological control of damping-off disease of *Trifolium alexandrinum* L. caused by *Pythium spinosum* Sawada var. *spinosum* using some soil fungi. Published online 28: 431-450.

Manamgoda DS, Rossman AY, Castlebury LA, Chukeatirote E and Hyde KD (2015) Taxonomic and phylogenetic re-appraisal of the genus *Curvularia* (Pleosporaceae): Human and plant pathogens. *Phytotaxa* 212: 175–198.

Manjunatha N, Maneet R, Kumar S, Tomar M, Vijay D, Maity A and Srinivasan R (2019) Morphological and molecular identification of stem rot pathogen in berseem (*Trifolium alexandrinum* L.). *Range Management and Agroforestry* 40(2): 262-268.

Manjunatha N, Vijay D, Kumar S, Wasnik VK, Maity A and Gupta CK (2017) Disease and pest management of forage crops in field and storage conditions. In *Trainer's Training on Forage Seed Production and Quality Control*, 16-18th March, ICAR—Indian Grassland and Fodder Research Institute, Jhansi, India 35-60.

Martin PH, Coulman BE and Peterson JF (1997) Genetics of resistance to alfalfa mosaic virus in red clover. *Canadian Journal of Plant Science*

Mattila TY, Kalko G, Hannuakkala A, Huhtala SP and Hakala K (2010) Prevalence, species composition, genetic variation and pathogenicity of clover rot (*Sclerotinia trifoliorum*) and *Fusarium* spp. in red clover in Finland. *European Journal of Plant Pathology* 126: 13-27.

Mikaliūnienė J, Lemežienė N, Danytė V and Supronienė S (2015) Evaluation of red clover (*Trifolium pratense* L.) resistance to Sclerotinia crown and root rot (*Sclerotinia trifoliorum*) in the laboratory and field conditions. *Zemdirbyste-Agriculture* 1: 102(2).

Mishra MD, Raychaudhuri SP, Ghosh A and Wilcoxson RD (1980) Berseem mosaic, a seed-transmitted virus disease. *Plant Disease* 64: 490-492.

Mishra RK and Gupta RP (2012) *In vitro* evaluation of plant extracts, bioagents and fungicides against purple blotch and stemphylium blight of onion. *Journal of Medicinal Plant Research* 45: 5658–5661.

Mohan C, Thind TS, Prem R and Arora JK (2003) Promising activities of triazoles and other fungicides against fruit rot of chilli and stemphylium blight of onion. *Plant Disease Reporter* 19: 200–203.

Mourão DDSC, Ferreira de Souza Pereira T, Souza DJD, Chagas Júnior AF, Dalcin MS, Veloso RA, Leão EU and Santos GRD (2017) Essential oil of *Cymbopogon citratus* on the control of the *Curvularia* leaf spot disease on maize. *Medicines* 4(3): 62.

Nassuth A and Bol JF (1983) Altered balance of the synthesis of plus and minus-strand RNAs induced by RNAs 1 and 2 of alfalfa mosaic virus in the absence of RNA 3. *Virology* 124: 75-85.

Norton MR and Johnstone GR (1998) Occurrence of alfalfa mosaic, clover yellow vein, subterranean clover red leaf, and white clover mosaic viruses in white clover throughout Australia. *Australian Journal of Agricultural Research* 49(4): 723-728.

Nyvall RF (ed) (2013) *Field Crop Diseases Handbook*. Springer Science and Business Media.

Omar AF, Hafez YM and Emeran AA (2015) Occurrence of a new leaf spot disease on the Egyptian clover (*Trifolium alexandrinum* L.) caused by *Stemphylium globuliferum* (Vestergren) in Egypt. *Egyptian Journal of Biological Pest Control* 25(1): 33–40.

Pande PP, Rathi AS, Avtar R, Kumar A (2008) Viability of sclerotia of *Sclerotinia trifoliorum* at different depth and duration in soil. *Forage Research* 34: 44–48.

Paull J (2011) The uptake of organic agriculture: A decade of worldwide development. *Journal of Social and Development Sciences* 2(3): 111–120.

Pineyro MJ, Albrecht KA, Mondjana AM and Grau CR (2002) First report of alfalfa mosaic virus in Kura clover (*Trifolium amgibuum*) in Wisconsin. *Plant Disease* 86(6): 695.

Pinto DA, Mantovani EC, Melo EDC, Sediyama GC and Vieira GHS (2014) Growth production and essential oil quality of lemongrass (*Cymbopogon citratus*) under different irrigation depths. *Rev Bras Plantas Med* 16: 54–61.

Purdy L (1979) *Sclerotinia sclerotiorum*: History, diseases and symptomatology, host range, geographic distribution, and impact. *Phytopathology* 69(8): 875–880.

Rathi AS, Avtar R, Jhorar BS (2007) Sources of multiple resistances against stem rot and root rot diseases in exotic and indigenous genotypes of Egyptian clover. *Forage Research* 32: 201–203.

Rathi AS, Niwas R, Avtar R and Pahuja SK (2010) Effects of weather variables on development of stem rot disease in berseem. *Forage Research* 36(3): 137-141.

Saira M, Rehman A, Gleason ML, Alam MW, Abbas MF, Ali S and Idrees M (2016) First Report of *Sclerotinia sclerotiorum* causing stem and crown rot of berseem (*Trifolium alexandrinum*) in Pakistan. *Plant Disease* 101(5): 835.

Santos GR, Brum RBCS, Castro HG, Gonçalves CG and Fidelis RR (2013) Effect of essential oils of medicinal plants on leaf blotch in Tanzania grass. *Rev Ciênc Agron* 44: 587–593.

Sarmento-Brum R, Castro H, Silva M, Sarmento R, Nascimento I and Santos GR (2014) Vegetable oils effect on the inhibition of mycelial growth of phytopathogenic fungi. *J Biotech Biodivers* 5: 63–70.

Sarmento-Brum RBC, Santos GR, Castro HG, Gonçalves CG, Júnior AFC and Nascimento IR (2013) Effect of essential oils of medicinal plants on the anthracnose of sorghum. *Biosci J* 29: 1549–1557.

Sharma VV and Murdia PC (1974) Utilization of berseem hay by ruminants. *Journal of Agricultural Science* 83(2): 289-293.

Siddiqui MA (1997) Effect of organic amendments together with clipping on the population of plant parasitic nematodes associated with berseem (*Trifolium alexandrinum*). *Indian Journal of Nematology* 27(1): 244-247.

Silow RA (1933) A systemic disease of red clover caused by *Botrytis anthophila* Bond. *Transactions of the British Mycological Society* 18(3): 239.

Singh C (2001) *Modern Techniques of Raising Field Crops*. New Delhi: Oxford & IBH Publishing Co. Pvt. Ltd., p. 523.

Singh H and Singh H (1995) Cultural control of stem rot of berseem caused by *Sclerotinia trifoliorum* Erikss. *Plant Disease Research* 10: 28–32.

Singh RP, Singh AK, Singh M and Singh RK (2020) Diseases in berseem and its management: A. *Journal of Pharmacognosy and Phytochemistry* 9(3): 2054–2057.

Singh VK, Singh HV and Singh P (2010) New record of root-knot nematode caused by *Meloidogyne incognita* infecting berseem in J and K, India. *Range Management and Agroforestry* 13(2): 154.

Sinha RC (1960) Red clover mottle virus. *Annals of Applied Biology* 48(4): 742–748.

Skipp RA and Christensen MJ (1990) Selection for persistence in red clover: Influence of root disease and stem nematode. *New Zealand Journal of Agricultural Research* 33(2): 319–333.

Smiley RW, Tayor PA, Clarke RG, Greenhalgh FC and Trutmann P (1986) Simulated soil and plant management effects on root rots of subterranean clover. *Australian Journal of Agricultural Research* 37: 633-645.

Smith RH (1919) A preliminary note concerning a serious nematode disease of red clover in the northwestern states. *Journal of Economic Entomology* 12(6): 460–462.

Souza Júnior ITS, Sales NLP and Martins ER (2009) Fungitoxic effect of concentrations of essential oils on *Colletotrichum gloeosporioides*, isolated from the passion fruit. *Biotemas* 22: 77–83.

Soytong K (2014) Bio-formulation of *Chaetomium cochliodes* for controlling brown leaf spot of rice. *Journal of Agricultural Technology* 10(2): 321–337.

Stanhill G (1990) The comparative productivity of organic agriculture. *Agriculture, Ecosystems and Environment* 30(1–2): 1–26.

Tann H and Soytong K (2016) Biological control of brown leaf spot disease caused by *Curvularia lunata* and field application method on rice variety IR66 in Cambodia. *AGRIVITA Journal of Agricultural Science* 39(1): 111–117.

Undersander DJ (1990) *Red Clover: Establishment, Management and Utilization*. University of Wisconsin–Extension.

Vleugels T, Baert J, Bockstaele EV (2013) Morphological and pathogenic characterization of genetically diverse Sclerotinia isolates from European red clover crops (*Trifolium pratense* L.). *Journal of Phytopathology* 161(4): 254–262.

Windham GL and Pederson GA (1988) Effects of *Meloidogyne incognita* on forage yields of four annual clover. *Plant Disease* 72: 152–154.

Yli-Mattila T, Kalko G, Hannukkala A, Paavanen-Huhtala S and Hakala K (2010) Prevalence, species composition, genetic variation and pathogenicity of clover rot (*Sclerotinia trifoliorum*) and *Fusarium* spp. in red clover in Finland. *European Journal of Plant Pathology* 126(1): 13–27.

Zaher EAM, Abada KAM and Zyton MAL (2013) Effect of combination between bioagents and solarization on management of crown-and stem-rot of Egyptian clover. *Journal of Plant Sciences* 1(3): 43-50.

6 Etiology and Management of Economically Significant Diseases of *Avena sativa*

Qaiser Shakeel, Muhammad Raheel, Rabia Tahir Bajwa, Ifrah Rashid, Hafiz Younis Raza, and Syeda Rashida Saleem

CONTENTS

DOI: 10.1201/9781003055365-6

6.1 HALO BLIGHT

6.1.1 Causal Organism: *Pseudomonas coronafaciens* pv. *coronafaciens*

6.1.2 Pathogen Biology

In 1996, halo blight of oats was observed for the first time in Norway and Sweden. The disease has been found in almost all oat-growing areas of the world and is the most common where oat hay crops are regularly grown and established with prolonged moist weather conditions that promote bacteria penetration and offer an appropriate environment for infection. Due to this disease, the most serious damage can be found under cool, moist, and humid conditions and sudden fluctuation in weather, that is, summer followed by heavy rains. Strains of *Pseudomonas coronafaciens* are known to be the chief pathogens of oats (Elliot, 1920; Griffiths and Peregrine, 1956; Wilkie, 1972; Paul and Smith, 1989; Persson and Sletten, 2001). Earlier, this Gram-negative bacterium, which is plant pathogenic, was categorized in the *Pseudomonas syringae* group, but according to the present research, yellow bud diseases of onion are also caused by *P. coronafaciens* strain. Subsequently, *P. coronafaciens* is considered separate from *P. syringae* and suggested as a new *Pseudomonas* species (Hwang et al., 2005; Dutta et al., 2018; Baltrus and Clark, 2020). The bacterium grows well at 22.5°C, while its extreme temperature is 33°C, and the minimum is 4°C. This rod-shaped bacterium is motile, with one to three polar flagella. The size of *P. coronafaciens* ranges from 0.4–0.6 × 1.5–2.5 μ. Its colonies on nutrient agar medium appear circular to irregular with whitish growth that can be uniformly smoothened, with slightly high margins (Tabei, 1964).

6.1.3 Disease Cycle

P. coronafaciens causing halo blight of oat is a seed-borne pathogen that can persist on crop residues or old oat straws. The disease is encouraged under high crop density when plants are becoming soft due to high nitrogen input. The bacterium develops an initial seedling infection on the seed surface. During the humid conditions of spring, the pathogen can spread voluntarily from plant to plant and leaf to leaf from these infections. The disease development is ultimately checked with the growing season's progression during warm, dry spring weather, resulting in new growth being pathogen free. Infection may take place through leaf stomata or through pores distributed at the leaf tip, as well as through wounds or injuries. Moreover, the disease disseminates through causal agents, including wind, rain splashes, and insects (usually aphids) (Martens *et al.*, 1985; Wallwork, 1992).

6.1.4 Symptoms

Halo blight disease symptoms are most common on leaves and appear on panicles, glumes, culms, or sheath.

Initially, minor, oblong, water-soaked lesions are formed, which later become reddish-brown with a light depressed center followed by ovoid yellow to pale green halo margin.

The size of lesions ranges from 1.5 to 2.5 mm in length and about 0.5 mm in width.

In some cases, small silvery measures of dry expel sticking at the center of lesions can be seen.

Halos are prominent at high temperatures, while at low temperatures, brownish to black-colored lesions appear. Severely infected leaves may become chlorotic, dry, fall, or die.

Infrequently, hulls are also found to show symptoms of tiny spots.

Under prolonged favorable conditions, the plant can be defoliated, or the pathogen may kill the plants, reaching the crown.

Symptoms of halo blight resemble lesions that appear due to *Septoria avenae* or *Drechslera* spp. Lesions or spots may merge, forming an uneven blotch (Wilkie, 1972; Martens et al., 1985; Harder and Haber, 1992; Wallwork, 1992; Wallwork, 2000; Persson and Sletten, 2001).

6.1.5 Management Strategies

In most temperate areas, particular measures are not generally required, as with the arrival of higher summer temperatures, little additional bacterial spread happens.

Rotation of crops and the use of clean or disease-free seeds is recommended.

Avoidance of infected crop residues reduces the disease level.

In mild infection, there is insignificant grain loss, but the quality of hay is significantly reduced.

As the pathogen is seed borne, avoid re-sowing of the seed from diseased crops.

Avoid sowing into infected stubble and burn or incorporate stubble if the problem is widespread.

Fungicides against bacterial diseases are neither registered nor effective.

The use of resistance varieties against halo blight disease will be useful. Breeding of resistant cultivar is not so difficult because resistance is a hereditary characteristic (Harder and Harris, 1973; Harder and Haber, 1992; Wallwork, 1992; Martens et al., 1985).

6.2 BACTERIAL STRIPE BLIGHT

6.2.1 Causal Organism

Pseudomonas syringae pv. *striafaciens*

6.2.2 Pathogen Biology

Bacterial stripe blight disease is also termed stripe blight. The disease was first reported and disease samples were collected in 1917 by A.G. Johnson at Urbana. The disease is most common in the United States but infrequently found in high plain areas and is less prevalent than halo blight disease. Bacterial blight disease is usually encouraged in a cool environment followed by sprinkler irrigation and moist and humid weather conditions. The optimum temperature range for the development of the pathogen ranges from 15–21°C. The bacterium disseminates or develops under these favorable conditions, but disease development ceases or is reduced with the rise in temperature followed by dry conditions (Elliott, 1927; Simons, 1949; Nyvall, 1989).

6.2.3 Disease Cycle

The disease cycle of *P. syringae* pv. *striafaciens* is similar to the *P. coronafaciens* pv. *coronafaciens*. Bacteria causing the bacterial blight are a seed-borne pathogen and can persist in infected seeds, crop residues, and weeds and epiphytically or pathogenically on collateral or alternate hosts for about two years. This infection is encouraged by wounds or injuries on leaf surfaces due to hail storms, heavy winds, sandblasting, feeding of insects, and regular rainfall (Elliott, 1927; Simons, 1949; Nyvall, 1989).

6.2.4 Symptoms

Small, sunken, water-soaked spots or lesions (usually 0.04 inches in diameter) initially appear on the leaves' surface, which later collapses to form blotches or stripes.

Lesions become rusty brownish and translucent with time.

A yellowish constricted margin surrounds these blotches, but the halo-like margin does not occur.

Bacteria ooze out from the lesions under humid conditions, which later appear as thin, colorless to whitish scales after drying out.

Symptoms generally appear on leaves but may rarely be observed on sheath, glumes, culms, or panicles.

The length of blotches or stripes ranges from one inch to a few inches, which may cover the length of leaves, reaching down the sheath.

In all development stages, from seedling to mature plants, lesions can be developed.

Newly emerging flowers become mottled whitish to brown and may appear sterile.

In case of severe infection, the pathogen can kill the entire plant.

The bacterial ooze can be seen under a microscope from the leaves' veins by observing lesions cut transversally.

Bacterial stripe blight can be differentiated from halo blight easily, as there is no halo development, long lesions like a stripe or streaks, and bacterial ooze presence (Elliott, 1927; Simons, 1949; Nyvall, 1989; Wallwork, 2000; Murray, 2007)

6.2.5 Management Strategies

Pseudomonas bacteria have a broad host range, so plantation of oats in a clean field (in the absence of crop debris) and crop rotation to non-host plants for at least two years may reduce pathogen inoculum.

Plowing of infected debris is recommended.

Injury of young leaves should be avoided.

Treatment of seed with a protectant fungicide may also be useful.

Sowing of good-quality seed, that is, pathogen-free seed, may also be useful.

Avoidance of overhead irrigation during cold and humid weather conditions is recommended.

Re-use of irrigation water should be restricted.

If available, resistant or tolerant varieties should be planted. (Nyvall, 1989; Byamukama et al., 2019)

6.3 BLACK CHAFF

6.3.1 Causal Organism

Xanthomonas compestris pv. *trnslucens*

6.3.2 Pathogen Biology

Black chaff disease is also known as bacterial leaf streak disease. This bacterial pathogen was first reported on barley and followed by wheat, rye, grasses, oats. A low level of seed infection cannot produce disease in the field because this pathogen's transmission rate is meager. However, the pathogen's virulence cannot be underestimated because it can vary based on environmental conditions; areas where sprinkler irrigation is adopted show significant yield loss due to this disease. The infection level is reduced following minimum grain loss in the late season when environmental conditions become dry. The infection may be promoted through natural injuries to induce primary lesions on seedlings. Small pores or injuries with water may be considered bacterial reservoirs, as free water availability is essential for the bacteria penetrating to the plant. The bacteria's path to reach the upper portion of plants from seeds involves conquering the coleoptile and entering the enclosed foliage, and before the leaf emerges from the coleoptile, the pathogen infects it. The pathogen of black chaff disease can survive in temperature ranges from 15–30°C, but the optimum temperature for the development of the pathogen is above 26°C (Jones et al., 1917; Smith et al., 1919; Reddy et al., 1924; Wallin, 1946; Forster et al., 1986; Schaad, 1988; Duveiller et al., 1991; Duveiller et al., 1993).

6.3.3 Disease Cycle

The pathogen overwinters in or on seed, soil, and infected crop debris. The seed is considered a primary source of inoculum and transmission of the disease to a large scale due to infected seed. The secondary source of inoculum includes disease spread over short distances through plant-plant contact, rain splashes, and insects. The pathogen can live on the surface of plants epiphytically during the growing season without showing any symptoms. Humidity is required for initial infection, but after the development of infection, humidity is not needed. The pathogen dies within 63–81 months depending on the storage environment, but in air-dried soil, free pathogen dies within 14 days (Forster and Schaad, 1987; Nyvall, 1989; Forster and Schaad, 1990; Duveiller et al., 1997).

6.3.4 Symptoms

After a consecutive rainy period, black chaff symptoms appear more predominantly on oat leaves than heads.

Brownish to blackish spots can be seen on the inner side of the glumes.

Honey-like slime oozes out, which later turns into a small, yellowish bead structure on the surface of infected areas under humid conditions.

Tiny water-soaked lesions can be seen on young leaves, seedlings, and sheaths of mature plants.

These lesions collapse and form glossy, translucent olive-green to yellowish-brown streaks.

Streaks are generally narrow, are restricted to the leaves' veins, and may reach the length of leaf blades.

Sometimes lesions can also appear as large blotches, resulting in the leaf shriveling and turning light brown.

Symptoms usually appear in the leaf center, as in the morning, dewdrops remain for a prolonged duration (Bamberg, 1936; Nyvall, 1989; Duveiller et al., 1997)

6.3.5 Management Strategies

Black chaff disease cannot be controlled easily, but the following measures reduce the inoculum.

Avoidance of sowing infected seeds and use of certified seeds only.

Following crop rotation with non-cereal crops.

In many areas, indexing the seeds is not generally followed, but there is a need to encourage it.

Control of insect attack to check the disease spread.

Plantation of genetically resistant varieties.

Regulation of appropriate irrigation interval so that before the next irrigation, the plant's crown dries completely.

Discarding infected seed before planting should be the primary control measure since sowing pathogen-free seed is the first logical step in avoiding an outbreak (McMullen, 1989; Duveiller et al., 1997; Hershman and Bachi, 2010).

6.4 BARLEY YELLOW DWARF VIRUS

6.4.1 Causal Organism: Barley Yellow Dwarf Leutovirus

6.4.2 Pathogen Biology

Barley yellow dwarf virus was first reported in the 1940s and is known to be the most critical disease compared with all other oat diseases. The barley yellow dwarf (BYD) virus belongs to the Luteovirus group. BYD disease is usually known as red leaf disease of oats. The pathogens have a broad host range, including oats, wheat, barley, rye, and about 150 species of grass. The severity of infection varies with plant age, virus involved, plant variety, and environmental conditions. The size of these hexagonal viruses responsible for BYD disease ranges from 25 to 28 nm in diameter.

Two proteins, that is, a minor read through and a significant protein, encapsulate a single-stranded RNA genome. BYD viruses are composed of these two proteins. The pathogen is restricted to the plant's phloem vessel and can be easily seen under an electron microscope residing in nuclei, cytoplasm, and infected vacuoles. Symptom expression is facilitated by high intensity of light under a cool environment ranging from 15–18°C. Under these conditions, aphids can be attracted to the infected host plants due to leaf discoloration (Simons, 1949; Mathews, 1982; D'Arcy and Burnett, 1995; Watkins and Lane, 2004; D'Arcy and Domier, 2005; Ransom et al., 2007; DAFWA, 2015).

6.4.3 Disease Cycle

Aphids act as a vector for transmission of the virus particle to the host plant's phloem vessel. After the discharge of single-stranded RNA by virion, the replication cycle of the virus gets started. This ssRNA acts as a messenger RNA (mRNA) and is termed +ve sense RNA. The ssRNA translates the protein complexes needed for making RNA copies of the virus. The -ve sense RNA strands serve as templates required to produce several full-length and +ve sense RNA copies. +ve sense ssRNA is the source for production of -ve sense RNA strands. These newly developed strands of +ve sense RNA are responsible for initiating more replication cycles within the host plants after migrating one cell to another. Aphids ingest structural proteins and full-length +ve sense RNA accumulated into virions and transfer the virus within the same plant to another cell or another healthy plant, where the virus replication cycle is again initiated. Newly produced virions are also transmitted within the infected plant through phloem but cannot initiate a new replication cycle. Following a 12-hour quiescent period, feeding is required for at least 15 minutes on the infected host plant to ingest and carry the virus particle. This virus's transmission mode is persistent or circulative, as the virus can be retained in the vector's body for several days or weeks or circulates in the vector's body cavity. About 23–25 various species of aphid are known to be vectors of this virus. The virus overwinters in cereal crops grown in the winter season when crops are grown to prevent soil erosion, like cover crops, fodder, and so on and grasses, including cocksfoot (*Dactylis glomerata* L.), little bluestem (*Andropogon scoparius* Mixch), and tall fescue (*Festuca arundinacea* Schreb), Moreover, bluegrass (*Poa pratensis* L.) acts as reservoirs for several virus strains responsible for causing BYD disease (Mathews, 1982; Miller and Rasochova, 1997; Oswald and Houston, 1951; Gray and Gildow, 2003; D'Arcy and Domier, 2005).

6.4.4 Symptoms

Older leaves show initial symptoms of this disease, including greenish-yellow blotches adjacent to the tips that may later turn tan, reddish, or purple depending on host plant variety and environmental conditions.

Leaves express discoloration within one to three weeks after infection, and water-soaked lesions may follow it.

Infected plants show early maturation and a short length of internodes resulting in stunted growth and low-weight seed production.

In severe cases, heads of the infected plants are unable to emerge because of stunted growth, while in mild cases, stunted growth may remain unnoticed if not compared with healthy ones.

The volume of infected plant roots is also reduced, due to which infected plants can be pulled out more easily than healthy ones.

Symptoms may vary depending on host plant variety, genotype, age of the plant, environmental conditions, and virus strain involved in causing infection and plant physiological condition.

Stiffness of leaves and their upright folding, reduction of tillering and flowering following their increased abortion, and unfilled kernels are the additional symptoms of this disease (Simons, 1949; Wallwork, 1992; Martens et al., 1985; Watkins and Lane, 2004; DAFWA, 2015)

6.4.5 Management Strategies

The best method is the prevention of this disease than curing it after infection occurs. The disease can be prevented efficiently by planting resistant or tolerant plant varieties. Tolerant varieties allow the virus to infect and multiply within the host but show negligible yield losses.

Another critical control measure is reducing the disease spreading by controlling or minimizing the aphid population.

According to the aphid migration season, alter the planting date, as old plants show more tolerance or resistance.

Suitable seed dressing can also be useful to distract aphids.

Use of insect traps and insecticides to reduce the aphid population.

Eradication of grassy plants or cereal volunteer is also required, as the virus residing on these may be carried or transmitted by aphids to healthy plants (D'Arcy and Burnett, 1995; Watkins and Lane, 2004; DAFWA, 2015).

6.5 OAT BLUE DWARF

6.5.1 Causal Organism: Oat Blue Dwarf Marafivirus

6.5.2 Pathogen Biology

The disease was reported for the first time in 1951 in the United States. The oat blue dwarf (OBD) virus is a phloem limited to the genus *Marafivirus* and family Tymoviridae. The OBD virus is a small oblong-shaped virus the size of which ranges from 28–30 nm in diameter. Among seven families, 18 different species are the host of the OBD virus. The occurrence of OBD has been found to be very little or about just 5% from various isolated fields, so the disease is not so economically essential, but if an outbreak occurs, this disease can be a severe threat to oat-producing areas because infected plants show little or no grain production (Boewe, 1960; Zeyen and Banttari, 1972; Edward and Weiland, 2010).

6.5.3 Disease Cycle

Aster leafhopper (*Macrosteles fascifrons*) is the only vector that transmits the OBD virus from plant to plant and from one place to another. To carry the virus, adults of leafhopper require two days to feed on the infected plant, while nymphs require one week of feeding. After acquiring the virus, leafhoppers require at least 15 minutes of feeding on a healthy plant to transmit the virus particle. Vectors require at least an incubation period of one day, but the vectors mostly transmit the virus between the 6th and 20th days. Usually, after the 66th day, the transmission of the virus ceases. The OBD virus is unable to be transmitted either through aphids or mechanically (Zeyen and Banttari, 1972; Conner and Banttari, 1979).

6.5.4 Symptoms

The flag leaves of infected plants become smaller and stiffer, exhibiting larger angles from culm than healthy ones.

Infected plants turn intense bluish-green and show stunted growth.

The development of enations can be seen on the leaf surface and veins of the infected culm.

Infected plants show very little or no seed formation in case of severe infection.

The number of tillers increases, reaching above the canopy.

The age of the infected plant is much greater compared to healthy ones (Simons, 1949; Boewe, 1960)

6.5.5 Management Strategies

The disease is not too prevalent, so management strategies are not required, nor is any strategy known (Simons, 1949; Boewe, 1960; Ransom et al., 2007).

6.6 OAT MOSAIC

6.6.1 Causal Organism: Oat Mosaic Bymovirus

6.6.2 Pathogen Biology

Oat mosaic virus is a soil-borne plant pathogen belonging to the genus *Bymovirus* and family Potyviridae. The pathogen was first reported in the United States in 1943. Two distinct strains of the virus are able to cause oat mosaic disease individually. The virus particle is cylindrical with a size ranging from 600–750 nm in length and 12–14 nm in width. Oat mosaic disease is limited to oats (*Avena sativa*) and cannot be found in other fodder or cereal crops. Mechanical transmission of this virus through sap inoculation is complicated but successful. The fungus *Polymyxa graminis* act as a vector for the transmission of this viral disease. In severe infection, yield loss can reach 25–50% in tolerant varieties, while no grains are produced in susceptible varieties (McKinney, 1946; Macfarlane et al., 1968; Monger et al., 2001; Clover et al., 2002; Catherall and Hayes, 2007).

6.6.3 Disease Cycle

Zoospores of *P. graminis* acquiring oat mosaic virus penetrate the roots of oats, where the infection is then initiated. Infection develops in the presence of free water in autumn. Free water is essential for the dispersal of fungal zoospores. Throughout the winter, the virus remains dormant after gaining entrance into the host plant. Symptom development on infected plants becomes visible during spring. The virus is susceptible to high temperatures, so infected foliage symptoms disappear as the summer period starts. The virus is protected under unfavorable environmental conditions and in the host plants' absence by the fungal vector, where the virus can persist for a very long period. Infection by the oat mosaic virus (OMV) is initiated in the roots of oats. After foliar symptoms disappear, stunting and yield loss are the only remaining OMV and oat golden stripe virus (OGSV) infection indications (Atkinson, 1945; Linford and McKinney, 1954; Hebert and Panizo, 1975; Catherall and Boulton, 1979; Catherall and Valentine, 1987).

6.6.4 Symptoms

In the case of severe infection, the production of small rosettes in infected plants can be noticed.

In mild infection cases, infected leaves show yellowish to light green narrow bands or streaks parallel to their axis, and occasionally necrotic mottling or blotching of leaves can also be seen.

The yellow streaks may turn brown or purple at the tips of the infected leaf.

Symptoms are usually restricted to the upper one to three leaves of the infected host plant.

Eyespot lesions may also be developed on infected plants by the second strain of this virus, which is less common.

Spindle-shaped spots appear at the leaves' borders with ash gray margins, and eventually the whole leaf becomes chlorotic (Simons, 1949; Elliot and Lommel, 1990).

6.6.5 Management Strategies

Planting resistant varieties is the only effective measure to check the disease. Oats with natural immunity are not known. Additionally, in the presence of a high disease inoculum, resistant varieties may also not resist the disease; an infection develops. However, different species show varying resistance levels. A variety named Falghum is a useful resistant cultivar derived by crossing one of the most resistant *Avena* spp. that is, *Avena byzantine*. The heritability level of resistance of this cultivar is very high, ranging from 0.59–0.88. However, the exact resistance mechanism against the fungal vector is unknown (Atkinson, 1945; McKinney, 1946; Hadden and Harrison, 1955; Toler and Herbert, 1963; Uhr and Murphy, 1992).

6.7 OAT NECROTIC MOTTLE

6.7.1 Causal Organism: Oat Necrotic Mottle Tritimovirus

6.7.2 Pathogen Biology

Oat necrotic mottle disease was first reported in 1966 in Southern Manitoba. The oat necrotic mottle virus belonging to the *Tritimovirus* genus is a rod-shaped flexuous particle with a 720-nm length and 11-nm diameter. The virus can be transmitted mechanically but is unable to be transmitted through aphids, unlike many viruses. The vector of this virus is unknown. Perennial grasses serve as natural reservoirs for the virus (Gill, 1976; Gill, 1980; Stenger and French, 2004).

6.7.3 Symptoms

Initial symptoms appear on young leaves as chlorotic streaking, which later turns into brownish to blackish necrotic regions.

Inclusions, like a pinwheel, develop on infected plant cells.
The stunted growth of the infected plant is noticeable.

6.7.4 Management Strategies

Control measures for oat necrotic mottle disease are unknown.

6.8 DOWNY MILDEW

6.8.1 Causal Organism

Scleropthora macrospora

6.8.2 Pathogen Biology

Scleropthora macrospora belongs to the class oomycete and infects a broad host range, such as different turf grasses, wheat, rice, maize, cereals, and oats. It is a water mold fungus, and for the establishment of infection, it requires saturated soil. *S. macrospora* is a member of the peronosporales and acts as an obligate parasite. It produces sporangia zoospores on the leaves of the infected plant. These spores then preserve in grasses as mycelium (Semeniuk and Mankin, 1964). The mycelium of *S. macrospora* is coenocytic and hyaline and has a phytophthora-like sporangial stage. At the sporangiophore's apices, the sporangia produced are single, lemon-shaped, and large (Thirumalachar et al., 1953; Channon, 1981).

6.8.3 Disease Cycle

All oomycetes produce oospores; *S. macrospora* is also included in oomycetes. By the fusion of an oogonium and antheridium, these sexual spores (oospores) are produced. By the germination of these oospores, sporangium is produced on the sporangiophores. The sporangium is lemon-shaped and pearly white. Downy mildew produces oospores in large numbers. These spores are resting spores, show a pale yellow color, and are smoothly walled. The spread of these spores from one field to another occurs through tillage implements, wind, plant residues, surface runoff water, soil, and seed grains. For germination, oospores require saturated soil and water and produce sporangia (lemon-shaped conidia) for one to two hours. Thirty to 90 motile zoospores are release after sporangial formation. These zoospores are motile and can swim a short distance in the water. Slender germ tubes are produced for penetration into host tissues after settling down. After penetration into the host plant, downy mildew grows systemically. The fungus is more abundant in those tissues, which are growing actively. The pathogen can survive at a wide range of soil temperature ranges from 6–31°C (with 10–25°C optimum temperature range). When there is a condition that remains for a long time, many zoospores are released from the sporangia. These zoospores of the pathogen act as a secondary inoculum after primary infection to become polycyclic. Many sporangia are produced from overwinter oospores in late spring. *S. macrospora* spores produced mycelium, which is used to absorb nutrients from the host plant (Singh, 1995).

6.8.4 Symptoms

Downy mildew infects more than 100 annual and perennial grasses and downy fungus growth, seldom visible on rice, wheat, oats, sorghum, and barley.

The plants affected by *S. macrospora* are mostly seen in standing water, and seedlings appear chlorotic and stunted.

The leaves of infected seedlings are thick and leathery. The head and leaves of the affected plant appear twisted.

The symptoms produced by downy mildew depend upon the disease severity. The symptoms produced by phenoxy herbicides or high concentrations of 2,4-D closely resemble the downy mildew caused by *S. macrospora*.

Infected plants produce excessive tillers. Many of them turn brown, wither, and then die after growing a few inches.

The infected plants look like dead clumps and are dense.

Infected plants are dwarfed, stunted, and deformed with wet or thickened, short, and upper leaves curled and twisted.

Some leaves turn yellow and fleshy as a whole or show stripes of yellow color.

Some affected plants die before maturity, and others are longer compared to healthy plants and remain green.

The characteristics symptoms of downy mildew are large, distorted, and twisted development of the head or panicle and floret proliferation.

The head of infected wheat plants is more open than the healthy plant, and its stem below the head deformed and thick.

The proliferation spikelet from a leafy structure in which the complete head is involved changes them into a brush-like form. If the crop is grown in an area affected by downy mildew disease in the previous year, it may build up in a severe form. The infected pant has thick leaf tissues (Thirumalachar et al., 1953; Singh, 1995).

6.8.5 Management Strategies

6.8.5.1 Sanitary and Cultural Methods

There is a relationship between the development of disease and wet soils. Therefore, adequate drainage is required to avoid soil saturation. The intensity of disease development has been reduced in rice crops by increasing mechanization. By using a mixture of different nitrogenous fertilizers, the intensity of *S. macrospora* disease can be reduced. However, fertilizer cannot reduce the number of infected plants but can mask disease symptoms. Rotation with non-cereal crops helps in controlling the disease. Remove all weeds from the field because they serve as an alternative host. Avoid low areas for cultivation. Take the seed from a disease-free field.

6.8.5.2 The Resistance of the Host Plant

Different resistant lines have been developed against *S. macrospora* for several hosts. When 27 oat lines are exposed to *S. macrospora* for natural infection under conditions favorable for disease development, the cultivar of the Golf Star *Stenotaphrum secundatum* has shown resistance against *S. macrospora*.

6.8.5.3 Chemical Control

To minimize the structures of the pathogen, preventive fungicides can be used. However, the use of these fungicides is not entirely successful. When metalaxyl fungicides are applied in the early season, they are very effective against yellow tuft disease. Keep in mind that fungicides are a preventive, not curative. Therefore, the farmer must recognize the early signs and symptoms to control the disease (Thirumalachar et al., 1953; Channon, 1981; Singh, 1995).

6.9 ERGOT

6.9.1 Causal Organism

Claviceps purporea

6.9.2 Pathogen Biology

The fungus of the genus *Claviceps* is unique because it attacks flowers of specific grasses, and except for flowers, no other part of the host plant is infected. Sphacelium is a specialized structure of the fungus. The ovary is replaced by sphacelium during the process of infection. This specialized structure then converts into another structure, which is known as a sclerotium. The sclerotium is characteristically hard and contains a thin outer layer (rind) around fungal tissues' compact mass. Sclerotia are four times larger than the host seed of *Claviceps* species. The spores which are produced during the process of ergot development are called conidia. The term ergot applies to all species of *Claviceps*. *Claviceps purporea* affects more than 200 grass species and is present in temperate climatic zones worldwide (Alderman et al., 1999).

6.9.3 Symptoms

Kernels or seeds in a panicle or spike are replaced by sclerotia (ergot bodies). These sclerotia have spur-like structures and are horny. These sclerotia are of dark purple-black color. The size of sclerotia is four times larger than standard kernels. Ergot sclerotia are produced instead of kernels by the affected plant. Therefore, the yield of grain is reduced. The heads affected by ergot sclerotia have empty florets and blasted kernels. These affected heads are usually light in weight and short compared to healthy ones. Ergot causes sterility, so there are certain grasses in Great Plains that cannot produce seed. The first sign of ergot disease expresses after the flowering stage when sugary slime (honeydew) comes from infected florets. During the flowering, honeydew or yellow droplets may appear on the head. The symptoms of ergot disease are often confused with bunt disease. The sclerotia of ergot disease are larger and challenging than bunt balls, and bunt balls are filled with powdery spores, but sclerotia are not (Orlando et al., 2017).

6.9.4 Disease Cycle

The sclerotium of ergot bodies contains a compact mass of fungal tissues, enveloped by a thin layer (rind). In that condition, sclerotium survives through winter in temperate areas. The sclerotia require 0–10°C temperature for one to two months to germinate. Temperate regions are suitable for adopting sclerotia, while warmer regions are not suitable for their survival. Sclerotia germinate in spring, before the flowering of grasses, and produce stalked fruiting bodies. Thread-like ascospores produce these fruiting bodies. These ascospores require high soil moisture and rainfall to produce and release. Ascospores release into the air and spread to grass flowers through air currents. The susceptible period of most grasses is from flowering to fertilization. After fertilization, the ovaries become resistant to infection. The period of susceptibility increases in cool temperatures, which delays pollination. Several conidia are produced within a week after infection. At the infection site, large drops accumulate by mixing conidia and sap. This stage is known as the honeydew stage. Secondary infection occurs by the spread of honeydew, with a rotten smell that attracts flies and other insects to the infected heads. The sclerotia appear within 14 days after infection. Infected grass and sclerotia mature at the same time. Their maturity is related to each other. *C. purprea* completes its life cycle at the maturity of grass. After completing the life cycle, it falls on the ground and survives winter (Warburton, 1911; Alderman et al., 1999).

6.9.5 Management Strategies

For the management of ergot, there are several different strategies. There is no method that is entirely successful in controlling ergot. That is why integrated pest management (IPM) is necessary to control ergot (Taylor et al., 2014).

6.9.5.1 Cultural Practices

6.9.5.1.1 Seed Cleaning

The method of seed cleaning is expensive and laborious. However, most ergot can be cleaned or removed from seed grain by flotation in brine (20% salt solution).

6.9.5.1.2 Sowing of Ergot-Free Seed

For the management of ergot, one should sow ergot-free seed. Certified seeds are ergot free, reducing the risk of disease.

6.9.5.1.3 Uniform Stand

Use seed which has a good germination percentage. Seed should be sown at the recommended and proper depth, following the recommended fertilizer quantity and agronomic practices. Ensure uniform stands within a region; otherwise, non-uniform stands increase the period of flowering. As a result, the spread of disease will increase from infected to healthy fields. The main shoots are less affected by ergot than late tillers, especially when the crop stand is thin. The reason behind this is that the main shoots have abundant pollens, while late tillers do not, so late tillers produce honeydew even after the maturation of the main crop. Then they cause secondary spread of disease. Fertilization and adequate seed density keep the crop well nourished and well developed, making the plant less vulnerable to ergot attack. Proper use of copper fertilization makes the crop less susceptible to ergot. There is much evidence showing that sterile pollen and small anthers are produced due to copper fertilizer deficiency. If pollination does not occur and the flower remains open for a long time, it becomes more vulnerable to ergot attack (Miedaner and Geiger, 2015; Orlando et al., 2017).

6.9.5.1.4 Sanitation

Before heading, cereals, weeds, and wild grasses should be removed from the field because they can be the primary source of ergot disease inoculum.

6.9.5.1.5 Crop Rotation

Crop rotation is significant for managing ergot with non-susceptible host plants because sclerotia survive only for one year.

6.9.5.1.6 Deep Plowing

When sclerotia are entirely buried in soil with deep plowing, then ascospores cannot be disseminated in the air from stromata. Now, some crops are sown directly into the stubble of the previous crop without any tillage practice. Crop rotation becomes necessary if deep plowing is not followed (Warburton, 1911).

6.9.5.1.7 Harvesting Method

Plants cultivated on field edges are affected more than others because spores are transported first at field edges through insects and wind.

6.9.5.1.8 Burning of Stubbles

In grass seed production, ergot is a significant disease. There is no option for crop rotation and tillage because many grass species belong to perennial crops. The United States has been following this practice for the management of some pests and diseases since 1940. Due to environmental concerns, many countries' legislation does not allow field burning (Alderman et al., 1999).

6.9.5.2 Chemical Control

In Germany, there is no fungicide officially recommended for *C. purporea*. The germination of conidia can be prevented by using strobilurins. Seventy to 90% growth of mycelium or ergot can be reduced by using tebuconazole and other azoles. In Zimbabwe, ergot infection in sorghum can be reduced using benomyl at 0.2% a.i/ha. In Brazil, three to four sprays of tebuconazole and propiconazole have been used for seed production of sorghum when applied at 250 g/ha. Generally, the disease can be controlled using fungicides if applied under dry conditions, low inoculum pressure, and before the establishment of disease (preventive).

Moreover, fungicides must be applied directly to stigma and head or panicle because fungicides do not move systemically into plant tissues. In the United States, sterol-inhibiting fungicides have been used at the flowering stage in Kentucky bluegrass to minimize ergot disease severity. These fungicides have been used for seed production in turfgrass and forage (Miedaner and Geiger, 2015).

Usually, fungicides can persist for one year. If fungicides treat the seed, it may delay the decomposition of sclerotia, and this period may increase to more than one year. The level of sclerotia can be reduced in harvested grains through multiple applications of fungicides. However, to control ergot disease, fungicides are not effective (Taylor et al., 2014).

6.10 *FUSARIUM* FOOT ROT

6.10.1 Causal Organisms

Fusarium culmorum

6.10.2 Pathogen Biology

Fusarium culmorum is a soil-borne fungus and has a highly competitive saprophytic ability. However, acting as a facultative parasite causes fusarium head blight and foot and root rot on different cereal crops. *F. culmorum* is also called a post-harvest pathogen because it causes disease on new grains which have not been stored and dried. Quality and quantity are both affected by *F. culmorum*. The mycelium of *F. culmorum is* tan or pale orange and whitish to yellow. Over time, their color turns brown-dark brown to reddish-brown. Some isolates may form a ring of spore masses under fluctuating temperature and light (Scherm et al., 2013).

6.10.3 Disease Cycle

Fusarium culmorum is considered a severe pathogen that causes foot and root rot in small grains of cereal crops. *F. culmorum* produces asexually by the formation of conidia, and its teleomorph stage is not known (i.e., ascospores are not produced). Conidia are the primary source of inoculum. In the winter season, host debris helps the chlamydospores to survive, whereas conidia formation does not occur naturally. *F. culmorum* is a soil-borne fungus and causes seedling death, which may be pre- or post-emergence, and infected seed helps the fungus survive. No evidence shows that seed-borne inoculum causes fusarium head blight (FHB). In the case of FHB, microconidia in crop residues and soil are disseminated through rain splashes or wind or by insects to its host. During anthesis, the host plant's air becomes susceptible to *F. culmorum*. It is reported that conidia move systemically to the crop head and contaminate the kernels with a mycotoxin. During the growing period, coleoptiles and primary and secondary roots, are infected by chlamydospores. As a result, fusarium foot rot develops. It is a monocyclic disease, and only fusarium foot rot contributes to the initial inoculum (Cook, 1980; Akgül and Erkilic, 2016).

6.10.4 Symptoms

Buckeye root rot and fusarium root rot (FRR) have similar symptoms. The mature plant shows the first symptoms as necrosis of leaves and chlorosis between the veins.

The disease in its severe form can cause the entire plant's death; leaves become brown and collapse. Up to 12 inches below the soil line, reddish dark brown lesions are visible on the primary lateral roots and taproot.

Vascular discoloration, which may spread 1–4 inches beyond lesions, can also be noticed.

The symptoms of *Fusarium* foot rot depend upon the time of infection. If the infection occurs during the early period after the sowing of seed, brown discoloration on roots, pseudostem, and coleoptiles results in pre- and post-emergence seedling death. If the infection starts at the later stages of the plant, it causes tiller absorption, and two to three internodes of the main stem contain brown lesions.

High humidity causes the sporulation of mycelium. The sporulation of mycelium causes reddish-brown discoloration on the nodes.

The disease becomes prominent when extended from late stem onward. As a result, lower nodes become dark brown.

At the basal portion of the stem, long streaks that are dark in color are also found.

An older plant can produce symptoms of real foot rot.

The stem's base becomes rotten and brown; thus, the head becomes white, and stems become lodged (Scherm et al., 2013).

6.10.5 Management Strategies

The disease can be controlled using IPM approaches such as resistant cultivars, cultural practices, biological control, and fungicides.

Among the cultural practices, plowing is better as compared to no-tillage or minimum tillage. The disease can be controlled by crop rotation with the non-cereal host plant, especially legumes, for three to four years.

Pre- and post-emergence mainly take place through *F. culmorum* when fungus colonizes in the seed. Therefore, the treatment of seeds with fungicides is an efficient approach.

The use of fungicides gives protection to the plant only at the early growing stage of the crop. That is why it has limited efficacy against *F. culmorum*.

The disease can be reduced up to 70% in the field by using the azole and strobilurin classes of fungicides.

The adoption of resistant cultivar is an ideal approach to control the disease.

The application of biological control can be an effective method. When antagonistic microbes and developing biological control products are sprayed, they also give good results against FRR.

To maintain soil fertility, apply potassium, nitrogen, and phosphorus regularly. It has been investigated that when ammonia is released in the slow form (urea or anhydrous ammonia), it suppresses root rot.

Avoid the application of nitrogen at a high rate (Cook, 1980; Akgül and Erkilic, 2016)

6.11 *FUSARIUM* HEAD BLIGHT

6.11.1 Causal Organisms

Fusarium graminearium

6.11.2 Pathogen Biology

The disease fusarium head blight is caused by *Fusarium graminearium* (anamorph). In most areas of the world, FHB is caused by *F. graminearum*. *F. graminearium* belongs to ascomycetes. The fungus contains a sac-like structure (known as an ascus) in which sexual spores are produced. The fungus produces two types of spores. Microconidia are produced at the asexual stage, while ascospores are produced at the sexual stage. The fungus *F. graminearium* also produces a mycotoxin called deoxynivalenol.

6.11.3 Disease Cycle

Crop residues such as straw and stalks help the pathogen to survive or overwinter. The fungus produces macroconidia (asexual spores) on infected crop residues. These asexual spores of *F. graminearum* are dispersed by wind or rain splashes. The development of the sexual stage (*Gibberella zeae*) of the fungus occurs on infected crop residues during favorable environmental conditions. The favorable environment for their development is warm, wet, and humid conditions. Perithecia of bluish-black color are produced on crop residues. After that, ascospores (sexual spores) are discharged into the air. Then these ascospores are transported a considerable distance with the help of air currents. When macroconidia and ascospores land on the head of their susceptible host, they develop infection. At the flowering stage of the host, anthers can be a primary site of infection. If infection occurs after the emergence of anthers, kernel development will not occur, and fungus will destroy the florets. When infection occurs in a later period of flowering, kernels are affected and show a wrinkled, wilted, and shriveled appearance. The development of kernels may not be affected by the disease if fungus colonizes in the late season. However, they may be contaminated with the mycotoxin of fungus. If these kernels are sown as a seed without any treatment, the next season may produce diseased seedlings. Disease incidence depends on the quantity of infected seed and soil conditions, which affects the development and growth of seedlings.

6.11.4 Management Strategies

Integrated pest management is required for the effective management of head blight.

6.11.4.1 Resistant Varieties

Among the commercially available varieties, none has high resistance against FHB. These commercial varieties have partial resistance to the disease. Encourage the farmer to select varieties that have some level of resistance against head blight disease.

6.11.4.2 Management of Residues

Excessive tillage should be encouraged after the cereal crop because excessive tillage helps manage crop residues, and their management reduces the quantity of overwintering inoculum. There are greater chances of disease occurrence in the oat crop when it is sown into any cereal crop residues. So, oat should be planted after less susceptible or non-hosts of FHB.

6.11.4.3 Use of High-Quality Seed

The use of high-quality seed can reduce the chances of head blight. Before seed sowing, conduct a germination test to evaluate seed vigor and clean it with FHB symptoms. Seeds treated with fungicides cannot protect the flower from disease infection, but they can minimize seedling blight (Wise et al., 2015).

6.11.4.4 Chemical Control

Fungicides have a partial effect against FHB and its toxins. A large number of fungicides have been used at the flowering stage in many areas around the world. Due to the fickle nature of FHB and the high cost of fungicide, farmers hesitate to use them. Many commercial fungicides are used for seed treatment to reduce seedling blight.

6.11.4.5 Biological Control

Research continues to find biocontrol against FHB pathogens to protect the environment from hazardous pollution. Organic cereal production is mostly possible through biocontrol agents. Some strains of bacteria show their role as a biocontrol agent against FHB. These strains are spore-producing, such as *Bacillus* species. There are some yeast species *(Cryptococcus flavescens)* which are also effective against FHB and its mycotoxin.

6.12 LEAF BLOTCH

6.12.1 Causal Organism

Drechslera avenae

6.12.2 Pathogen Biology

Helminthosporium blotch is a disease of oats and *Hordeum* species. It is caused by *Drechslera avenae*. It is an asexual stage of *Pyrenophora avenae* reported from all oat-growing regions of the world. The formation of the sexual stage takes place on crop debris in the spring season. *Pyrenophora aveane* belongs to the family Pleosporaceae. In Germany and the southern United States, leaf blotch of oats has occurred in epidemic form as the most severe infection (responsible for about 30–40% yield loss) following crown rust of oat. *P. aveane* is also found as a seed-borne fungus in Sweden and Finland, causing about 10% crop losses. *Pyrenophora chaetomioides* is thought to be the leading cause of oat diseases in Brazil. Furthermore, there is no evidence of toxins produced by this fungus on infected oat crops, leaves, or grains that can interrupt the utilization of infected crops as fodder (Gough and McDaniel, 1974; Harder and Haber, 1992; Blum, 1997).

6.12.3 Disease Cycle

The fungi overwinter in dead or living leaves, seeds, stubbles of other crops, and crown. Helminthosporium produces its spores (ascospores and conidia) in the older lesions of living or dead grasses. These spores are produced in early spring and summer, and again in fall. These spores are transported to new growth of grasses by rain splashes and air currents. When conditions are humid in summer, spring, and fall, seedlings, leaves, crowns, and leaf sheaths become a fungus. The seed is the primary source of this disease. Conidia are produced in bulk on lower leaves of the oat plant, which serve as secondary inoculum and are transmitted to other plant parts, particularly to the panicles, hence completing the pathogen's life cycle. Furthermore, there is no evidence regarding a secondary host of this pathogen (Shaner, 1981; Blum, 1997; Rosa et al., 2003).

6.12.4 Symptoms

Leaf sheaths and leaves contain small to large spots (lesions), which are usually elongated and oval to oblong.

The lesion can be brown to purple-black or yellowish tan, dark brown, or medium brown, mostly with a yellowish boundary around them.

The size of the lesions on the forage grasses is 1/2 inch, while on forage sorghum and Sudan grass, the size of these lesions may extend up to 6 inches.

The older lesion center is surrounded by gray to straw-colored and purplish to brown or dark red to brown color boundaries.

A net blotch symptom may occur on fescues and other grass.

The stem and leaf surface has purple to brown tiny lines and a netlike appearance.

Due to the merging of some lesions, streaks are formed on affected stems and leaves.

The leaf becomes yellow or brown, withers, and dies due to the extension of remaining lesions. Plants infected by *Helminthosporium* look scalded. Lower leaves are infected first, and infection then progress to upper young leaves.

Near the collar of the leaf blade, leaf spots become more prominent.

When the pathogen attacks rhizomes and stolons, severe damage appears (Ivanoff, 1963; Luke et al., 1957)

6.12.5 Management Strategies

Leaf spot disease in oats has a drastic effect on the nutritive value and plant yield. Various management strategies are adopted against leaf spot disease. Avoidance plays a key role and is thought to be very useful for managing leaf spot disease. However, the following management practices are those which reduce the chance of disease development.

6.12.5.1 Genetic Resistance

The choice of a resistant variety undoubtedly plays a significant role in managing leaf spot diseases. Oats of genotypes IL85–6467, IL86–4189, and Maldwyn are referred to as resistant against *P. chaetomioides* (Harder and Haber, 1992; Frank and Christ, 1988; Sabesta et al., 1996).

6.12.5.2 Enhancing Soil Fertility

Disease development can also be found in resistant varieties when growing conditions are not good. In this regard, enhancing soil fertility is essential. Once the stand is established, it is essential to pay keen attention to management strategies.

Maintaining soil fertility is very important by taking a soil sample, and within it, the fertility level is adjusted. Observations have revealed a considerable reduction in nutrients in the ratio 4:1:3 for N, P_2O_4, and K_2O, respectively, on the hay harvest. The potassium level is significant for leaf spot resistance, and it has to be maintained. So, each season, 75% as much potash as nitrogen should be applied. The application of potash should be either in splits or single, depending on the location. However, in sandy soil, the split application is preferred. A soil test is also significant for controlling leaf spot by maintaining the required level of potash in the soil after analyzing that particular plant tissue, which, based on dry matter, indicates less than 1.8% level of potassium in the soil.

6.12.5.3 Management with Irrigation Water

The fungus grows pretty well when it is more in contact with a wet and relatively hot environment, so controlling the level of moisture in dryland is quite a challenging task. However, in this regard, irrigation water can do some management against growth of the fungus. Leaf spot disease requires many hours for the leaf to be wet to thrive better. So, by cutting off the supply of irrigation water from mid-afternoon until the next morning, leaves dry and ultimately resist the fungus growing over them.

6.12.5.4 Burning of Thatch from the Field

Thatch burning is essential for the control of leaf spot disease because it can hold nutrients tightly and create a suffocating environment by producing humid conditions as it holds water. Thatch has dry, decayed material that acts as the spore reservoir. So, its burning is necessary for resisting leaf spot disease. Besides, forestry groups can help by consulting about the precautionary measures that should be taken.

6.12.5.5 Keeping the Proper Cutting Interval

Timely cutting is essential for the resistance of leaf spot disease in Bermuda grass. It tends to maintain a high level of nutrients in hay. Potassium mobility in plants can infect older leaves to younger ones when facing deficiency in nutrients. The predisposition of older leaves will produce leaf spots disease due to not maintaining a proper cutting interval.

6.13 POWDERY MILDEW

6.13.1 Causal Organisms

This disease is caused by:

Erysiphe graminis
Blumeria graminis
Oidium monilioide

6.13.2 Pathogen Biology

Powdery mildew of cereals and grasses is widely distributed worldwide, mostly in China, South America, Europe, and Central Asia. There is no quarantine for this disease because of its air-borne dispersal. The fungus (*Erysiphe graminis* f.s.p *avena*) is an obligate parasite and cannot be grown on dead or artificial media. So, only living cells get attacked by the pathogen. Germination of conidia occurs from about 3–31°C, while the optimum temperature is 15°C with 95% relative humidity. Free water inhibits conidial germination. Powdery mildew is encouraged by sowing the crop in the autumn season, using excessive nitrogen fertilizer, growing crops close together under humid conditions, and using non-resistant varieties (Zillinsky, 1983; Martinelli, 1990).

6.13.3 Disease Cycle

Powdery mildew overwinters primarily as mycelium on volunteers and autumn-sown crops. The cleistothecia are produced in late summers and are resistant to high temperatures and dryings to survive for a long time without any host. Under humid conditions, cleistothecia release the ascospores that further initiate the infection.

6.13.4 Symptoms

Powdery mildew of oat shows similar symptoms to powdery mildew of other cereal crops.

Powdery mildew appears as white gray pustules on the aerial part of the cereals and grasses (like leaves, stem, and ears).

The upper surface of the leaves gets infected and is easily visible. Initial symptoms on plant tissues may appear as chlorotic flecks, followed by white, beige, or gray patches that produce conidia masses that give a powdery appearance.

Whenever the plant is shaken, the cloud or the dust of conidia burst out from the patches. Ascomata may or may not form, but it is embedded in the mildew colonies as tiny dark-colored dots when it forms (Martinelli, 2001).

6.13.5 Management Strategies

6.13.5.1 Cultural Control

Volunteer cereals, crop debris, and stubbles should be removed and burnt, as these may serve as a source of overwintering for the chasmothecia. Close plantation should be avoided. Use a limited source of nitrogen fertilizer, as lush green crops encourage mildew infection.

6.13.5.2 Chemical Control

Powdery mildew is controlled by different sorts of fungicides like Morpholines, Triazole, Falcon 460 EC, Simeconazole, and a noble Triazole.

6.14 CROWN RUST OF OATS

6.14.1 Causal Organisms

Puccinia coronata f.sp. *avenae*

6.14.2 Pathogen Biology

Crown rust is the most damaging oat disease distributed in all oat-growing regions of the world. Warm temperature and high humidity favor this disease, with an optimal temperature of 20–25°C. The pathogen is ceased or inhibited at temperatures above 30°C. This disease is widely distributed worldwide and causes considerable losses to crop in terms of low yield and kernel weight. Its epidemic can cause 10–40% of losses. Various physiological, biochemical, and structural changes are induced in the host plant due to this fungal infection (Simons, 1985; Harder and Haber, 1992).

6.14.3 Disease Cycle

Puccinia coronata f.sp. *avenae* is macrocyclic rust with five infectious stages that involve both sexual and asexual phases. The asexual stages occur only in oat plants, while the sexual stages occur in alternative hosts. The asexual phase involves repeated cycles and continuous sporulation every two weeks. At this stage, urediniospores germinate on the leaves with the favorable condition of mild temperature, adequate moisture, and short exposure to light. Appressoria are formed once germination occurs; penetration pegs are produced to give fungus access to the stoma. The substomatal vesicle is formed in the stomatal cavity from where the hyphae originate required for nutrient uptake; later on, within 7–10 days, a fungal colony is formed on the leaves of a set of urediospores. Uredinia, later on, produces orange-yellow oblong pustules that show characteristic symptoms of the infection. The sexual phase involves both the oat and alternative host (*Rhamnus* spp.). Late in the cropping season, when blooming starts, infection differentiates into teliospores. These dikaryotic, thick-walled structures undergo meiosis and produce haploid basidiospores that later infect the leaves of *Rhamnus* spp. When the fungus is fully developed, pycniospores are produced on the adaxial surface of leaves that act as gametes and fuse with hyphae to produce the fungus's dikaryotic stage (Browning, 1973; Harder and Haber, 1992; Agrios, 1997; Martinelli, 2000).

6.14.4 Symptoms

On susceptible oat varieties, orange-yellow round to oblong pustules appear on leaves.

Pustules vary in size and can be larger than 5 mm. The infection primarily occurs on the leaf and later on leaf sheaths and floral structures.

On resistant oat varieties, small pustules and flecks are formed with chlorotic halos and necrosis.

Most *Rhamnus* species have aecial and pycnial stages on the leaves. Aecial structure varies in size and shows hypertrophy (Browning, 1973; Simons, 1985; Harder and Haber, 1992).

6.14.5 Management Strategies

The use of resistant varieties should be adopted. Avoid planting near its alternative host. Fungicides with pyraclostrobin or azoxystrobin or a combination of propiconazole and trifloxystrobin can control this pathogen effectively.

6.15 STEM RUST OF OATS

6.15.1 Causal Organism

Puccinia graminis f.sp. *avenae.*

6.15.2 Pathogen Biology

Oat is the major crop for several countries and is grown in winter. There are two major diseases of oats, crown rust and stem rust. All oat species, including wild species, get infected by stem rust disease. It is found in all oat-growing areas of the world but mostly occurs in several areas of Canada and can cause 100% devastation in Australia. Warm temperature and moist conditions enhance this disease. The optimal temperature that favors this disease is 15–30°C. The latent period is 7–10 days under favorable temperature (Martens et al., 1985; Wallwork, 1992).

6.15.3 Disease Cycle

Stem rust initiates its sexual stage on *Berberis vulgaris* L. In North America, this disease is not widespread because they have limited alternative hosts. Without an alternative host, its distribution and epidemiology are similar to crown rust (Leonard and Martinelli, 2004).

6.15.4 Symptoms

Oat stem rust mainly affects the leaf sheath and stem, but leaf blades and spikes may also get infected.

Urediniospores are embedded in the pustules that rupture the epidermis and release masses of reddish-brown spores.

The pustules that form in stem rust are larger than the crown rust's pustules; mainly, the pustules are oval and elongated with torn epidermal tissue along the margins.

They continue to produce pustules until the plant reaches maturity. Later on, teliospores are produced either in the same uredia or some different fruiting body called telia.

6.15.5 Management Strategies

Destroy and burn volunteer oats and wild oats for at least four weeks before seeding.

Proper cleaning and crop rotation in the fields should be promoted.

Early planting and harvesting of the oat crop make them less impacted by this disease.

Remove alternative plants like barberry plants.

Resistant varieties should be adopted, and proper fungicides like Triazole should be used.

6.16 ROOT ROT OF OAT

6.16.1 Causal Organisms

Root rot is caused by many pathogens, including:

Bipolaris sorokiniana.
Cochliobolus sativus
Fusarium spp.
Pythium debaryanum
Pythium irregular
Pythium ultimum

6.16.2 Pathogen Biology

Root rot is not a conspicuous disease, but when it gets severe, it causes great devastation. The infected plants produce a limited number of tillers with small seeds. Warm temperature and drought or drying conditions favor the disease. Nutritional deficiency in the soil encourages infection. Excessive use of nitrogen fertilizer and compact soil favors the disease.

6.16.3 Disease Cycle

Different soil-borne fungi are involved in this disease. Spores of the fungi germinate in the soil and infect the seedling; later on, spores are produced on the plant's infected tissues and disseminated by wind, water, cultural practices, and infected seeds. As the plant matures, new spores are formed in the new growing season that cause further infection. Spores may remain viable for many years in the soil.

6.16.4 Symptoms

Patchy pattern disease appearance in the field is the first indication of root rot disease.

The disease progresses from root or sub-crown internodes toward aerial parts like leaves.

Seedlings may die before or just after the emergence even if they were slightly damaged.

This disease's main symptoms are sudden death of emerging seedlings, small root growths, stunted and chlorotic growth, and decaying of the crown or root area.

In severe cases, the tillers contain no seeds.

Dark reddish-brown decay occurs on the sub-crown internode region.

6.16.5 Management Strategies

Irrigate the field properly to remove soil compaction.

Stubbles and soil debris should be pulled out and burned to reduce infection.

An adequate quantity of nitrogen, phosphorous, and potassium should be used to increase the crop's resistance, as excessive quantities may favor the disease.

Seed should be sown in well-drained soil with an appropriate depth of 3–4 cm, as uniform sowing may reduce the infection.

Crop rotation must be done with non-cereal crops, with the removal of grassy weeds.

Commercial seed treatment fungicide should be used.

6.17 *RHIZOCTONIA* ROOT ROT OF SOYBEAN

6.17.1 Causal Organism

Rhizoctonia solani

6.17.2 Pathogen Biology

This disease mainly causes damage to the seedling and older plants. Superficial lesions are produced in this disease and can kill and stunt the plant and cause significant yield losses. Warm soil and high humidity encourage this disease. The optimal temperature for this disease is about 26–32°C, optimal soil moisture ranges from 30–60%, and optimal soil pH is above 6.6. Nutrient and water deficiencies also favor the disease.

6.17.3 Disease Cycle

Rhizoctonia solani is a saprophytic soil-borne pathogen, and it can survive in the soil for a long time in the form of sclerotia or mycelium. Overwintering sclerotia or mycelium may serve as the primary inoculum for the infection. The diseased plant may die, but the pathogen may spread to the root system and cause lateral root rots if the plant survives. Later on, decay starts on lateral roots; as a result, plants become susceptible to drought and heat.

6.17.4 Symptoms

Rusty-brown and dry, sunken lesions are the main symptom of *Rhizoctonia* root rot, appearing on the stem and roots near the soil line.

Decaying of lateral roots occurs.

Seedlings and older plants become stunted, girdled, and chlorotic and may wilt.

Superficial symptoms may appear, but in the case of adequate moisture, these symptoms will not occur.

6.17.5 Management Strategies

Tillage practices should be promoted in fields so that soil drainage will be better in return; it reduces the disease's conducive environment. Avoid nematode and herbicidal injuries. Resistant varieties should be planted. Strobilurin fungicide should be adopted for seed treatment.

REFERENCES

Agrios GN (1997) *Plant Pathology* (4th eds) New York NY, USA: Academic Press, pp. 635.

Akgül DS and Erkilic A (2016) Effect of wheat cultivars, fertilizers, and fungicides on *Fusarium* foot rot disease of wheat. *Turkish Journal of Agriculture and Forestry* 40(1): 101–108.

Alderman S, Frederickson D, Milbrath G, Montes N, Narro-Sanchez J and Odvody G (1999) A laboratory guide to the identification of *Claviceps purpurea* and *Claviceps africana* in grass and sorghum seed samples. *By the Seed Trade* 112.

Atkinson RE (1945) A new mosaic chlorosis of oats in the Carolinas. *Plant Disease Reporter* 29: 86–89.

Baltrus DA and Clark M (2020) Complete genome sequence of *Pseudomonas coronafaciens* pv. *oryzae* 1_6. *Microbiology Resource Announcements* 9(3).

Bamberg RH (1936) Black chaff disease of wheat. *Journal of Agricultural Research* 52: 397–417.

Blum MMC (1997) *Pyrenophora avenae*: Ocorrência, Inóculo, Patogenicidade e Sobrevivência Dissertação (Mestrado–Fitotecnia)—Programa de Pós-Graduação em Agronomia, Faculdade de Agronomia, Universidade Federal do Rio Grande do Sul, *Porto Alegre*. pp. 111.

Boewe GH (1960) Diseases of wheat, oats, barley and rye. *Illinois Natural History Survey* Circular 48.

Browning JA (1973) Oats: Continental Control Program. In: Nelson RR (eds) *Breeding Plants for Disease Resistance*. Pennsylvania, USA: Pennsylvania State University Press, pp. 155–180.

Byamukama E, Ali S and Caffe M (2019) Bacterial Blight Developing in Oats. SDSU Extension. Available at https://extension.sdstate.edu/bacterial-blight-developing-oats.

Catherall P and Hayes J (2007) Oat mosaic virus. *Plant Pathology* 19(2): 78–81.

Catherall PL and Boulton RE (1979) Reaction of some winter oat cultivars to oat mosaic and oat tubular viruses. *Plant Pathology* 28: 57–60.

Catherall PL and Valentine J (1987) Resistances to oat mosaic virus in autumn-sown oats. *Annals of Applied Biology* 111: 483–487.

Channon AG (1981) Downy mildew of Brassicas. In: Spencer DM (ed) *The Downy Mildews*. London: Academic Press, pp. 321–339.

Clover G, Ratti C and Rubies-Autonell C (2002) Detection of European isolates of oat mosaic virus. *European Journal of Plant Pathology* 108: 87–91.

Conner RL and Banttari EE (1979) Storage of the oat blue dwarf virus in whole leafhoppers. *Canadian Journal of Plant Pathology* 1(2): 111–112.

Cook RJ (1980) *Fusarium* foot rot of wheat and its control in the Pacific Northwest. *Plant Disease* 64(12): 1061–1066.

DAFWA (2015) Managing barley yellow dwarf virus and cereal yellow dwarf virus in cereals. Department of Primary Industries and Regional Development. Available at www.agric.wa.gov.au/barley/managing-barley-yellow-dwarf-virus-and-cereal-yellow-dwarf-virus-cereals?nopaging=1

D'Arcy CJ and Burnett PA (eds) (1995) *Barley Yellow Dwarf: 40 Years of Progress*. American Phytopathological Society Press.

D'Arcy CJ and Domier LL (2005) Luteoviridae. In: Mayo MA, Maniloff J, Desselberger U, Ball LA and Fauquet CM (eds) *Virus Taxonomy: VIIIth Report of the International Committee on Taxonomy of Viruses*. New York., NY: Academic Press.

Dutta B, Gitaitis R, Agarwal G, Coutinho T and Langston D (2018) *Pseudomonas coronafaciens* sp. *nov*., a new phytobacterial species diverse from *Pseudomonas syringae*. *PLoS One* 13(12).

Duveiller E, Bragard C and Maraite H (1991) Bacterial diseases of wheat in the warmer areas—reality or myth? In D. Saunders (eds) *Wheat for the Non-Traditional Warm Areas: A Proceedings of the International Conference* 29 July-3 August 1990. Foz do Iguaçu, Brazil: CIMMYT, pp. 189–202.

Duveiller E, Bragard C and Maraite H (1997) Bacterial leaf streak and black chaff caused by *Xanthomonas translucens. The Bacterial Diseases of Wheat: Concepts and Methods of Disease Management.* Mexico, DF: CIMMYT 25–47.

Duveiller E, van Ginkel M and Tijssen M (1993) Genetic analysis of resistance to bacterial leaf streak caused by *Xanthomonas campestris* pv. *undulosa* in bread wheat. *Euphytica* 66: 35–43.

Edwards MC and Weiland JJ (2010) First infectious clone of the propagatively transmitted oat blue dwarf virus. *Archives of Virology* 155 (4): 463–470.

Elliot C (1920) Halo-blight of oats. *Journal of Agricultural Research* 19: 139–172.

Elliot LG and Lommel SA (1990) Identification and partial characterization of oat mosaic and oat golden stripe viruses co-infecting oats in the Southeastern United States. *Schriftenreihe der Deutschen Phytomedizinischen Gesellschaft (Germany, FR).*

Elliott C (1927) Bacterial stripe blight of oats. *Journal of Agricultural Research* 35(9).

Forster RL and Schaad NW (1987) Tolerance levels of seed borne *Xanthomonas campestris* pv. *translucens*, the causal agent of black chaff of wheat. *Plant Pathogenic Bacteria.* Dordrecht: Springer, pp. 974–975.

Forster RL and Schaad NW (1990) Longevity *of Xanthomonas campestris* pv. *translucens* in wheat seed under two storage conditions. In: Klement AZ (eds) *Proceedings of the 7th International Conference of Plant Pathogenic Bacteria, Budapest, Hungary*, Part. Budapest: Akadémiai Kiadó, pp. 329–331.

Forster RL, Mihuta-Grimm L and Schaad NW (1986) Black chaff of wheat and barley. *Current Information Series* 784: 2. University of Idaho, College of Agriculture.

Frank JA and Christ BJ (1988) Rate-limiting resistance to *Pyrenophora* leaf blotch in spring oats. *Phytopathology* 78: 957–960.

Gill CC (1976) Serological properties of oat necrotic mottle virus. *Phytopathology* 66: 415–418.

Gill CC (1980) Some properties of the protein and nucleic acid of oat necrotic mottle virus. *Canadian Journal of Plant Pathology* 2(2): 86–89.

Gough FJ and McDaniel ME (1974) Occurrence of oat leaf blotch in Texas in 1973. *Plant Disease Reporter* 58: 80–81.

Gray S and Gildow FE (2003) Luteovirus-aphid interactions. *Annual Review of Plant Pathology* 41: 539–566.

Griffiths DJ and Peregrine WTH (1956) Halo blight of oats. *Plant Pathology* 5: 95–97.

Hadden SJ and Harrison HF (1955) Occurrence of oat mosaic in the lower coastal plain of South Carolina. *Plant Disease Reporter* 8: 628–630.

Harder DE and Haber S (1992) Oat diseases and pathologic techniques. In: Marshall HG and Sorrells ME (eds) *Oat Science and Technology*, pp. 307–425.

Harder DE and Harris DC (1973) Halo blight of oats in Kenya. *East African Agricultural and Forestry Journal* 38(3): 241–245.

Hebert TT and Panizo CH (1975) Oat mosaic virus. CMI/AAB. *Descriptions of Plant Viruses* 145(3).

Hershman DE and Bachi PR (2010) UK cooperative extension service. *Plant Pathology* Fact sheet.

Hwang MS, Morgan RL, Sarkar SF, Wang PW and Guttman DS (2005) Phylogenetic characterization of virulence and resistance phenotypes of *Pseudomonas syringae. Applied and Environmental Microbiology* 71: 5182–5191.

Ivanoff SS (1963) The cause of spikelet drop of oats. *Plant Disease Reporter* 3: 206–207.

Jones LR, Johnson AG and Reddy CS (1917) Bacterial blight of barley. *Journal of Agricultural Research* 11: 625–643.

Leonard KJ and Martinelli JA (2004) Virulence of oat crown rust in Brazil and Uruguay. *Plant Disease* (in press).

Linford MB and McKinney HH (1954) Occurrence of *Polymyxa graminis* in roots of small grains in the United States. *Plant Disease Reporter* 38(10): 711–713.

Luke HH, Wallace AT and Chapman WH (1957) A new symptom incited by the oat leaf blotch pathogen, *Helmintosporium avenae. Plant Disease Reporter* 2: 109–110.

Macfarlane I, Jenkins JEE and Melville SC (1968) A soil-borne virus of winter oats. *Plant Pathology* 17(4): 167–170.

Martens JW, Seaman WL and Atkinson TG (eds) (1985) Diseases of field crops. In Canada: An illustrated compendium. Ontario, Canada: *Canadian Phytopathology Society* Harrow, pp. 160.

Martinelli JA (1990) Induced resistance in barley (*Hordeum vulgare* L.) to powdery mildew (*Erysiphe graminis* DC: Fr. f. sp. *hordei* Em. Marchal) and its potential for crop protection. Doctoral dissertation, University of Cambridge.

Martinelli JA (2000) Major diseases on oats in South America. In: Cross J. (ed) *Proceedings of the Sixth International Oat Conference*, 13–16 November 2000. Canterbury, New Zealand: Lincoln University, pp. 277–283.

Martinelli JA (2001) Oídio de cereais. In: Stadnik MJ and Rivera MC (eds) *Oídios*. Embrapa Meio Ambiente: Jaguariúna, SP, Brazil, pp. 195–216.

Mathews REF (1982) Classification and nomenclature of plant viruses. *Intervirology* 17(1–3): 76–91.

McKinney HH (1946) Mosaics of winter oats induced by soil-borne viruses. *Phytopathology* 36: 359–369.

McMullen M (1989) *Black Chaff and False Black Chaff of Wheat.* NDSU Extension service. 749.

Miedaner T and Geiger HH (2015) Biology, genetics, and management of ergot (*Claviceps* spp.) in rye, sorghum, and pearl millet. *Toxins* 7(3): 659–678.

Miller WA and Rasochova L (1997) Barley yellow dwarf viruses. *Annual Review Phytopathology* 35: 167–190.

Monger WA, Clover GR and Foster GD (2001) Molecular Identification of oat mosaic virus as a bymovirus. *European Journal of Plant Pathology* 107: 661–666.

Murray GM (2007) Review of diseases of oats for hay: Current and future management, *Part II: Identification and control options for the diseases of importance.* Canberra: Rural Industries Research and Development Corporation.

Nyvall RF (1989) Diseases of Oats. In: *Field Crop Diseases Handbook*. Boston, MA: Springer.

Orlando B, Maumené C and Piraux F (2017) Ergot and ergot alkaloids in French cereals: Occurrence, pattern and agronomic practices for managing the risk. *World Mycotoxin Journal* 10(4): 327–338.

Oswald JW and Houston BR (1951) A new virus disease of cereals transmissible by aphids. *Plant Disease Reporter* 35: 471–475.

Paul VH and Smith IM (1989) Bacterial pathogens of Gramineae: Systematic review and assessment of quarantine status for the EPPO region 1. *EPPO Bulletin* 19 (1): 33–42.

Persson P and Sletten A (2001) Halo blight of oats in Scandinavia. In *Plant Pathogenic Bacteria* 265–268.

Ransom JK, McMullen MS and Meyer DW (2007) Oat production in North Dakota. Extension service, North Dekota State University, Fargo, North Dekota 58105. A-891 (Revised).

Reddy CS, Godkin J and Johnson AG (1924) Bacterial blight of rye. *Journal of Agricultural Research* 28: 1039–1040.

Rosa CRE da, Martinelli JA, Federizzi LC and Bocchese CAC (2003) Quantificação de conídios produzidos por *Pyrenophora chaetomioides* em folhas mortas de *Avena sativa* em condições de campo. *Fitopatologia Brasileira* 28(3): 319–321.

Sabesta J, Zwartz B, Harder DE, Corazza L, Roderick H and Stojanović S (1996) Incidence and resistance of oats to fungus diseases in Europe in 1988–1994. *Ochrana Rostlin* 32(2): 103–113.

Schaad NW (ed) (1988) Laboratory Guide for Identification of Plant Pathogenic Bacteria. Moscow ID, USA, pp. 164.

Scherm B, Balmas V, McKinney F, Spanu, Pani G, Delogu G, Pasquali M and Migheli Q (2013) Pathogen profile *Fusarium culmorum*: Causal agent of foot and root rot and head blight on wheat. *Molecular Plant Pathology* 14(4): 323–341.

Semeniuk G and Mankin CJ (1964) Occurrence and development of *Sclerophthora macrospora* on cereals and grasses in South Dakota. *Phytopathology* 54(4): 409–416.

Shaner G (1981) Effect of environment on fungal leaf blights of small grains. *Annual Review of Phytopathology* 19: 273–296.

Simons MD (1949) *Oat Diseases and their Control* (No. 343). US Department of Agriculture.

Simons MD (1985) Crown rust. In: Roelfs AP and Bushnell WR (eds) *The Cereal Rusts: Diseases, Distribution, Epidemiology and Control*. New York, NY, USA: Academic Press, pp. 132–172.

Singh SD (1995) Downy mildew of pearl millet. *Plant Disease* 79(6): 545–550.

Smith EF, Jones LR and Reddy CS (1919) The black chaff of wheat. *Sci.* 50: 48.

Stenger D and French R (2004) Complete nucleotide sequence of Oat necrotic mottle virus: A distinct Tritimovirus species (family Potyviridae) most closely related to Wheat streak mosaic virus. Arch. Virol. 149: 633–640.

Tabei H (1964) Bacterial halo-blight of oats in Japan. *Japanese Journal of Phytopathology* 29(1): 20–24.

Taylor P, Menzies JG and Turkington TK (2014) An overview of the ergot (*Claviceps purpurea*) issue in western Canada: Challenges and solutions. *Canadian Journal of Plant Pathology* 37(1): 40–51.

Thirumalachar MJ, Shaw CG and Narasimhan MJ (1953) The sporangial phase of the downy mildew on *Eleusine coracana* with a discussion of the identity of *Sclerospora macrospora* sacc. *Bulletin of the Torrey Botanical Club* 80(4): 299–307.

Toler RW and Hebert TT (1963) Reaction of oat varieties, *Avena* species, and other plants to artificial inoculation with the soil borne oat mosaic virus. *Plant Disease Reporter* 47: 58–62.

Uhr DV and Murphy JP (1992) Heritability of oat mosaic resistance. *Crop Science* 32: 345–348.

Wallin JR (1946) Parasitism of *Xanthomonas translucens* (JJ and R) Dowson on grasses and cereals. *Iowa State College Journal of Science* 20: 171–193.

Wallwork H (1992) *Cereal Leaf and Stem Diseases*. Barton, Australia: Grains Research and Development Corporation, p. 102.

Wallwork H (20000) *Cereal Leaf and Stem Diseases*. GRDC.

Warburton CW (1911) Ergot on oats. *Botanical Gazette* 51(1): 64–64.

Watkins JE and Lane LC (2004) Barley yellow dwarf disease of barley, oats, and wheat. In *NebGuide*. USA: University of Nebraska.

Wilkie JP (1972) Halo blight of oats in New Zealand. *New Zealand Journal of Agricultural Research* 15: 461–468.

Wise K, Woloshuk C and Freiji A (2015) Diseases of Wheat: *Fusarium* head blight (Head Scab). *Purdue Extension, BP-33-W* 1–4.

Zeyen RJ and Banttari EE (1972) Histology and ultrastructure of oat blue dwarf virus infected oats. *Canadian Journal of Botany* 50(12): 2511–2520.

Zillinsky FJ (1983) *Common Diseases of Small Grain Cereals: A Guide to Identification*. Mexico, DF: CIMMYT 141.

USEFUL STUDY LINKS

https://en.wikipedia.org/wiki/Sclerophthora_macrospora
https://wheat.pw.usda.gov/cgibin/GG3/report.cgi?class=pathology;name=Downy+Mildew
https://en.wikipedia.org/wiki/Sclerophthora_macrospora
www.plantwise.org/KnowledgeBank/datasheet/49243

https://wiki.bugwood.org/HPIPM:Claviceps_purpurea
www.apsnet.org/edcenter/disandpath/fungalasco/pdlessons/Pages/Ergot.aspx
https://en.wikipedia.org/wiki/Fusarium_culmorum
www.seminis-us.com/resources/disease-guides/tomatoes/fusarium-foot-rot/
https://ahdb.org.uk/knowledge-library/fusarium-and-microdochium-in-cereals
www.agroatlas.ru/en/content/diseases/Avenae/Avenae_Pyrenophora_avenae/index.htl
http://ipm.illinois.edu/diseases/series300/rpd309/

7 Hordeum vulgare Diseases, Etiology, and Management

Rana Muhammad Sabir Tariq, Tariq Mukhtar, Tanveer Ahmad, Shahan Aziz, Zahoor Ahmad, Sajjad Akhtar, and Asghar Ali

CONTENTS

DOI: 10.1201/9781003055365-7

7.1 ECONOMICALLY IMPORTANT DISEASES OF BARLEY

7.1.1 Viral Diseases

7.1.1.1 Barley Mosaic Virus

7.1.1.1.1 Occurrence

It has been known in Japan since 1940, but the first report of the barley yellow mosaic virus of infection in Europe was 1978 in Germany (Hill and Fivans, 1980; Plumb et al., 1986). It occurs where winter barley is grown in all fields of the United Kingdom. It was first investigated in Britain in 1980. It has been reported in England (Adams et al., 1987), the United States (Carroll and Mayhew, 1976), Greece (Katis et al., 1997), Spain (Achon et al., 2003), China (Zheng et al., 1999), Japan (Adams et al., 1986), East Asia (You and Shirako, 2012), Poland (Jeżewska and Trzmiel, 2009), Spain (Achon et al., 2005), and Korea (Lee et al., 2006).

7.1.1.1.2 Causal organism

Barley yellow mosaic virus (BaYMV)

7.1.1.1.3 Vector

Polymyxa graminis (zoosporic fungal pathogen)

7.1.1.1.4 Symptoms

At the initial stage, yellow patches appear on leaves. These symptoms appear in the crop from February. Sometimes it looks the same as waterlogging and acidic soil effects. The small leaf patches changes to streaky yellow, including long chlorotic spots on new small leaves of a few plants, brown at the top of leaves, and often necrosis and curling in leaves, leading to a pointy, sharp look (Katis et al., 1997; Jeżewska and Trzmiel, 2009).

7.1.1.1.5 Management

There is no resistant variety available, but researchers have found some resistant genes (rym1–6) that are efficient to the virus at the cellular level and movement level. Still, none of them entirely reduce infection (You and Shirako, 2013).

Continuous cropping of barley-soybean and three years of fallow land reduce disease severity to a medium level (Zheng et al., 1999), and pausing barley cultivation for a year reduces the amount of *Polymyxa graminis*, a vector in root and soil for BaYMV and Barley mild Mosaic virus (Park et al., 2010).

There is no chemical and biological control available. Use a fungal vector-free area to reduce infection.

7.1.1.2 Barley Yellow Dwarf

7.1.1.2.1 Occurrence

The barley yellow dwarf virus has been known in Kenya since 1983 but was first reported in 1990 (Wangai, 1990). It is known all over the world in Asia, Africa, Europe,

Australia, North America, South America, and Antarctica. It mostly occurs in wet summers. and high rainfall regions of the cropping area but may occur in all regions.

7.1.1.2.2 Causal Organism

Barley yellow dwarf virus (BYDVs)

7.1.1.2.3 Vector

More than 25 species of aphid as vector have been reported for BYDVs (Henry and Dedryver, 1991; Halbert and Voegtlin, 1995). The most critical vectors include *Rhopalosiphum padi, R. maidis, R. rufiabdominalis, Sitobion avenae, Metopolophium dirhodum*, and *Schizaphis graminum*.

7.1.1.2.4 Symptoms

BYDVs cause yellowing and stunting in barley. The strains and ecological conditions can be easily confused and look similar to nutritional, environmental, and abiotic factors. Symptoms on leaves include depigmentation from top to base and from edge to center. Different colors can appear depending on the plant. In barley, the leaf changes to dark yellow. Reddening or yellowing appears in different species of grasses, and some of them are symptomless. Plant infertility and failure of germination capability of grains occur in the most severe cases (D'arcy, 1995; Chay et al., 1996).

7.1.1.2.5 Management

Use biological control agents to control vectors. The parasitoid *Aphidius rhopalosiphi., Ephedrus plagiato*, and *Praon volucre* are used as biological control agents against *Metopolophium dirhodum* and *M. dirhodum* (Farrell and Stufkens, 1990; Dean, 1974).

There is no chemical pesticide available to control the virus. Chemical treatment of the barley yellow dwarf virus can only be through the control of its vector. Before leaf emergence, use alpha-cypermethrin or beta-cyfluthrin spray (12.5 g a.i./ha). Use imidacloprid for seed treatment (70 g a.i./ha), with two foliar sprays with alpha-cypermethrin (Gray et al., 1996; McKirdy and Jones, 1996). Remove regrowths. The stubble of cereals acts as a supply of viruses and vectors (Plumb et al., 1986). Delayed sowing may minimize BYDVs (McKirdy and Jones, 1997).

The *Yd2* gene provides resistance against barley yellow dwarf virus disease in barley. Three Ethiopian varieties (C.I. 3906–1, C.I. 5809, and C.I. 9658) have been studied, and resistance in each variety is identified as incompletely dominant to the *Yd2* gene (Schaller et al., 1964; Collins et al., 1996).

7.1.2 Fungal Diseases of Barley

7.1.2.1 Black Point of Barley (Kernel Blight)

7.1.2.1.1 Occurrence

Black point of barley is a significant disease of barley that occurs worldwide. It is an important disease because discoloration affects the price of grains. However, the frequency of infecting pathogens is variable for different regions, as shown in Table 7.1.

7.1.2.1.2 Etiology

This disease is caused by multiple factors: *Alternaria* spp., *Cochliobolus sativus (Bipolaris sorokiniana)*, and *Fusarium* spp.

TABLE 7.1
Worldwide Occurrence of Black Point in Barley

Pathogen	Country	Reference
Alternaria sp.	Australia	Cook and Dubé, 1989.
"	Brazil	Mendes et al., 1998.
"	Canada	Ginns, 1986.
"	Iran	Poursafar et al., 2018.
"	Mexico	Alvarez, 1976.
"	Russia	Gannibal, 2004.
Alternaria alternate	USA (Illinois)	Boewe, 1964.
Bipolaris sorokiniana (*Cochliobolus sativus*)	Australia	Sivanesan, 1987.
"	Austria	Richardson, 1990.
"	Brazil	Richardson, 1990. Mendes et al., 1998. Zhong and Steffenson, 2001.
"	Canada	Ginns, 1986. Richardson, 1990. Woudenberg et al., 2013.
"	China	Tai, 1979. Zhong and Steffenson, 2001.
"	Czechoslovakia	Richardson, 1990.
"	Denmark	Richardson, 1990.
"	Egypt	Sivanesan, 1987.
"	Finland	Richardson, 1990.
"	Germany	Sivanesan, 1987. Richardson, 1990.
"	Italy	Richardson, 1990.
"	Japan	Zhong and Steffenson, 2001.
"	Kenya	Nattrass, 1961.
"	Morocco	Rehman et al., 2020.
"	New Zealand	Pennycook, 1989. Sivanesan, 1987. Richardson, 1990.
"	South Africa	Crous et al., 2000.
"	Thailand	Richardson, 1990.
"	UK	Sivanesan, 1987. Richardson, 1990.
"	Russia	Richardson, 1990.
"	Zambia	Sivanesan, 1987.
"	Zimbabwe	Sivanesan, 1987.

7.1.2.1.3 Symptoms

- Barley grains are typically infected with these fungi in the process of the dough stage.
- If warmer climate predominates for few days or a week just before the harvesting stage, the rate of infection will rise rapidly.
- Dark brown to black tips appear on the embryo of the grain.
- Other parts of the grain usually are darkened.
- When associated with spot blotch, more of the grain surface is generally darkened, and the grain may be dried up (AHDB, 2020).

7.1.2.1.4 Management

- Avoid sowing susceptible varieties. Avoid late sowing of long-season varieties.
- Use registered fungicides at the recommended rate and appropriate time
- Plant resistant varieties.
- Properly rotate crops every three years.
- High rates of nitrogen application increase the prevalence of this disease, so reduce rates whenever possible (Neate and McMullen, 2005).

7.1.2.2 Head Blight of Barley

7.1.2.2.1 Occurrence

Head blight of barley is another acute disease of barley. It also prevails worldwide in various countries (as shown in Table 7.2). It was declared the worst disease of barley since the outbreak of the 1950s by the United States Department of Agriculture. In 1990, Canada faced significant losses due to this disease. It is also prevalent in the United States (AHDB, 2020).

Etiology: This disease is caused by fungal pathogen *Fusarium graminearum.*

7.1.2.2.2 Perpetuation

It is introduced to the field using diseased kernels or an air-carrying inoculum. The infection spreads rapidly through rain and wind.

7.1.2.2.3 Symptoms

- Common symptoms are found during the dry season.
- Water-soaked spots appear on glumes.
- The base of the glume shows discoloration.
- Spikelets are also affected.
- Sometimes a whitish or brownish fungal mass appears on the base of glumes.
- Kernel development is also affected.
- Kernels reduce their size, and discoloration is also observed.
- Inside the kernels, a floury appearance develops.
- The symptoms of barley head blight resemble those of root, crown, and blight diseases.
- Grain becomes scabby (Nganje et al., 2004).

TABLE 7.2
Occurrence of Head Blight in Barley

Serial No.	Country	Reference
1.	Argentina	Castanares et al., 2016.
2.	Australia	Simmonds, 1966.; Cook and Dubé, 1989.; Schilling et al., 1995. Wright et al., 2010.
3.	Brazil	Mendes et al., 1998; Castanares et al., 2016.
4.	USA (California)	Sprague, 1950.
5.	Canada	Ginns, 1986.
6.	China	Tai, 1979.; Zhuang, 2005.
7.	Denmark	Kristensen et al., 2005.
8.	England	Richardson, 1990.
9.	Germany	Schilling et al., 1995.

7.1.2.2.4 Control

- Use resistant varieties against head blight.
- Under severe disease conditions, even tolerant species will need proper care to achieve adequate control.
- Protective fungicides are particularly required at the flowering stage to stop spore (conidia) infection.
- Stop watering before flowering until after planting. This reduces the distribution of spores from crop residues.
- Weeds serve as alternative hosts; hence, the destruction of the alternative host is the best disease escape strategy.
- The use of certified seeds provided after treatment with fungicide is required.
- Replacement of crops on non-grain crops or soil transplantation will reduce the inoculum.
- Circulation of maize, wheat, or barley is at high risk of FHB and should be avoided.
- Destroy maize ranges for farming to reduce the risk of FHB. Consistent varieties will need to be sprayed with fungicide (Nganje et al., 2004).

7.1.2.3 Leaf and Spot Blotches

Occurrence: This problem is reported due to multiple pathogens almost all over the world. Popular causes of leaf and spot blotches in barley are *Bipolaris* sp., *Pyrenophora*, and *Septoria* sp., as described in Table 7.3.

7.1.2.3.1 Epidemiology and Secondary Cycles

The fungus overwinters in barley straw and stubble, in the soil or on the seed. Spores are produced on barley debris in the warm climate and dispersed through the wind,

TABLE 7.3
Worldwide Occurrence of Leaf and Spot Blotch of Barley

Pathogen	Country	Reference
Bipolaris sorokiniana (Cochliobolus sativus)	Australia	Sivanesan, 1987.
"	Austria	Richardson, 1990.
"	Brazil	Richardson, M.J., 1990.; Mendes et al., 1998.; Zhong and Steffenson, 2001.
"	Canada	Ginns, 1986.; Richardson, 1990.; Woudenberg et al., 2013.
"	China	Tai, 1979.; Zhong and Steffenson, 2001.
"	Czechoslovakia	Richardson, 1990.
"	Denmark	Richardson, 1990.
"	Egypt	Sivanesan, 1987.
"	Finland	Richardson, 1990.
"	Germany	Sivanesan, 1987. Richardson, 1990.
"	Italy	Richardson, 1990.
"	Japan	Zhong and Steffenson, 2001.
"	Kenya	Nattrass, 1961.
"	Morocco	Rehman et al., 2020.
"	New Zealand	Pennycook, 1989.; Sivanesan, 1987.; Richardson, 1990.
"	South Africa	Crous et al., 2000.
"	Thailand	Richardson, 1990.
"	UK	Sivanesan, 1987.; Richardson, 1990.
"	Russia	Richardson, 1990.
"	Zambia	Sivanesan, 1987.
"	Zimbabwe	Sivanesan, 1987.
Pyrenophora teres (Helminthosporium gramineum)	Australia	Shivas, 1989.
	Brazil	Richardson, 1990.
	Canada	Sivanesan, 1987. Ginns, 1986.
	China	Sivanesan, 1987.
	Cyprus	Sivanesan, 1987.
	Czechoslovakia	Richardson, 1990.
	Denmark	Richardson, 1990.

(Continued)

TABLE 7.3
(Continued)

Pathogen	Country	Reference
	Egypt	Richardson, 1990.
	England	Richardson, 1990.
	Ethiopia	Sivanesan, 1987.
	Germany	Sivanesan, 1987.
	India	Sivanesan, 1987.
	Iraq	Sivanesan, 1987.
	Israel	Sivanesan, 1987.
	Kenya	Sivanesan, 1987.; Nattrass, 1961.
	Korea	Cho and Shin, 2004.
	Libya	Sivanesan, 1987.
	Mexico	Alvarez, 1976.
	Nepal	Sivanesan, 1987.
	New Zealand	Pennycook, 1989.
	Pakistan	Sivanesan, 1987.
	Romania	Sivanesan, 1987.
Septoria hordei	Poland	Mulenko et al., 2008.
Septoria sp.	Australia	Sampson and Walker, 1982.
	Poland	Mulenko et al., 2008.
Setosphaeria rostrata– *(Exserohilum rostratum)*	Pakistan	Richardson, 1990.

and water and insects may also be involved in the dispersal of fungal spores. Barley yield can be infected from the inoculum on the seed or infected debris left in the soil. Warm temperatures above 20°C and humidity also favor the development of spot blotch disease.

7.1.2.3.2 Symptoms

Small dark chocolate-colored spots appear on the infected area of leaves; eventually, it merges with other necrotic spots, forming a large patch on leaves. These blotches are irregular in shape, surrounded by a yellow lining. Severely infected leaf dries up completely, leaving the characteristic brown lesion. Brown spots also appear on glumes.

7.1.2.3.3 Control and Management: This Disease Is Managed in the Following Ways

- Use of resistant varieties.
- Use of clean seeds.
- Application of foliar fungicide.

- Rotation with non-cereal crops.
- Destruction of the debris of infected plants in the field.
- Treating the infected seed with an effective fungicide (AHDB, 2020).

7.1.2.4 Wilt of Barley

7.1.2.4.1 Occurrence

In 1989, this fungus was isolated from the leaves of the barley plant. It was identified as *Verticullium dahliae*. This fungus has a broad host range, including wheat, potato, cotton, and tomato.

7.1.2.4.2 Etiology and Infection Cycle

- Verticillium wilt of barley is caused by a soil fungus, *Verticillium dahliae*. *Verticillium dahliae* fungus lives in the soil as small, darkened structures that are called microsclerotia. These microsclerotia can live in the soil for many years.
- When the roots of any of susceptible plant grow close to microsclerotia, the fungus *Verticillium dahliae* germinates and infects the roots of the plant through cuts, wounds, or natural openings of root parts of the plant.
- The fungus spreads into the branches via the vascular system of the plant. When the xylem of the plant is infected, it blocks the system, and water cannot reach the leaves.
- *Verticillium dahliae* can also spread to the other plants with equipment/tools used in the field (Mathre, 1989).

7.1.2.4.3 Symptoms

Wilting is the most common sign of plants affected by *Verticullium dahliae*. But some of the common symptoms are:

- Chlorosis of leaves and stripes along the leaf margin.
- Stunting and yellowing of the leaves.
- Necrosis of leaves occurs in the later stages.

7.1.2.4.4 Management/Control

We can manage *Verticullium* wilt by

- Planting resistant varieties.
- Soil fumigation.
- It is important to clear plant debris to decrease the spread of disease in the field.
- Solarization is also very effective against *Verticillium*.

7.1.2.5 Smut of Barley

Etiology: This disease is caused by *Ustilago hordei*.

7.1.2.5.1 Occurrence

The disease is of frequent occurrence in the barley-growing tracts of the Indian subcontinent. This disease is most common in hilly regions. It is seed- and soil-borne and can be carried in contaminated machinery. Smut spores are released during harvest and contaminate clean seed, machinery, and soil.

TABLE 7.4
Worldwide Occurrence of Barley Smut in Barley

Serial No.		
1.	Algeria	Zundel, 1953. Zambettakis, 1971.
2.	Argentina	Zundel, 1953. Hirschhorn, 1986.
3.	Australia	Zundel, 1953. Sampson and Walker, 1982. Simmonds, 1966. Vanky and Shivas, 2008.
4.	Brazil	Zundel, 1953.
5.	Bulgaria	Denchev, 2001.
6.	Canada	Zundel, 1953.
7.	China	Zundel, 1953. Tai, 1979. Guo, 1988. Teng, 1996. Zhuang, 2001.
8.	Egypt	Zundel, 1953.
9.	Europe	Vanky, 1994.
10.	India	Zundel, 1953.
11.	Iran	Ershad, 2001. Stoll et al., 2005. Vanky and Abbasi, 2013.
12.	Mexico	Alvarez, 1976.
13.	Morocco	Zundel, 1953.
14.	Nepal	Dahal et al., 1992. Singh, 1968.
15.	New Zealand	Zundel, 1953. Pennycook, 1989. McKenzie and Vanky, 2001. Vanky and McKenzie, 2002.
16.	Norway	Zundel, 1953.
17.	Pakistan	Ahmad, 1956. Ahmad et al., 1997.
18.	Palestine	Zundel, 1953.
19.	Poland	Zundel, 1953. Mulenko et al., 2008.
20.	Russia	Karatygin, 2012.
21.	South Africa	Gorter, 1977.
22.	Sweden	Zundel, 1953.

Serial No.		
23.	Taiwan	Zundel, 1953.
24.	Tunisia	Zambettakis, 1971.
25.	Turkey	Zundel, 1953.
26.	USA	Bakkeren et al., 2000.
27.	Zimbabwe	Whiteside, 1966. Zambettakis, 1971. Vanky and Vanky, 2002.

Database source: (Farr and Rossman, 2020).

7.1.2.5.2 Etiology

Externally seed-borne disease. Mycelium is intercellular and dikaryotic. The hyphae are septate and branched. Mycelium grows in the stem as well as leaf, but generally, it is confined to the stem. Teliospore (smut spore) is round to elliptical brown, and the spore wall is smooth.

7.1.2.5.3 Classification

Kingdom: Fungi, Division: Basidiomycota, Class: Ustilaginomycetes, Order: Ustilaginales, Family: Ustilaginaceae, Genus: *Ustilago*, Species: *Ustilago hordei*

Favorable Environment: Overwintered smut covers the seed as teliospore. When the temperature of the soil is 15–25°Celsius, spore germinates in low soil moisture.

7.1.2.5.4 Symptoms

- The infected plant's head appears normal as healthy plants. On leaf sheaths, blackened ears emerge, by which one can recognize the infected plant.
- It becomes visible at ear emergence.
- Smutted ears emerge at the same time as those of healthy plants but remain shorter and are usually retained within the sheath for a longer time.
- Every ear on a diseased hill and every grain in an infected ear is affected.
- A black powdery mass of teliospore replaces the grain.
- A persistent membrane covers the black spores.
- Occasionally, smut sori may also develop in leaf blades (AHDB, 2020).

7.1.2.5.5 Control

The disease only affects barley seeds. Hence, seed treatments control the disease best.

Cultural Practices:

- Buy certified seed or test home-saved seed for the presence of the pathogen.
- Use healthy seeds.

- Seed treatments help protect seed and seedling diseases.
- Uproot and burn affected plants.
- Grow resistant varieties.

7.1.2.5.6 Chemical Control

- Treat the seeds with Agrosan, Ceresan, or Spergon to successfully control the disease.

7.1.2.6 Root Rot

7.1.2.6.1 Occurrence

The economic importance of this disease is not yet entirely determined; however, it is caused by a group of economically important pathogens, including *Rhizoctonia* sp., *Fusarium* sp., *Pythium* sp., and *Sclerotium* sp., as described in Table 7.5.

7.1.2.6.2 Symptoms

- The disease is characterized by an increasingly apparent lack of vigor as the plant develops, with a reduction in tillering, stem elongation, green color, and head size.
- Some highly susceptible varieties of barley are vulnerable to complex of common root rot attacking basal stem tissues as well as roots.
- Chlorotic streaks frequently occur in barley leaves.
- Symptoms of this disease are too variable for a complete diagnosis. These differences may result from environmental effects or inherent differences in host materials.
- The browning symptom is associated with severe root pruning, together with succulent foliage developed under cool, moist conditions followed by leaf desiccation brought on by hot, dry weather.
- The root systems of diseased plants vary from nearly clear with only slight yellowing, mainly in the stele, to marked yellow and even brown.
- Casual examination may indicate that the roots are healthy, when a detailed examination reveals local necrotic areas and general mild discoloration of the stele (Bruehl, 1953).

7.1.2.6.3 Control

- Success has been seen with phosphatic fertilizers in the control of *Pythium* root rot in ammonium.
- Phosphate. Barley was treated with ammonium phosphate treble superphosphate and ammonium nitrate. The increases in grain production were 73% from ammonium phosphate, 19% from treble superphosphate, and 114% from ammonium nitrate.
- Reduce tillage that makes the condition worse.
- Improve phosphorus nutrition.
- Use fungicides for seed treatment (Bruehl, 1953; Neate and McMullen, 2005).

TABLE 7.5
Worldwide Incidence of Root Rot of Barley

Pathogen	Country	Reference
Rhizoctonia sp.	Australia	Simmonds, 1966.
Rhizoctonia solani	California (USA)	French, 1989.
	Iran	Choupannejad et al., 2017.
	Mexico	Alvarez, 1976.
Rhizoctonia oryzae	USA (North Dakota)	Sprague, 1950.
Pythium aristosporum	**Canada**	Ginns, 1986.
Pythium hypogeum (*Globisporangium hypogeum*)	**USA**	Sprague, 1950.
Sclerotium rolfsii	**Australia**	Shivas, 1989. Simmonds, 1966.
	Brazil	Mendes et al., 1998.
	California (USA)	French, 1989.
	China	Tai, 1979.
	Thailand	Giatgong, 1980.
	Zimbabwe	Whiteside, 1966.
Fusarium acuminatum	Brazil	Mendes et al., 1998.
	California (USA)	Sprague, 1950.
	Canada	Ginns, 1986. Richardson, 1990.
Fusarium equiseti	Argentina	Richardson, 1990.
	Brazil	Mendes et al., 1998.
	California (USA)	French, 1989.
	Canada	Richardson, 1990.
	Czech Republic	Xia et al., 2019.
	Denmark	Kristensen et al., 2005.
	Estonia	Loiveke, 2006.
	Kenya	Nattrass, 1961.
	Norway	Kristensen et al., 2005.
	Poland	Mulenko et al., 2008.
	Sweden	Kristensen et al., 2005.

7.1.2.7 Take-All Disease

7.1.2.7.1 Occurrence

This is an important disease of barley that may also occur in rice and wheat. It has been reported in various regions of the world, as described in Table 7.6.

TABLE 7.6
Worldwide Occurance of Take-All Disease of Barley

Pathogen	Country	Reference
Gaeumannomyces graminis	Australia	Sampson and Walker, 1982. Shivas, 1989.
	Brazil	Mendes et al., 1998.
	Canada	Ginns, 1986.
	China	Tai, 1979.
	Korea	Cho and Shin, 2004.
	Virginia	Roane and Roane, 1994.
Gaeumannomyces tritici	California, USA	French, 1989.
	Iran	Yosefvand et al., 2016.
	New Zealand	Pennycook, 1989.
	South Africa	Gorter, 1977. Crous et al., 2000.
	UK	Hernandez-Restrepo et al., 2016.

7.1.2.7.2 Etiology

Take-all disease is caused by *Gaeumannomyces graminis* and *Gaeumannomyces tritici*, the soil-borne fungi. They survive from season to season in soil or debris of crops that can become the host and weed host. Take-all mostly occurs in early fall sown crops.

7.1.2.7.3 Symptoms

- At the very first stage, patches of poor growth can appear on the crops that can't be clearly defined.
- Then discoloration of the crown, root, and stem base occurs.
- The blackening of the root is one of the most prominent symptoms of this disease.
- These blackened roots are 'rat-tail' in shape.
- In mature plant patches, whiteheads and stunted growth of different areas of the plant can be seen above the ground.

7.1.2.7.4 Management and Control

An effective strategy for take-all is not to allow the fungus to multiply in the soil by excluding hosts. Grass containing low-level pasture will also help in reducing take-all disease. Minimum tillage has a significantly increased breakdown of residue that reflects the management for that (AHDB, 2020).

7.1.2.8 Powdery Mildew of Barley

7.1.2.8.1 Occurrence

This disease is caused by a fungus, *Blumeria gramineae*. This disease prevails all where barley is grown all over the world (Akhkha, A., 2009).

TABLE 7.7
Worldwide Occurrence of Powdery Mildew of Barley

Serial no.	Country	Reference
	Bulgaria	Fakirova, 1991.
	China	Zhuang, 2001. Zheng and Yu, 1987. Zhuang, 2005.
	Europe	Braun, 1995.
	India	Sarbhoy and Agarwal, 1990. Paul and Thakur, 2006. Gautam and Avasthi, 2018.
	Israel	Voytyuk et al., 2009.
	Korea	Shin, 2000. Cho and Shin, 2004.
	New Zealand	Boesewinkel, 1979.
	Russia	Rusanov and Bulgakov, 2008.

7.1.2.8.2 Insect Vector

The most important vector is a wooly aphid (Eriosomantinae). The vector of transmission of barley powdery mildew is other sucking insects.

7.1.2.8.3 Symptoms

- The infected plant shows spots on the leaves and stems resembling white powder.
- Sometimes the lower leaves are most affected by this disease.
- The spore masses developed on leaves (leaf, sheath, and heads) of powdery mildew are buff or gray white color.
- Browning and yellowing appear on dead leaves (Akhkha, A., 2009).

7.1.2.8.4 Control

- Resistant varieties.
- If the disease is severe in field/crop nitrogen fertilizer, reducing rates gives some control.
- Crop rotation helps reduce the disease.
- Clean plowing is helps to control the disease.
- Foliar fungicide application is beneficial under severe disease pressure (Akhkha, A., 2009).

7.1.3 Bacterial Diseases of Barley

7.1.3.1 Basal Glume Rot

7.1.3.1.1 Causing Agent

Pseudomonas syringae pv. *atrofaciens*

7.1.3.1.2 Host

Wheat, barley, rye and other cereal crops.

7.1.3.1.3 Introduction

Basal glume rot is a disease generally reported in America, Europe, Canada, New Zealand, South Africa, Belgium, and many other countries and distributed worldwide. In 1982, basal glume rot was reported for the first time in Belgium. Basal glume rot leads to yield loss but has low economic loss, so basal glume rot has no economic importance. The disease generally attacks at the heading stage (AHDB, 2020; Neate and McMullen, 2005).

7.1.3.1.4 Etiology

Basal glume rot is caused by a bacterium, *Pseudomonas syringae* pv. *atrofaciens*. The bacterium has a rod shape. *Pseudomonas syringae* pv. *atrofaciens* are Gram-negative bacteria. Basal glume rot is a seed-borne disease, but the bacterium is an epiphyte (lives on plant surface) also. In extreme conditions, the bacterium can survive. The bacteria release effector proteins through a type-III secretion system, which causes disease or pathogenicity in the plant.

7.1.3.1.5 Epidemiology

The disease is nourished in moist, wet, and high-humidity conditions. The favorable temperature for the disease is 12–25°C, which means the bacteria needs a cool environment to cause disease. The bacterium enters the plant through natural openings like stomata, wounds, hydathodes, and abrasions. It enters the natural openings of the plant through pilli or filaments. Basal glume rot generally develops at the grain filling or heading stage.

7.1.3.1.6 Transmission

The pathogen survives on the plant's surface until conditions favor the pathogen. The disease develops more in cool and wet conditions. Infection can be transmitted from one plant to another through water (rain or irrigation). The pathogen also survives on weeds (unwanted plants, e.g., grasses). It also lives on plant debris. It can spread quickly by water, but in some cases, it can be spread by wind blowing and also by insects. When the host is not present, it survives on grasses and weeds. It survives in the soil in overwintering (Neate and McMullen, 2005; AHDB, 2020).

7.1.3.1.7 Symptoms

The general symptoms of basal glume rot are

- Symptoms generally appear on glumes.
- Water-soaked lesions appear at the base of glumes.
- Dark green spots also develop at the base of glumes after some time. These spots change color and turn into black- and dark brown–colored spots.
- Water-soaked lesions also appear on spikelets.
- Glumes show a translucent appearance (light passes through glumes).
- In favorable conditions, a white, grayish bacterial ooze appears on glumes.
- Discoloration is observed on the stem.

- Irregular water-soaked spots and lesions develop on leaves.
- The symptoms of basal glume rot have a resemblance to black chaff, frost damage, and glume blotch (Neate and McMullen, 2005; AHDB, 2020).

7.1.3.1.8 Control

- Crop rotation.
- Use resistant varieties.
- Use treated seeds.
- Use cultural practices.
- Avoid over-irrigation.
- Avoid irrigation at the heading stage.
- Uproot infected plants.
- Remove plant debris from the field.
- Remove weeds from the field.
- Plow the field before sowing seeds (Neate and McMullen, 2005; AHDB, 2020).

7.1.3.2 Bacterial Blight and Black Chaff

7.1.3.2.1 Occurrence

Bacterial leaf streak has been reported in the United States, Canada, Mexico, Argentina, Bolivia, Brazil, Paraguay, Peru, and Uruguay (Mehta, 1990; Mohan and Mehta, 1985; Frommel, 1986; Tessi, 1949). The disease is known on wheat in China (Sun and He, 1986), Pakistan (Akhtar and Aslam, 1985, 1986), and Iran (Alizadeh et al., 1995) and on triticale in India (Richardson and Waller, 1974). In the Near and Middle East, it affects durum (*T. turgidum* var. *durum* L.) and bread wheat in irrigated areas of Syria (Mamluk et al., 1990), Israel (CIMMYT, 1977), Turkey (Demir and Üstün, 1992), and Yemen (Bragard et al., 1995). In Africa, BLS has been found in Kenya (Burton, 1931), Ethiopia (Korobko et al., 1985), South Africa (Smit and Van A. Bredenkamp, 1988), Tanzania (Bradbury, 1986), Libya, Madagascar (Bragard et al., 1995), and Zambia (Bragard et al., 1995). In Australia, BLS has been recorded on wheat and rye in New South Wales (Noble, 1935).

7.1.3.2.2 Etiology

Kingdom: Bacteria, Class: Probacteria, Family: Lysobacteraceae, Genus: *Xanthomonas*, Species: *Xanthomonas translucens* pv. *translucens*

- *Xanthomonas translucens* pv. *translucens* is a Gram-negative bacterium that causes worldwide disease of bacterial blight and black chaff in barley and other cereal crops.

7.1.3.2.3 Favorable Condition

- Initially, wet and humid weather are favorable conditions for bacterial blight and black chaff of barley.
- Humidity isn't needed if the bacterium enters the plant.
- The most important cause of bacterial growth is temperature.
- It grows optimally at 78°F or above.
- Bacterial blight doesn't require moisture for disease infection and its development (Neate and McMullen, 2005).

7.1.3.2.4 Symptoms

- Brown–purple patches seen on upper leaves and below the stems.
- Symptoms start as minor linear water-soaked patches, which later become brown.
- Wounds become irregular and glossy surfaced, and the length of the leaf may extend.
- The lesions produce honey-like emissions in moist conditions.
- In barley, a light-green ring surrounds the wounds.
- A tint of orange appears on the leaf (Neate and McMullen, 2005).

7.1.3.2.5 Control

- We can rotate with broadleaf species.
- The seed is considered a primary inoculum, which means they can survive in unfavorable conditions. Thus, the use of certified pathogen-free seeds is the best way to reduce injuries and increase yield.
- The disinfection of seeds is the best method to remove the disease by using physical and chemical techniques.
- We can also use fungicidal seed treatment on infected seeds.
- We can also destroy infected stubble.
- We can minimize infected crop debris (Neate and McMullen, 2005).

7.1.4 Nematodes of Barley

7.1.4.1 Root-Knot Nematode

7.1.4.1.1 Occurrence

The root-knot nematode of barley, also known as *Meloidogyne naasi*, is not a very common species but is present on small grains and many kinds of grass. It is an introduced species mostly found on creeping grasses in the northern part of the United States (Allen et al., 1970).

7.1.4.1.2 Host

- Barley
- Oats
- Rice
- Colonial and creeping bentgrass
- Orchardgrass
- Rye
- Wheat
- Ryegrass
- Number of weed grasses

The only important crop hosts outside of the grass family are soybean and sugar beet.

7.1.4.1.3 Symptoms

It causes severe yield losses in small grains, causing necrosis, stunting of plants, hyperplasia, disorganization of xylem elements, reduced tilling, and small seeds of a crop.

At the beginning of spring, the stunting and yellowing of plants starts to occur. Plant parts tend to disappear in favorable growing conditions. Small galls or knots appear on the roots of the plant (Allen et al., 1970).

7.1.4.1.4 Control

- Crop rotation.
- Incorporation of organic matter.
- Use of resistant varieties.
- Early planting.
- Temperature control.
- Heat treatment of propagation material.
- Heat treatment of soil (Allen et al., 1970).

7.2 ABIOTIC STRESS ON BARLEY

Different unpredictable abiotic stresses are responsible for more reduction in the growth and yield of crops due to their timing and intensity, which are emerging threats nowadays and good topics for all researchers who study plants. Due to changing climatic conditions, the different abiotic stresses increase day by day, which affects the growth and physiological processes of plants, and the results are decreasing the yield of crop plants under these different stresses (Ahmad et al., 2020). Global climate change and the increasing occurrence of droughts (water and heat stresses) of varying severity, duration, and scale in various parts of the world can have adverse effects on crop growth and grain productivity (Ashraf, 2010). Abiotic stresses resulting from extreme water deficit or salinity are responsible for reducing transpiration and leaf pigments related to crop productivity. According to a worldwide approximation, abiotic stresses cause yield losses in various crops by an average of 50%, of which low temperature, salinity, drought, and high temperature constitutes about7%, 9%, 10%, and 20% (Choudhury et al., 2017). The role of different stresses such as high temperature, increasing water shortage, and increasing salt concentration, which are responsible for affecting the growth, physiology, and yield of barley plants, are described in detail in the following.

7.2.1 Drought Stress and Barley

Across the world, drought is becoming more common, causing distressing effects on the production of crops (Ahmad et al., 2018, 2020). In various developing countries, drought represents a restraint in the production of agriculture, and it also results excessive losses of yield. The unavailability of water related to heat stress is responsible for reducing the growth of plants with another stresses (Ahmad et al., 2017). In the semi-arid Mediterranean region today, water stress is an important threat during grain filling (N. Samarah and Alqudah, 2011). Barley is considered a very important winter cereal crop; drought stress reduces the grain yield of barley. Grain yield was correlated positively with leaf gross photosynthetic rate and negatively with leaf osmotic potential. Various research has revealed that leaf photosynthetic rate is reduced by drought stress (high temperature) during the development of seeds (N. H. Samarah,

2005; Al-Sayaydeh et al., 2019), the duration of seed production is reduced (Sánchez-Díaz et al., 2002; N. H. Samarah, 2005), and the barley harvesting period is boosted, which causes a severe drop in yield components and grain yield (Forster et al., 2004).

Moreover, all physiological processes that occur in plants are affected by drought stress, such as seed development causing decreased vigor and seed germination. Chloupek et al., (2003) showed the result of three out of seven years of studies; during unsuitable weather conditions, barley crop decreased its production of seed vigor ranging from 61–86% compared to the remaining four years (exceeded 94%). N. Samarah and Alqudah (2011) concluded that low germination in the small seed size category (composed of hard, shriveled, and misshapen seeds) also reduced vigor.

7.2.2 Salinity Stress and Barley

Increased salinization is a big danger for sustainable farming, and it is expected that more than 50% of agricultural soil will be affected by the year 2050 (Vinocur and Altman, 2005). Over the world, about 800 mha of land is affected by various salt and constitutes greater than 6% of the whole land area across the world (FAO, 2015). Among the 6%, 189 Mha is located in the Middle East, which makes it the largest such area (Wicke et al., 2011). Poor methods of irrigation and expected change in climate in the future will result in increasing such area (Hayes et al., 2015). Mahmood (2011) stated that barley is considered one of the best salt-tolerant crops. Still, under a high-stress situation, the development and growth of barley are badly affected by osmotic and ionic stresses. The varieties of barley which are grown in those problematic areas have a high tendency to tolerate soil salinity. In cereals, salt tolerance is recognized as linked to controlling the shoot content of sodium ions, such as tolerant lines that have been used for more competent systems for excluding Na from inside the plant (Dubcovsky et al., 1996). Barley is considered one of the best crops in maintaining its growth even in high concentrations of accumulation of sodium ions in its leaves. Barley's intensity of high tissue tolerance can include sequestration of sodium ions into intracellular vacuoles and the synthesis of compatible solutes that gather in the cytoplasm to stabilize the osmotic potential of the vacuolar $Na+$ (Munns, 2002). The toxicity of sodium ions $Na+$ affects the growth of the plant and an enlarged sodium ion/potassium ion ratio, hence dislocation of potassium ions by sodium ions in the cell of the plant, which affects the movement of hydrogen ions and plasma membrane activities. Under salt-induced water scarcity situations in leaves, a significant variation has been reported in soluble carbohydrates, total phenolic compounds, $Na+/K+$ ratio, antioxidant activity and leaf photosynthetic pigments. Generally, water or osmotic stress is stabilized by plants' metabolic adaptation which increases synthesis of osmoprotectants, for example, soluble sugar and proline, which contributes to osmoregulation and also protects the structure of diverse membranes and biomolecules (Hare et al., 1998).

7.2.3 Heat Stress and Barley

In barley, it is observed that temperature stress affects vigor and development and growth. Heat stress is the chief abiotic stress that is linked with high surrounding temperatures that maintain crop productivity and growth (Shao et al., 2015).

Temperature, when increased from optimal, adversely affects crop growth and decreases yield, causes early evaporation and unestablished plants due to loss of moisture, and disturbs the optimal period such as jointing and tillering, which causes deprived formation. It is also found that at the reproductive stage, high temperature causes an accelerated grain filling period, premature flowering, size, and decreased grain number and finally decreases the yield of the crop (Wahid et al., 2007). Heat stress (high temperature) disturbs geographic distribution, the time of the growing season (Dawson et al., 2015), and the malting quality of barley (Savin et al., 1996). Heat stress depends on various factors such as rate and duration of high temperature and stress intensity (temperature in degrees) (Wahid et al., 2007). Heat stress affects barley in various ways such as decreased yield and the components of yield, that is, low concentration of final starch, probably due to inalterable inactivation of sucrose synthase, the number of tillers per plant, thousand-grain weight, and spike length (Wahid et al., 2007). Similar to some of other cereal crops, barley is also very sensitive to heat stress, especially at meiosis and panicle growth, producing abnormal pollens and leading to whole infertility (Sakata et al., 2000). However, yield losses vary and are linked with phase and time at which the plant met with inherent genetic tolerance, abnormal temperature, and the mechanism involved in bettering the stress and availability of moisture present in the soil, which plays a role in evaporative freezing of the canopy. In Iran, field research was conducted under heat stress; after anthesis, 17% decreased in grain yield in barley, and parameters such as deposition in the developing bran, translocation of photosynthates to the grain, and starch synthesis were affected significantly (Modhej et al., 2015). Ugarte et al. (2007) conducted field trials and concluded that a decrease in grain yield produced by heat stress caused serious damage when temperature rose during stem elongation. Field research was conducted in which a panel of 138 spring barley genotypes was used which proved that a reduction in grain yield up to 55.8% was due to heat stress (Ingvordsen et al., 2015).

7.2.4 Management of Abiotic Stress in Barley

7.2.4.1 Agronomic Approach

Agronomic field-based management strategies can be incorporated with resilient breeding material (genotypes) to improve various abiotic stresses in barley. A study has shown that greater water storage prolonging the availability of moisture leads to 1.0, 1.7, and 6.3 times higher grain yield under no tillage than CT in three years of barley mono-cropping. Moreover, no-till results in higher N uptake, N grain content, and yield-based water use efficiency. The study strongly argues that the integration of no-till along with pig slurry (agronomic rates) significantly improved the grain yield and nutrient and water use efficiency of barley compared to conventional management (intensive tillage with mineral-based N fertilization) under rainfed conditions (Plaza-Bonilla et al., 2017). A five-year study reported that a no-till malt barley–pea (*Pisum sativum* L.) cropping system sustained barley malt yield and quality under a temperate dryland scenario of a short crop growing season owing to lower precipitation with a cool climate. This resilient system prevents soil erosion and nitrogen leaching and escalates soil organic matter compared to no-till continuous malt barley, no-till malt barley–fallow, and conventional till malt barley–fallow (Sainju et al., 2013).

7.2.4.2 Plant Growth-Promoting Rhizobacteria

Under stress scenarios, plant growth may be improved by using microbial-based inoculation, that is, plant growth-promoting rhizobacteria (PGPR), including generally bacterial strains that have potential to synthesize growth-promoting substances for plants, that is, enzymes, phytohormones, siderophores, and vitamins along with ethylene synthesis inhibitors, organic phosphate mineralizers, atmospheric nitrogen fixers, and also increase resilience to various stresses (drought, metal toxicity, salinity, etc.) (Turan et al., 2013). The mechanism of growth improvement by these microbes regulating hormonal and nutritional balance in plants amplified production of plant growth regulators, enhanced solubilization of nutrient for the ease in their uptake and hydration, and imparted resistance against pathogens (Boostani et al., 2014).

7.2.4.3 Exogenous Application of Different Chemicals

Under combined salinity and drought stress, application of 50 μM SA or 10 mM KNO3 ameliorated oxidative stress in barley by maintaining low MDA and Na+/K+ ratio in leaves (Fayez and Bazaid, 2014). Li et al. (2008) treated barley leaves with 50 μM SNP (sodium nitroprusside), an nitric oxide (NO) donor, lessened the harm of salt stress, imitated by reduced ion leak, and carbonyl, MDA (malondialdehyde), and H_2O_2 (hydrogen peroxide) content. The existence of a nitric oxide donor enhanced the actions of ascorbate peroxidases (APXs), Catalase, and Superoxide dismutase. Meanwhile, the addition of sodium nitroprusside (SNP) enhanced ferritin accretion at the protein level, which indicates that NO directly controlled ferritin accumulation. These findings concluded that NO can efficiently buffer germination from the harm of salt stress by improving the actions of antioxidant enzymes to reduce the extreme reactive O_2 (oxygen) species triggered by salt stress through enhancing the accumulation of ferritin to chelate a greater number of Fe^{+2} (ferrous ions). A comparative study showed that salicylic acid (SA) performed best with greater amelioration effects compared to cycocel and jasmonic acid under salinity conditions (Pakar et al., 2016). Application of synthetic zeolite with 5% significantly ameliorated salinity stress and increased concentration of calcium (Ca^{2+}), iron (Fe^{2+}), and manganese (Mn^{2+}) by about 5%, 19%, and 10%, respectively, with improved nutrient balance and water holding capacity (Al-Busaidi et al., 2008).

7.2.4.4 Genetic Approach

Barley crops can survive with various types of stresses with the involvement of genes. The natural tolerance of barley to abiotic stresses is mainly due to its quick flower initiation, which guarantees that good production of grain or seed, maturation, and pollination happens in an optimal time duration. The most important genes which originate in barley crops that affect the duration of flowering are mostly associated with circadian clock vernalization and photoperiod (Gutiérrez et al., 2011a, 2011b), such as circadian clock gene Ppd-H1, which controls photoperiod-related output genes, whereas HvLux1 and HvCEN control the duration of flowering (Campoli et al., 2012; Gürel et al., 2016).

7.3 COMMON WEEDS IN BARLEY AND THEIR MANAGEMENT

7.3.1 Introduction

A weed is any undesirable, unpleasant, unwanted, prolific, noxious, persistent, and harmful plant, or it is a plant that grows by itself in the field. Crop yield and quality are the main objectives of cultivation. New production technology is used to achieve these objectives. Weed control is part of the increasing yield and quality of produce. Weed losses are not visible, but they case huge losses. Every year weeds cause damage of 6–7 billion rupees to agricultural produce. Weeds cause more yield losses (50–60%) in Kharif than rabi season due to rainfall, which hinders intercultural practices (www.slideshare.net/anilbhosale13/agro-359-weed-notes1).

Common weeds of barley include the following.

7.3.2 Broad-Leaved Weeds

7.3.2.1 Wild Radish

Wild radish (*Raphanus raphanistrum*) is inexpensive and may cause damage to yield up to 10–90%. If eaten by animals, it may cause health problems for them. It is a substitute host for several diseases and pests. It is a winter- and spring-growing annual plant. It may develop up to 1.5 meters tall. Its heart-shaped cotyledons are hairless, bearing longer stems. The first young leaf has irregular lobes at edges, and the base of the leaf blade has fully developed lobes.

7.3.2.1.1 Management

7.3.2.1.1.1 Hoeing It is difficult to remove wild radish when it is fully established in an area. Dry season care ought to be confined to defiled enclosures and careful checking of harvest seeds and produce entering clean regions. Different plants can be evacuated by hand before seeding.

Suppose rains in summer result in the germination of wild radish. At that point, shower with 750 mL/ha Tigrex when the plants are small and allow sheep to feed on the plants that are not exposed to wild radish seed for 14 days. If it is not possible to allow sheep to feed, shower with 500 mL/ha of glyphosate (450g/L) in addition to 10g/ha of Eclipse in addition to 1% oil.

7.3.2.1.1.2 Chemical Control Initial control is necessary to get the highest yield in barley. A spray of diflufenican-based products (e.g., Brodal) at the three-leaf stage of barley gives four to five times higher yield than spray at the tillering stage. It will inhabit the soil for longer periods and destroy the weeds, which grow at a later stage. Metosulam (e.g., Eclipse) plus oil should be used to kill weeds before flowering, while triasulfuron (e.g., Logran) plus oil should be used after flowering. At the hard batter phase of grains (barley), 2,4-D may be used to diminish the germination ability of seed, and radish plants should be buried in the soil so the collector may ignore and move over them. Green radish can be dried up with diquat before harvesting.

7.3.2.2 Doublegee

It is an annual herb with a steady taproot and a long, soft, and smooth stem. The cotyledons are also smooth, long, and club shaped. Subsequent leaves are alternate, smooth, and triangular with undulating margins. During the initial stages of growth, leaves are egg shaped and form a flat rosette, but in the well-established crop, leaves may be semi-erect. The stem is circular and arises from the rosette. It can grow up to 600 mm. White, small, and subtle flowers are produced in clusters. Each flower bears hard woody achenes, with three sharp spines arising from the top of the stem (Western Australian Department of Agriculture and Food, 2006).

TABLE 7.8
Strategies for Integrated Weed Management of Doublegee

Strategy	% control	Range	Comments
Crop choice and sequence	80	0–95	Cheaper and easier to control in cereals. Avoid crops that don't have a good herbicide control option
Herbicide-tolerant crops	90	50–95	Very useful for the non-cereal phase of rotations
Inversion plowing	90	80–99	Use once on intractable infestations only and then don't deep cultivate for many years
Autumn tickle	40	20–60	Depends on seasonal break. More effective when used in conjunction with a follow-up herbicide treatment or cultivation
Knockdown (non-selective) herbicides for fallow and pre-sowing control	75	50–80	Use robust rates
Pre-emergent herbicides	75	50–80	It can be variable depending on the season. Subsequent crop choice may be limited after treatment
Selective post-emergent herbicides	90	70–95	Spray when young and actively growing. Repeat if required
Grazing—actively managing weeds in pastures	70	50–90	Doublegee is palatable to the stock until the formation of the spiny achenes. Useful for suppression and reduction of seed production, enabling favorable pasture species to compete actively

7.3.2.3 Capeweed

Capeweed (*Arctotheca calendula*) is an annual herb, has no stem, and grows horizontally on land. It germinates in autumn and winter. Its cotyledons are hairless and club shaped. The first two leaves develop in pairs. Leaves are spear-shaped and have semicircles on the margins. Young leaves develop separately, are highly lobed, and are round at the apex. Leaves store water to withstand xeric conditions and have white hairs on both sides. Capeweed seeds are easily dispersed by wind between fields. It is widely distributed in meadows (Western Australian Department of Agriculture and Food, 2006).

7.3.2.3.1 Management

Prosulfuron is a sulfonylurea herbicide that kills some annual broadleaved weeds (OMA, 2013; Senseman, 2007). Sulfonylurea herbicides prevent the synthesis of branched-chain amino acids. These amino acids are valine, leucine, and isoleucine (Senseman, 2007). Protein synthesis of broadleaved weeds does not take place in the absence of these amino acids. Vulnerable plants are unable to continue their growth and eventually die. The application of low doses of sulfonylurea herbicides is effective to control broadleaved weeds (Senseman, 2007). Roots and shoots of vulnerable plants absorb sulfonylurea quickly; then, it moves throughout the plant. Vulnerable plants show reduced growth and a red to purple color of leaves showing necrosis and chlorosis. Symptoms first appear in leaves and then move on to the whole plant (OMA, 2013; Senseman, 2007). Sulfonylurea herbicides are environmentally friendly and less toxic to aquatic species, birds, and mammals (OMA, 2013; Senseman, 2007; Soltani et al., 2014)

7.3.3 Grass Weeds

7.3.3.1 Wild Oats

Wild oats (*Avena fatua* and *A. ludoviciana*) represent a high cost to cropping. Wild oats are highly competitive and, when left uncontrolled, can reduce wheat yields by up to 80%. The highest yield loss occurs when wild oat plants emerge at the same time as the crop. Wild oats have a large ligule with no auricles, and the leaves tend to be hairy with a slightly bluish hue. The emerging sheet is rolled around the barley seedlings anticlockwise. Seeds are dark but may be cream colored. Wild oats have a growth habit of growing in distinct patches with plant density ranging from low to moderate (up to 100 plants/m^2). At the seedling stage, it resembles bromegrass. It has rolled sheets, and a smaller number of hairs on leaves and brome species also have hairy leaf sheets of tubular shape.

7.3.3.1.1 Management

Integrated weed management, including cultural practices, non-chemical strategies, and herbicides, is necessary to manage wild oats efficiently. Post-emergent herbicides (Group A, fops and dims; Group B, sulfonylurea; and Group Z, flamprop-m-methyl) should be used in barley only after checking their resistance level against wild oats. Group B (imidazolinone) class herbicides are still useful for the control of wild oats.

Winter fallow and crop rotation with summer crops are also effective in controlling wild oats. Before the seed shedding of wild oats, cutting for forage or brown manuring is also useful in the control of weeds.

7.3.3.2 Annual Ryegrass

Annual ryegrass (*Lolium rigidum*) is smooth, lush green with narrow leaves and a shiny backside of the leaf blade. Each leaf has a broader ligule, large auricles, and an emerging leaf. At the base, leaves are reddish-purple and release a transparent fluid when crushed. Mature plants can grow straight up to 900 mm, and flowering stems are flat with 300 mm length. Each spikelet bears three to nine flowers. Flat seeds are 1 mm wide, 4–6 mm long, and straw colored. Seed embryos are frequently visible through external layers. They are held safely to the flower stem, and huge power is required to isolate them either as individual seeds or as a component of the flower stem.

7.3.3.2.1 Management

Barley, due to its ability of more tillering and underground root competition, is a stronger competitor with weeds than wheat, canola, and pulses. Barley was commonly unaffected by ryegrass in earlier-field examination, and it limits the seed production of ryegrass more effectively than corn or peas. Barley yields were not influenced by sowing time, but early sowing improves the grain quality. Crop rotation with barley will be progressively fruitful in diminishing ryegrass densities. Collection of seeds of ryegrass at harvest time reduces the number of weed seeds in the soil available for germination in the next season. It may also reduce the chances of invasion by other weeds. Burning of stubble does not affect the viability of ryegrass seeds significantly, but it can be used as a management strategy to weed control.

REFERENCES

Achon MA, Marsinach M, Ratti C and Rubies-Autonell C (2005) First report of barley yellow mosaic virus in barley in Spain. *Plant Disease* 89(1): 105–105.

Achon MA, Ratti C and Rubies-Autonell C (2003) Occurrence of barley mild mosaic virus on barley in Spain. *Plant Disease* 87(8): 1004–1004.

Adams MJ, Swaby AG and Jones P (1987) Occurrence of two strains of barley yellow mosaic virus in England. *Plant Pathology* 36(4): 610–612.

Adams MJ, Swaby AG and Macfarlane I (1986) The susceptibility of barley cultivars to barley yellow mosaic virus (BaYMV) and its fungal vector, *Polymyxa graminis*. *Annals of Applied Biology* 109(3): 561–572.

AHDB (2020) *Wheat and Barley Disease Management Guide*. Agriculture and Horticulture Development Board, UK. Available at https://ahdb.org.uk/cereal-dmg (Accessed on 29th July, 2020).

Ahmad S (1956) Ustilaginales of West Pakistan. *Mycol. Pap.* 64: 1–17.

Ahmad S, Iqbal SH and Khalid AN (1997) *Fungi of Pakistan*. Sultan Ahmad Mycological Society of Pakistan, 248 pages.

Ahmad Z, Anjum S, Waraich EA, Ayub MA, Ahmad T, Tariq RMS, Ahmad R and Iqbal MA (2018) Growth, physiology, and biochemical activities of plant responses with foliar potassium application under drought stress—a review. *Journal of Plant Nutrition*. https://doi.org/10.1080/01904167.2018.1459688.

Ahmad Z, Ejaz AW, Celaleddin B, Akbar H and Murat E (2020) Enhancing drought tolerance in wheat through improving morpho-physiological and antioxidants activities of plants by the supplementation of foliar silicon. https://doi.org/10.32604/phyton.2020.09143.

Ahmad Z, Ejaz AW, Rashid A and Muhammad S (2017) Modulation in Water relations, chlorophyll contents and antioxidants activity of maize by foliar phosphorus application under drought stress. *Pakistan Journal of Botany* 49(1): 11–19.

Akhkha A (2009) Barley powdery mildew (*Blumeria graminis* f. sp. hordei): Interaction, resistance and tolerance Egypt. *J. Exp. Biol. Bot.* 5: 1–2.

Akhtar MA and Aslam M (1985) Bacterial stripe of wheat in Pakistan. *Rachis.* 4: 49.

Akhtar MA and Aslam M (1986) *Xanthomonas campestris* pv. *undulosa* on wheat. *Rachis.* 5: 34–37.

Al-Busaidi A, Yamamoto T, Inoue M, Eneji AE, Mori Y and Irshad M (2008) Effects of zeolite on soil nutrients and growth of barley following irrigation with saline water. *Journal of Plant Nutrition.* https://doi.org/10.1080/01904160802134434.

Alizadeh A, Barrault G, Sarrafi A, Rahimian H and Albertini L (1995) Identification of bacterial leaf streak of cereals by their phenotypic characteristics and host range in Iran. *Europ. J. Plant Pathol.* 101: 225–229.

Allen M, Hart WH and Baghot K (1970) Crop rotation controls barley root knot nematode at Tulelake. Available at http://calag.ucanr.edu/download_pdf.cfm?article=ca.v024n07p4 (Retrieved 30th July 2020).

Al-Sayaydeh R, Al-Bawalize A, Al-Ajlouni Z, Akash MW, Abu-Elenein J and Al-Abdallat AM (2019) Agronomic evaluation and yield performance of selected barley (*Hordeum vulgare* L.) landraces from Jordan. *International Journal of Agronomy.* https://doi.org/10.1155/2019/9575081.

Alvarez MG (1976) Primer catalogo de enfermedades de plantas Mexicanas. *Fitofilo* 71: 1–169.

Ashraf M (2010) Inducing drought tolerance in plants: Recent advances. *Biotechnology Advances.* https://doi.org/10.1016/j.biotechadv.2009.11.005.

Bakkeren G, Kronstad J and Lévesque CA (2000) Comparison of AFLP fingerprints and ITS sequences as phylogenetic markers in *Ustilaginomycetes. Mycologia* 92: 510–521.

Boesewinkel HJ (1979) Erysiphaceae of New Zealand. *Sydowia* 32: 13–56.

Boewe GH (1964) Some plant diseases new to Illinois. *Plant Disease Reporter* 48: 866–870.

Boostani HR, Mostafa C, Abdol AM and Naeimeh E (2014) mechanisms of plant growth promoting rhizobacteria (PGPR) and *Mycorrhizae* fungi to enhancement of plant growth under salinity stress: A review. *Scientific Journal of Biological Sciences.* https://doi.org/10.14196/SJBS.V3I10.1262.

Bradbury JF (1986) *Guide to Plant Pathogenic Bacteria.* Farnham House, Slough, UK: Mycological Institute, CAB International, 332 pp.

Bragard C, Verdier V and Maraite H (1995) Genetic diversity among *Xanthomonas campestris* strains pathogenic for small grains. *Appl. Environ. Microbio.* 61: 1020–1026.

Braun U (1995) The powdery mildews (*Erysiphales*) of Europe. Gustav Fischer Verlag, 337 pages.

Bruehl GW (1953) *Pythium Root Rot of Barley and Wheat.* Technical Bulletin No. 1084. USA: USDA.

Burton GJL (1931) Annual report of the senior plant breeder 1931. *Kenya Dept. Agric. Ann. Rep.* 176–209.

Campoli C, Shtaya M, Davis SJ and Korff MV (2012) Expression conservation within the circadian clock of a monocot: Natural variation at barley Ppd-H1 affects circadian expression of flowering time genes, but not clock orthologs. *BMC Plant Biology.* https://doi.org/10.1186/1471-2229-12-97.

Carroll TW and Mayhew DE (1976) Occurrence of virions in developing ovules and embryo sacs of barley in relation to the seed transmissibility of barley stripe mosaic virus. *Canadian Journal of Botany* 54(21): 2497–2512.

Castanares E, Dinolfo MI, Del P, Pan D and Stenglein SA (2016) Species composition and genetic structure of *Fusarium graminearum* species complex populations affecting the main barley growing regions of South America. *Pl. Pathol.* 65: 930–939.

Chay CA, Smith DM, Vaughan R and Gray SM (1996) Diversity among isolates within the PAV serotype of barley yellow dwarf virus. *Phytopathology* 86(4): 370–377.

Chloupek O, Hrstková P and Jurecka D (2003) Tolerance of barley seed germination to cold- and drought-stress expressed as seed vigour. *Plant Breeding.* https://doi.org/10.1046/j.1439-0523.2003.00800.x.

Cho WD and Shin HD (2004) *List of Plant Diseases in Korea.* 4th ed. Korean Society of Plant Pathology, pp. 779.

Choudhury FK, Rosa MR, Blumwald E and Mittler R (2017) Reactive oxygen species, abiotic stress and stress combination. *Plant Journal.* https://doi.org/10.1111/tpj.13299.

Choupannejad R, Sharifnabi B, Fadaei T and Gholami J (2017) *Rhizoctonia solani* AG4 associated with foliar blight symptoms on barley in Iran. *Australasian Plant Disease Notes* 12: 2.

CIMMYT (1977) Israel. In *CIMMYT Report on Wheat Improvement*, Mexico, DF, pp. 237.

Collins NC, Paltridge NG, Ford CM and Symons RH (1996) The Yd2 gene for barley yellow dwarf virus resistance maps close to the centromere on the long arm of barley chromosome 3. *Theoretical and Applied Genetics* 92(7): 858–864.

Cook RP and Dubé AJ (1989) *Host-pathogen Index of Plant Diseases in South Australia.* South Australian Department of Agriculture, pp. 1–142.

Crous PW, P Plumb ips AJL and Baxter AP (2000) *Phytopathogenic Fungi from South Africa.* University of Stellenbosch, Department of Plant Pathology Press, 358.

D'arcy CJ (1995) Symptomatology and host range of barley yellow dwarf. *Barley Yellow Dwarf* 40: 9–28.

Dahal G, Amatya P and Manandhar H (1992) Plant diseases in Nepal. *Rev. Pl. Pathol.* 71: 797–807.

Dawson IK, Joanne R, Wayne P, Brian S, Thomas WTB and Robbie W (2015) Barley: A translational model for adaptation to climate change. *New Phytologist.* https://doi.org/10.1111/nph.13266.

Dean GJ (1974) Effects of parasites and predators on the cereal aphids *Metopolophium dirhodum* (Wlk.) and *Macrosiphum avenae* (F.) (Hem., Aphididae). *Bulletin of Entomological Research* 63(3): 411–422.

Demir G and Üstün N (1992) Studies on bacterial streak disease (*Xanthomonas campestris* pv. *translucens* (Jones et al.) dye of wheat and other gramineae. *J. Turk. Phytopathol.* 21: 33–40.

Denchev CM (2001) Class Ustomycetes (Orders Tilletiales, Ustilaginales, and Graphiolales). *Fungi Bulgaricae* 4: 1–286.

Dubcovsky J, María GS, Epstein E, Luo MC and Dvořák J (1996) Mapping of the K+/Na+ discrimination locus Kna1 in Wheat. *Theoretical and Applied Genetics.* https://doi.org/10.1007/BF00223692.

Ershad D (2001) Smut fungi reported from Iran. *Rostaniha* (Bot. J. Iran) 1 (Suppl).

Fakirova VI (1991) Fungi Bulgaricae, 1 tomus, ordo Erysiphales. In Aedibus Academiae Scientiarum Bulgaricae Serdicae 154.

FAO (2015) Food Outlook. *Global Information and Early Warning System on Food and Agriculture.* https://doi.org/10.1044/leader.PPL.19102014.18.

Farr DF and Rossman AY (2020) *Fungal Databases*, U.S. National Fungus Collections, ARS, USDA. Available at https://nt.ars-grin.gov/fungaldatabases/ (Accessed on 31 July 2020).

Farrell JA and Stufkens MW (1990) The impact of *Aphidius rhopalosiphi* (Hymenoptera: Aphidiidae) on populations of the rose grain aphid (*Metopolophium dirhodum*) (Hemiptera: Aphididae) on cereals in Canterbury, New Zealand. *Bulletin of Entomological Research* 80(4): 377–383.

Fayez KA and Bazaid SA (2014) Improving drought and salinity tolerance in barley by application of salicylic acid and potassium nitrate. *Journal of the Saudi Society of Agricultural Sciences* 13(1): 45–55. https://doi.org/10.1016/j.jssas.2013.01.001.

Forster BP, Ellis RP, Moir J, Talame V, Sanguineti MC, Tuberosa R, This D, et al. (2004) Genotype and phenotype associations with drought tolerance in barley tested in North Africa. *Annals of Applied Biology*. https://doi.org/10.1111/j.1744-7348.2004.tb00329.x.

French AM (1989) California Plant Disease Host Index. Sacramento: Calif. Dept. Food Agric., 394 pages.

Frommel MI (1986) *Xanthomonas campestris pv.* translucens, causal agent of bacterial streak of wheat (*Triticum aestivum*). Montevideo, Uruguay, Dirección de Sanidad Vegetal (in Spanish).

Gannibal FB (2004) Small-spored species of the genus *Alternaria* on grasses. *Mikol. Fitopatol* 38: 19–28.

Gautam AK and Avasthi S (2018) Diversity of powdery mildew fungi from north western Himalayan Region of Himachal Pradesh—a checklist. *Pl. Pathol & Quarantine* 8(1): 78–99.

Giatgong P (1980) *Host Index of Plant Diseases in Thailand.* 2nd ed. Bangkok, Thailand: Mycology Branch, Plant Pathology and Microbiology Division, Department of Agriculture and Cooperatives, 118.

Ginns JH (1986) Compendium of plant disease and decay fungi in Canada 1960–1980. *Res. Br. Can. Agric. Publ.* 1813: 416.

Gorter GJMA (1977) Index of plant pathogens and the diseases they cause in cultivated plants in South Africa. *Republic South Africa Dept. Agric. Techn. Serv. Pl. Protect. Res. Inst. Sci. Bull.* 392: 1–177.

Gray SM, Bergstrom GC, Vaughan R, Smith DM and Kalb DW (1996) Insecticidal control of cereal aphids and its impact on the epidemiology of the barley yellow dwarf luteoviruses. *Crop Protection* 15(8): 687–697.

Guo L (1988) The genera *Ustilago* and *Liroa* in China. *Mycosystema* 1: 211–240.

Gürel F, Öztürk ZN, Uçarlı C and Rosellini D (2016) Barley genes as tools to confer abiotic stress tolerance in crops. *Frontiers in Plant Science*. https://doi.org/10.3389/fpls.2016.01137.

Gutiérrez L, Alfonso CM, Ariel J, Castro, Jarislav VZ, Mark S and Patrick MH (2011a) Association mapping of malting quality quantitative trait loci in winter barley: Positive signals from small germplasm arrays. *The Plant Genome*. https://doi.org/10.3835/plantgenome2011.07.0020.

Gutiérrez L, Marcos AC, Castro AJ, Von Zitzewitz J, Schmitt M and Hayes PM (2011b) Association mapping of malting quality quantitative trait lci in winter barley: Positive signals from small germplasm arrays. *The Plant Genome*. https://doi.org/10.3835/plantgenome2011.07.0020.

Halbert S and Voegtlin D (1995) Biology and taxonomy of vectors of barley yellow dwarf viruses. *Barley Yellow Dwarf* 40: 217–258.

Hare PD, Cress WA and Van Staden J (1998) Dissecting the roles of osmolyte accumulation during stress. *Plant, Cell and Environment*. https://doi.org/10.1046/j.1365-3040.1998.00309.x.

Hayes JE, Pallotta M, Garcia M, Öz MT, Rongala J and Sutton T (2015) Diversity in boron toxicity tolerance of Australian barley (*Hordeum vulgare* L.) genotypes. *BMC Plant Biology*. https://doi.org/10.1186/s12870-015-0607-1.

Henry M and Dedryver CA (1991) Occurrence of barley yellow dwarf virus in pastures of western France. *Plant Pathology* 40(1): 93–99.

Hernandez-Restrepo M, Groenewald JZ, Elliott ML, Canning G, McMillan VE and Crous PW (2016) Take-all or nothing. *Stud. Mycol.* 83: 19–48.

Hill SA and Fivans EJ (1980) Barley yellow mosaic virus. *Plant Pathology* 29: 197–199.

Hirschhorn E (1986) *Las ustilaginales de la flora Argentina*. Provincia de Buenos Aires: Comision de Invest. Cientificas, pp. 530.

Ingvordsen CH, Gunter B, Michael FL, Pirjo PS, Jens DJ, Marja Jalli, Ahmed J, et al. (2015) Significant decrease in yield under future climate conditions: Stability and production of 138 spring barley accessions. *European Journal of Agronomy*. https://doi.org/10.1016/j.eja.2014.12.003.

Jeżewska M and Trzmiel K (2009) First report of barley yellow mosaic virus infecting barley in Poland. *Plant Pathology* 58(4).

Karatygin IV (2012) Smut fungi in European part of Russia. A preliminary checklist. *Mikol. Fitopatol.* 46: 41–53.

Katis N, Tzavella-Klonari K and Adams MJ (1997) Occurrence of barley yellow mosaic and barley mild mosaic bymoviruses in Greece. *European Journal of Plant Pathology* 103(3): 281–284. Available at https://link.springer.com/article/10.1023/A:1008689308869.

Korobko AP, Wondimagegne E and Anisimoff BV (1985) *Bacterial Stripe and Black Chaff of Wheat in Ethiopia*. Ambo, Ethiopia: Scientific Phytopathological Laboratory, pp. 5.

Kristensen R, Torp M, Kosiak B and Holst-Jensen A (2005) Phylogeny and toxigenic potential is correlated in *Fusarium* species as revealed by partial translation elongation factor 1 alpha gene sequences. *Mycol. Res.* 109: 173–186.

Lee KJ, Choi MK, Lee WH and Rajkumar M (2006) Molecular analysis of Korean isolate of barley yellow mosaic virus. *Virus Genes* 32(2): 171–176.

Li QY, Hong BN, Jun Y, Meng BW, Hong BS, De-Zhi D, Xiao XC, Jiang PR and Yong CL (2008) Protective role of exogenous nitric oxide against oxidative-stress induced by salt stress in barley (*Hordeum vulgare*). *Colloids and Surfaces B: Biointerfaces*. https://doi.org/10.1016/j.colsurfb.2008.04.007.

Loiveke H (2006) Incidence of *Fusarium* spp. on several field crops in Estonia and their toxicity towards *Bacillus stearothermophilus*. *Agron. Res.* 4: 273–280.

Mahmood K (2011) Salinity tolerance in barley (*Hordeum vulgare* L.): Effects of varying NaCL, K+/Na+ and NaHCO3 levels on cultivars differing in tolerance. *Pakistan Journal of Botany*.

Mamluk OF, Al-Ahmed M and Makki MA (1990) Current status of wheat diseases in Syria. *Phytopath. Medit.* 29: 143–150.

Mathre, D (1989) Pathogenicity of an isolate of *Verticullium dahliae* from barley. 1st ed. [ebook] Bozeman: Department of Pathology, Montana State University, pp. 4.

McKenzie EHC and Vanky K (2001) Smut fungi of New Zealand: An introduction, and list of recorded species. *New Zealand J. Bot.* 39: 501–515.

McKirdy SJ and Jones RAC (1996) Use of imidacloprid and newer generation synthetic pyrethroids to control the spread of barley yellow dwarf luteovirus in cereals. *Plant Disease* 80(8): 895–901.

McKirdy SJ and Jones RAC (1997) Effect of sowing time on barley yellow dwarf virus infection in wheat: Virus incidence and grain yield losses. *Australian Journal of Agricultural Research* 48(2): 199–206.

Mehta YR (1990) Management of *Xanthomonas campestris* pv. *undulosa* and *hordei* through cereal seed testing. *Seed Sci. Tech.* 18: 467–476.

Mendes MAS, da-Silva VL, Dianese JC, et al. (1998) Fungos em Plants no Brasil. Brasilia: Embrapa-SPI/Embrapa-Cenargen, 555 pages.

Modhej A, Farhoudi R and Afrous A (2015) Effect of post-anthesis heat stress on grain yield of barley, durum and bread wheat genotypes. *Journal of Scientific Research and Development.*

Mohan SK and Mehta YR (1985) Studies on *Xanthomonas campestris* pv. *undulosa* in wheat and triticale in Paraná State. *Fitopatologia Brasileira* 10: 447–453.

Mulenko W, Majewski T and Ruszkiewicz-Michalska M (2008) A Preliminary Checklist of Micromycetes in Poland. W. Szafer Institute of Botany, Polish Academy of Sciences 9: 752.

Munns R (2002) Comparative physiology of salt and water stress. *Plant, Cell and Environment.* https://doi.org/10.1046/j.0016-8025.2001.00808.x.

Nattrass RM (1961) Host lists of Kenya fungi and bacteria. *Mycol. Pap.* 81: 1–46.

Neate S and McMullen M (2005) *Barley Disease Handbook.* USA: Department of Plant pathology, North Dakota State University.

Nganje WE, Kaitibie S, Wilson WW, Leistritz FL and Bangsund DA (2004) Economic impacts of fusarium head blight in wheat and barley: 1993–2001. Agribusiness and Applied Economics Report No 538. North Dakota State University, USA.

Noble RJ (1935) Australia: Notes on plant diseases recorded in New South Wales for the year ending 30th June 1935. *Int. Bull. Plant Prot.* 12: 270–273.

Ontario Ministry of Agriculture (OMA) (2013) Food, and Rural Affairs, Guide to Weed Control, Publication 75, Ontario Ministry of Agriculture, Food, and Rural Affairs, Toronto, Canada.

Pakar N, Pirasteh-Anosheh H, Emam Y and Pessarakli M (2016) Barley growth, yield, antioxidant enzymes, and ion accumulation affected by PGRs under salinity stress conditions. *Journal of Plant Nutrition.* https://doi.org/10.1080/01904167.2016.1143498.

Park JC, Noh TH, Kim MJ, Lee SB, Park CS, Kang CS and Kim TS (2010) Effect of cropping system on disease incidence by soil-borne bymovirus in barley and on density of the vector, *Polymyxa graminis. Research in Plant Disease* 16(2): 115–120.

Paul YS and Thakur VK (2006) *Indian Erysiphaceae.* Jodhpur: Scientific Publishers (India), pp. 134.

Pennycook SR (1989) *Plant diseases recorded in New Zealand.* 3 Vol. Auckland: Pl. Dis. Div., D.S.I.R.

Plaza-Bonilla D, Cantero-Martínez C, Bareche J, Arrúe JL, Lampurlanés J and Álvaro-Fuentes J (2017) Do no-till and pig slurry application improve barley yield and water and nitrogen use efficiencies in rainfed Mediterranean conditions? *Field Crops Research.* https://doi.org/10.1016/j.fcr.2016.12.008.

Plumb RT, Lennon EA and Gutteridge RA (1986) The effects of infection by barley yellow mosaic virus on the yield and components of yield of barley. *Plant Pathology* 35(3): 314–318.

Poursafar A, Ghosta Y, Orina AS, Gannibal PB, Javan-Nikkhah M and Lawrence DP (2018) Taxonomic study on *Alternaria* sections *Infectoriae* and *Pseudoalternaria* associated with black (sooty) head mold of wheat and barley in Iran. *Mycol. Progr.* 17(3): 343–356.

Rehman S, Gyawali S, Amri A and Verma RPS (2020) First report of spot blotch of barley caused by *Cochliobolus sativus* in Morocco. *Plant Disease* 104(3): 988.

Richardson MJ (1990) *An Annotated List of Seed-Borne Diseases.* 4th ed. Zurich: International Seed Testing Association, 387+ pages.

Richardson MJ and Waller JM (1974) Triticale diseases in CIMMYT trial locations. In *Triticale: Proc. Int. Symp.*, El Batan, Mexico, Monograph 024e. Ottawa, Canada: International Development Research Center, pp. 193–199.

Roane CW and Roane MK (1994) Graminicolous fungi of Virginia: Fungi associated with Cereals. *Virginia J. Sci.* 45: 279–296.

Rusanov VA and Bulgakov TS (2008) Powdery mildew fungi of Rostov region. *Mikol. Fitopatol.* 42: 314–322.
Sainju UM, Lenssen AW and Barsotti JL (2013) Dryland malt barley yield and quality affected by tillage, cropping sequence, and nitrogen fertilization. *Agronomy Journal*. https://doi.org/10.2134/agronj2012.0343.
Sakata T, Takahashi H, Nishiyama I and Higashitani A (2000) Effects of high temperature on the development of pollen mother cells and microspores in barley *Hordeum vulgare* L. *Journal of Plant Research*. https://doi.org/10.1007/pl00013947.
Samarah N and Alqudah A (2011) Effects of late-terminal drought stress on seed germination and vigor of barley (*Hordeum vulgare* L.). *Archives of Agronomy and Soil Science*. https://doi.org/10.1080/03650340903191663.
Samarah NH (2005) Effects of drought stress on growth and yield of barley. *Agronomie*. https://doi.org/10.1051/agro:2004064.
Sampson PJ and Walker J (1982) *An Annotated List of Plant Diseases in Tasmania*. Department of Agriculture Tasmania, 121 pages.
Sánchez-Díaz M, García JL, Antolín MC and Araus JL (2002) Effects of soil drought and atmospheric humidity on yield, gas exchange, and stable carbon isotope composition of barley. *Photosynthetica*. https://doi.org/10.1023/A:1022683210334.
Sarbhoy AK and Agarwal DK (1990) *Descriptions of Tropical Plant Pathogenic Fungi*. Set 1. New Delhi: Malhotra Publ. House.
Savin R, Stone PJ and Nicolas ME (1996) Responses of grain growth and malting quality of barley to short periods of high temperature in field studies using portable chambers. *Australian Journal of Agricultural Research*. https://doi.org/10.1071/AR9960465.
Schaller CW, Qualset CO and Rutger JN (1964) Inheritance and linkage of the Yd2 gene conditioning resistance to the barley yellow dwarf virus disease in barley 1. *Crop Science* 4(5): 544–548.
Schilling AG, Moller EM and Geiger HH (1995) Molecular differentiation and diagnosis of the cereal pathogens *Fusarium culmorum* and *F. graminearum*. *Sydowia* 48: 71–82.
Senseman SA (2007) *Herbicide Handbook*. 9th ed. Champaign, Ill, USA: Weed Science Society of America, 2007.
Shao H, Wang H and Tang X (2015) NAC transcription factors in plant multiple abiotic stress responses: Progress and prospects. *Frontiers in Plant Science*. https://doi.org/10.3389/fpls.2015.00902.
Shin HD (2000) *Erysiphaceae of Korea*. Suwon, Korea: National Institute of Agricultural Science and Technology, 320 pages.
Shivas RG (1989) Fungal and bacterial diseases of plants in Western Australia. *J. Roy. Soc. W. Australia* 72: 1–62.
Simmonds JH (1966) *Host Index of Plant Diseases in Queensland*. Brisbane: Queensland Department of Primary Industries, 111.
Singh SC (1968) Some parasitic fungi collected from Kathmandu Valley (Nepal). *Indian Phytopathol.* 21: 23–30.
Sivanesan A (1987) Graminicolous species of *Bipolaris*, *Curvularia*, *Drechslera*, *Exserohilum* and their teleomorphs. *Mycol. Pap.* 158: 1–261.
Smit IBJ and Van ABT (1988) Items from South Africa: International nurseries. *Ann. Wh. Newsl.* 34: 84.
Soltani N, Brown LR, Cowan T and Sikkema PH (2014) Weed management in spring seeded barley, oats, and wheat with prosulfuron. *Int. J. Agron.* 2014. https://doi.org/10.1155/2014/950923.
Sprague R (1950) *Diseases of Cereals and Grasses in North America*. New York: Ronald Press Company, pp. 538.

Stoll M, Begerow D and Oberwinkler F (2005) Molecular phylogeny of *Ustilago*, *Sporisorium*, and related taxa based on combined analyses of rDNA sequences. *Mycol. Res.* 109: 342–356.

Sun F and He L (1986) Studies on determinative techniques for resistance of wheat to black chaff (*Xanthomonas translucens f.* sp. *undulosa*). *Acta Phytophylacica Sinica* 13: 109–115.

Tai FL (1979) *Sylloge Fungorum Sinicorum*. Peking: Sci. Press, Acad. Sin., pp. 1527.

Teng SC (1996) *Fungi of China*. Ithaca, NY: Mycotaxon, Ltd., pp. 586.

Tessi JL (1949) *Current status of work on* Xanthomonas translucens *var.* cerealis. Presented at the 4th Wheat, Oats, Barley and Rye Meeting, Castellar, Argentina, pp. 200.

Turan M, Medine G, Ramazan Ç and Fikrettin Ş (2013) Effect of plant growth-promoting rhizobacteria strain on freezing injury and antioxidant enzyme activity of wheat and barley. *Journal of Plant Nutrition*. https://doi.org/10.1080/01904167.2012.754038.

Ugarte C, Calderini DF and Slafer GA (2007) Grain weight and grain number responsiveness to pre-anthesis temperature in wheat, barley and triticale. *Field Crops Research*. https://doi.org/10.1016/j.fcr.2006.07.010.

Vanky K (1994) *European Smut Fungi*. Stuttgart: Gustav Fischer Verlag, pp. 570.

Vanky K and Abbasi M (2013) Smut fungi of Iran. *Mycosphere* 4: 363–454.

Vanky K and McKenzie EHC (2002) *Smut Fungi of New Zealand. Fungi of New Zealand*, Vol. 2. Hong Kong: Fungal Diversity Press, pp. 256.

Vanky K and Shivas RG (2008) *Fungi of Australia. The smut fungi*. Canberra: ABRS; Melbourne: CSIRO Publishing, pp. 267.

Vanky K and Vanky C (2002) An annotated checklist of *Ustilaginomycetes* in Malawi, Zambia and Zimbabwe. *Lidia* 5: 157–176.

Vinocur B and Altman A (2005) Recent Advances in engineering plant tolerance to abiotic stress: Achievements and limitations. *Current Opinion in Biotechnology*. https://doi.org/10.1016/j.copbio.2005.02.001.

Voytyuk SO, Heluta, VP, Wasser, SP, Nevo E, Takamatsu S and Volz PA (2009) Biodiversity of the Powdery Mildew Fungi (Erysiphales, Ascomycota) of Israel: Vol. 7. Biodiversity of Cyanoprocaryotes, Algae and Fungi of Israel. Koeltz Scientific Books, pp. 290.

Wahid A, Gelani S, Ashraf M and Foolad MR (2007) Heat tolerance in plants: An overview. *Environmental and Experimental Botany*. https://doi.org/10.1016/j.envexpbot.2007.05.011.

Wangai AW (1990) Effects of barley yellow dwarf virus on cereals in Kenya. *World Perspectives on Barley Yellow Dwarf Virus*. Mexico, DF: CIMMYT, pp. 391–393.

Western Australian Department of Agriculture and Food (2006) Weeds and integrated weed management (IWM) options for barley production. Available at http:/Avww.agrie.wa, gov.au/content/FCP/CER/BAR/PW/barley weeds manag.htm.

Whiteside JO (1966) A revised list of plant diseases in Rhodesia. *Kirkia* 5: 87–196.

Wicke B, Smeets E, Dornburg V, Vashev B, Gaiser T, Turkenburg W and Faaij A (2011) The global technical and economic potential of bioenergy from salt-affected soils. *Energy and Environmental Science*. https://doi.org/10.1039/c1ee01029h.s.

Woudenberg JHC, Groenewald JZ, Binder M and Crous PW (2013) *Alternaria* redefined. *Stud. Mycol.* 75: 171–212.

Wright DG, Thomas GJ, Loughman R, Fuso-Nyarko J and Bullock S (2010) Detection of *Fusarium graminearum* in wheat grains in Western Australia. *Australasian Plant Disease Notes* 5: 82–84.

Xia JW, Sandoval-Denis M, Crous PW, Zhang XG and Lombard L (2019) Numbers to names—restyling the *Fusarium incarnatum-equiseti* species complex. Persoonia 43: 186–221.

Yosefvand M, Abbasi S, Chagha-Mirza K and Bahram-Nezhad S (2016) Genetic diversity analysis of *Gaeumannomyces graminis var. tritici* in Kermanshah Province of Iran using RAPD markers. *Pl. Pathol and Quarantine* 6(2): 207–218.

You Y and Shirako Y (2012) Influence of amino acid at position 132 in VPg on replication and systemic infection of barley yellow mosaic virus. *Virus Research* 166(1–2): 121–124.

You Y and Shirako Y (2013) Evaluation of host resistance to barley yellow mosaic virus infection at the cellular and whole-plant levels. *Plant Pathology* 62(1): 226–232.

Zambettakis C (1971) Recherches sur les Ustilaginales d'Afrique. *Bull. Soc. Mycol. France* 86: 305–692.

Zheng R and Yu Y (1987) *Flora Fungorum Sinicorum. Vol. 1. Erysiphales.* Beijing: Science Press, pp. 552.

Zheng T, Cheng Y, Chen JP, Antoniw JF and Carroll MJ (1999) The occurrence of barley mild mosaic virus (BaMMV) in China and the nucleotide sequence of its coat protein gene. *Journal of Phytopathology* 147(4): 229–234.

Zhong S and Steffenson BJ (2001) Virulence and molecular diversity in *Cochliobolus sativus.* Phytopathology 91: 469–476.

Zhuang WY (2001) *Higher Fungi of Tropical China.* Ithaca, NY: Mycotaxon, Ltd., 485 pages.

Zhuang WY (2005) *Fungi of Northwestern China.* Ithaca, NY: Mycotaxon, Ltd., 430 pages.

Zundel GL (1953) The Ustilaginales of the World. *Contr. Dept. Bot. Pennsylvania State Coll. School Agric.* 176: 1–410.

8 Rapeseed and Mustard
Diseases, Etiology, and Management

Muhammad Zunair Latif, Imran Ul Haq, Siddra Ijaz, and Anjum Faraz

CONTENTS

DOI: 10.1201/9781003055365-8

8.1 FUNGAL CAUSES

8.1.1 *Alternaria* Blight

Alternaria blight is one of the most important and destructive diseases of rapeseed and mustard; it is reported globally to cause considerable economic losses. Disease severity varies between years and sites depending on local seasonal conditions and is favored by warm, humid conditions during spring. The disease is also known as black spot, gray leaf spot, dark leaf spot, and pod drop. The fungus infects various plant parts, but the damage it causes to leaves affects the plant's photosynthetic potential and is associated with the premature shattering of infected pods (Nyvall, 2013; Thomma, 2003; Meena et al., 2016).

8.1.1.1 Cause

The disease is usually caused by *A. brassicae*, a dominant invasive species, and occasionally by *A. alternata*, *A. brassicola*, and *A. raphai.* The fungus, *A. brassicae*, can infect host species at all growth stages, including seed. The fungus can overwinter on crop debris, susceptible Brassica weeds, seed plants, or cuttings. Windborne conidia are important sources of initial crop infections. These conidia remain intact on plants, penetrate host tissues when they encounter sufficient moisture, and cause a lesion. Infected seeds may result in pre or post-emergence damping-off of seedlings. Seed infections and infestations with fungal propagules are the principal means of distribution of the pathogen. Conidia produced on these seedlings may cause infection of hypocotyls and cotyledons of surviving seedlings. The chief source of *Alternaria* infections includes epidermis penetration, stomatal openings, and mechanical wounding. *A. brassicae* enters into host plants through stomatal openings, while *A. brassicola* directly invades plant tissues. Fungal hyphae develop on the epidermis beneath the leaf waxes. The development of fungal infection structures and disease symptoms primarily depends upon incubation temperature and relative humidity. The optimum growth temperature for hyphal growth of *A. brassicae* and *A. brassicola* is 18–24°C and 20–30°C, respectively. Conidia produced on infected leaves are the secondary source of inoculum. Rain- and wind-transported fungal spores are also important means of infection. During harvesting and when removing infected debris from fields, conidia can spread up to 1800 m. The pathogen is influenced by weather conditions such as warm, humid, wet seasons and high rainfall conditions that favor disease development and lead to severe black spots. Minimum (8–12°C) and maximum (18–27°C) temperatures with an average relative humidity greater than 92% favor disease initiation on leaves, while maximum (20–30°C) temperature, nine hours of sunshine, and ten hours of leaf wetness are ideal for pod infection (Humpherson-Jones, 1989; Chattopadhyay and Bagchi, 1994; Mehta et al., 2002; Nyvall, 2013; Nowicki et al., 2012; Aneja and Agnihotri, 2016).

8.1.1.2 Morphology

A. brassicae: Fungal colonies on PDA medium are moderate, fast-growing, amphigenous, usually ash gray, fluffy, and circular in the beginning and dark greenish olive on maturity with abundant sporulation. Mycelium is septate, branched, and light brown at the initial stage, which becomes darker in later growth stages. Conidiophores are usually simple, arise in a group of two to ten with 0–8 septations and three to five geniculate conidiophores that are olive-brown, measuring 34.5–184.5 μm length and 4.7 μm diameter. The size of conidia is 86.4–240.5 × 15.5–30.0 μm. Beak is short, cylindrical, 10–130 × 3–8 μm and pale brown. *A. brassicicola*: Mycelium; septate, olive-gray to grayish black, conidiophores; olivaceous, septate, and branched measuring 35–45 × 5–8 μm, conidia; cylindrical to oblong and dark. Fungal colonies on synthetic media are black and sooty with distinct zonation. *A. brassicicola*: Mycelium is initially cottony whitish and turns greenish gray to dark olive in advanced stages, conidiophores are septate and olive-brown, 29–160 × 4–8 μm in size; conidia are muriform, olive-brown to dark and obclavate (Chadar et al., 2016; Kolte, 2018).

8.1.1.3 Host Range

Most *Alternaria* species are common saprophytes; some are plant pathogens, while others are well-known post-harvest pathogens. Pathogenic *Alternaria* spp. have a tremendous economic impact on various agronomic crops, including cereals, oilseed crops, ornamentals, vegetables such as broccoli, carrot, cauliflower, head cabbage, Chinese cabbage, and potato and fruits like apple, citrus, and tomato. Alternaria blight of mustard is distributed throughout Bangladesh, Canada, Europe, India, Mexico, Pakistan, Poland, and the United States (Thomma, 2003; Nowicki et al., 2012; Meena et al., 2016; Ahmed et al., 2018; Latif et al., 2018).

8.1.1.4 Symptoms

Disease symptoms appear on all above-ground parts and at any growth stage. Due to successful pathogen attack, small brown to black pin-head spots of varying sizes ranging from 0.5 cm to several cm in diameter appears on leaves, stems, and silique.

FIGURE 8.1 *Alternaria* blight lesions showing black fungal growth on the lower surface of the leaf.

These spots with characteristic concentric rings sometimes have a yellow chlorotic halo. Under favorable conditions, these lesions become covered with brownish to black downy-like bloom of germinating hyphae, providing diagnostic criteria for disease identification (Fig 8.1). Infected stems exhibit elongated lesions with pointed ends. Circular, dark brown-black lesions are also visible on siliquae. Over time, several lesions will coalesce, covering a large portion of the leaf and silique, resulting in leaf fall and eventual death. Infected pods may split open prematurely and develop small-sized shrunken seeds. Infected plants respond to infections via quick browning of cell walls, particularly in para-stomatal cells. Spots produced by *A. brassicola* are darker and less regular than those caused by *A. brassicae* (Nyvall, 2013; Nowicki et al., 2012; Chadar et al., 2016).

8.1.1.5 Management

Alternaria blight can be best managed by implementing suitable integrated disease management (IDM) strategies, including agricultural management practices. An IDM plan composed of seed treatment with propiconazole, balanced use of fertilizer, acceptable cultural practices such as removing the lower four leaves, and limited use of fungicides such as propiconazole effectively managed Alternaria blight in mustard. Agronomic practices such as timely sowing, balanced nutrition, suitable phytosanitary measures, and crop rotation are highly meaningful to prevent/reduce disease development. In Bangladesh, it was found that mustard crops planted before the first week of November can escape this disease. Soil mixing of cow dung and gypsum during tillage operations proved helpful to reduce disease incidence. Disease incidence can be significantly reduced either by using certified disease-free seeds or by pre-sowing seed treatment with fungicides. Timely application of fungicides, usually at the appearance of first disease symptoms, is a practical approach to minimize yield losses. Breeding programs must be introduced for the selection of resistant germplasm. Biological control agents like *Trichoderma harzianum*, *T. viridae*, *Pseudomonas fluorescens*, and bulb extract of *Allium sativum* help manage the disease. One percent garlic bulb extract used either as seed dresser or as foliar spray restricts disease incidence. There are several fungicides, including amistar, bion, captan, chlorothalonil, selected copper fungicides, Duter, fludioxonil, imazalil, iprodione, mancozeb, maneb, rovral, and thiram, that have varying degrees of efficacy against Alternaria species (Ayub et al., 1996; Ansari et al., 2004; Meena et al., 2011, 2004a; Nowicki et al., 2012; Singh et al., 2014; Ahmed et al., 2018; Gupta et al., 2018).

8.1.2 *Sclerotinia* Stem Rot

Sclerotinia stem rot, locally known as 'Polio disease' in India, is another important disease and a challenging threat to rapeseed and mustard production. The disease is also named stem blight, stalk break, stem canker, white blight, and white rot based on the disease's symptomology. The disease was previously of minor importance, but it has become the primary fungal disease in most rapeseed and mustard growing countries, especially in the Indian sub-continent. It results in heavy crop losses. The disease has caused severe yield and crop losses in Canada, China, Finland, Florida, Germany, India, and New Zealand. Infection occurs at different growth stages and on different plant parts, resulting in severe yield losses, up to 40–80% in some cases, and a significant reduction in oil content and quality. Disease losses may vary concerning the growth stage. For example, early flower infection will produce little or no seed, while infections at a late flowering stage may result in little yield reduction and a percent portion of plants infected. It has been commonly observed in temperate and sub-tropical regions with cool and wet seasons, but disease outbreaks may also occur in irrigated fields of drier areas and less commonly from semi-arid regions where unfavorable disease conditions prevail (Jamalainen, 1954; Kruger, 1976; Morrall et al., 1976; Roy and Saikia, 1976; Purdy, 1979; Chauhan et al., 1992; Kang and Chahal, 2000; Chattopadhyay et al., 2003; Ghasolia et al., 2004; Shukla, 2005; Saharan and Mehta, 2008; Bharti et al., 2019).

8.1.2.1 Cause

A necrotrophic fungus, *Sclerotinia sclerotiorum*, is responsible for this disease. It is a ubiquitous, soil-borne, and aggressive plant pathogenic fungus. It survives as sclerotia, a hard, asexual, resting structure. Mature sclerotia have three layers: 1) rind, the outer black protective layer composed of hyphae; 2) cortex, a thin intermediate region constituted by pseudoparenchymatous cells; and finally 3) medulla, light-colored interior composed of prosenchymatous tissue. Sclerotial development, a complex process regulated by MAPK and PKA signal transduction pathways, is completed in three macroscopically distinct phases 1) initiation, 2) development, and 3) maturation and is thought to be different both morphologically and biochemically (Willetts and Wong, 1971; Colotelo, 1974; Arseniuk and Macewicz, 1994; Chen and Dickman, 2005; Chen et al., 2004; Harel et al., 2005; Erental et al., 2008).

8.1.2.2 Disease Cycle

The spread of the disease is facilitated by various sexual and asexual stages of this pathogen. The fungus overwinters in the soil as sclerotia, which enters the soil via infected debris or contaminated seeds. According to a study, about 432 sclerotia were found in 1 kilogram of seed. During harvesting and threshing processes, sclerotia, which remain on the soil surface, are buried in the soil by succeeding tillage operations. The disease cycle begins with the germination of overwintering sclerotia. Sclerotial germination is of two types 1) carpogenic and 2) myceliogenic. However, carpogenic germination is the most common method that results in apothecia production (fleshy-colored discs with a diameter of 4 to 8 mm) and usually requires a wetness period of one or two weeks before germination of the sclerotia. Ascospores produced by the apothecia due to sexual reproduction are forcibly discharged into the surroundings and land on susceptible plants or carried by wind a distance from few centimeters to several kilometers. Each apothecium can produce 2–30 million ascospores, covered with sticky mucilage, over several days. In the laboratory, dried, frozen *Sclerotinia* ascospores can remain viable for several years, while freshly produced spores can survive 5–21 days depending upon relative humidity. *S. sclerotium* is a genetically variable fungus, and sclerotia belonging to different geographical regions have different carpogenic germination. Soil moisture and temperature are key factors that affect *S. sclerotium*. In the other type of sclerotial germination, myceliogenic, mycelium is produced by the sclerotia, and infection occurs at or below the ground. Germination of sclerotia occurs in the presence of exogenous nutrients that invade dead organic matter, form mycelium, and then infect living host tissues. Several factors, including soil type, crop residues, amount and frequency of irrigation, and environmental conditions, affect the sclerotia and apothecial development's germination and survival. Maximum production of apothecia has been observed in sandy loam soils and the least number of apothecia in sandy soils. Secondary infection takes place when healthy tissues come in contact with the infected ones. Continuous sclerotial germination and discharge of ascospores ensure suitable infection potential over three to four weeks (Neergaard, 1958; Abawi and Grogan, 1975; Wu, 1988; Carpenter et al., 1999; Clarkson et al., 2004; Saharan and Mehta, 2008; Mehta et al., 2009).

8.1.2.3 Morphology

Hyphae are hyaline, septate, multinucleate, and thin walled. Mycelium is white to tan in culture media. Round, semi-spherical to irregularly shaped sclerotia is embedded in the mycelial net, measuring 2–10 × 3–15 mm in size. Apothecia, produced on stipe measuring 20–80 mm in length, are light yellowish-brown, cup-shaped and variable in size (2–11 mm). Asci are inoperculate, cylindrical, narrow, and rounded at the apex, usually with eight ascospores, and are arranged on the periphery of the ascocarp. Ascopores are hyaline, uniform, and ellipsoid, with smooth walls 10.2–14.0 × 6.4–7.7 μ in size and eight chromosomes (Whetzel, 1945; Kosasih and Willets, 1975).

8.1.2.4 Host Range

S. sclerotiorum has a worldwide distribution with a broad host range of over 500 plant species, including 278 genera in 75 families, and is pathogenic to around 400 plant species. The pathogen affects not only dicotyledonous crops like chickpea, dry pea, edible dry bean, lentils, rapeseed and mustard, soybean, sunflower, and many vegetables but also several notable monocotyledonous crop species like onion, tulip, and so on. The disease is prevalent in many countries across the world, including Australia, Brazil, Canada, Denmark, Finland, France, Germany, India, Nepal, Sweden, United Kingdom, and the United States (Boland and Hall, 1994; Bolton et al., 2006; Ordonez-Valencia et al., 2015; Sharma et al., 2016).

8.1.2.5 Symptoms

Sclerotinia rot disease signs are visible as cottony mycelial growth at different plant growth stages and are similar to leaves, stems, and siliques. Although the disease attacks all above-ground plant parts, the stem portion is more frequently affected. The initial sign of the mycelial infection is forming elongated, water-soaked lesions at the base of the stem or internodes of infected plants, later on becoming bleached and necrotic and subsequently developing patches of fluffy mycelium. These lesions are later covered by white mycelial growth, and the affected stems become soft, giving a whitish look to the diseased plants from a distance. Mycelial growth leads to girdling of the stem, after which the plant wilts and ultimately dies. This type of infection usually appears in patches. Carpogenic germination of sclerotia under the dense, cool, and wet plant canopy occurs mostly at the flowering stage. Ascosporic infection is generally uniform and often occurs at the leaf axil where petals lodge; infection also occurs on leaves and produces shot hole symptoms.

In some cases, infection is restricted to a smaller portion of pith, results in stunted plant growth and premature ripening instead of the sudden collapse of diseased plants, and can be easily identified in the field from a distance. Numerous grayish-white to black spherical sclerotia can be seen either in the pith or on the affected stems' surface. At the seed maturation stage, the plants tend to lodge so that the silique touches the ground (Fig 8.2). However, such plants do not develop stem or aerial infection, showing rotting of the silique with abundant fungal growth. The sclerotia may develop just above the soil level when conditions are conducive to their development or nutrition sufficient for mycelial growth. Occasionally during ascospore liberation, the disease becomes airborne, and large, oval to round-shaped holes can be seen on the leaves as shot hole symptoms. Lesions on leaves are of irregular shape and grayish, which are often associated with adhering petals. These ascospores are also carried by the petals, which generally fall in the leaves' axil where they germinate and cause aerial infection, resulting in drying of the infected branch. These infected branches often produce shriveled silique (Christias and Lockwood, 1973; Saharan et al., 2005; Bolton et al., 2006; Saharan and Mehta, 2008).

8.1.2.6 Management

Management of the disease is difficult, inconsistent, and uneconomical due to the broad host range of the fungus and long-term survival ability of fungal propagules, sclerotia, in the soil and is a significant challenge for plant pathologists. The disease can be best managed by integrating cultural, genetic, biological, and chemical approaches. Cultural practices such as five-year crop rotation with non-host crops,

FIGURE 8.2 Mustard pods infected by *Sclerotinia* stem rot.

zero tillage, less use of nitrogenous fertilizers, and soil amendment with bougainvillea leaves and vermin compost help reduce the number of apothecia, lesions length, seedling mortality, and disease intensity. Host resistance is a very economical and sustainable disease-controlling strategy against this disease. Oilseed breeders have made several attempts to find resistant germplasm for the effective management of stem rot disease, but a resistant genotype to *S. sclerotium* has not been found yet. However, partial resistance has been identified in genotypes from Australia, China, and India. Biological Control: The soil microbial community, mainly fungi and bacteria, has a vital role in limiting the inoculum build-up of the fungus *S. sclerotium* because the fungus overwinters in the soil as sclerotia. Among BCAs, *Bacillus* spp., *Coniothyrium minitans*, *Gliocladium virens*, *Pseudomonas* spp., and *Trichoderma* spp. have a natural potential to parasitize this fungus. Chemical Control: Application of systemic fungicides has provided varying degrees of disease control. Fungicidal spray is ineffective when the disease comes at the late pod formation stage. Antracol, captan, carbendazim, mancozeb, phenylpyrrole, and thiophenate methyl are effective fungicides to control the fungus's mycelial growth in the laboratory. Seed treatment with carbendazim is helpful in the reduction of disease incidence and intensity. Foliar spray of benomyl, captan, carbendazim, and hexaconazole has significantly reduced disease incidence (Adams and Ayers, 1979. Pereira et al., 1996; Budge and Whipps, 2001; Nelson et al., 2001; Shivpuri et al., 2001; Buchwaldt et al., 2003; Gupta et al., 2004; Savchuk and Fernando, 2004; Zhao et al., 2004; Mehta et al., 2005; Li et al., 2008; Singh et al., 2010; Rathi et al., 2012; Rakesh et al., 2016).

8.1.3 White Rust

Among white rusts, diseases caused by *Albugo candida* are most severe and widespread on Cruciferae and have long been a popular subject for mycologists and pathologists. White rust is a common disease of rapeseed and mustard, causing heavy yield losses in different geographical areas worldwide depending upon the nature and severity of the disease. The disease mostly appears along with downy mildew and causes significant damage from combined infection. It can cause about 20–90% annual losses all over the world. The disease is causing annual yield losses of 5–10% in Australia, 20% in Canada, 23–89.8% in India, 30–60% in Manitoba, and 60–90% in Pakistan. The disease appears in early January on the lower surface of the leaves and later moves quickly to the upper surface and other aerial plant parts (Bernier, 1972; Bains and Jhooty, 1980; Kolte, 2002; Mishra et al., 2009; Asif et al., 2017; Bisht et al., 2018).

8.1.3.1 Cause

White rust of crucifers is a fungal infection disease caused by an oomycete, *Albugo candida*, an obligate parasite that belongs to the family Albuginaceae in the order Peronosporales. As an obligate parasite, it can survive in soil and plant debris in the form of oospores for up to 20 years. Morphology: Isolates of *A. candida* from different species or cultivars of Brassica and different geographical regions may vary in shape, size, texture, sporangia production, zoospores, and pustules, as well as aggressiveness. Mycelium is intercellular, Sori; white to pale, prominent, deep-seated, often confluent, and variable in size and shape; sporangiophores are thick walled, non-septate, hyaline, and clavate and measure 30–45 × 15–18 μm in diameter. Sporangia are hyaline, thin walled, globose to oval, and 12–18 μm in diameter. Oospores are globose and brownish, epispores; thick walled, verrucose to tuberculate. Disease cycle: The pathogen survives through the mycelium, in resting spores (sporangia and oospores) in decaying tissues, or as seed contaminant. Some spores germinate in the spring and infect cotyledons and leaves. The infection progress and formation of sori take place on the stem or lower surface of leaves. These spores are discharged into the air, from where they spread to other plant parts or nearby plants, resulting in a secondary infection on leaves, stems, or flower buds. Infected buds produce stag heads, which may be broken during harvesting and threshing, causing contamination of the seeds with resting spores. Epidemiology: Many factors, such as amount of initial inoculum, weather conditions, and time of the disease appearance will collectively determine epidemics of the disease. Cool and wet weather favors disease development. Optimum temperature and RH for successful infection is 12–22°C and 60–90%, respectively (Saharan and Verma, 1992; Sullivan et al., 2002; Gupta et al., 2002; Mishra et al., 2009; Armstrong, 2007; Siddaramaiah, 2007; Meena et al., 2011, 2014b).

8.1.3.2 Host Range

The fungus *A. candida* has a broad host range, from cultivated crops to non-cultivated weeds. It infects 241 plant species belonging to 63 genera in the Cruciferae family only. White rust is most prevalent in Australia, Brazil, Canada, China, Fiji, Germany, India, Japan, Korea, New Zealand, Palestine, Pakistan, Romania, Turkey, the United Kingdom, and the United States (Biga, 1955; Farr et al., 1989; Saharan and Verma, 1992; Choi et al., 2011; Asif et al., 2017; Prasad et al., 2017).

8.1.3.3 Symptoms

The disease is a local and systemic infection. It first appears on leaves, which predisposes it to attack by downy mildew fungus, *Peronospora parasitica*. Initial symptoms start as local infection with white to creamy-yellow sporangial pustules (sori) and variable size and shape on the lower surface of leaves (Fig 8.3; Fig 8.4), stems, and tender shoots from the seedling stage onward, and infected tissues become necrotic and die. These pustules are initially separate but later coalesce, covering the whole plant organ. Following stem and pod infection, raised green blisters become white during wet weather, but on maturity, they turn brown, hard, and dry. Systemic infection is characterized by unequal growth and distortion of inflorescence, giving a stag-headed look, as well as flower sterility (Fig 8.5).

FIGURE 8.3 Powdery white pustules on the lower side of leaf infected by white rust.

FIGURE 8.4 Mustard leaf showing small, scattered, and white circular colonies.

FIGURE 8.5 Staghead deformation associated with white rust.

Abnormal swelling, malformation of infected tissues, twisting, and extensive distortion of inflorescence give a stag-headed look to the plants. Leaves, stems, and inflorescence axis become thickened. The systemic infection phase increases fruit shattering, and infected plants produce fewer pods and fewer seeds (Saharan and Mehta, 2002; Mishra et al., 2009).

8.1.3.4 Management

IDM practices are the best option for the management of white rust as well as to enhance yield. Use of healthy seeds, removal of weed hosts, crop rotation with at least three years, and growing resistant cultivars help disease control. Field application of *T. viridae* and botanicals such as neem oil, garlic, onion, and kikar causes a significant disease reduction. Seed treatment with antagonistic fungus *T. harzianum* and fungicides Apron, Bravo, Metalaxyl, and Manzat are useful in preventing leaf and stag headed phase. Chemical Control: Copper-based fungicides have provided success in controlling the leaf phase but limited success in controlling the stag-headed phase. Benlate, Calixin, Difolatan, Dithane, difenoconazole, hexaconazole, Miltox, Thiovit, Mancozeb, metalaxyl, propiconazole, Ridomil, and sulfex are effective fungicides (Saharan and Verma, 1992; Bhatia and Gangopadhyay, 1996; Meena et al., 2011; Asif et al., 2017; Mohan et al., 2017; Prasad et al., 2017; Gairola and Tewari, 2019; Wangkhem et al., 2020).

8.1.4 Blackleg

Blackleg is the most severe disease of canola worldwide, particularly in Australia. It is also known as root and collar rot, dry rot, and stem canker. Areas with extensive canola production are more vulnerable to blackleg disease. Although this disease is not common, yield losses of 50% and greater up to 90% in some cases have been recorded (Rimmer and van der Berg, 2007; Nyvall, 2013; Kumar et al., 2017).

8.1.4.1 Cause

The disease is caused by *Leptosphaeria maculans*, anamorph: *Phoma lingam*. Two strains of this fungus have been involved in blackleg disease in rapeseed and mustard. The prevalent strain, weakly virulent, infects at maturity and is known to cause minute damage, while the other strain is highly virulent and sporulates early in the spring. A third strain of the fungus has also been reported on stinkweed in Canada but has no role in this disease. The pathogen survives, potentially producing spore types, pseudothecia, and pycnidia on canola stubbles, for one to four years and as mycelium in seed for years. In spring, during wet weather or high RH periods, ascospores are produced in fruiting bodies known as perithecia. Sometimes ascospores are released at a flowering stage, which is highly susceptible to the disease. Large quantities of airborne ascospores are released from these fruiting bodies, capable of spreading several kilometers. This spore liberation is highly triggered by autumn and winter rainfalls. The fungus is also carried over by seeds, further spread by farm machinery, irrigation water, rain splashes, and wind. On successful infection, cotyledons and newly emerged leaves develop off-white lesions within two weeks of spores landing. These lesions produce pycnidia, which releases rain-splashed spores. Production of pycnidia continues throughout the growing season, but perithecia only develop on plant maturity. Blackleg disease is more severe at 18°C and is favored by wet and windy conditions. An increase in temperature and light intensity and prolonged leaf wetness favors the maximum formation of fruiting bodies and ultimate spore production and thus increases infection severity (Vanniasingham et al., 1989; Rimmer and van der Berg, 2007; Khangura et al., 2007; GRDC, 2019; Nyvall, 2013).

8.1.4.2 Host Range

A variety of brassica crops serve as the host plants. It can also occur on *Raphanus* and *Sinapis* spp. as well as brassica weeds. Blackleg disease is found in almost all cruciferous growing countries. However, it is most severe in Australia, Canada, and Europe (Rimmer and van der Berg, 2007; Nyvall, 2013; Peng et al., 2020).

8.1.4.3 Symptoms

The pathogen infects leaves, petioles, pods, stems, and roots; however, younger plant tissues are more susceptible than older tissue. On leaves, the lesions are pale, irregular with a purple border and black dots, pycnidia. Stem lesions are elongated and extend to the soil line, causing black rot or lower stem dry rot. Stem girdling and canker formation over a large area can be seen on severely infected plants. The fungus can grow within the crown's vascular systems, where it causes the crown to rot. It can also cause partial blockage of vascular bundles and ultimate plant death. Seed pods rarely develop disease symptoms. Disease symptoms may also be observed in plant roots. Less severe infections limit water and nutrient flow within the plant, while severe infections may result in stunted growth, premature ripening, production of light and shriveled seeds, and even premature plant death (Rimmer and van der Berg, 2007; GRDC, 2013; Nyvall, 2013)

8.1.4.4 Management

Eradicate susceptible weeds and volunteer crucifers, adopt field sanitation practices, use certified seed, rotate crop with non-host crops such as cereals for a minimum of three to five years, use tolerant cultivars, and avoid rapeseed and mustard planting in or adjacent to a site where the disease previously occurred. Seed treatment with Fluopyram and fluquinconazole or hot water treatment at 50°C for 15 to 30 min effectively control the disease. Additionally, foliar application of Prosaro (a mixture of prothioconazole and tebuconazole) and Thiram 50 WP represents a promising tool for the management of blackleg disease in Australia and Canada (du Toit and Derie, 2005; Guo et al., 2005; Nyvall, 2013; Van de Wouw et al., 2017; GRDC, 2019; Peng et al., 2020).

8.1.5 Powdery Mildew

Powdery mildews are a widespread and economically significant group of diseases. Powdery mildew of rapeseed and mustard has become an epidemic disease in India, occurring in 17 states, causing yield losses up to 17%. Losses due to this disease have been estimated at 10–90% from different *Brassicas*. The disease is most destructive when it occurs at early growth stages (Dange et al., 2002; Meena et al., 2018).

8.1.5.1 Cause

Erysiphe cruciferarum, the cause of the disease, is an obligate biotrophic fungus in the family Erysiphaceae. Airborne fungal spores of the pathogen overwinter on infected plant debris or weeds, which provides initial inoculum for the polycyclic disease. The pathogen does not require water on the leaf surface for successful infection to occur. Moderate to dry weather, usually warm climate, and low RH favor pathogen growth and disease development. High RH is required for spore germination. The disease occurs in late summer and autumn. Morphology: Mycelium: amphigenous, septate, and white; Conidia: hyaline, cylindrical, with size 25–45 × 12–16 μm; Perithecia: scattered, globose, initially yellowish-orange and brown to dark brown or black on maturity, and 90–130 μm in diameter; Appendages: myceloid, slightly hyaline to faintly colored and rarely branched; Asci: oval to Pyriform, 3–12 in number, usually 6–8; Ascospore: Ovoid, 2–7 in number measuring 16–22 × 11–14 μm (Micali et al., 2008; Kumar et al., 2015).

8.1.5.2 Host Range

Powdery mildew fungi infect more than 125 crucifer hosts, including cereals, cucurbits, oilseeds, and vegetables, in more than 25 countries of the world. The disease is prevalent in the United States, United Kingdom, Turkey, Sweden, Pakistan, Japan, Germany, France, and India (Saharan et al., 2005; Micali et al., 2008).

8.1.5.3 Symptoms

The fungus affects all foliar plant parts. The disease is most common on leaves' upper surfaces compared to lower surfaces and other plant parts. The disease first appears on the oldest leaves as small, scattered, white circular colonies. Under favorable environmental conditions, these colonies coalesce and cover the entire leaf area. Infected leaves turn pale and fall prematurely. Infected buds may also fail to open. It also affects developing green silique (Singh, 2000; Enright and Cipollini, 2007).

8.1.5.4 Management

The disease can be prevented by growing resistant cultivars. Furthermore, Iprodione +carbendazim and ZnSO4 + sulfur are effective fungicides for powdery mildew management (Kumar et al., 2015).

8.1.6 Downy Mildew

This is a common disease of rapeseed and mustard, and when it occurs together with white rust, it is referred to as stag heads. The disease poses a major threat to seedlings but does not cause significant yield losses. Infection occurs under cool and moist conditions when leaves or cotyledons contact infected leaves or soil (Nyvall, 2013; Kumar et al., 2017).

Cause: *Peronospora parasitica* causes the disease. The pathogen survives systemically as mycelium in infected perennials and soil and crop residues as oospores for more extended periods. Infection occurs when oospores germinate and produce *sporangia*. Production of *sporangia* occurs in the evening, while *sporangia* are discharged during the daytime. Penetration of windblown sporangia takes place through leaves and flowers. Systemically infected plants produced wind-borne conidia. Conidia produced on the underside of infected leaves are thought to be the primary source of secondary inoculum. Cool and wet weather favors infection, and new infections may develop within three to four days under ideal weather conditions (GRDC, 2013; Nyvall, 2013; Kumar et al., 2017).

Host range: Turnip crops, radish, kohlrabi, kale, cauliflower, cabbage, brussels sprouts, broccoli, and brassica are important host plants for downy mildew fungus. The disease has a worldwide distribution (Satou and Fukimoto, 1996).

8.1.6.1 Symptoms

Disease symptoms can be observed on leaves, stems, and pods. First symptoms of the disease emerge as light green or slightly chlorotic or purple lesions of variable size on the upper leaf surface of seedlings. Infected cotyledons are likely to die prematurely, and severe infections may also result in seedling mortality due to excessive defoliation. Leaf spots become enlarged and turn yellow. White moldy growth may appear on the upper surfaces of leaves. This type of growth also occurs on blisters and stag heads. Individual spots merge to form large blotches of an irregular shape. These necrotic lesions can cause part of the leaf to dry (GRDC, 2013; Nyvall, 2013; Kumar et al., 2017).

Management: Use of healthy seeds, growing resistant or tolerant cultivars, removal of volunteer hosts and cruciferous weeds, crop rotation with non-susceptible crops, and a spray of copper-based fungicides are suitable management options for this disease (GRDC, 2013; Nyvall, 2013).

8.1.7 Damping-Off

Damping-off or seedling blight attacks at early growth stages when conditions are unfavorable for seedling growth. Problems occur when seeds are sown dry and weather conditions become cool and damp or close to the autumn break. Heavy, wet soils and cool, wet weather increase the chances of higher disease incidence and severity. Yield loss is infrequent (GRDC, 2013; Nyvall, 2013).

Cause: Fungi such as *Rhizoctonia* spp., *Pythium* spp., *Phytophthora* spp., *Leptosphaeria* sp., *Fusarium* spp., and *Alternaria* spp. are frequently associated with damping-off disease. These fungi are common soil inhabitants and weak pathogens that affect young succulent tissues under unfavorable growth conditions and reduce germination or post-emergence damping-off in seedlings. Cool and moist weather is required to cause seedling infection, while warm weather favors mature plant infection (GRDC, 2013; Nyvall, 2013).

8.1.7.1 Host Range

Damping-off fungi have a broad host range and can even overwinter on non-brassica hosts. The disease is generally distributed in all rapeseed and mustard-growing countries (Nyvall, 2013; Kumar et al., 2017).

8.1.7.2 Symptoms

Disease symptoms vary, ranging from pre-emergence rot to post-emergence damping-off. Infested seeds fail to germinate. Post-emergence damping-off is characterized by constriction, tapering, and formation of brown lesions on hypocotyl. Seedlings may not grow any further at the one- to four-leaf stages. Infected plants show stunted growth and early flowering and may mature prematurely. Pre- and post-emergence occurs in patches. Affected leaves show discoloration and turn orange, purple, or yellow. Taproots turn dark and shriveled at soil level (GRDC, 2013; Nyvall, 2013).

Management: Seed treatment with fungicides, cultural practices such as soil tillage, and long-term crop rotation can reduce damping-off damage. Biological control includes *T. harzianum*, and a few strains of *Pseudomonas fluorescens* are well known to limit damping-off fungi. Fungicides including benodanil, carboxin, carbothiin, cyproconazole, fenpropimorph, iprodione, lindane, pencycuron, thiram, and thiabendazole have proved effective (GRDC, 2013; Kumar et al., 2017).

8.1.8 Clubroot

Clubroot is an economically significant disease of crucifers. The disease is cosmopolitan and causes significant yield losses in Australia (GRDC, 2013).

Cause: The fungus *Plasmodiophora brassicae*, an obligate biotroph, is the cause of the disease. The pathogen can survive in soils as resting spores for years even without a susceptible host. Infection is restricted to roots and may occur at any growth stage. Spores germinate and discharge small motile spores that penetrate and form fungal colonies inside root cells. Resting spores mature in the infected roots and are released into the soil. Infested soils and seedling transplants cause infection to paddocks. The life cycle of the associated fungus is completed in two phases. In the first phase, root hairs are attacked, while the second phase affects the cortex of hypocotyl and roots. The fungus causes infection in warmer months and spreads through irrigation water, infested soils, farm machinery, and equipment (GRDC, 2013).

Host range: Brussels sprouts, broccoli, cabbage, canola, cauliflower, and mustard are the primary host plants. The disease occurs worldwide (GRDC, 2013).

8.1.8.1 Symptoms

As a result of a root infection, cells become enlarged and divide rapidly and induce swollen galls on the roots of infected plants. At first, galls are firm and white, but they become soft, turn grayish brown, and decay. These galls interfere with water and nutrient uptake. Consequently, leaves become yellow, wilted, and lodged (GRDC, 2013; Kumar et al., 2017).

8.1.8.2 Management

Avoid plantation of cruciferous crops in infested soils; crop rotation; reduced soil tillage; good drainage; weed eradication; sanitation practices; maintenance of high pH by using lime; resistant cultivars; and fungicides such as fluazinam, Flusulfamide, cyazofamid; and soil fumigation with metham sodium are useful in the pathogen and disease inhibition (Donald and Porter, 2009).

8.1.9 Foot Rot

The disease is similar to brown girdling root rot but apparently differs in symptomology and is sometimes referred to as late root rot. Cause: The disease's primary cause is *Rhizoctonia solani*, but *Fusarium roseum* also causes it. *R. solani* survives as mycelium and sclerotia in residue, whereas *F. roseum* overwinters as chlamydospores, mycelium, and often as perithecia of its teleomorph, *Gibberella zeae*, in the soil and crop residue, respectively. Other fungi associated with this disease complex include *Pythium debaryanum* and *P. polymastum*, *Leptosphaeria maculans*, *F. tricinctum*, *F. solani*, and *A. alternata*. Host range: It has a broad host range and is generally distributed among rapeseed and mustard-growing areas. Symptoms: Prematurely ripened plants may occur singly or in patches. Infected stems have hard, brown lesions encircled by a black border at the base. Salmon-pink spore masses can be seen on these lesions during moist conditions. Discolored tissues may develop white mycelial growth. Severe infections cause stem girdling and ultimate plant death. Management: Always use clean and healthy seeds, eradicate volunteer plants and cruciferous weeds, and rotate rapeseed and mustard with cereals (Nyvall, 2013).

8.1.10 Brown Girdling Root Rot

This disease is also called *Rhizoctonia* root rot. It is similar to foot rot disease because of the number of similar pathogenic fungi. Cause: The primary cause of this root rot complex is *R. solani*, but it is sometimes also caused by *P. ultimum* and *F. roseum* as well. Distribution: Alberta and British Columbia. Symptoms: Pre- and post-emergence damping-off may occur. Tap root develops brown to red-brown lesions. A single lesion or several lesions may coalesce to girdle the taproot. Management: Management options are the same as those listed for foot rot (Nyvall, 2013).

8.1.11 White Leaf Spot and Gray Stem

The disease is also known as black blight, black stem ring spot, and ring spot. It usually does not cause yield losses. Cause: It is caused by *Mycosphaerella brassicicolai*. The fungus survives as thick-walled mycelium and perithecia in the residue. In the spring, wind-borne conidia produce residue from the resting structure. Formation of perithecia takes place at low night temperatures and with heavy dew. In Europe and the United States, the disease is spread by ascospores. Conidia produced in cool and moist weather act as a source of secondary infection. Host range: Has a wide host range and is distributed in Australia, Canada, Europe, and the United States. In Australia, it infects canola seedlings. Symptoms: Initial spots on leaves are circular and whitish. Lesions on stem and pods are elongated, large, purple to gray lesions with lighter centers speckled with numerous perithecia. These lesions coalesce, causing blacking of the large portion of infected plants. Management: Always use healthy and cleaned seeds, rotate crops with non-cruciferous crops, eradicate weed hosts, and balance use of fertilizer to reduce crop stress (GRDC, 2013; Nyvall, 2013).

8.1.12 Pod Drop

Pod drop is caused by *A. alternata* and *Cladosporium* sp. Drying or dead petals are infected, which provide nutrition to the fungi to grow into the pedicel. The disease has been reported in Canada. Symptoms: Blackened pedicel can be seen below or at its point of attachment to the pod. Normal filling of severely infected pods is common. Pedicel becomes so weakened that pods drop to the pedicel's soil, and tips show typical blackening (Nyvall, 2013).

8.1.13 Storage Molds

Several fungi, including *Alternaria* spp., *Aspergillus* spp., *Cephalosporium acremonium*, and *Cladosporium cladosproioides*, are involved in storage molds (Nyvall, 2013).

8.2 VIRAL CAUSES

Three viruses, turnip mosaic virus (TuMV), cauliflower mosaic virus (CaMV), and beet western yellow virus (BWYV), affect canola in Australia and Alberta. Among these, BWYV is most common, with estimated yield losses of 34% in Europe and up to 46% in Western Australia. It lowers oil quality, while the other two may also cause significant losses but usually tend to occur at low incidences. TuMV is transmitted either mechanically or by green peach aphids and others (Hertel et al., 2004; GRDC, 2013; Nyvall, 2013).

Cause: BWYV, family Luteovirus, is a persistent virus carried by the body of aphids, by the green peach aphid, and transmitted during feeding. Virus-acquiring aphids remain infective through several molts and even for the rest of their lives, but effectiveness decreases over time. It has a broad host range, infecting 150 species of dicotyledonous, including canola, chickpea, faba bean, field bean, shepherd's purse, soybean subterranean clover, wild radish, and wild turnip. TuMV, family Potyvirus, is a non-persistent virus transmitted by various aphid species and does not persist for more than four hours. It usually infects brassicaeous and leguminous hosts. CaMV, genus *Caulimovirus*, is a semi-persistent virus transmitted by many aphid species and retained for up to four days. It has a limited host range restricted to Brassicaceae. Volunteer canola host plants and weeds act as the survival places for these viruses outside the growing season, spread and transmitted by aphids. Autumn is considered most critical for infection. Four species of aphids are generally associated with canola and mustard 1) green peach aphid, *Myzus persicae*, transmits BWYV, TuMV, and CaMV; 2) turnip aphid, *Lipaphis erysimi*, transmits TuMV and CaMV; 3) cabbage aphid, *Brevicoryne brassicae*, transmits BWYV, TuMV, and CaMV; and 4) cowpea aphid, *Aphis craccivora*, transmits BWYV, TuMV, and possibly CaMV (Hertel et al., 2004; GRDC, 2013).

Symptoms: BWYV causes stunted growth and purpling of lower leaves; CaMV causes stunting, mottling, and chlorotic ring spots; and TuMV causes severe mosaic symptoms, stunting, tip necrosis, and premature death (Hertel et al., 2004; GRDC, 2013).

8.2.1 Management

Control broadleaf weeds, retain stubble at sowing to cover the ground, late sowing, and use of varieties carrying strain-specific resistance genes will help minimize infection. Seed dressing with imidacloprid insecticides helps kill green peach aphid (Hertel et al., 2004; GRDC, 2013).

8.3 MYCOPLASMAL CAUSES

Aster yellows are of minor importance. Several leafhoppers and dicotyledonous plants are important sources of the survival of this mycoplasma. Cause: *Macrosteles fascifrons* primarily transmit these, along with *M. laevis*, *Endria inimical*, and aster leafhopper. Symptoms: Plants suffer systematic infection, but symptoms can only be observed on inflorescences. Pods produced by infected plants become sterile and distorted and look like small, blue green, hollow, and bladder-like normal seed pods. Management: No practical management is available (Nyvall, 2013).

8.4 BACTERIAL CAUSES

Bacterial wilt and rot caused by *Erwinia carotovora* pv. *atroseptica* occurs in Mexico. Symptoms: The central part of infected stems shows partial or complete rotting and wilting associated with the molted appearance of the leaf. Rot initiates at the point of infection, caused by larvae of *Hylemyia* sp. usually at the ground level or where branches occur. Management: No management strategy is reported (Nyvall, 2013).

Black rot, a commonly occurring disease in Canada, is caused by seed-borne *Xanthomonas campestris*. The pathogenic bacteria survive in the infected residue. Bacteria enter the plant through stomata, usually when splashed by water. Symptoms: Leaf tissues become chlorotic with dark color veins in the chlorotic area (Nyvall, 2013).

8.4.1 Virus-Like Disorder

It is distributed in India. Cause: Unknown, but most probably of genetic or physiologic origin. Infected leaves of diseased plants are darker green than normal leaves and show stunting and twisted leaves. Affected plants produce fewer flowers and seeds. Longitudinal fissures can be observed on aerial parts (Nyvall, 2013).

REFERENCES

Abawi GS and Grogan RG (1975) Source of primary inoculum and effects of temperature and moisture on infection of beans by *Whetzelinia sclerotiorum*. *Phytopathology* 65: 300–309.

Adams PB and Ayers WA (1979) Ecology of *Sclerotinia* species. *Phytopathology* 69: 896–899.

Ahmed M, Tonu NN, Hornaday K, Aminuzzaman FM, Chowdhury MS and Islam MR (2018) Effect of chemical seed treatment and BAU-Biofungicide on *Alternaria* blight (*Alternaria brassicae*) of mustard. *Agricultural Sciences* 9(5): 566–576.

Aneja JK and Agnihotri A (2016) *Alternaria* blight of oilseed brassicas: Epidemiology and disease control strategies with special reference to use of biotechnological approaches for attaining host resistance. *Journal of Oilseed Brassica* 1(1): 1–10.

Ansari AM, Monjil MS and Hossain I (2004) Efficacy of bion and amistar in controlling *Alternaria* blight and yield of mustard variety agrani. *Journal of the Bangladesh Agricultural University* 2: 1–8.

Armstrong T (2007) Molecular detection and pathology of the oomycete *Albugo candida* (white rust) in threatened coastal cresses. Science & Technical Pub., Department of Conservation.

Arseniuk E and Macewicz J (1994) Scanning electron microscopy of sclerotia of *Sclerotinia trifoliorum* and related species. *Journal of Phytopathology* 141: 275–284.

Asif M, Atiq M, Sahi ST, Ali S, Nawaz A, Ali Y, . . . and Saleem A (2017) Effective management of white rust (*Albugo candida*) of rapeseed through commercially available fungicides. *Pakistan Journal of Phytopathology* 29(2): 233–237.

Ayub A, Dey TK, Jahan M, Ahmed HU and Alam KB (1996) Foliar spray of fungicides to control *Alternaria* blight of mustard. *Journal of the Bangladesh Agricultural University* 6: 47–50.

Bains SS and Jhooty JS (1980) Mixed infections by *Albugo candida* and *Peronospora parasitica* on *Brassica juncea* inflorescence and their control. *Indian Phytopathology* 32(2): 268–271.

Bernier CC (1972) Diseases of rapeseed in Manitoba in 1971. *Canadian Plant Disease Survey* 52: 108.
Bharti O, Pandey RK, Singh R and Singh RK (2019) A potential menace: Stem rot in mustard. *International Journal of Chemical Studies* 7(3): 4672–4678.
Bhatia JN and Gangopadhyay S (1996) Studies on chemical control of white rust disease of mustard. *International Journal of Pest Management* 42(1): 61–65.
Biga MLB (1955) Review of the species of the genus *Albugo* based on the morphology of the conidia. *Sydowia* 9: 339–358.
Bisht KS, Rana M, Upadhyay P and Awasthi RP (2018) Disease assessment key for white rust disease caused by *Albugo candida*, in rapeseed-mustard. *Plant Pathology Journal* 17: 11–18.
Boland GJ and Hall R (1994) Index of plant hosts of *Sclerotinia. Canadian Journal of Plant Pathology* 16: 93–108.
Bolton DM, Thomma PHJB and Nelson DB (2006) *Sclerotinia sclerotiorum* (Lib.) de Bary: Biology and molecular traits of a cosmopolitan pathogen. *Molecular Plant Pathology* 7: 1–16.
Buchwaldt L, Yu FQ, Rimmer SR and Hegedus DD (2003) Resistance to *Sclerotinia sclerotiorum* in a Chinese *Brassica napus* cultivar. In: *Proceedings of 8th International Congress of Plant Pathology*, Christchurch, New Zealand, p. 289.
Budge SP and Whipps JM (2001) Potential for integrated control of *Sclerotinia sclerotiorum* in glasshouse lettuce using *Coniothyrium minitans* and reduced fungicide application. *Phytopathology* 91: 221–227.
Carpenter MA, Frampton C and Stewart A (1999) Genetic variation in New Zealand populations of the plant pathogen *Sclerotinia sclerotiorum. New Zealand Journal of Crop and Horticultural Sciences* 27: 13–21.
Chadar LK, Singh RP, Singh RK, Yadav RR, Mishra MK, Pratap N and Vishnoi RK (2016) Studies on *Alternaria* blight of rapeseed-mustard (*Brassica juncea* L.) caused by *Alternaria brassicae* (Berk.) Sacc. and its integrated management. *Plant Archives* 16(2): 897–901.
Chattopadhyay AK and Bagchi BN (1994) Relationship of disease severity and yield due to leaf blight of mustard and spray schedule of mancozeb for higher benefit. *Journal of Mycopathological Research* 32(2): 83–87.
Chattopadhyay C, Meena PD, Sastry RK and Meena RL (2003) Relationship among pathological and agronomic attributes for soil-borne diseases of three oilseed crops. *Indian Journal of Plant Protection* 31: 127–128.
Chauhan LS, Singh J and Chandra DR (1992) Assessment of losses due to stem rot of yellow sarson. In: *Proceedings of National Symposium*, University of Allahabad, Allahabad, 19–21 November, pp. 65–66.
Chen C and Dickman MB (2005) cAMP blocks MAPK activation and sclerotial development via Rap-1 in a PKA-independent manner in *Sclerotinia sclerotiorum. Molecular Microbiology* 55: 299–311.
Chen C, Harel A, Gorovoits R, Yarden O and Dickman MB (2004) MAPK regulation of sclerotial development in *Sclerotinia sclerotiorum* is linked with pH and cAMP sensing. *Molecular Plant Microbiology* 17: 404–413.
Choi YJ, Park MJ, Park JH and Shin HD (2011) White blister rust caused by *Albugo candida* on Oilseed rape in Korea. *Plant Pathology Journal* 27: 192.
Christias C and Lockwood JL (1973) Conservation of mycelia constituents in four sclerotium forming fungi in nutrient deprived conditions. *Phytopathology* 63: 602–605.
Clarkson JP, Phelps K, Whipps JM, Young CS, Smith JA and Watling M (2004) Forecasting Sclerotinia disease on lettuce: Toward developing a prediction model for carpogenic germination of sclerotia. *Phytopathology* 94: 268–279.
Colotelo N (1974) A scanning electron microscope study of developing sclerotia of *Sclerotinia sclerotiorum. Canadian Journal of Botany* 52: 1127–1130.

Dange KK, Patel RL, Patel SI and Patel KK (2002) Assessment of losses in yield due to powdery mildew disease in mustard under north Gujarat conditions. *Journal of Mycology and Plant Pathology* 32: 249–250.
Donald EC and Porter IJ (2009) Integrated control of clubroot. *Journal of Plant Growth Regulation* 28: 289–303.
du Toit LJ and Derie ML (2005) Evaluation of fungicide seed treatments for control of black leg of cauliflower. *Fungicide & Nematicide Tests* 60.
Enright SM and Cipollini D (2007) Infection of powdery mildew *Erysiphe cruciferarum* (Erysiphaceae) strongly affects growth and fitness of *Alliaria petiolata* (Brassicaceae). *American Journal of Botany* 94: 1813–1820.
Erental A, Dickman MB and Yarden O (2008) Sclerotial development in *Sclerotinia sclerotiorum*: Awakening molecular analysis of a "dormant" structure. *Fungal Biology Reviews* 22(1): 6–16.
Farr DF, Bills GF, Chamuris GP and Rossman AY (1989) *Fungi on Plants and Plant Products in the United States.* St. Paul, USA: APS Press.
Gairola K and Tewari AK (2019) Management of white rust (*Albugo candida*) in Indian mustard by fungicides and garlic extract. *Pesticide Research Journal* 31(1): 60–65.
Ghasolia RP, Shivpuri A and Bhargava AK (2004) Sclerotinia rot of Indian mustard (*Brassica juncea*) in Rajasthan. *Indian Phytopathology* 57: 76–79.
Grain Research and Development Coop (2013) *Diseases of Canola and Their Management—The Back Pocket Guide.* Available at https://grdc.com.au/resources-and-publications/all-publications/bookshop/2013/02/grdc-bpg-canoladiseases (accessed on August 08 2020).
Grain Research and Development Coop (2019) *Canola—What Disease Is That and Should I Apply a fungicide*? Available at https://grdc.com.au/resources-and-publications/grdc-update-papers/tab-content/grdc-update-papers/2019/02/canola-what-disease-is-that-and-should-i-apply-a-fungicide (Accessed on 21 September 2020).
Guo XW, Fernandoa WGD and Entz M (2005) Effect of crop rotation and tillage on blackleg disease on canola. *Canadian Journal of Plant Pathology* 27: 53–57.
Gupta AK, Raj R, Kumari K, Singh SP, Solanki IS and Choudhary R (2018) Management of major diseases of Indian mustard through balanced fertilization, cultural practices and fungicides in calcareous coils. In: *Proceedings of the National Academy of Sciences, India Section B: Biological Sciences* 88(1): 229–239.
Gupta K and Saharan GS (2002) Identification of pathotype of *Albugo candida* with stable characteristic symptoms on Indian mustard. *Journal of Mycology and Plant Pathology* 32: 46–51.
Gupta R, Awasthi RP, and Kolte SJ (2004) Effect of nitrogen and sulphur on incidence of Sclerotinia rot of mustard. *Indian Phytopathology* 57: 193–194.
Harel A, Gorovits R and Yarden O (2005) Changes in protein kinase A activity accompany sclerotial development in *Sclerotinia sclerotiorum. Phytopathology* 95: 397–404.
Hertel K, Schwinghamer M and Bambach R (2004) *Virus Diseases in Canola and Mustard.* Bill Noad Agnote DPI, 495.
Humpherson-Jones FM (1989) Survival of *Alternaria brassicae* and *Alternaria brassicicola* on crop debris of oilseed rape and cabbage. *Annals of Applied Biology* 115(1): 45–50.
Jamalainen EA (1954) Overwintering of cultivated plants under snow. *FAO Plant Prot Bull* 2: 102.
Kang IS and Chahal SS (2000) Prevalence and incidence of white rot of mustard incited by *Sclerotinia sclerotiorum* in Punjab. *Plant Disease Research* 15: 232–233.
Khangura R, Speijers J, Barbetti MJ, Salam MU and Diggle AJ (2007) Epidemiology of blackleg (*Leptosphaeria maculans*) of canola (*Brassica napus*) in relation to maturation of pseudothecia and discharge of ascospores in Western Australia. *Phytopathology* 97: 1011–1021.
Kolte SJ (2002) Diseases and their management in oilseed crops new paradigm. In: Rai M, Singh H and Hedge DM (eds) Oilseeds and Oils: Research and Development Needs. Hyderabad, India: Indian Society of Oilseeds Research, pp. 244–252.

Kolte SJ (2018) *Diseases of Annual Edible Oilseed Crops: Volume II: Rapeseed-Mustard and Sesame Diseases.* CRC press.

Kosasih BD and Willetts HJ (1975) Ontogenetic and histochemical studies of the apothecium of *Sclerotinia sclerotiorum. Annals of Botony* 39: 185–191.

Kruger W (1976) Important root and stalk diseases of Rape in Germany. *Gesunde Pflanzen* 28: 7.

Kumar K, Tiwari Y and Trivedi L (2017) Major fungal diseases of oilseed brassica and their preventive measures: A brief review. *International Journal of Botany Studies* 2(5): 105–112.

Kumar S, Singh D, Yadav SP and Prasad R (2015) Studies on powdery mildew of rapeseed-mustard (*Brassica juncea* L.) caused by *Erysiphe cruciferarum* and its management. *Journal of Pure Applied Microbiology* 9(2): 1481–1486.

Latif MZ, Haq IU, Khan SH, Habib A, Khawaja S, Shehroz A, Nazir N and Ramzan I (2018) Evaluation of rapeseed-mustard germplasm against *Alternaria* leaf blight under local field conditions. *Pakistan Journal of Phytopathology* 30(2): 163–168.

Li CX, Liu SY, Sivasithamparam K and Barbetti MJ (2008) New sources of resistance to Sclerotinia stem rot caused by *Sclerotinia sclerotiorum* in Chinese and Australian *Brassica napus* and *Brassica juncea* germplasm screened under Western Australian conditions. *Australasian Plant Pathology* 38: 149–152.

Meena PD, Awasthi RP, Chattopadhyay C, Kolte SJ and Kumar A (2016) Alternaria blight: A chronic disease in rapeseed-mustard. *Journal of Oilseed Brassica* 1(1): 1–1.

Meena PD, Awasthi RP, Godika S, Gupta JC, Kumar A, Sandhu PS, Sharma P, Rai PK, Singh YP, Rathi AS, Prasad R, Rai D and Kolte SJ (2011) Eco-friendly approaches managing major diseases of Indian mustard. *World Applied Sciences Journal* 12(8): 1192–1195.

Meena PD, Meena RL, Chattopadhyay C and Kumar A (2004a) Identification of critical stage for disease development and biocontrol of *Alternaria* blight of Indian mustard (*Brassica juncea*). *Journal of Phytopathology* 152(4): 204–209.

Meena PD, Mehta N, Rai PK and Saharan GS (2018) Geographical distribution of rapeseed-mustard powdery mildew disease in India. *Journal of Mycology and Plant Pathology* 48(3): 284–302.

Meena PD, Verma PR, Saharan GS and Borhan MH (2014b) Historical perspectives of white rust caused by *Albugo candida* in oilseed brassica. *Journal of Oilseed Brassica* 5: 1–41.

Mehta N, Hieu NT and Sangwan MS (2009) Influence of soil types, frequency and quantity of irrigation on development of Sclerotinia stem rot of mustard. *Journal of Mycology and Plant Pathology* 39: 506–510.

Mehta N, Sangwan MS and Saharan GS (2005) Fungal diseases of rapeseed-mustard. In: Saharan GS, Mehta N and Sangwan MS (ed) *Diseases of Oilseed Crops.* New Delhi, India: Indus Publishing Co., pp. 15–86.

Mehta N, Sangwan MS, Srivastava MP and Kumar R (2002) Survival of *Alternaria brassicae* causing *Alternaria* blight rapeseed-mustard. *Journal of Mycology and Plant Pathology* 32(1): 64–67.

Micali C, Gollner K, Humphry M, Consonni C, and Panstruga R (2008) The powdery mildew disease of *Arabidopsis*: A paradigm for the interaction between plants and biotrophic fungi. *The Arabidopsis Book/American Society of Plant Biologists* 6.

Mishra KK, Kolte SJ, Nashaat NI and Awasthi RP (2009) Pathological and biochemical changes in *Brassica juncea* (mustard) infected with *Albugo candida* (white rust). *Plant Pathology* 58: 80–86.

Mohan M, Mehta N and Avtar R (2017) Integrated management of foliar diseases of Indian mustard. *Forage Research* 42(4): 274–278.

Morrall RAA, Dueck J, McKenzie DL and McGee DC (1976) Some aspects of *Sclerotinia sclerotiorum* in Saskatchewan, 1970–75. *Canadian Plant Disease Survey* 56: 56.

Neergaard P (1958) Mycelial seed infection of certain crucifers by *Sclerotinia sclerotiorum* (Lib) de bary. *Plant Disease Report* 42: 1105–1106.

Nelson BD, Christianson T and McClean P (2001) Effects of bacteria on sclerotia of *Sclerotinia sclerotiorum*. In: *Proceedings of XI International Sclerotinia Workshop*. York, UK: Central Science Laboratory, 8–12 July, p. 39.

Nowicki M, Nowakowska M, Niezgoda A and Kozik E (2012) Alternaria black spot of crucifers: Symptoms, importance of disease, and perspectives of resistance breeding. *Vegetable Crops Research Bulletin* 76: 5–19.

Nyvall RF (2013). Diseases of rapeseed (canola) and mustard. In Nyvall RF (ed) *Field Crop Diseases Handbook*. New York: Springer Science and Business Media.

Ordonez-Valencia C, Ferrera-Cerrato R, Quintanar-Zuniga RE, Flores-Ortiz CM, Guzman GJM, Alarcon A, . . . and Garcia-Barradas O (2015) Morphological development of sclerotia by *Sclerotinia sclerotiorum*: A view from light and scanning electron microscopy. *Annals of microbiology* 65(2): 765–770.

Peng G, Liu X, McLaren DL, McGregor L and Yu F (2020) Seed treatment with the fungicide fluopyram limits cotyledon infection by *Leptosphaeria maculans* and reduces blackleg of canola. *Canadian Journal of Plant Pathology* 1–13.

Pereira JCR, Chaves GM, Matsouok K, Silva AR and Vale FXR (1996) Integrated control of *Sclerotinia sclerotiorum*. *Fitopatholgia Brasileira* 21: 254–260.

Prasad R, Kumar D and Kumar V (2017) Symptoms and management of white rust/blister of Indian mustard (*Albugo candida*). *International Journal of Current Microbiology and Applied Sciences* 6: 1094–1100.

Purdy LH (1979) *Sclerotinia sclerotiorum*: History, diseases and symptomatology, host range, geographic distribution, and impact. *Phytopathology* 69: 875–880.

Rakesh R, Rathi AS, Kumar A and Singh H (2016) Evaluation of fungicides for the control of Sclerotinia stem rot of Indian mustard caused by *Sclerotinia sclerotiorum* (Lib.) de Bary. *Journal of Natural and Applied Sciences* 8: 441–444.

Rathi AS, Sharma S and Singh D (2012) Efficacy of carbendazim as prophylactic control of Sclerotinia rot in Indian mustard. In: *1st National Brassica Conference on "Production Barriers and Technological Options in Oilseed Brassica*" held at CCS HAU, Hisar from March 02–03, 2012. Abstracts p. 130.

Rimmer SR and van der Berg CGJ (2007) Black leg (phoma stem canker). Pp. 19–22 In: Rimmer SR, Shattuck VI and Buchwaldt L (eds) *Compendium of Brassica Diseases*. St. Paul, MN: APS Press, 117 pp.

Roy AK and Saikia UN (1976) White blight of mustard and its control. *Indian Journal of Agricultural Science* 46: 197.

Saharan GS and Mehta N (2002) Fungal diseases of rapeseed-mustard. *Diseases of field crops* 193–228.

Saharan GS and Mehta N (2008) *Sclerotinia Diseases of Crop Plants: Biology, Ecology and Disease Management*. The Netherlands: Springer Science+Busines Media BV, p. 485.

Saharan GS, Mehta N and Sangwan MS (2005) Development of disease resistance in rapeseed-mustard. *Diseases of Oilseed Crops* 561–617.

Saharan GS and Verma PR (1992) *White Rusts: A Review of Economically Important Species*. Ottawa, ON, CA: IDRC.

Satou M and Fukimoto F (1996) The host range of downy mildew, *Peronospora parasitica*, from *Brassica oleracea*, cabbage and broccoli crops. *Japanese Journal of Phytopathology* 62(4): 393–396.

Savchuk S and Fernando WGD (2004) Effect of timing of application and population dynamics on the degree of biological control of *Sclerotinia sclerotiorum* by bacterial antagonists. *FEMS Microbiology Ecology* 49: 379–388.

Sharma P, Meena PD, Verma PR, Saharan GS, Mehta N, Singh D and Kumar A (2016) *Sclerotinia sclerotiorum* (Lib) de Bary causing Sclerotinia rot in oilseed brassicas: A review. *Journal of Oilseed Brassica* 1(2): 1–44.

Shivpuri A and Gupta RBL (2001) Evaluation of different fungicides and plant extracts against *Sclerotinia sclerotiorum* causing stem rot of mustard. *Indian Phytopathology* 54: 272–274.

Shukla AK (2005) Estimation of yield losses to Indian mustard (*Brassica juncea*) due to Sclerotinia stem rot. *Journal of Phytological Research* 18: 267–268.

Siddaramaiah AL (2007) Epidemiological studies of white rust, downy mildew and *Alternaria* blight of Indian mustard (*Brassica juncea* (Linn.) Czern. and Coss.). *African Journal of Agricultural Research* 2(7): 305–308.

Singh JP, Singh HK, Singh RB and Singh AK (2014) Major foliar diseases of rapeseed-mustard and their integrated management. *International Journal of Agricultural and Statistical Sciences* 10(1): 9–13.

Singh R, Singh D, Salisbury P and Barbetti MJ (2010) Field evaluation of Indian and exotic oilseed *Brassica napus* and *B. juncea* germplasm against Sclerotinia stem rot. *Indian Journal of Agricultural Science* 80: 1067–1071.

Singh UP (2000) Pea-powdery mildew-an ideal pathosystem. *Indian Phytopathology* 53(1): 1–9.

Sullivan MJ, Damicone JP and Payton ME (2002) The effects of temperature and wetness period on the development of spinach white rust. *Plant Disease* 86: 753–758.

Thomma BP (2003) *Alternaria* spp.: From general saprophyte to specific parasite. *Molecular Plant Pathology* 4(4): 225–236.

Van de Wouw AP, Elliott VL, Chang S, Lopez-Ruiz FJ, Marcroft SJ and Idnurm A (2017) Identification of isolates of the plant pathogen *Leptosphaeria maculans* with resistance to the triazole fungicide fluquinconazole using a novel in-planta assay. *PLoS One* 12: 11.

Vanniasingham VM and Gilligan CA (1989) Effects of host, pathogen and environmental factors on latent period and production of pycnidia of *Leptosphaeria maculans* on oilseed rape leaves in controlled environments. *Mycological Research* 93: 167–174.

Wangkhem B, Devi KS and Chanu WT (2020) Integrated disease management IDM of white rust of rapeseed and mustard. In: *AgriCos e-Newsletter*. Available at www.researchgate.net/publication/341788078 (Accessed 05 December 2020).

Whetzel HH (1945) Synopsis of the genera and species of Sclerotiniaceae, a family of somatic inoperculate discomycetes. *Mycologia* 37: 648–714.

Willetts HJ and Wong AL (1971) Ontogenetic diversity of sclerotia of *Sclerotinia sclerotiorum* and related species. *Transactions of the British Mycological Society* 57: 515–524.

Wu H (1988) Effects of bacteria on germination and degradation of sclerotia of *Sclerotinia sclerotiorum*(Lib) de Bary. M.Sc. thesis, North Dakota State University, Fargo, North Dakota.

Zhao J, Peltier AJ, Meng J, Osborn TC and Grau CR (2004) Evaluation of Sclerotinia stem rot resistance in oilseed *Brassica napus* using a petiole inoculation technique under greenhouse conditions. *Plant Disease* 88: 1033–1039.

9 Impact of Environmental and Edaphic Factors on Winter Fodders and Remedies

Zaffar Malik, Muhammad Asaad Bashir, Ghulam Hassan Abbasi, Bushra, Muhammad Ali, Muhammad Waqar Akhtar, Muhammad Adil, Muhammad Irfan, Freeha Sabir, and Maqshoof Ahmad

CONTENTS

DOI: 10.1201/9781003055365-9

9.1 INTRODUCTION

The word 'fodder' refers to any agricultural foodstuff that has been used for raising domestic livestock, including cattle, sheep, horses, chicken, and pigs. The difference between fodder and feed is that fodder contains any food (plant material) that has been cut and carried to the animals. Fodders are delicious, juicy, and nutritious and have an adequate amount of carbohydrates but lack proteins essential for animal production and health. Most animal feed is derived from plant material. However, some manufacturers also add ingredients to process feeds that have animal origins. Since ancient times, leguminous crops have been utilized as animal feed and contributed to the agriculture world in fodder and soil fertility. However, with the beginning of the agricultural revolution (1500–1850) in Europe, leguminous crops dominated as a food source (fodder) for animals and gained a central position in agricultural development. Legume crops mostly perform two functions, symbiotic N_2 fixation and provision of high-protein fodder to livestock by maintaining grain-based rotation productivity (Sumberg, 2002). The cultivation of leguminous crops such as *Mucuna pruriens* (a trailing green manure species) started in East and West Africa (Nigeria and Uganda) and has been widely used in Asia (Boonman, 1993). Leguminous species from South America showed noticeable results in Australia and later were introduced in Africa's sub-Saharan region.

Fodder leguminous crops serve as a binding agent in recent farming models (Sumberg, 2002). There is a variety of winter leguminous crops that has been cultivated extensively to meet domestic as well as commercial livestock demands. These crops include oats, senji, shaftal, vetch, berseem, and lucerne. Barley is also reported to grow well in winter with annual legume-cereal cropping systems for forage purposes (Kardağ and Büyükburc, 2003). Cereal production, along with climbing vetches, helps vetches with light interception and facilitates mechanical harvesting.

Similarly, mixture cropping of legumes and grains can improve both crops' nutritional value and enhance soil fertility. The intercropping or mixture cropping of winter fodder (vetch, oats, barley, and wheat) is often preferred; however, this cropping method may affect the growth, yield, and quality of individual forage (Lithourgidis et al., 2006). Farmers prefer oats, barley, wheat, and triticale to provide climbing frames for legumes and increase fodder production (Roberts et al., 1989). Recently, sustainable land use has gained importance in forming a link between agricultural and livestock production through innovation and modern techniques. The utilization of land to grow fodder rather than human food is a controversial topic, but it is believed that livestock plays an essential role in small-scale agricultural sustainability and nutritional productivity (Schulz et al., 2001). There are also some crops (maize) that can be grown to fulfill human and animal needs. In many cases, fodder production (for animals) with the main crop (human food) is considered valuable, as it builds organic matter in soil and improves soil fertility.

Winter fodder can be grown on various soils with variable fertility, texture, and tillage methods. Apart from sustainable land use, a number of several environmental and edaphic factors affect winter fodder growth and production. The environmental factors that reduce fodder production are irregular precipitation, drought, salinity, temperature, waterlogging, drainage, light intensity, and various inorganic and organic pollutants (Van Der Heijden et al., 2008). It has also been observed that edaphic factors (texture, structure, pH, redox potential (Eh), soil porosity, color, and soil water holding capacity) considerably affect the yield and quality of winter fodder (Pang et al., 2011).

9.2 GROWTH CONDITIONS AND TYPES OF WINTER FODDER

Winter fodder includes many crops and legumes (alfalfa, berseem, barley, clover, and rapeseed). Several strategies have been compiled to improve the quality and quantity of winter fodder, and herbal legumes are introduced. Legumes provide not only rich protein feed but also fix atmospheric nitrogen through symbiotic association. These legumes (N_2 fixation) can reduce the dependence on expensive chemical fertilizers and enhance soil fertility for future crops. Apart from animal feed, winter fodder also adds much-required organic matter to the soil. In many countries, winter fodder is also intercropped with cash crops like sugarcane and wheat. A variety of winter fodder is used nowadays considering the local environmental conditions and soil properties. A few of these crops will be discussed in the following.

9.2.1 Alfalfa

Alfalfa, also known as lucerne (*Medicago sativa*), is an essential forage species (Babinec et al., 2003). It is a perennial flowering plant belonging to a leguminous family, Fabaceae. This fodder is native to the Middle East and is also cultivated in South Africa, North America, Australia, and South Asia. It is considered one of the vital fodder crops in many countries. Alfalfa is used for grazing, hay, silage, and green manuring. At earlier stages, alfalfa resembles clover, with trifoliate leaves. Alfalfa dominates in warm temperate climates but can also be grown in tropical and sub-tropical areas. The history of alfalfa used as forage dates back to ancient Greeks and Romans. Its sprout is used in South Indian cuisine. It is considered an excellent fodder with high protein content and energy. Alfalfa reduces the addition of supplements in the ration and is recommended for high-yielding dairy cows (Undersander et al., 2011).

9.2.2 Berseem

Berseem (*Trifolium alexandrinum*) is an annual, sparsely haired, standing pasture legume, 30 to 80 cm tall (Nichols et al., 2007; Heuzé et al., 2016). This fodder has shallow roots and stems with alternate leaves of small bearings (405 cm long and 2–3 cm wide). Berseem forms yellowish-white dense oval clusters of 2 cm diameter.

It is cross-pollinated and bears purple-red seed in a fruit pod (Ijaz et al., 2019). Berseem is a variable species divided into four classes according to its branching behavior and successive productivity. The highly branched and productive types are Miskawi and Kahrubi (Ijaz et al., 2019; Suttie, 1999a). Berseem is a fast-growing high-quality forage, mainly used as green chopped forage; its comparable feed value is often compared with alfalfa. However, unlike alfalfa, it has never been reported to cause swelling. It is a moisture-loving crop with moderate resistance to drought but can be grown in alkaline soil. Berseem can be sown in early autumn to provide feed before and during the colder months (Suttie, 1999a). High temperatures after winter improve its vegetative growth (Hannaway and Larson, 2004; Suttie, 1999b). It can produce seeds in high quantity under favorable environments. Berseem is often mixed with oats to make silage or mixed with chopped straw. Grazing is possible, although not as common as cutting. Berseem is also recommended for green manuring.

9.2.3 Oat

Oat (*Avena sativa* L) is a cereal and forage crop grown for animal and human consumption (Sapre et al., 2018). Oat is a secondary crop (originating from the weed of primary cereal domestication) native to the Middle East and Europe. Oat is an annual plant and is planted in autumn or spring. It is best grown in temperate regions with low heat requirements. It is a multipurpose crop (grain, forage, and rotation crop) with high protein content and is often compared to soy protein. Oat is a very nutritious cereal crop, with eminent oil, protein, and bioactive fiber β-glucan (Gorissen et al., 2018; Yue et al., 2020). Oat leaf production, initiation, and appearance depend on the temperature and day length. Oat roots, after germinating from radicle, penetrate in coleorhiza and develop as primary and secondary roots. Adventitious roots originate from stem internodes and develop on either side of tillers. These adventitious roots later form the permanent root system (Brouwer and Flood, 1995). Oat is less resistant to salinity compared to wheat and barley and moderately resistant to drought.

9.2.4 Barley

Barley (*Hordeum vulgare* L) belongs to the grass family and is a prominent cereal grain grown in temperate climates. Barley is self-pollinated, diploid species native to Western Asia and Northeast Africa. Barley is an important feed grain in many areas of the world (Edney et al., 2002; Newman and Newman, 2006). The recorded history of barley consumption by humans dates back 10,000 years. Barley is also a multipurpose crop and can be utilized for fodder, fermented material for beer and beverages, and healthy foods (Badr and El-Shazly, 2012). Barley is a widely adopted crop and is sown in temperate (summer crop) and tropical regions (winter crop). This fodder is sensitive to salinity and drought. It has been observed that drought reduce lamina length and width, whereas stomatal density was decreased under salt stress in barley (Wyka et al., 2019; Kiani-Pouya et al., 2020). Overall, barley is a vital fodder, and half of the United States' barley production is used as livestock feed.

9.2.5 Mustard

Brassica juncea L., commonly known as brown mustard or Indian mustard, is a mustard plant species that belongs to the Brassicaceae family. This plant is native to the Himalayas and is commercially cultivated in India, the United Kingdom, Denmark, and the United States. It is widely used in the Indian sub-continent as vegetables and for oil and fodder at the domestic level. It is a rich source of minerals, vitamins, and antioxidants (Kapoor et al., 2014). Mustard is a high biomass-producing plant with ample oil content used in biofuel production (Sridhar et al., 2005). Besides its edible uses, the mustard plant is also used as a green manure crop and for phytoremediation of metal-polluted soil (Smrithi et al., 2012).

9.2.6 Clover

Clover, also called 'trefoil,' is the common name for plants of the genus (*Trifolium*) derived from the Latin word tres meaning 'three,' and *folium* mean 'leaf,' and it consists of about 300 flowering plants in the legume or pea family Fabaceae. The genus has a worldwide distribution and the highest diversity in the temperate northern hemisphere, but many species exist in South America, Asia, and Africa, including high-altitude areas in tropical mountains. They are annual, biennial, or short-lived perennial herbs. Clover can be evergreen. The leaves are trifoliate (rarely four-leaf), five-leaf, or seven-leaf, with stipules attached to the petiole, and small red, purple, white, or yellow flower heads with dense spikes; sepals are packed with small seed pods. Other related genera often referred to as clover include *Melilotus* (sweet clover) and *Medicago* (alfalfa).

9.3 ECONOMIC VALUE OF WINTER FODDER

Fast-growing fodder legumes can be intercropped or used as a sole crop that can improve soil physical and biological conditions, suppress weeds, and provide high economic benefits. Besides, various field and environmental options are also available for effective weed management in fodder legumes. Pakistan has various types of livestock. It provides food security by providing milk, meat, and self-employment for men and women and plays a vital role in reducing poverty among small farmers. Its contribution to gross national product is about 11.5% and to agricultural gross domestic product is about 55%. The livestock population in Pakistan (buffalo, cattle, goat, sheep, donkey, camel, horse) is 163 million heads (economic survey 2010–11). The livestock population is growing at a rate of 4.2% per year, so its fodder demand is also increasing. A regular supply of adequate and nutritious fodder is essential for the promotion and development of livestock. Fodder crops are the primary and cheapest source of livestock feed. However, insufficient feed production is the main restricting factor in the country's livestock production. Every ten years, the feed area is reduced by about 2%.

9.4 GROWTH CONDITIONS FOR WINTER FODDER

During winter, low temperatures (<20°C) will limit legume yield, and summer temperatures exceeding 40°C could be harmful (Johansen et al., 2000). In eastern parts of the Indo-Gangetic plains, legumes face boron (B) and calcium (Ca) deficiency and acidic soil limitation when not growing on alluvial soil. In alkaline soils with excessive moisture, iron (Fe) deficiency may be a problem (Reddy and Krishna, 2009). In the dry season, disease attack in rabi legumes is less severe than that on rainy season legumes due to more favorable canopy microclimatic conditions during the rainy season. However, rabi and summer legumes are affected by the same soil-borne diseases, viruses, and insects.

Black gram and mung bean are short-day tropical crops (Johansen et al., 2000). Crop phenology is affected by day length, precipitation, temperature, and humidity. For example, longer days extend the crop stages, such as delayed flowering and maturity, leading to irregularities in pod maturity. These variations are mainly dependent on genotype, soil, and environmental interaction in subtropical regions. However, recently, some short-term evolutionary genotypes have been relatively unresponsive to daylight (Ramakrishna et al., 2000). Growth of rainy season fodder legumes is strictly inhibited by insect pests (*Maruca lestulalis)* geyer, (*Helicoverpa armigera)* pod borer, legume pod borer, and *Melanagromyza* spp. (pod fly); waterlogging; and damage caused by rain to growing pods. However, some genotypes may be appropriate for rabi cultivation (Raza et al., 2005).

Numerous essential tropical fodder legumes are usually suitable to grow in the Indo-Gangetic plain, with long days (from April to October). These are typically short-day crops with growth stages stimulated by declining temperatures in short day lengths in winter. One exception is peanuts (*Arachis hypogaea* L), in which some genotypes are favored by short-day environments (Collomb et al., 2004). These tropical legumes have an optimal temperature for most growth and development processes within the range of 20–40°C; this is the usual temperature limit for growing legumes in long-day environments.

9.5 SOCIO-ECONOMIC BENEFITS OF WINTER FODDERS

Several biological, non-biological, and socio-economic restrictions explain fodder legumes' poor performance relative to significant grain and cash crops. Previous research to improve grain production of legumes in the Indo-Gangetic plain is relatively limited and lacks focus. Methods to develop, assemble, and evaluate and improved production technologies have recently been developed. Previous research was mainly focused on a variety of development. Factors such as limited inherent yield potential, susceptibility to pests and diseases, sensitivity to water level, and soil and environmental changes lead to substantial changes in grain and legume production year after year. There are two critical periods of fodder shortage, one in winter (November to January) and the other in summer (May to June). Currently, the output of fodder crops exceeds 10.3% of the total crop area of 22.6 million hectares. Punjab's area share is 82.56%, Sindh 11.50%, non-timber forest products 4.48%, and Baluchistan 1.46%. It is estimated that the area of various fodder crops in the country is 2.31 million hectares, and the annual feed output is 51.92 million tons. The average feed output is 22.5 tons ha^{-1} (Pakistan Bureau of Agricultural Statistics, 2009–10), which is too low to meet half of its current livestock population maintenance needs.

In terms of nutrition, the estimated deficit is 15% to 30% of the demand. If expressed in digestible protein, the shortage is even more significant. Two major livestock production systems in Pakistan are rural households and large livestock. The first depends upon raising fodder crops and livestock feed. In later systems, mostly small ruminants are raised in pastures. More than half of animal feed comes from fodder and crop residues; one-third comes from pastures, wastelands, and canal bank roadside grazing; and the rest is from crops and their by-products. Different sources to feed livestock in Pakistan are 51% from fodder and crop residues; 38% from forage/grazing; 6% from cereal by-products; 3% from post-harvest grazing; and 2% from oilcake, meals, and animal protein.

The main fodder crops grown in winter include berseem, lucerne, oats, barley, and mustard. These are corn, sorghum, S.S. hybrids, millet and cow beans, and guar in summer. The less volatile areas decreased from 2.6 million hectares in 1976–77 to 2.31 million hectares in 2009–10. However, the total output of the corresponding year increased from 45.10 million tons to 51.92 million tons. Similarly, due to feeding research scientists' concerted efforts through improved production technology and research and development (R&D), the yield per unit area has also increased from 17.4 tons/ha (about 30%) to 22.5 tons/ha. Recently, through the cultivation of multispecies feed crops (such as S.S. hybrids, lucerne, a mixture of grains and pulses, and mortgrass), the green feed shortages that occurred during the two deficit periods have been solved. However, the availability of improved seeds for forage crops is one of the main limiting factors for producing forage crops. It is estimated that only improved seeds can grow 5% to 10% in the fodder crop area. Considering the problem of increasing seed yield, this involves many interconnected systems: agro-climatic conditions, adaptation of specific crops to the environment, and socio-economic and political factors, including prices and sales, crop management, and production.

9.6 CONSTRAINTS IN THE PRODUCTION AND GROWTH OF WINTER FODDER

The use of agricultural land to grow fodder rather than human food can be controversial; some types of feeds, such as corn (maize), can also serve as human food; and those that cannot be used as human food, such as grassland grass, may be grown on land that can be used for crops consumed by humans. Legumes are grown on various soils with variable textures; low-energy tillage methods restrict planting to a shallower depth by producing a compact layer in the soil (plow pan). This may hinder the root growth of fodder, especially in clay soils. Besides, intensive tillage reduces soil organic matter, resulting in a reduced organic combination of soil particles. These conditions are not suitable for fodder legume crop establishment. Cloddy soils result in poor germination of legumes due to reduced seed-soil interaction and reduced moisture availability to the seeds, while moist and dryness tend to form dense seedbeds with high soil strength, preventing it from appearing physical constraints impeding emergence. Compact layer development will be improved in the dry land planting system, suitable for legume crops. According to previous research, alfalfa has increased starch and sugar content during the day, but the variability in results may be due to the difference in the analytical method, plant growth stage, and environmental impacts. Sowing time is critical, affecting productivity and legumes' success by changing the plant environment, temperature, photoperiod, and soil moisture availability. Low crop stand is one of the main factors restricting low legume yield, especially under the country's rice and wheat planting systems. The grain yield of legumes can be significantly increased by increasing plant density.

Application of micronutrients such as boron (B), iron (Fe), zinc (Zn), molybdenum (Mo), and manganese (Mn) on legumes increased plant physiological parameters. In chickpea, improved nodulation, root growth, and yield were observed with the application of 25 kg zinc sulfate ($ZnSO_4$) ha^{-1} (Singh et al., 2013), and improved Fe, Zn, and P uptake was observed (Dravid and Goswami, 1987). Lentils are extremely sensitive to zinc deficiency. Application of 12.5–15 kg $ZnSO_4$ ha^{-1} to the soil can increase legume yield. $ZnSO_4$ foliar sprays with a combination of lime can also effectively correct zinc deficiency in legumes.

Droughts of varying intensities have been experienced in rainfed regions at different growth stages of legumes. Response to limited irrigation has been observed in many grain legumes; common bean (*Phaseolus vulgaris* L.) responds more than pea. An adequate supply of irrigation is the most critical factor in the successful growth of mung bean as a catch crop in fallow months. In most cases, pod initiation is considered the most critical growth stage in legumes. However, the moisture and soil type determine the demand for subsequent irrigation. In contrast, excessive water or waterlogging will reduce the rhizosphere's oxygen concentration, affecting biological nitrogen fixation (BNF) activity and nutrient utilization, ultimately reducing yield. Therefore, proper drainage should be provided for better germination and legume growth, especially in lowland areas.

During the early stages of growth, legumes are weaker competitors to weeds, and as a result, legumes suffer from heavy yield losses. Weeds cause higher production losses in pea, black gram, and chickpeas by reducing their yield by 44%, 50%, and 42%, respectively. The nature and extent of crop weed competition are affected by several factors, such as crop type, planting system, planting time, plant population, water supply, and soil fertility conditions. Legume-weed competition adversely affects crop yield and BNF. Legumes such as mung bean, black gram, cowpea, and soybean are short statured and early maturing and can be intercropped in broad stature crops, for example, pigeon pea, in order to reduce the weed threat.

9.7 ENVIRONMENTAL FACTORS AFFECTING WINTER FODDER GROWTH AND YIELD

9.7.1 Drought Stress

Environmental conditions directly influence winter fodder growth, and weather factors like sunlight, rainfall, snowfall, and hail considerably affect their yield. These factors control crop development and wellbeing, yearly harvest yield, and yield of every crop that specific season. (Howden et al., 2007; Kang et al., 2009; Lehmann, 2013; Paudel et al., 2014; Liang et al., 2017). Expected atmospheric changes contain more successive climate boundaries, including dry spells, which will influence numerous parts of the life cycle, water assets, and wellbeing besides thriving of the occupants and farming (Karl et al., 2009). Less water availability resulting from drought conditions gives low yields, negatively affecting crops' growth (Karl et al., 2009). Texas is a state where water availability is a big issue, making it a water-deficit state. Their susceptibility is being faced by a sharply increasing population (Singh and Mishra, 2011).

Drought is one of the climatic factors, a kind of environmental condition resulting from no or less rainfall (Paulo and Pereira, 2006). It can be observed in the areas of high and low rainfall. This phenomenon is a compound that is not readily known. It is observed from time to time in random seasons with random intensity. Drought is defined in agriculture as a long-term water shortage for the soil in any specific area which affects any specific crop species or all at a specific time. It becomes the main reason for the deleterious conditions and, ultimately, low crop yields.

The Czech Republic is a country where wheat is the most commonly cultivated crop in winter. Wheat surpassed a land area of 830,000 ha in 2014–2015, and it yielded 5,274,000 tons (ČSÚ, 2016). The Czech Republic faced drought in the year 2015, one of the most dangerous droughts ever. Frequently observed drought might be a primary issue in the near future (Daňhelka et al., 2015). With a rise in temperature by 2°C or more (above the levels of the end of the 20th century), unfavorable climatic changes for the main crops begin, which negatively affect production, reducing the variability of annual yield in many areas (IPCC, 2014).

Stress factors are conditions resulting in a short- or long-term change in plant structure and function. If the degree of response to the stressor does not surpass the plant's genetically defined tolerance, the plant assumes that the conditions have changed, while long-term exposure to the stressor will lead to permanent harm at high intensity in cells, tissues, and organs, even causing the death of the entire plant in extreme cases (Chaves and Oliveira, 2004). Water shortages significantly limit crop growth, development, and in particular, yield. The growth and rate of cell division in the leaves are initially inhibited by moderate water stress; as the water content drops below the tissue's saturation point, the growth rate decreases. Prolonged stress can trigger plant metabolism disruptions, especially in the plant's photosynthetic activity. The photosynthesis rate decreases under drought, possibly due to decreased RuBisCo activity due to less stomatal conductivity and decreased CO_2 availability (Hura et al., 2007; Kalaji and Loboda, 2010). With various approaches, the physiological processes that take place in plants under stress can be studied. The photosynthesis method, one of the essential processes in plants but at the same time responsive to stressors, is investigated by the most comprehensive techniques. (Kalaji and Loboda, 2010). Several authors note a decrease in photosynthesis strength and transpiration in forage grasses subjected to drought stress (Olszewska, 2006, 2009; Xu and Zhou, 2005). The largest decrease in CO_2 uptake under water stress was observed for *F. pratensis*, according to Olszewska et al. (2010), of the tested grass species (*Lolium perenne, Dactylis glomerata, Festuca pratensis, Phleum pratense*, and *Arrhenatherum elatius*). In *D. glomerata*, the reduction was minimal and negligible. Plants show the mechanisms of the battle against drying out under conditions of water scarcity. Second, they increase abscisic acid (ABA) content and synthesize stress proteins that protect and participate in cell membrane osmoregulation (Farooq et al., 2009). The accelerated level of ABA in the cells contributes to a decrease in transpiration rate. Furthermore, abscisic acid has an inhibitory effect on shoot growth, but at the same time, it promotes root growth and development, which helps to resolve stress significantly (Farooq et al., 2009).

9.7.2 Temperature Stress

In the sense of potential climate change, complex environmental constraints affecting plant growth and development have become a significant concern. Farming today is under tremendous environmental pressure across the world. In terms of global temperature measurements, by the end of the 21st century, the mean atmospheric temperature is predicted to rise by 1–6°C (De Costa, 2011). Such a global temperature increase will depend on the magnitude of the stress and affect agricultural production due to elevated temperatures, drought, salinity, floods, and toxicity of minerals. High-temperature heat stress is expressed as an increase in air temperature above a threshold for a reasonable period to cause irreversible damage or damage to crops in general (Teixeira et al., 2013). Heat stress causes the plant's water conditions to change (Hasanuzzaman et al., 2012, 2013), a reduction in photosynthetic potential (Hasanuzzaman et al., 2012, 2013), (Almeselmani et al., 2012; Ashraf and Harris, 2013), reduced metabolic activities (Farooq et al., 2011) and hormonal changes (Krasensky and Jonak, 2012), production of ROS (Wang et al., 2011), increased production of ethylene (Hays et al., 2007), decreased growth of pollen tubes, and increased mortality of pollen (Oshino et al., 2011) of wheat.

Heat stress affects many plant processes, leading to morphophysiological changes in wheat plants, impeding growth processes, and eventually leading to significant yield losses (McClung and Davis, 2010; Grant et al., 2011). Depending on the degree and length of the temperature and the growth phases in which the stress is confronted, plant responses to heat stress vary considerably (Ruelland and Zachowski, 2010). In many crops, including wheat, heat stress's primary effect is to prevent seed germination and insufficient steadiness (Johkan et al., 2011; Hossain et al., 2013). An average temperature of about 45°C has extreme effects on embryonic wheat cells, decreasing crop mass through the germination and emergence of seeds (Essemine et al., 2010). By encouraging leaf aging and abscission and reducing photosynthesis, heat stress primarily affects plant meristems and decreases plant growth (Kosová et al., 2011). Less biomass is created by a warm environment than plants are grown at optimum or low temperatures. Day and night temperatures of about 30 and 25°C can have significant implications for leaf growth and productive stem development in wheat, respectively (Rahman et al., 2009). However, heat stress is more harmful to wheat development in the reproductive process (Nawaz et al., 2013). Increasing the average temperature by one degree during the cultivation process will lead to a severe loss of wheat yield (Bennett et al., 2012; Yu et al., 2014). In embryonic wheat embryos, high-temperature stress degenerates mitochondria, changes protein expression profiles, and lowers ATP storage and oxygenation, resulting in higher seed loss frequency correlated with semen mass and vigor growing conditions (Balla et al., 2012; Hampton et al., 2013). By speeding seed growth and shortening the filling times of the grains in wheat, a temperature rise of 1 to 2°C lowers the seed mass (Nahar et al., 2010).

The growth and production of wheat plants are affected by heat stress. In parallel to adapting applicable farming techniques, these effects can be primarily controlled by generating suitable plant species (Asseng et al., 2011; Chapman et al., 2012). Many attempts have been made to generate heat-tolerant genotypes using the information acquired so far about wheat plants' responses to heat stress. The two most necessary strategies can be pursued for sustainable wheat production in heat-stressed areas are 1) Application by molecular and biotechnological breeding of transgenic or genetically modified wheat genotypes along with traditional breeding methods and 2) triggering various sustainable agriculture management practices, interacting so far with thermal stress management under field trials.

9.7.3 Salinity Stress

The fodder beet is a halophyte that can be successfully cultivated around the world on salt-affected soils. It not only meets the requirements of dairy cattle and other livestock for forage, but it has also been useful for the productive use of moderately salt-damaged land, as no other crops can be cultivated. It can be grown from August to September in regions such as Pakistan and India to provide forage during the full winter season when no other forage crop can sustain it. Compared to other fodder crops, food beets can tolerate salinity; thus, they can be successfully grown in saline soils worldwide. Millions of hectares of marginal and saline soils have reduced the size of agricultural production (Wang et al., 2012).

Salinity interferes with all processes related to plant maturation (Kaya and Day, 2008). The mechanism of photosynthesis is also impaired by the adverse effects of salinity (Netondo et al., 2004), chlorophyll content reduction, and protein synthesis due to the insufficient supply of CO_2 (Jamil et al., 2007; Khan et al., 2007). Salinity is significant abiotic stress that adversely decreases plant development and growth (Majeed et al., 2010). Salinity slows the growth and output of plants (Al-Karaki, 2000). Ionic toxicity, nutrient imbalance, and decreased water capacity are the three main effects of salt stress on plant growth (Flowers and Flowers, 2005). Salt stress also changes the stability of the cell membrane (Mansour et al., 2005). Salt stress affects plant morphology and physiology (Akram et al., 2010; Ahmad, 2010). For example, sodium at higher concentrations is toxic, and the selective absorption of Na^+ ions in the cell membrane can be a significant factor in determining the vulnerability or resistance of crops to salinity (Khan et al., 2007). Osmotic tolerance involves all physiological changes in plants through the production and transfer of osmoprotectants such as amino acids (e.g., proline) and sugars (Rhodes et al., 2002), advantageous selective uptake pathways of K^+, and K^+ translocation in shoots according to various K^+ requirements of conveyors and networks. The removal mechanism is primarily designed to reduce the amount of toxic Na^+ in the root and shoot cytoplasm and preserve a high K^+/Na^+ ratio (Almeida et al., 2017). Plants with tissue tolerance reduce the concentration of Na^+ in the cytoplasm when the salt concentration in the leaves is high, thus preventing adverse effects on cell metabolism by storing large quantities of salts in vacuoles and other cellular spaces. (Adem et al., 2014; Roy et al., 2014). Research has concentrated mainly on the hypothesis of excluding Na+ and preserving the K^+/Na^+ ratio as the primary cause of salinity tolerance in recent decades (Munns, 2005). Previous findings, however, show a more significant influence of osmotic tolerance and tissue tolerance as the primary components of salt tolerance (Genc et al., 2016; Li et al., 2017). In general, by combining these mechanisms to maintain osmotic and ionic stress caused by salinity stress, plants respond to salinity stress.

9.7.4 Waterlogging

Water and oxygen are the essential components of all living things, including plants. Plants photosynthesize in the leaves while capturing water and minerals in the soil. More than 300,000 species of plants exist in the world. Many of them deal with congestion in part of their useful life. The resulting flooding is so severe that biochemical and morphological adaptations have been made during evolution (Cook, 1999) to allow for the succession of sporadically or permanently inundated areas on this green planet (Andrews, 1996). Although water is chemically benign, specific physical properties interfere with the free exchange of gases. Therefore, when immersed in water (Jackson and Richard, 2003) or even when only the ground is flooded, it can injure and destroy plants. (Vartapetian and Jackson, 1997).

Vegetable crops are more sensitive to waterlogging than field crops in terms of yield. It is estimated that about 10% of all irrigated agricultural land in the world is regularly affected by flooding, reducing crop productivity by 20% (Jackson, 2004). Besides, many rain-fed areas are also prone to temporary flooding, such as many parts of Bangladesh. Around 10 million hectares of land are frequently affected by floods during India and Bangladesh's monsoon season (Huke, 1982). These floods have severe consequences for crop productivity on much of the arable land. The effects of waterlogging on winter crops are generally considered to be reduced growth and chlorosis of older leaves (Ellington, 1986) due to the deterioration of physiological activities. Flooded plants have a higher concentration of reactive oxygen species (ROS) than typical plants.

A variety of cellular molecules; metabolic reactions; and metabolites such as proteins, lipids, pigments, photosynthesis, PS II quality, DNA, and so on are harmed by these ROS (Achraf, 2009). Calvin cycle inhibition is caused by an increased cell level of hydrogen peroxide (Ashraf and Akram, 2009). ROS are free radicals with one or more electrons that have not been combined. It is not a stable configuration, so radicals react to other cellular molecules to create more free radicals (Foyer and Halliwell, 1976; Hideg, 1997). Persistent plants are exposed to flooding conditions that could result in root disorders by inducing specific changes in photosynthesis's biochemical reactions, limiting photosynthetic ability. The restricted activity of ribulose bisphosphate carboxylase (RuBPC), phosphoglycolate, and glycolate oxidase was included in these biochemical modifications (Yordanova and Popova, 2001). Breakdown of the chloroplast membrane restricts photosynthetic electrons' movement and photosystem II efficiency (Titarenko, 2000).

In submerged soils, water strongly limits the diffusion of gases through the soil pores and thus does not meet emerging roots' needs. The primary cause of damage to the roots and shoots supplying them is a slower supply of oxygen. Aerobic microorganisms and roots readily absorb a small amount of dissolved oxygen (approximately 3%) during the initial stages of flooding. The diffuse escape and oxidative breakdown of gases such as ethylene or carbon dioxide released by roots and soil microbes are also prevented by flooding, requiring them to accumulate. Accumulated ethylene can slow down root proliferation, while carbon dioxide can severely damage certain species' roots, such as soybeans (*Glycine max*), but not rice (Boru et al., 2003). Bicarbonate ions can be produced by the trapped carbon dioxide that can increase high calcium content, contributing to iron and chlorosis absence (Jackson, 2004). The production of these potentially hazardous soil conditions would be accelerated by warm temperatures and a sufficient organic matter supply. In maize, an engorgement-induced reduction in the activity of superoxide dismutase (SOD) oxygen-processing enzymes has also been documented.

In comparison, when subjected to different degrees of waterlogging, an increase in various enzyme antioxidant activity was observed in maize seedlings (Bin et al., 2010). A rise in SOD, catalase, peroxidase (POD), and ascorbate peroxidase activities was shown by the submerged pigeon pea (Kumutha et al., 2009). These examples show that flooded plants use an antioxidant defense mechanism to counteract the harmful effects of ROS-induced oxidative stress. Plant growth regulator (PGR) endogenous levels such as gibberellins (GAs) and cytokinins (CKs) are limited in the roots. This raises the ABA and ethylene bud concentrations, contributing respectively to stomatal closure and the early onset of aging. The levels of auxin and aminocyclopropane-1-carboxylic acid (ACC), a substrate of ethylene biosynthesis, are also demonstrated to inhibit and increase under flood stress.

9.7.5 Inorganic and Organic Pollutants

An important class of contaminants in the atmosphere is polycyclic aromatic hydrocarbons (PAHs), primarily through incomplete combustion of anthropogenic fuels. PAHs are toxic compounds, and many mutagens are carcinogenic (Lin et al., 2001). Because of their environmental persistence, long-term toxicity, and vast quantities, PAHs merit more consideration than other contaminants, according to current knowledge (Aina et al., 2006; Parrish et al., 2005). Plants take up dissolved PAHs through the root system (Kacálková and Tlustoš, 2011), and these particles are accumulated on the waxy cuticles of leaves and the passage gases through the stomata; plants mainly consume these PAHs from polluted soil (Gao and Zhu, 2004). Collins et al., 2006). The deposition of suspended solids, which also induces soil pollution, is possibly the primary source of PAHs for plants. However, when sewage sludge, chicken manure, compost, or composted waste compounds are used, PAH concentrations in agricultural soils may increase dramatically (Abad et al., 2005). All stages of plant growth can be affected by PAHs, from germination to reproduction. According to Baud-Grasset et al., 1993, the critical stages of growth are mainly the early stages of ontogeny, seed germination, and root system formation. The rising

PAH load can be a deciding factor for plant species' future habitat diversity and can decide the potential economic return (Kummerová et al., 1997). In plants exposed to PAHs, biochemical and physiological changes are reported, and they can be identified earlier than anatomical and morphological changes (Babu et al., 2001). It is known that plants react to the presence of stressors by changing their phytohormone development (Mok and Mok, 2001; Kummerová et al., 2010). This phase is triggered by variations in the levels of enzymes involved in the biosynthesis of phytohormones and the control of enzymes associated with the degradation or deactivation of xenobiotics (Vaseva-Gemisheva et al., 2005). The gaseous ethylene phytohormone (C_2H_4) has profound effects on plants during their life cycle (Petruzzelli et al., 2000), and one of the first plant responses to the presence of stressors is an increase in production. The detection and biosynthesis of ethylene are highly regulated, as seen in different vegetative tissues, seedlings, ripe fruit, and aging flowers (Johnson and Ecker, 1998).

The available heavy metals for plant uptake are present in the soil solution as soluble components or are readily solubilized by root exudates (Gadd, 2001). While plants need certain heavy metals for their growth and maintenance, excessive quantities of these metals may become toxic to plants. The ability of plants to absorb essential metals also helps them produce other non-essential metals (Petruzzelli et al., 2000). Since metals cannot be broken down, they harm the plant, both directly and indirectly, when the plant's concentrations reach the optimum level. Inhibition of cytoplasmic enzymes and damage to çellular structures due to oxidative stress is a direct toxic effect induced by high metal concentrations (Assche and Clijsters, 1990). The substitution of essential nutrients at the cation exchange locations of plants is an example of an indirect toxic effect (Tzfira and Citovsky, 2003).

Moreover, the detrimental effect of heavy metals on soil micro-organisms' growth and activities may also indirectly influence plants' growth. For example, because of a high concentration of metals, a reduction in the number of useful soil microorganisms may decrease the decomposition of organic material, leading to decreased soil nutrients.

Due to heavy metal interaction with soil microorganisms, enzyme activities useful for plant metabolism may also be disrupted. Such toxic direct and indirect effects cause a decrease in plant growth, often leading to plant death (Schilcher, 1977). One of the significant feed grains grown in winter all over Pakistan is oats (*Avena sativa* L.) It is ideal for a wide range of types of soil, heights, and conditions of rain. It can tolerate waterlogging better than other grains (Alemayehu, 1997). The production of roughage in Pakistan is around 52–54% lower than the actual need (Bhatti, 1988). Due to many constraints, this unsatisfactory yield of crops is most relevant (Oad et al., 2004).

9.7.6 Light Intensity

Light is one of the most significant environmental factors for plant growth as the key energy source (Naoya et al., 2008). For plant growth, morphogenesis, and other physiological responses, light intensity and quality are crucial (Fukuda et al., 2008; Li and Kubota, 2009). The parameters of leaf anatomy, physiology, and morphology are highly influenced by changes in the light spectrum (Hogewoning et al., 2010; Macedo et al., 2011). The blue spectrum has been shown to increase the palisade's epidermis and mesophyll cell thickness, while the red spectrum reduces the abaxial surface and spongy tissue thickness (Macedo et al., 2011). In the growth of plants, the magnitude of light is also an essential factor. Plants grown in low light also tend to be more sensitive than plants grown in high light to photoinhibition. In general, rises in net photosynthesis (Pn) are associated with increases in light intensity. The high intensity of the radiation, however, resulted in a reduction in net photosynthesis. If it is not possible to rapidly dissipate the excess light energy absorbed by the photosynthetic apparatus, it can decrease photosynthetic efficiency and lead to photoinhibition and even damage to the photosynthetic reaction core. For example, high light stress can efficiently inhibit photosystem I, and excessive light often inhibits photosystem II repair (Takahashi and Murata, 2008).

Fodder beet in the coastal areas of several European countries is successively grown as a fodder crop. The crop is used to provide adequate food for livestock (Niazi et al., 2000). It improves milk production due to the higher water and sugar content and is an ideal feed for dairy cows. Fodder beet, combined with straw, are usually chopped and fed. Environmental variables such as environment and soil affect plant development and growth. Knowing the minimum, optimum, and maximum temperatures and light intensity required for plant growth and development is helpful (Aréchiga and Carlos, 2000; Håkansson et al., 2002). These variables evaluate the species of plants from which to grow. The growth rate of plants in the early stages of development is a significant determinant of productivity in an area. The proportion between the leaf and plant total weight is the leaf weight ratio. Slow-growing species can accumulate considerable quantities of secondary compounds such as lignin and phenols, increasing the leaf dry weight.

The broadleaf area was negatively correlated with high light intensity and temperature positively (Evans, 1989). Light intensity affects plants' dry matter content, and the proportion of the leaf surface decreases with increasing light intensity (Picken et al., 1986). The net assimilation rate is one of the most critical growth parameters. The net output efficiency of the assimilation device is defined here (Bruggink and Heuvelink, 1987). The relative growth rate indicates an increase in dry weight from the initial weight over a time period (Gardner et al., 1994). In terms of relative growth rate, plant species may differ widely. Habitat-related changes in abiotic variables such as temperature, water, light, nutrients, or biotic variables, such as competition and disease, can cause this (Poorter, 1989). Warmer temperatures accelerate development and early growth but can adversely affect the final biomass due to canopy aging and increased respiration maintenance (Demmers-Derks et al., 1998). Stomata are critical channels for exchanging water and air with the external environment, and by increasing the motive force of protons, the light intensity affects the conductivity of the stomata (Figure 9.1). Furthermore, stomatal growth also appears to be connected to light intensity (Lee et al., 2007).

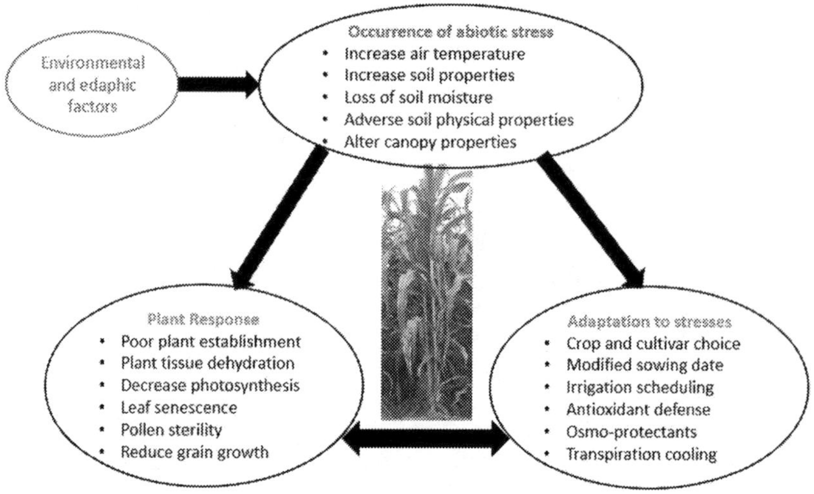

FIGURE 9.1 The relation between environmental and edaphic factors on fodder production and plant modification mechanism.

9.8 EFFECT OF EDAPHIC FACTORS ON THE GROWTH OF WINTER FODDER

9.8.1 Effect of Texture and Structure on the Growth of Winter Fodder

Soil factors like soil type and management strategies affect the productivity of crops. It is imperative to know soil conditions (texture, CEC, pH) for specific fodder production. As it mediates many biological and physical procedures in soils, soil structure is a vital soil property to be assessed; for instance, it determines porosity and infiltration and thus the availability of water to plants and vulnerability to soil erosion. Since soil structure also affects agrochemical losses and C and N gas loss sequestration, it is necessary to preserve soil structure to reduce agricultural practices' effect on the environment. Soil structure can sustain plant and animal life, as it is a prerequisite for soil's proper functioning. It regulates various significant soil properties and processes, such as retention and soil water movement, exchange of gases, and soil erosion.

Furthermore, soil structure also directly influences root penetration, organic matter, nutrient undercurrents, and ultimately crop yield (Bronick and Lal, 2005; Rabot et al., 2018). For soil structure, both the pore and solid-phase perspectives are imperative, as they are directly involved in soil structure formation (Rabot et al., 2018). During soil formation, the soil structure changes continuously; that is, the interaction of various biota (soil fauna, plants, microbes) with the weathered mineral material shapes the soil structure, which depends on the parent material climatic changes specific to the site.

It is not easy to disentangle the various biotas' relative contribution to soil structure formation, especially in controlled systems. Rising roots or burrowing earthworms reorganize soil particles' spatial arrangements as they align individual minerals and organic substances of different types and sizes, which may result in soil compaction along with the bio pores (Kautz, 2015). Biopores are cylindrical, have low tortuosity and high vertical continuity, and extend along with the entire soil profile. Thus, penetration and unique flow phenomena are substantially affected (Koestel and Larsbo, 2014; Naveed et al., 2013). It has been suggested that root growth depends more on an established pore network in dense soils and with growing depth (Gao et al., 2016), which results in a close relationship between new root growth, water extraction, and old bio pores (Stirzaker et al., 1996). Microbes modify the pore wall's surface by secreting extracellular polymeric substances (EPSs), leading to creating habitat covers on the micro scale (Colica et al., 2014). EPSs stabilize the soil structure by forming the 'glue' to bind mineral particles in the soil together with plant polysaccharides (Totsche et al., 2018; Watteau et al., 2006). The soil quality is mainly referred to as the soil aggregate distribution and stability measurements. The increased microbial activity level due to the addition of organic matter can increase micro-aggregate binding to form water-stable macro-aggregates. The major factors in soil aggregation include fibrous roots, polysaccharides, and fungal hyphae (Tisdall and Oades, 1982). The addition of organic matter in the soil enhances soil aggregation, ultimately increasing seed germination and plant vigor (Van Noordwijk et al., 1993).

9.8.2 Soil Organic Matter and Availability of Nutrients for Fodder Crop

The ultimate goal of sustainable agricultural growth is to increase the excellent quality of crop production while reducing its adverse effects on the environment (Tilman et al., 2002). The volume and consistency of soil organic matter represent and influence primary productivity, as they are the key constituents of biogeochemical cycles and major nutrient components. Because of their local availability as a source of multiple nutrients and the capacity to boost soil characteristics, organic materials hold great promise (Khaliq et al., 2006). Organic materials' input is needed to improve fertility and soil quality, especially in low-input agricultural systems (Naureen et al., 2005). A significant determinant of carbon and nutrient cycles in the biosphere is soil organic matter (SOM), as:

1. It is the primary source of nutrients for plant growth (after microbial decomposition).
2. It contributes to soil quality (erosion resistance, soil structure) (Herrick and Wander, 1998).
3. It also represents the biosphere-atmosphere system's major carbon reservoir (Falkowski et al., 2000).

Given the current global warming issues, it is important to consider when soils are either a source or a drain for atmospheric warming (Smith et al., 2000). Organic matter deposition in soil results from soil biota (soil fauna and microorganisms) on the available plant material. It has been shown that SOM mineralization can be exacerbated by the inclusion of fresh organic matter (FOM) such as green manure or straw in soil (Wu et al., 1993). The fertilizer requirement differs for each of the crops because of their ability to draw mineralized soil nutrients and differential output capacity. Oat (*Avena sativa* L.) is a vital feed crop among fodders and needs more attention to increase good-quality fodder production. Due to its outstanding routine of fast regrowth, oat is a widespread fodder crop. It requires abundant fertilizers to develop succulent and high-quality herbage (Hukkeri et al., 1977). A balanced fertilizer supply ensures that all nutrients are used efficiently. Inorganic fertilizers supply the critical plant nutrients, but heavy doses of inorganic fertilizers cause water contamination, soil depletion, and so on.

Among rabi fodder crops, oat requires more nutrient supply for proper growth and development. This is consistent with previous studies showing that the time and quantity of N release were positively affected by the previous crop (Dawson et al., 2008). By applying balanced fertilizer, especially with phosphorus fertilizer, the output of rabi fodder such as oat and maize fodder could be increased, as its quality can be significantly improved with this nutrient. The rise in fodder yield is due to the rise in plant height, leaf area per plant, and stem diameter with P application (Ayub et al., 2002). Similarly, (Alias et al., 2003) reported that an increase in the P level to 125 kg/ha significantly increased Stover maize growth; a further increase in the P rate (150 kg ha^{-1}) caused a reduction in Stover yield (Eijk et al., 2006). For better animal output, quality fodder is needed. Crude fiber, crude protein, acid detergent fibre, ash, and neutral detergent fiber content (percent) are the parameters used to describe the quality of forage (Soest, 1985). It is essential to add P fertilizer to improve both green fodder quality and quantity (Dhillon et al., 1998). P application controls Mg^{2+} absorption, preventing tetany disease in animals (Lock et al., 2000).

9.8.3 pH and Eh Affecting Growth and Yield of Fodder Crops

Soil pH is an indicator of the ionized H (H+) ion concentration in the soil solution. The single most insightful calculation that can be made to assess soil characteristics is possibly soil pH. The availability and uptake of nutrients influence soil pH by plants. In many areas of the world, plant growth can be significantly decreased due to excess soil acidity. Nutrient supply is impacted by soil pH for crop plants, thus impacting plant growth (Fageria et al., 1997). Due to the complex existence of different soil processes and these processes' interactions with plants and microorganisms, soil pH is also highly evolving (Adams, 1984). pH, which can be highly affected by plant cultivation and fertilizers, is one of the most significant factors determining soil fertility. Two common mechanisms by which plants change the rhizosphere to attain nutrients are acidification and root exudate release. Soil pH has been decreased by about one unit in Australia due to continuous clover cultivation for more than 30 years. This reduction in soil pH is primarily due to legume crop roots, which release protons in the soil (Schubert et al., 1990). For most crop plants' growth, including winter fodder, a pH range of 5.5–8.5 is recommended, but the optimum pH ranges from 5.5–6.5.

A pH of 5.5–6.0 or higher is required for winter fodder beet, and it is known to be responsive to low soil sodium levels.

9.8.4 Soil Porosity, Color, and Water Availability for Fodder Crops

Soil porosity is affected by soil compaction and earthworm behavior (Douglas et al., 1992; Lavelle, 1988) and offers valuable information on the potential for partly anaerobic soil conditions. In soil nutrient cycling, the mineralization of organic material can be accelerated by earthworm activity (Lavelle, 1988). Both soil cultivation and biological activity (root penetration) generate macropores and affect soil functions such as gas, water and nutrient transport, and organic matter storage (Peng et al., 2015). A factor that decreases soil fertility is known to be soil compaction. Many researchers assume that the increased use of chemical fertilizers compensates for decreased soil fertility. Compaction allows soil particles and aggregates to be rearranged, which reduces pore spaces, leading to a reduction in large pores, an increase in apparent specific weight, and total porosity. Subsequently, this outcome changes in the soil properties such as decreased air movement, decreased permeability, and conductivity of water, increased impedance to root and shoot growth during emergence, and changes in the soil's thermal properties. All these variations disturb the crop growth and development with potential effects on yields. The lower bulk density was due to higher soil organic matter content, increased root growth, better aggregation, and increased micropore volume. Fertilizer and manure application to the soil adds organic matter in the soil, enhances root growth, and increases aggregation and micropores' volume (Schjønning et al., 1994). Depending on the quantity of materials applied, total soil porosity increases with the application of fertilizer and compost (Celik et al., 2004)

The soil's water holding capacity is mainly regulated by the number and size of pores and the soil's actual surface area (Haynes and Naidu, 1998). Water is an essential natural resource, and its growing scarcity has led to the emergence of numerous problems relating to its efficient use, sustainability, and management. Just 2.7% of the world's water is available as freshwater, of which only 30% can be used to meet human and livestock requirements. It is projected that demand for good-quality water may increase several times and may even surpass its supply due to the ever-growing population, urbanization, industrialization, and increasing demand for food (ENVIS Centre, Punjab, 2005). Historically, the growth of high-quality livestock forage in arid and semiarid lands has been restricted due to reduced accessibility of water. Almost 50% of the global land area is considered sub-humid, arid, semiarid, or desert, resulting in low agronomic productivity (Zika and Erb, 2009). The water demand for winter fodder is very strong, for example, 140 cm of water a year is needed for berseem. Oat can be grown as a substitute crop in areas where irrigation water is not adequate for berseem, as it needs less water for growth than other fodder crops. Good yields are provided by four to five irrigations. In general, for winter fodder, irrigation immediately after each cut is mandatory. Water application to the soil will increase the availability of nutrients and transform nutrients or fertilizers (Li et al., 2009). This is evidence of crop N uptake, which increases with an increase in available water (Clarke et al., 2001). Organic nitrogen mineralization is proportional to soil water, affecting nitrogen movement and plant absorption (Song et al., 2006). As phosphorus in the soil is relatively immobile, soil moisture controls soil phosphorus mobility by mass diffusion (Yang et al., 2011). Research indicates that injecting K with water through the sub-surface drip will bring this nutrient to the soil's surface and more in-depth into the soil profile (Balkcom, 2009). As opposed to drip irrigation, another study found a possible leaching risk for applied K under furrow irrigation (Hebbar et al., 2004). Among the winter fodders, oat and ryegrass were less susceptible to saline water, while Indian clover was most sensitive to salt load.

The reduction in the yield and dry matter of various crops, including green fodders, depends on a variety of factors, such as the type of soil, inherent resistance of a crop to the saline environment, the need for crop water, climate change, and the stage of crop growth at which it was subjected to an aggressive environment (Yadav et al., 2007).

9.9 MANAGEMENT STRATEGIES

As discussed earlier, several environmental and edaphic factors affect fodder production. These factors contribute to low fertility, which is a limiting factor to sustaining livelihoods. Therefore, it is necessary to increase fodder production. the following management strategies can be adopted for sustainable production of fodder crops without further degradation of natural resources according to the current situation, problems, and potential for developing fodder resources. 1) Identification and propagation of high-yielding varieties of fodder crops should be introduced to develop the fodder's quality. This necessity can be attained by acquiring certified seed from the government and private sector or imported from other countries. There should also be training and potential of seed production of domestic fodder crops at the farmer level. Despite providing good-quality fodder seed in the government sector, many private seed companies have entered the country recently, and these companies should also have introduced good-quality fodder seed. 2) Growing multi-cut fodder crops such as hybrid fodder, lucerne, berseem, mixtures of legumes and cereals, and mott grass should be introduced. This approach will reduce fodder shortage and improve fodder availability throughout the year. 3) Soil amelioration practices also improve the establishment and growth of fodder to cover fodder production requirements. Our soil's production potential has been lost in remarkably inaccessible areas, which contributes to severe consequences. Most fodder-producing farmers lack technical information and production technologies and are unaware of soil fertility conservation. 4) Public awareness of the reduced fodder production problem by practical extension work needs to be generated. Support of the extension department needs to be utilized and deliver awareness with technical research findings. 5) Management, production, and exploitation of fodder resources should be introduced at the farmer level and developed as baseline information. 6) Socio-economic and political factors, including prices and marketing, should also be controlled. 7) Training/capacity-building for farmers, students, researchers, and extension workers should be organized. Moreover, there should be technical support and services to students for internship and research programs. 8) Concerned institutes need to be well equipped with the workforce, technical support facilities, required funds, and other physical infrastructure to conduct research projects for fodder production. 9) Research and development on fodder production should combine to use existing natural resources and introduce new technologies for enhancing fodder production. 10) Breeding of fodder crops should improve quality, yield, and resistance against adverse environmental conditions, that is, drought, temperature, diseases, and so on. Research and development should be conducted simultaneously and introduced to fodder-producing farmers to improve fodder production rapidly. Therefore, the best fodder species identification, productivity evaluation, and nutritive value should be considered a long-term strategy for fodder production and its effective exploitation.

9.10 CONCLUSION

The livestock sector largely contributes to milk and meat production. Fodder crops are the fundamental source of livestock feed. Berseem, lucerne, oats, barley, and mustard are the main winter fodder crops. Several environmental and edaphic factors influence the production of fodder crops. Improved management strategies can play an essential role in fodder production by regulating adverse factors.

REFERENCES

Abad E, Martinez K, Planas C, Palacios O, Caixach J and Rivera J (2005) Priority organic pollutant assessment of sludges for agricultural purposes. *Chemosphere* 61(9): 1358–1369.

Adams F (1984) Crop response to lime in the southern United States. *Soil Acidity and Liming* 12: 211–265.

Adem GD, Roy SJ, Zhou M, Bowman JP and Shabala S (2014) Evaluating the contribution of ionic, osmotic, and oxidative stress components towards salinity tolerance in barley. *BMC Plant Biology* 14(1): 1–13.

Ahmad B (2010) Effect of salinity and N sources on the activity of antioxidant enzymes in canola (*Brassica napus* L.). *Journal of Food, Agriculture & Environment* 8(2): 350–353.

Aina R, Palin L and Citterio S (2006) Molecular evidence for benzo [a] pyrene and naphthalene genotoxicity in *Trifolium repens* L. *Chemosphere* 65(4): 666–673.

Akram M, Ashraf MY, Ahmad R, Waraich EA, Iqbal J and Mohsan M (2010) Screening for salt tolerance in maize (*Zea mays* L.) hybrids at an early seedling stage. *Pak. J. Bot* 42(1): 141–154.

Alemayehu M (1997) Conservation-based forage development for Ethiopia. *Conservation-based forage development for Ethiopia.*

Alias A, Usman M, Ullah E and Waraich EA (2003) Effects of different phosphorus levels on the growth and yield of two cultivars of maize (*Zea mays* L.). *Int. J. Agric. and Biolo.* 5(4): 632–634.

Al-Karaki GN (2000) Growth, water use efficiency, and sodium and potassium acquisition by tomato cultivars grown under salt stress. *Journal of Plant Nutrition* 23(1): 1–8.

Almeida DM, Oliveira MM and Saibo NJ (2017) Regulation of Na+ and K+ homeostasis in plants: Towards improved salt stress tolerance in crop plants. *Genetics and Molecular Biology* 40(1): 326–345.

Almeselmani M, Deshmukh PS and Chinnusamy V (2012) Effects of prolonged high temperature stress on respiration, photosynthesis and gene expression in wheat (*Triticum aestivum* L.) varieties differing in their thermotolerance. *Plant Stress* 6(1): 25–32.

Andrews CJ (1996) How do plants survive ice? *Annals of Botany* 78(5): 529–536.

Aréchiga M and Vázquez-Yanes C (2000) Cactus seed germination: A review. *Journal of Arid Environments* 44(1): 85–104.
Ashraf M (2009) Biotechnological approach of improving plant salt tolerance using antioxidants as markers. *Biotechnology Advances* 27(1): 84–93.
Ashraf M and Akram NA (2009) Improving salinity tolerance of plants through conventional breeding and genetic engineering: An analytical comparison. *Biotechnology Advances* 27(6): 744–752.
Ashraf MHPJC and Harris PJ (2013) Photosynthesis under stressful environments: An overview. *Photosynthetica* 51(2): 163–190.
Assche F and Clijsters H (1990) Effects of metals on enzyme activity in plants. *Plant, Cell & Environment* 13(3): 195–206.
Asseng S, Foster IAN and Turner NC (2011) The impact of temperature variability on wheat yields. *Global Change Biology* 17(2): 997–1012.
Ayub M, Nadeem MA, Sharar MS and Mahmood N (2002) Response of maize (*Zea mays* L.) fodder to different levels of nitrogen and phosphorus. *Asian J. Plant Sci* 1(4): 352–355.
Babinec J, Kozova Z and Zapletalova E (2003) The characteristics of some lucerne (*Medicago sativa* L.) varieties. *Czech Journal of Genetics and Plant Breeding* 39(Special issue): 188–193.
Babu TS, Marder JB, Tripuranthakam S, Dixon DG and Greenberg BM (2001) Synergistic effects of a photooxidized polycyclic aromatic hydrocarbon and copper on photosynthesis and plant growth: Evidence that *in vivo* formation of reactive oxygen species is a mechanism of copper toxicity. *Environmental Toxicology and Chemistry: An International Journal* 20(6): 1351–1358.
Badr A and El-Shazly H (2012) Molecular approaches to origin, ancestry and domestication history of crop plants: Barley and clover as examples. *Journal of Genetic Engineering and Biotechnology* 10(1): 1–12.
Balkcom K (2009) *Effects of Subsurface Drip Irrigation on Chemical Soil Properties and Cotton Yield*. Doctoral dissertation.
Balla K, Karsai I, Bencze S and Veisz O (2012) Germination ability and seedling vigour in the progeny of heat-stressed wheat plants. *Acta Agronomica Hungarica* 60(4): 299–308.
Baud-Grasset F, Baud-Grasset S and Safferman SI (1993) Evaluation of the bioremediation of a contaminated soil with phytotoxicity tests. *Chemosphere* 26(7): 1365–1374.
Bennett D, Izanloo A, Reynolds M, Kuchel H, Langridge P and Schnurbusch T (2012) Genetic dissection of grain yield and physical grain quality in bread wheat (*Triticum aestivum* L.) under water-limited environments. *Theoretical and Applied Genetics* 125(2): 255–271.
Bhatti MB (1988) *National Perspectives for Fodder Crops in Pakistan*. Islamabad: NARC, 1–2.
Bin T, Xu SZ, Zou XL, Zheng YL and Qiu FZ (2010) Changes of antioxidative enzymes and lipid peroxidation in leaves and roots of waterlogging-tolerant and waterlogging-sensitive maize genotypes at seedling stage. *Agricultural Sciences in China* 9(5): 651–661.
Boonman G (1993) *East Africa's Grasses and Fodders: Their Ecology and Husbandry*.
Boru G, van Ginkel M, Trethowan RM, Boersma L and Kronstad WE (2003) Oxygen use from solution by wheat genotypes differing in tolerance to waterlogging. *Euphytica* 132(2): 151–158.
Bronick CJ and Lal R (2005) Soil structure and management: A review. *Geoderma* 124(1–2): 3–22.
Brouwer J and Flood RG (1995) Aspects of oat physiology. In *The Oat Crop*. Dordrecht: Springer, pp. 177–222.
Bruggink GT and Heuvelink E (1987) Influence of light on the growth of young tomato, cucumber and sweet pepper plants in the greenhouse: Effects on relative growth rate, net assimilation rate and leaf area ratio. *Scientia Horticulturae* 31(3–4): 161–174.

Celik I, Ortas I and Kilic S (2004) Effects of compost, mycorrhiza, manure and fertilizer on some physical properties of a chromoxerert soil. *Soil and Tillage Research* 78(1): 59–67.

Chapman SC, Chakraborty S, Dreccer MF and Howden SM (2012) Plant adaptation to climate change—opportunities and priorities in breeding. *Crop and Pasture Science* 63(3): 251–268.

Chaves MM and Oliveira MM (2004) Mechanisms underlying plant resilience to water deficits: Prospects for water-saving agriculture. *Journal of Experimental Botany* 55(407): 2365–2384.

Clarke AE, Bacic A and Lane AG (2001) *U.S. Patent No. 6,271,001.* Washington, DC: U.S. Patent and Trademark Office.

Colica G, Li H, Rossi F, Li D, Liu Y and De Philippis R (2014) Microbial secreted exopolysaccharides affect the hydrological behavior of induced biological soil crusts in desert sandy soils. *Soil Biology and Biochemistry* 68: 62–70.

Collins C, Fryer M and Grosso A (2006) Plant uptake of non-ionic organic chemicals. *Environmental Science & Technology* 40(1): 45–52.

Collomb M, Sollberger H, Bütikofer U, Sieber R, Stoll W and Schaeren W (2004) Impact of a basal diet of hay and fodder beet supplemented with rapeseed, linseed and sunflowerseed on the fatty acid composition of milk fat. *International Dairy Journal* 14(6): 549–559.

Cook CD (1999) The number and kinds of embryo-bearing plants which have become aquatic: A survey. *Perspectives in Plant Ecology, Evolution and Systematics* 2(1): 79–102.

ČSÚ (2016) Definitivní údaje o sklizni zemědělských plodin—2015. In *Český Statistický Úřad* [Online]. Available at www.czso.cz/csu/czso/definitivni-udaje-o-sklizni-zemedelskych-plodin-2015. [05–05–2016].

Daňhelka J, et al. (2015) *Drought in the Czech Republic in 2015* [Online]. Prague: ČHMÚ. Available at http://portal.chmi.cz/files/portal/docs/meteo/ok/SUCHO/zpravy/en_drought2015.pdf. [05–05–2016].

Dawson JC, Huggins DR and Jones SS (2008) Characterizing nitrogen use efficiency in natural and agricultural ecosystems to improve the performance of cereal crops in low-input and organic agricultural systems. *Field Crops Research* 107(2): 89–101.

De Costa WAJM (2011) A review of the possible impacts of climate change on forests in the humid tropics. *J Natl Sci Found Sri* 39: 281–302.

Demmers-Derks H, Mitchell RAC, Mitchell VJ and Lawlor DW (1998) Response of sugar beet (*Beta vulgaris* L.) yield and biochemical composition to elevated CO_2 and temperature at two nitrogen applications. *Plant, Cell & Environment* 21(8): 829–836.

Dhillon NS, Dresi TS and Brar BS (1998) Extractants for assessing P availability to berseem in flood-plain soils of Punjab. *Journal of the Indian Society of Soil Science* 46(3): 402–406.

Douglas JT, Koppi AJ and Moran CJ (1992) Changes in soil structure induced by wheel traffic and growth of perennial grass. *Soil and Tillage Research* 23(1–2): 61–72.

Dravid MS and Goswami NN (1987) Effect of FYM, P and Zn on dry matter yield and uptake of P, K, Zn, Fe by rice under non-saline and saline soil conditions. *J Nucl Agric Biol* 16: 201–205.

Edney MJ, Rossnagel BG, Endo Y, Ozawa S and Brophy M (2002) Pearling quality of Canadian barley varieties and their potential use as rice extenders. *Journal of Cereal Science* 36(3): 295–305.

Eijk D, Janssen BH and Oenema O (2006) Initial and residual effects of fertilizer phosphorus on soil phosphorus and maize yields on phosphorus fixing soils.

Ellington A (1986) Effects of deep ripping, direct drilling, gypsum and lime on soils, wheat growth and yield. *Soil and Tillage Research* 8: 29–49.

ENVIS Centre (2005) State of Environment, Punjab. Report by Punjab State Council for Science and Technology, Chandigarh, Punjab, pp. 315. epinasty in geranium under red light condition. *J. Hort. Sci.* 115: 176–182.

Essemine J, Ammar S and Bouzid S (2010) Impact of heat stress on germination and growth in higher plants: Physiological, biochemical and molecular repercussions and mechanisms of defence. *Journal of Biological Sciences* 10(6): 565–572.

Evans JR (1989) Photosynthesis and nitrogen relationships in leaves of C 3 plants. *Oecologia* 78(1): 9–19.

Fageria NK, Baligar VC and Jones C (1997) *Growth and Mineral Nutrition of Field Crop.* New York: Marcel Dakker. Inc, 1001.

Falkowski P, Scholes RJ, Boyle EEA, Canadell J, Canfield D, Elser J, . . . and Steffen W (2000) The global carbon cycle: A test of our knowledge of earth as a system. *Science* 290(5490): 291–296.

Farooq M, Bramley H, Palta JA and Siddique KH (2011) Heat stress in wheat during reproductive and grain-filling phases. *Critical Reviews in Plant Sciences* 30(6): 491–507.

Farooq M, Wahid A, Kobayashi N, Fujita DBSMA and Basra SMA (2009) Plant drought stress: Effects, mechanisms and management. *Sustainable Agriculture*: 153–188.

Flowers TJ and Flowers SA (2005) Why does salinity pose such a difficult problem for plant breeders? *Agricultural Water Management* 78(1–2): 15–24.

Foyer CH and Halliwell B (1976) The presence of glutathione and glutathione reductase in chloroplasts: A proposed role in ascorbic acid metabolism. *Planta* 133(1): 21–25.

Fukuda N, Fujita M, Ohta Y, Sase S, Nishimura S and Ezura H (2008) Directional blue light irradiation triggers epidermal cell elongation of abaxial side resulting in inhibition of leaf epinasty in geranium under red light condition. *Scientia Horticulturae* 115(2): 176–182.

Gadd GM (2001) Metal transformations. In *British Mycological Society Symposium Series*, Vol. 23, pp. 359–382.

Gao W, Hodgkinson L, Jin K, Watts CW, Ashton RW, Shen J, . . . and Whalley WR (2016) Deep roots and soil structure. *Plant, Cell & Environment* 39(8): 1662–1668.

Gao Y and Zhu L (2004) Plant uptake, accumulation and translocation of phenanthrene and pyrene in soils. *Chemosphere* 55(9): 1169–1178.

Gardner DK, Lane M, Spitzer A and Batt PA (1994) Enhanced rates of cleavage and development for sheep zygotes cultured to the blastocyst stage *in vitro* in the absence of serum and somatic cells: Amino acids, vitamins, and culturing embryos in groups stimulate development. *Biology of Reproduction* 50(2): 390–400.

Genc Y, Oldach K, Taylor J and Lyons GH (2016) Uncoupling of sodium and chloride to assist breeding for salinity tolerance in crops. *New Phytologist* 210(1): 145–156.

Grant RF, Kimball BA, Conley MM, White JW, Wall GW and Ottman MJ (2011) Controlled warming effects on wheat growth and yield: Field measurements and modeling. *Agronomy Journal* 103(6): 1742–1754.

Gorissen SH, Crombag JJ, Senden JM, Waterval WH, Bierau J, Verdijk LB and van Loon LJ (2018) Protein content and amino acid composition of commercially available plant-based protein isolates. *Amino Acids* 50(12): 1685–1695.

Håkansson I, Myrbeck Å and Etana A (2002) A review of research on seedbed preparation for small grains in Sweden. *Soil and Tillage Research* 64(1–2): 23–40.

Hampton JG, Boelt B, Rolston MP and Chastain TG (2013) Effects of elevated CO_2 and temperature on seed quality. *The Journal of Agricultural Science* 151(2): 154–162.

Hannaway DB and Larson C (2004) *Berseem Clover* (*Trifolium alexandrinum* L.). Oregon State University, Species Selection Information System.

Hasanuzzaman M, Hossain MA, da Silva JAT and Fujita, M (2012) Plant response and tolerance to abiotic oxidative stress: Antioxidant defense is a key factor. In *Crop Stress and Its Management: Perspectives and Strategies*. Dordrecht: Springer, pp. 261–315.

Hasanuzzaman M, Nahar K, Alam MM, Roychowdhury R and Fujita M (2013) Physiological, biochemical, and molecular mechanisms of heat stress tolerance in plants. *Int J Mol Sci*. 14: 9643–9684.

Haynes RJ and Naidu R (1998) Influence of lime, fertilizer and manure applications on soil organic matter content and soil physical conditions: A review. *Nutrient Cycling in Agroecosystems* 51(2): 123–137.

Hays DB, Do JH, Mason RE, Morgan G and Finlayson SA (2007) Heat stress induced ethylene production in developing wheat grains induces kernel abortion and increased maturation in a susceptible cultivar. *J Plant Sci* 172: 1113–1123.

Hebbar SS, Ramachandrappa BK, Nanjappa HV and Prabhakar M (2004) Studies on NPK drip fertigation in field grown tomato (*Lycopersicon esculentum* Mill.). *European Journal of Agronomy* 21(1): 117–127.

Herrick JE and Wander MM (1998) Relationships between soil organic carbon and soil quality in cropped and rangeland soils: the importance of distribution, composition, and soil biological activity. *Relationships Between Soil Organic Carbon and Soil Quality in Cropped and Rangeland Soils: The Importance of Distribution, Composition, and Soil Biological Activity*: 405–425.

Heuzé V, Tran G, Boudon A, Bastianelli D and Lebas F (2016) *Berseem (Trifolium alexandrinum).*

Hideg E (1997) Free radical production in photosynthesis under stress conditions. In: Pessarakli M (ed) *Handbook of Photosynthesis.* New York: Marcel Dekker, pp. 911–930.

Hogewoning SW, Trouwborst G, Maljaars H, Poorter H, van Ieperen W and Harbinson J (2010) Blue light dose—responses of leaf photosynthesis, morphology, and chemical composition of *Cucumis sativus* grown under different combinations of red and blue light. *Journal of Experimental Botany* 61(11): 3107–3117.

Hossain A, Sarker MAZ, Saifuzzaman M, Teixeira da Silva JA, Lozovskaya MV and Akhter MM (2013) Evaluation of growth, yield, relative performance and heat susceptibility of eight wheat (*Triticum aestivum* L.) genotypes grown under heat stress. *International Journal of Plant Production* 7(3): 615–636.

Howden SM, Soussana JF, Tubiello FN, Chhetri N, Dunlop M and Meinke H (2007) Adapting agriculture to climate change. *Proceedings of the National Academy of Sciences* 104(50): 19691–19696.

Huke RE (1982) *Rice Area by Type of Culture: South, Southeast, and East Asia.* Int. Rice Res. Inst.

Hukkeri SB, Shukla NP and Rajput RK (1977) Effect of levels of soil moisture and nitrogen on the fodder yield of oat on two types of soils [India]. *Indian Journal of Agricultural Sciences.*

Hura T, Hura K, Grzesiak M and Rzepka A (2007) Effect of long-term drought stress on leaf gas exchange and fluorescence parameters in C 3 and C 4 plants. *Acta Physiologiae Plantarum* 29(2): 103–113.

Ijaz F, Riaz U, Iqbal S, Zaman Q, Ijaz MF, Javed H, . . . and Ahmad I (2019) Potential of rhizobium and PGPR to enhance growth and fodder yield of berseem (*Trifolium alexandrinum* L.) in the presence and absence of tryptamine. *Pakistan Journal of Agricultural Research* 32(2): 398–406.

IPCC. 2014. Summary for Policymakers. In: Field CB, et al. (eds) *Climate Change 2014: Impacts, Adaptation and Vulnerability. Part A: Global and Sectoral Aspects. Contribution of working Group II to the Fifth Assessment Report of the Intergovernmental Panel on Climate Change.* United Kingdom and New York, NY, USA: Cambridbe University Press, Cambridge, pp. 1–32.

Jackson MB (2004) The impact of flooding stress on plants and crops.

Jackson MB and Ricard B (2003) Physiology, biochemistry and molecular biology of plant root systems subjected to flooding of the soil. *Root Ecology* 193–213.

Jamil M, Lee KJ, Kim JM, Kim HS and Rha ES (2007) Salinity reduced growth PS2 photochemistry and chlorophyll content in radish. *Scientia Agricola* 64(2): 111–118.

Johansen C, Ali M, Gowda CLL, Ramakrishna A, Nigam SN and Chauhan YS (2000) Regional opportunities for warm season grain legumes in the Indo-Gangetic plain.

Johkan M, Oda M, Maruo T and Shinohara Y (2011) Crop production and global warming. *Global Warming Impacts-case Studies on the Economy, Human Health, and on Urban and Natural Environments* 139–152.

Johnson PR and Ecker JR (1998) The ethylene gas signal transduction pathway: A molecular perspective. *Annual Review of Genetics* 32(1): 227–254.

Kacálková L and Tlustoš P (2011) The uptake of persistent organic pollutants by plants. *Central European Journal of Biology* 6(2): 223–235.

Kalaji MH and Łoboda T (2010) *Chlorophyll Fluorescence in the Studies of the Physiological Condition of Plants*. Warsaw: SGGW, 116.

Kang Y, Khan S and Ma X (2009) Climate change impacts on crop yield, crop water productivity and food security—a review. *Prog Natl Sci* 19(12): 1665–1674.

Kapoor D, Kaur S and Bhardwaj R (2014) Physiological and biochemical changes in *Brassica juncea* plants under Cd-induced stress. *BioMed Research International* 2014.

Karadağ Y and Büyükburç U (2003) Effects of seed rates on forage production, seed yield and hay quality of annual legume-barley mixtures. *Turkish Journal of Agriculture and Forestry* 27(3): 169–174.

Karl TR, Melillo JM, Peterson TC and Hassol SJ (eds) (2009) *Global Climate Change Impacts in the United States*. Cambridge University Press.

Kautz T (2015) Research on subsoil biopores and their functions in organically managed soils: A review. *Renewable Agriculture and Food Systems* 30(4): 318–327.

Kaya MD and Day S (2008) Relationship between seed size and NaCl on germination, seed vigor and early seedling growth of. *African Journal of Agricultural Research 3*(11): 787–791.

Khaliq A, Abbasi MK and Hussain T (2006) Effects of integrated use of organic and inorganic nutrient sources with effective microorganisms (EM) on seed cotton yield in Pakistan. *Bioresource Technology* 97(8): 967–972.

Khan MN, Siddiqui MH, Mohammad F, Khan MMA and Naeem M (2007) Salinity induced changes in growth, enzyme activities, photosynthesis, proline accumulation and yield in linseed genotypes. *World J Agric Sci* 3(5): 685–695.

Kiani-Pouya A, Rasouli F, Rabbi B, Falakboland Z, Yong M, Chen ZH, . . . and Shabala S (2020) Stomatal traits as a determinant of superior salinity tolerance in wild barley. *Journal of Plant Physiology* 245: 153108.

Koestel J and Larsbo M (2014) Imaging and quantification of preferential solute transport in soil macropores. *Water Resources Research* 50(5): 4357–4378.

Kosová K, Vítámvás P, Prášil IT and Renaut J (2011) Plant proteome changes under abiotic stress—contribution of proteomics studies to understanding plant stress response. *Journal of Proteomics* 74(8): 1301–1322.

Krasensky J and Jonak C (2012) Drought, salt, and temperature stress-induced metabolic rearrangements and regulatory networks. *Journal of Experimental Botany* 63(4): 1593–1608.

Kummerová M, Slovak L and Holoubek I (1997) Growth response of spring barley to short-or long-period exposures to fluoranthene. *Rostlinna Vyroba-UZPI (Czech Republic).*

Kummerová M, Váňová L, Fišerová H, Klemš M, Zezulka Š and Krulová J (2010) Understanding the effect of organic pollutant fluoranthene on pea *in vitro* using cytokinins, ethylene, ethane and carbon dioxide as indicators. *Plant Growth Regulation* 61(2): 161–174.

Kumutha D, Ezhilmathi K, Sairam RK, Srivastava GC, Deshmukh PS and Meena RC (2009) Waterlogging induced oxidative stress and antioxidant activity in pigeonpea genotypes. *Biologia Plantarum* 53(1): 75–84.

Lavelle P (1988) Earthworm activities and the soil system. *Biology and Fertility of Soils 6*(3): 237–251.

Lee SH, Tewari RK, Hahn EJ and Paek KY (2007) Photon flux density and light quality induce changes in growth, stomatal development, photosynthesis and transpiration of *Withania somnifera* (L.) Dunal. plantlets. *Plant Cell, Tissue and Organ Culture* 90(2): 141–151.

Lehmann N (2013) *How climate change impacts on local cropping systems: A bioeconomic simulation study for western Switzerland* (Doctoral dissertation, ETH Zurich).

Li Q and Kubota C (2009) Effects of supplemental light quality on growth and phytochemicals of baby leaf lettuce. *Environmental and Experimental Botany* 67(1): 59–64.

Li Q, Yang A and Zhang WH (2017) Comparative studies on tolerance of rice genotypes differing in their tolerance to moderate salt stress. *BMC Plant Biology* 17(1): 1–13.

Li SX, Wang ZH, Malhi SS, Li SQ, Gao YJ and Tian XH (2009) Nutrient and water management effects on crop production, and nutrient and water use efficiency in dryland areas of China. *Advances in Agronomy* 102: 223–265.

Liang XY (8). others. 2017. Determining climate effects on US total agricultural productivity. In: Proceedings of the National Academy of Sciences doi (Vol. 10).

Lin CH, Huang X, Kolbanovskii A, Hingerty BE, Amin S, Broyde S, . . . and Patel DJ (2001) Molecular topology of polycyclic aromatic carcinogens determines DNA adduct conformation: A link to tumorigenic activity. *Journal of Molecular Biology* 306(5): 1059–1080.

Lithourgidis AS, Vasilakoglou IB, Dhima KV, Dordas CA and Yiakoulaki MD (2006) Forage yield and quality of common vetch mixtures with oat and triticale in two seeding ratios. *Field Crops Research* 99(2–3): 106–113.

Lock TR, Kallenbach RL, Blevins DG, Reinbott TM, Crawford Jr RJ, Massie MD and Bishop-Hurley GJ (2000) Phosphorus fertilization of tall fescue may prevent grass tetany. *Better Crops* 84(3): 12–13.

Macedo AF, Leal-Costa MV, Tavares ES, Lage CLS and Esquibel MA (2011) The effect of light quality on leaf production and development of *in vitro*-cultured plants of *Alternanthera brasiliana* Kuntze. *Environmental and Experimental Botany* 70(1): 43–50.

Majeed A, Nisar MF and Hussain K (2010) Effect of saline culture on the concentration of Na+, K+ and Cl-in *Agrostis tolonifera*. *Curr. Res. J. Biol. Sci* 2(1): 76–82.

Mansour MMF, Salama KHA, Ali FZM and Abou Hadid AF (2005) Cell and plant responses to NaCl in *Zea mays* L. cultivars differing in salt tolerance. *Gen. Appl. Plant Physiol* 31(1–2): 29–41.

McClung CR and Davis SJ (2010) Ambient thermometers in plants: From physiological outputs towards mechanisms of thermal sensing. *Current Biology* 20(24): R1086-R1092.

Mok DW and Mok MC (2001) Cytokinin metabolism and action. *Annual Review of Plant Biology* 52(1): 89–118.

Munns R (2005) Genes and salt tolerance: Bringing them together. *New Phytologist* 167(3): 645–663.

Munns R, James RA and Läuchli A (2006) Approaches to increasing the salt tolerance of wheat and other cereals. *Journal of Experimental Botany* 57(5): 1025–1043.

Nahar K, Ahamed KU and Fujita M (2010) Phenological variation and its relation with yield in several wheat (*Triticum aestivum* L.) cultivars under normal and late sowing mediated heat stress condition. *Notulae Scientia Biologicae* 2(3): 51–56.

Naoya F, Mitsuko F, Yoshitaka O, Sadanori S, Shigeo N and Hiroshi E (2008) Directional blue light irradiation triggers epidermal cell elongation of abaxial side resulting in inhibition of leaf epinasty in geranium under red light condition. *Scientia Horticulturae* 115: 176–182.

Nichols PGH, Loi A, Nutt BJ, Evans PM, Craig AD, Pengelly BC, . . . and You MP (2007) New annual and short-lived perennial pasture legumes for Australian agriculture—15 years of revolution. *Field Crops Research* 104(1–3): 10–23.

Naureen Z, Yasmin S, Hameed S, Malik KA and Hafeez FY (2005) Characterization and screening of bacteria from rhizosphere of maize grown in Indonesian and Pakistani soils. *Journal of Basic Microbiology: An International Journal on Biochemistry, Physiology, Genetics, Morphology, and Ecology of Microorganisms* 45(6): 447–459.

Naveed M, Moldrup P, Arthur E, Wildenschild D, Eden M, Lamandé M, . . . and de Jonge LW (2013) Revealing soil structure and functional macroporosity along a clay gradient using X-ray computed tomography. *Soil Science Society of America Journal* 77(2): 403–411.

Nawaz A, Farooq M, Cheema SA and Wahid A (2013) Differential response of wheat cultivars to terminal heat stress. *International Journal of Agriculture and Biology* 15(6).

Netondo GW, Onyango JC and Beck E (2004) Sorghum and salinity: II. Gas exchange and chlorophyll fluorescence of sorghum under salt stress. *Crop Science* 44(3): 806–811.

Newman CW and Newman RK (2006) A brief history of barley foods. *Cereal Foods World* 51(1): 4–7.

Niazi BH, Rozema J, Broekman RA and Salim M (2000) Dynamics of growth and water relations of fodderbeet and seabeet in response to salinity. *Journal of Agronomy and Crop Science* 184(2): 101–110.

Oad FC, Buriro UA and Agha SK (2004) Effect of organic and inorganic fertilizer application on maize fodder production. *Asian J. Plant Sci* 3(3): 375–377.

Olszewska M (2006) Effect of water stress on physiological processes, leaf greenness (SPAD index) and dry matter yield of *Lolium perenne* and *Dactylis glomerata. Polish Journal of Natural Sciences (Poland).*

Olszewska M (2009) Response of cultivars of meadow fescue (*Festuca pratensis* Huds.) and Timothy (*Phleum pratense* L.) grown on organic soil to moisture deficiency. *Acta Scientiarum Polonorum-Agricultura* 8(1): 37–46.

Olszewska M, Grzegorczyk S, Olszewski J and Baluch-Malecka A (2010) Porównanie reakcji wybranych gatunków traw na stres wodny. *Łąkarstwo w Polsce* 13.

Oshino T, Miura S, Kikuchi S, Hamada K, Yano K, Watanabe M and Higashitani A (2011) Auxin depletion in barley plants under high-temperature conditions represses DNA proliferation in organelles and nuclei via transcriptional alterations. *Plant, Cell & Environment* 34(2): 284–290.

Pang ZP, Yang N, Vierbuchen T, Ostermeier A, Fuentes DR, Yang TQ, . . . and Wernig M (2011) Induction of human neuronal cells by defined transcription factors. *Nature* 476(7359): 220–223.

Parrish ZD, Banks MK and Schwab AP (2005) Assessment of contaminant lability during phytoremediation of polycyclic aromatic hydrocarbon impacted soil. *Environmental Pollution* 137(2): 187–197.

Paudel B, Acharya BS, Ghimire R, Dahal KR and Bista P (2014) Adapting agriculture to climate change and variability in Chitwan: Long-term trends and farmers' perceptions. *Agricultural Research* 3(2): 165–174.

Paulo AA and Pereira LS (2006) Drought concepts and characterization: Comparing drought indices applied at local and regional scales. *Water International* 31(1): 37–49.

Peng X, Horn R and Hallett P (2015) Soil structure and its functions in ecosystems: Phase matter & scale matter. *Soil & Tillage Research* 146(Part A): 1–3.

Petruzzelli L, Coraggio I and Leubner-Metzger G (2000) Ethylene promotes ethylene biosynthesis during pea seed germination by positive feedback regulation of 1-aminocyclopropane-1-carboxylic acid oxidase. *Planta* 211(1): 144–149.

Picken AJF, Stewart K and Klapwijk D (1986) Germination and vegetative development. In *The Tomato Crop.* Dordrecht: Springer, pp. 111–166.

Poorter H (1989) Plant growth analysis: Towards a synthesis of the classical and the functional approach. *Physiologia Plantarum* 75(2): 237–244.

Rabot E, Wiesmeier M, Schlüter S and Vogel HJ (2018) Soil structure as an indicator of soil functions: A review. *Geoderma* 314: 122–137.

Rahman MA, Chikushi J, Yoshida S and Karim AJMS (2009) Growth and yield components of wheat genotypes exposed to high temperature stress under control environment. *Bangladesh Journal of Agricultural Research* 34(3): 360–372.

Ramakrishna A, Gowda CLL and Johansen C (2000) Management factors affecting legumes production in the Indo-Gangetic Plain.

Raza WASEEM, Yousaf SOHAIL, Niaz ABID, Rasheed MK and Hussain IQBAL (2005) Subsoil compaction effects on soil properties, nutrient uptake and yield of maize fodder (*Zea mays* L.). *Pakistan Journal of Botany* 37(4): 933.

Reddy DV and Krishna N (2009) Precision animal nutrition: A tool for economic and eco-friendly animal production in ruminants. *Livestock Research for Rural Development* 21(3): 36.

Rhodes D, Nadolska-Orczyk A and Rich PJ (2002) Salinity, osmolytes and compatible solutes. In *Salinity: Environment-Plants-Molecules.* Dordrecht: Springer, pp. 181–204.

Roberts CA, Moore KJ and Johnson KD (1989) Forage quality and yield of wheat-vetch at different stages of maturity and vetch seeding rates. *Agronomy Journal* 81(1): 57–60.

Roy SJ, Negra¨ o S and Tester M. 2014. Salt resistant crop plants. *Current Opinion in Biotechnology* 26: 115–124.

Ruelland E and Zachowski A (2010) How plants sense temperature. *Environmental and Experimental Botany* 69(3): 225–232.

Sapre S, Gontia-Mishra I and Tiwari S (2018) *Klebsiella* sp. confers enhanced tolerance to salinity and plant growth promotion in oat seedlings (*Avena sativa*). *Microbiological Research* 206: 25–32.

Schilcher H (1977) Biosynthese des (–)–α–bisabolols und der bisabololoxide. *Planta Medica* 31(04): 315–321.

Schjønning P, Christensen BT and Carstensen B (1994) Physical and chemical properties of a sandy loam receiving animal manure, mineral fertilizer or no fertilizer for 90 years. *European Journal of Soil Science* 45(3): 257–268.

Schubert S, Schubert E and Mengel K (1990) Effect of low pH of the root medium on proton release, growth, and nutrient uptake of field beans (*Vicia faba*). In *Plant Nutrition—Physiology and Applications.* Dordrecht: Springer, pp. 443–448.

Schulz S, Carsky RJ and Tarawali SA (2001) Herbaceous legumes: The panacea for West African soil fertility problems? *Sustaining Soil Fertility in West Africa* 58: 179–196.

Singh AK, Meena MK, Bharati RC and Gade RM (2013) Effect of sulphur and zinc management on yield, nutrient uptake, changes in soil fertility and economics in rice (*Oryza sativa*)—lentil (*Lens culinaris*) cropping system. *Indian Journal* of *Agricultural Scien ces* 83(3): 344–348.

Singh VP and Mishra AK (2011) Report as of FY2010 for 2009TX334G: "Hydrological Drought Characterization for Texas under Climate Change, with Implications for Water Resources Planning and Management". *Department of Biological and Agricultural Engineering, Texas A & M University.*

Smith P, Powlson DS, Smith JU, Falloon P and Coleman K (2000) Meeting Europe's climate change commitments: Quantitative estimates of the potential for carbon mitigation by agriculture. *Global Change Biology* 6(5): 525–539.

Smrithi A, Bhaigyabati T and Usha K (2012) Bioremediation potential of *Brassica juncea* against textile disposal. *Research Journal of Pharmaceutical, Biological and Chemical Sciences* 3(2): 393–400.

Soest PJ (1985) Composition, fiber quality, and nutritive value of forages.

Song HX and Li SX (2006) Root function in nutrient uptake and soil water effect on NO_3–-N and NH_4+-N migration. *Agricultural Sciences in China* 5(5): 377–383.

Sridhar BM, Diehl SV, Han FX, Monts DL and Su Y (2005) Anatomical changes due to uptake and accumulation of Zn and Cd in Indian mustard (*Brassica juncea*). *Environmental and Experimental Botany* 54(2): 131–141.

Stirzaker RJ, Passioura JB and Wilms Y (1996) Soil structure and plant growth: Impact of bulk density and biopores. *Plant and Soil* 185(1): 151–162.

Sumberg J (2002) The logic of fodder legumes in Africa. *Food Policy* 27(3): 285–300.

Suttie JM (1999a) Berseem clover (*Trifolium alexandrinum*). *A searchable catalogue of grass and forage leg-umes, FAO.*

Suttie JM (1999b) Temperate Forage Legumes. In: Frame J, Charlton JFL and Laidlaw AS. Wallingford, UK: Cab International (1998), pp. 327, £ 27.50. ISBN 0–85199–2145. *Experimental Agriculture* 35(1): 111–114.

Takahashi S and Murata N (2008) How do environmental stresses accelerate photoinhibition? *Trends in Plant Science* 13(4): 178–182.

Teixeira EI, Fischer G, Van Velthuizen H, Walter C and Ewert F (2013) Global hot-spots of heat stress on agricultural crops due to climate change. *Agricultural and Forest Meteorology* 170: 206–215.

Tilman D, Cassman KG, Matson PA, Naylor R and Polasky S (2002) Agricultural sustainability and intensive production practices. *Nature* 418(6898): 671–677.

Tisdall JM and Oades JM (1982) Organic matter and water-stable aggregates in soils. *Journal of Soil Science* 33(2): 141–163.

Titarenko T (2000) Test parameters of revealing the degree of fruit plants tolerance to the root hypoxia caused by flooding of soil. *Plant Physiol. Biochem* 38(Suppl 115).

Totsche KU, Amelung W, Gerzabek MH, Guggenberger G, Klumpp E, Knief C, . . . and Kögel-Knabner I (2018) Microaggregates in soils. *Journal of Plant Nutrition and Soil Science* 181(1): 104–136.

Tzfira T and Citovsky V (2003) The *Agrobacterium*-plant cell interaction. Taking biology lessons from a bug.

Undersander D, Cosgrove D, Cullen E, Rice ME, Renz M, Sheaffer C, . . . and Sulc M (2011) *Alfalfa Management Guide*. American Society of Agronomy, Incorporated.

Van Der Heijden MG, Bardgett RD and Van Straalen NM (2008) The unseen majority: Soil microbes as drivers of plant diversity and productivity in terrestrial ecosystems. *Ecology Letters* 11(3): 296–310.

Van Noordwijk M, Schoonderbeek D and Kooistra MJ (1993) Root–soil contact of field-grown winter wheat. In *Soil Structure/Soil Biota Interrelationships*. Elsevier, pp. 277–286.

Vartapetian BB and Jackson MB (1997) Plant adaptations to anaerobic stress. *Annals of Botany* 79(suppl_1): 3–20.

Vaseva-Gemisheva I, Lee D and Karanov E (2005) Response of *Pisum sativum* cytokinin oxidase/dehydrogenase expression and specific activity to drought stress and herbicide treatments. *Plant Growth Regulation* 46(3): 199–208.

Wang Q, Wu C, Xie B, Liu Y, Cui J, Chen G and Zhang Y (2012) Model analysing the antioxidant responses of leaves and roots of switchgrass to NaCl-salinity stress. *Plant Physiology and Biochemistry* 58: 288–296.

Wang X, Cai J, Jiang D, Liu F, Dai T and Cao W (2011) Pre-anthesis high-temperature acclimation alleviates damage to the flag leaf caused by post-anthesis heat stress in wheat. *Journal of Plant Physiology* 168(6): 585–593.

Watteau F, Villemin G, Burtin G and Jocteur-Monrozier L (2006) Root impact on the stability and types of micro-aggregates in silty soil under maize. *European Journal of Soil Science* 57(2): 247–257.

Wu J, Brookes PC and Jenkinson DS (1993) Formation and destruction of microbial biomass during the decomposition of glucose and ryegrass in soil. *Soil Biology and Biochemistry* 25(10): 1435–1441.

Wyka TP, Bagniewska-Zadworna A, Kuczyńska A, Mikołajczak K, Ogrodowicz P, Żytkowiak M, . . . and Adamski T (2019) Drought-induced anatomical modifications of barley (*Hordeum vulgare* L.) leaves: An allometric perspective. *Environmental and Experimental Botany* 166: 103798.

Xu ZZ and Zhou GS (2005) Effects of water stress and high nocturnal temperature on photosynthesis and nitrogen level of a perennial grass *Leymus chinensis*. *Plant and Soil* 269(1): 131–139.

Yadav RK, Singh SP, Lal D and Kumar A (2007) Fodder production and soil health with conjunctive use of saline and good quality water in ustipsamments of a semi-arid region. *Land Degradation & Development* 18(2): 153–161.

Yang L, Zhang YL, Fu-Sheng LI and Lemcoff JH (2011) Soil phosphorus distribution as affected by irrigation methods in plastic film house. *Pedosphere* 21(6): 712–718.

Yordanova RY and Popova LP (2001) Photosynthetic response of barley plants to soil flooding. *Photosynthetica* 39(4): 515–520.

Yu Q, Li L, Luo Q, Eamus D, Xu S, Chen C, . . . Nielsen DC (2014) Year patterns of climate impact on wheat yields. *International Journal of Climatology* 34(2): 518–528.

Yue J, Gu Z, Zhu Z, Yi J, Ohm JB, Chen B and Rao J (2020) Impact of defatting treatment and oat varieties on structural, functional properties, and aromatic profile of oat protein. *Food Hydrocolloids* 112: 106368.

Zika M and Erb KH (2009) The global loss of net primary production resulting from human-induced soil degradation in drylands. *Ecological Economics* 69(2): 310–318.

10 Application of Precision Agriculture

Mitigating the Effect of Climate Change on Winter Fodders

Muhammad Kaleem Sarwar, Imran Ul Haq, Siddra Ijaz, and Nabeeha Aslam Khan

CONTENTS

DOI: 10.1201/9781003055365-10

10.1 CLIMATE CHANGE

'Climate' means the average weather conditions such as temperature, rainfall, humidity, and wind at a certain point on the earth's surface. 'Climate change' is the variability in climatic conditions persisting over an extended time. Climate change, influenced by natural factors, that is, volcanic activity, the sun's energy output, the earth's orbit, and other external/internal processes, is long-term natural climate change. It is due to this natural climate change that the earth experiences regular ice ages and warmer periods. Since the end of the most recent ice age, human civilization has flourished during the current stable warm climate period, about 20 millennia ago.

However, apart from natural climate change, scientific observations through the latest instruments and models indicate that human activities change the earth's climate. This anthropogenic climate change results from the greenhouse effect, which over time has resulted in 'global warming.' Global warming—Earth's average temperature increase—drives other observable climate changes such as changes in pattern and distribution of rainfalls, droughts, floods, heatwaves, and storms. Activities responsible for anthropogenic climate change are embedded in modern human life, and scientists believe it will have devastating consequences on the earth's ecosystem (Riedy, 2016).

10.1.1 Climate Change and Agriculture

Among the 17 sustainable goals of the UN agenda, zero hunger, food security, improved nutrition, and sustainable agriculture are its foundation. Anthropogenic climate change has a deleterious impact on agriculture and food security, making the challenge of ending hunger more challenging to achieve.

FAO estimates that food production must be increased by 60% in the next decade to feed the world population; as 161 million children are stunted, 800 million people are malnourished, 500 million are obese, and 2 billion lack essential nutrients for healthy living. In 2015, the UN reported that 836 million of the world population is living in a state of indigence. According to IFAD, 70% of the extreme poverty population in rural areas depends directly or indirectly on agriculture. In developing countries of Asia and Africa, 500 million small landholding farmers support a population of 2 billion and produce almost 80% of the consumed food (UN, 2015; Gitz et al., 2016; IFAD, 2011).

10.1.2 Impact of Climate Change on Fodder Productivity

Slight changes in prevailing climatic conditions profoundly impact plants and animals of that region. A few may become less productive or disappear, whereas others may become more productive. The impact of climate change at a specific stage of plant growth, in some cases, is easy to predict, such as the impact of a heatwave. Contrarily, the impact of climate change on the whole ecosystem is complex and challenging to predict, as different elements react and interact differently. This can be explained by taking into consideration the impact of increased carbon dioxide levels on crop yield. Under laboratory conditions, crop plants respond positively to an increase in atmospheric CO_2, but weeds also react favorably. Hence, under field conditions, an increase in cultivated crop yield directly depends on weed competition. Likewise, pests and diseases following climatic changes spread to areas that are biologically and institutionally less prepared for them, resulting in potentially severe damages to crop growth and yield. The risk to agricultural production is directly linked with food safety, security, and nutrition of people who depend entirely on agriculture for food and livelihood. This impact on crop growth and production is not limited to the rural population but can also be observed in distant populations in increased prices and trade disruption. Thus, the impact of climate change is a cascade of risks affecting the ecosystem, crop production, economics, food security, and nutrition (Riedy, 2016). The global average temperature is likely to increase by 0.3–0.7°C from 2016 to 2035, and its impact is predicted to be larger on land than over the oceans (Kirtman et al., 2013).

Alfalfa is grown over 583,561 ha in Mexico and covers more than 57% of the total forage surface area. During a recent study, the impact of global warming on alfalfa production was estimated in Mexico's irrigated areas. It was concluded that global warming could reduce alfalfa's potential productive area from 24.7% to 42% in 2070. It is expected that the mean reduction in rainfall by 2050 in the region of Mexico will be 7.2%, and this water deficit is expected to increase further. Given the high water requirement of alfalfa, fodder crops may be replaced with those with less water requirements, such as maize (Medina-Garcia et al., 2020).

An overwhelming number of farmers (77%) in Pakistan reported that climate change has drastically reduced fodder production and is responsible for livestock diseases. As discussed earlier, an increase in temperature and greenhouse gas levels and altered frequency of precipitation result from climate change. These changes have a negative impact on fodder and forage production. This results in reduced livestock production, especially milk production (Rahut and Ali, 2018).

10.2 CRUCIAL CLIMATIC VARIATIONS FOR FODDER YIELD

Climate change events have dramatically increased over the past few decades and are responsible for various abiotic stresses on plants and reduced crop yields. It is hard to calculate the exact loss in crop yield due to abiotic stresses, but it is believed that it significantly influences crop production. Loss in yield due to global warming, drought, and other climate change factors is directly proportional to the extent of damage to the total area of crop cultivation. Climate changes are expected to increase crop yield in NW Europe and decrease in the Mediterranean area. Wheat yield has already been affected by high temperatures/heatwaves. When temperature extremes combine with the drought, it disrupts the rubisco enzyme and consequently stops the photosynthesis process, such as in maize, sorghum, and barley. The negative impact of heat stress on antioxidant enzymes has also been reported in maize. Water deficiency at flower initiation and inflorescence stage of crop badly affect crop yield by causing sterility (in cereals), reduction in grain set, fertilization process, and anthesis (in rice) (Raza et al., 2019).

Water scarcity is expected to increase in various world regions under ongoing climate change, and it will pose a challenge for climate adaptation. Scientific data show that glaciers have considerably contracted, and dry land area has doubled since the

1970s. Climate change model simulations show that water availability will be modified, especially during crucial dry periods in regions irrigated by mountain rivers. The Himalayas are the source of 40% of the world's irrigation water. Rivers receiving their water from flows of the Himalayas will be heavily impacted due to global warming. Freshwater availability has significantly decreased in regions of low latitudes, especially in agricultural and heavily irrigated regions of the sub-continent, China, and Egypt (Elbehri and Burfisher, 2015).

Water stress reduces fodder crop growth and yield. In the case of severe drought, fodder crop stops its growth, and if prolonged, above-ground green foliage may die, leading to insufficient forage supply and livestock production. Moderate drought conditions may increase fodder crop digestibility due to increased leaf to stem ratio. Water stress in alfalfa drastically reduces yield by 49% and increases leaf to stem ratio and digestibility by 18% and 8%, respectively. This trend was also observed in other forage crops and grasses (Magan Singh et al., 2017).

An agricultural production system is impacted by direct effects caused by changes in physical characteristics, such as temperature and water availability, whereas indirect effects are caused by changes in other critical species, that is, insect pests and pollinator vectors. Direct effects, contrary to indirect effects, are easily simulated and modeled. Indirect effects are complicated due to various parameters and links which interact with each other in various unpredictable ways.

Climatic changes have affected crop yields and quality, such as in wheat, rice, and maize, globally (Lobell et al., 2011). It is evident from globally consistent climate change crop models that food production systems will be fundamentally altered by climate change, especially in tropical regions at low latitudes. In some regions, warmer temperatures with a more extended growth period will benefit crop yield; however, other environmental parameters, such as soil quality, will hinder its expansion. It is also predicted that climate change will exacerbate existing imbalance and crop yield variability between the developed and developing world (Elbehri et al., 2015).

In a consolidated study to determine the impact of climatic changes on agriculture, it was found that by the end of the 21st century, global climate change will reduce the yield of maize, wheat, rice, and soybean by 20–45%, 5–50%, 20–30%, and 30–60%, respectively. The European Union and the United States of America can face water shortage and frequent heat events during the growing wheat and maize season, respectively, but these regions have the flexibility for climate adaptation (Müller and Elliott, 2015).

10.3 IMPORTANCE OF LIVESTOCK IN AGRICULTURE

Among the agriculture-dependent population, one-third keep livestock, and almost 800 million of them earn <$2/day. Livestock is the key to food security and has a more than 40% share in global agricultural GDP. In a mixed agricultural system, apart from being a protein and energy source, livestock provides draught power, organic fertilizers, and transport. Livestock consumes crop by-products and residues and serves as a buffering agent in times of financial crises. Additionally, income from livestock serves rural communities by contributing to education and health.

Climate change impacts livestock productivity in various ways, most notably on animal production and yield of fodder crops. Increased temperature, reduced rainfall, and drought have a detrimental impact on fodder crop production, for example, up to 60% loss of green fodder production during summer 2003 in France. Climate change also affects fodder quality and feed safety through increased lignification, shrub cover, secondary metabolites, and fungal infestation. Variability in rainfall and temperature results in decreased fodder and feed crop yields, compromised fodder quality, production system changes, and pasture compositions (Gitz et al., 2016; Escarcha et al., 2018).

10.4 CLIMATE IMPACT ON QUALITY AND FEEDING VALUE OF FODDER

Global warming considerably impacts the crop calendar by affecting the growing season of fodder or forage crops. Multiple cropping systems involving fodder crops under drought stress reduce fodder crops' production potential, especially in high evapotranspiration, low humidity, and heatwaves. Drought in multiple cropping systems involving fodder crops under heat stress worsens the issue. Additionally, most fodder crops, particularly alfalfa and berseem, are susceptible to drought and high temperature, and increased daylight average temperature decreases their respective yields.

Climate change reduces fodder crop production and lowers feeding value, resulting in reduced livestock performance. The feeding value of a forage crop depends on physical and chemical properties affected by soil quality, plant parts being fed, fodder species, and growth stage at harvesting. It is determined through various field experiments that a fodder crop's feeding value is higher in spring, which falls gradually from summer to winter due to dry and rainy tropics. Fodder crops produce higher yields when grown at a temperature lower than optimal, whereas higher temperature than optimal causes lower yields due to early blooming and maturation.

Climatic changes responsible for high temperature significantly influence fodder crop quality and adaptation in a specific environment. The optimum growth temperature for winter fodder crops is 20°C, and for summer, fodders range from 30–35°C. Winter fodder crops during photosynthesis produce highly digestible sugars; contrarily, summer fodder crops have an increased development rate due to high temperature, which results in reduced leaf to stem ratio, lower digestibility, and compromised forage quality. The digestibility of fodder crop reduces by 0.3–0.7% units with each degree temperature increase. Fodder crop quality in a tropical climate is lower than temperate fodder crops, associated with higher fiber content and lower dry matter digestibility. Likewise, the intake value of tropical fodder crops is usually less than temperate fodder crops.

Levels of atmospheric carbon dioxide have increased from 338 ppm in 1980 to 398 ppm in 2014. According to various climate models, it is expected that at the end of the 21st century, its level could reach up to 421 to 936 ppm, depending on RCP. These elevated levels of CO_2 decrease nitrogen content up to 9% and contrarily increase carbohydrate content up to 30% in fodder crops. Although carbohydrate contents tend to increase due to elevated CO_2 levels, these have high instability within plant tissues. Moderate heat stress cause early maturation of crop plants, low water content level in plant tissue, increased step/leaf ratio, and lignin, which hinders polysaccharide digestion in livestock; it acts as a physical barrier against enzymatic activity (Bertrand Dumont et al., 2015).

In 2018, a detailed study was conducted to assess modern forage crops' adaptability—meadow fescue, tall fescue, timothy, red clover, festulolium—to climate change in the northern Europe region. All cultivars of meadow fescue were sensitive to low rainfall, especially during the hardening period. It was found that lower winter temperatures consistently reduced the forage crop yield from 10–25%. Comparatively, high winter temperatures boost forage yield up to 20%. Few cultivars are resistant to high temperatures and moderate drought conditions, but high temperatures right after the first cut can drastically reduce cultivar yield. In tall fescue cultivars, cold temperatures of less than 5°C and low rainfall reduced forage yield in all cultivars. Cultivars of festulolium also showed sensitivity to low winter temperatures (–15°C) and high temperatures after the first cut, leading to reduced forage yield (Hanna et al., 2018).

Climate change impact data on fodder production is not widely available, but few countries have recorded it for years. Lowland areas with reduced precipitation in the United Kingdom have greatly affected herbage yield, particularly in dry seasons. It is also evident from research that enhanced levels of CO_2 change crop response to temperature and water. Likewise, weeds, pests, and diseases of fodder crops also respond to climatic changes and may adversely affect forage yield. For example, an increase in temperature favor growth and consequent damages by clover stem nematode. Reported and predicted adverse effects of climate change on fodder production adversely impact livestock production and animal economic rearing. Developing world countries, especially in Africa and Asia, already have deficits in dry and green fodder and concentrate on meeting livestock demands. These shortages under climate change are further aggravated, with destructive effects on already-wavering agricultural production. Adverse consequences of climate change will be inflicted on fodder production in areas with high temperatures, declining precipitation, drought, and evapotranspiration due to global warming. In 1987, drought lowered fodder and water availability in India and affected 168 million livestock animals, particularly cattle. Gujrat state was hard hit, with the worst drought reported as having killed more than 50% of cattle. In 2001, drought damaged 7.8 million ha of crop area; consequently, it affected 40 million cattle due to fodder shortage; fodder availability declined from 144 to 127 million tons. Frequent droughts force pastoral families to migrate. For example, Gujrat's drought forced almost 50% of the pastoral families to migrate (Sirohi and Michaelowa, 2007).

10.5 FOOD SECURITY AND SAFETY THROUGH PRECISION AGRICULTURE

The agriculture sector's mechanization in the 20th century increasingly replaced labor with machinery that enabled farmers to manage larger crop fields. In the mid-20th century, the green revolution reduced crop losses and improved farm yields via genetically improved crop varieties, synthetic fertilizers, and pesticides. These innovations enabled farmers to grow crops on larger and well-managed fields with high profits and lower costs. Precision agriculture includes standardized practices for temporal, site-specific, and practical farm management. Precision agriculture is applicable and useful for all cropping and livestock farms, small and large, organic and conventional, developed and developing. Advanced precision farming technologies are not fully adopted worldwide, mainly due to high costs and complex operating systems. Challenges of food security and safety, along with the protection of the environment from adverse impacts of agriculture, have attracted policymakers' attention in the context of recent climate change. In response, environmental protection policies are becoming stricter. However, these policies and steps are not meant to jeopardize food production and the agricultural economy. Hence, sustainable agricultural production requires dealing with this nexus of food security, environmental safety, and agricultural sector economy. Though not a panacea, precision agriculture can effectively contribute to achieving sustainable agricultural production (Finger et al., 2019).

10.6 RECENT FARMING AND POLICYMAKING IN THE CONTEXT OF CLIMATE CHANGE

Precision agriculture is the 'farm management system in which the right treatment is applied at the right time in the right way for a particular issue in the right place.' This system is designed to achieve sustainable whole farm production efficiency and profitability while reducing unintended impacts on the environment. As described earlier, this system is not limited to a sustainable cropping system but the whole agricultural production system. In this approach, resource application decisions and on-farm agronomic practices are compatible with soil type and crop requirements at a specific stage. Precision agriculture improves the climate resilience of the agricultural system while ensuring FAO's food security goal. It is widely acknowledged that better decisions should exhibit economies of scale in the agriculture sector. In an experiment, it was found that precision agriculture was profitable in 68% of cases. With expanding agricultural markets, profitability is getting tighter and urges farmers to adopt cost-effective technologies. Apart from increasing profitability, farmers adopt precision agriculture strategies to increase production, ensuring more considerable economic benefits.

Precision agriculture (PA) is an environmentally friendly approach, as along with increasing agricultural profits and production, it focuses on minimizing adverse impacts on the environment. PA is getting more attention in the 21st century, as it directly links to key drivers related to climate change. There are pieces of evidence that PA strategies reduce environmental degradation, groundwater contamination, nitrate leaching, and carbon footprint due to increased fuel use efficiency. Precision agriculture in an EU meeting was proposed as an effective way to reduce water and soil contamination with agro-chemicals. PA technologies in dairy farming have been reported to reduce labor costs and improve production and animal welfare (Zhang et al., 2002; Zarco-Tejada et al., 2014).

10.7 FARMING AND INTEGRATION OF RECENT TECHNOLOGIES

Precision agricultural strategies are effectively implemented by combining sensor technologies and mapping variables. This helps farmers to perform suitable management actions, such as land preparation, sowing, irrigation, fertilization, herbicide and pesticide application, and harvesting at the right time and place. Global navigation satellite system (GNSS) technology is widely used for geo-positioning related tasks (auto-steer system) and yield mapping. GNSS has enabled more expanded and accurate machinery guidance systems; as a result, farm operations are completed on time without any error or fatigue. With the help of variable rate technology, improved and efficient application of synthetic chemicals is possible, resulting in cost-effective farm production with a reduced impact on the environment. Airborne and satellite platforms (e.g., remotely piloted aerial systems) generate crop imagery that helps determine crop canopy conditions (disease incidence, stresses, crop biomass, and chlorophyll content) and their variability at different locations in the field. The crop plants' nutritional status during the growth period can also be measured by using handheld devices. Diagnostic tools are also helpful in determining appropriate harvest time and crop yield. Yield-monitoring tools play a critical role in mechanical and handpicked harvesting systems, particularly for perishable crops like vegetables and fruits. Diagnostic tools are combined with near-infrared spectroscopy to determine crop quality traits, such as moisture content and protein content. Furthermore, forage crop quality is assured by frequent monitoring of fiber, moisture, and protein content near harvest to determine the optimum harvesting time. These are beneficial technologies but require knowledge and skill to determine weather and soil properties' effects on crop yield (Zude-Sasse et al., 2016; Finger et al., 2019).

Agricultural production is required to feed the ever-increasing world population while reducing fossil fuel dependence and negative impacts on climate. Precision agriculture technologies improve the productivity and efficiency of cropping and livestock system. It can be achieved by using PA approaches combined with the latest breeding and genetic approaches, cultural practices, and agronomic management. Precision agriculture provides a broad spectrum of climate-related benefits through a targeted approach for fertilizers and pesticide application, reduced water losses, and greenhouse gas emissions. However, the extent of these beneficial effects in the real-world is highly variable. Guidance systems and controlled traffic technologies have been reported to reduce fuel use and expenditures up to 6% and 25%, respectively. These benefits are more for large-scale crop fields, as they provide co-benefits of fewer N losses, soil compaction, erosion, and runoff. Case studies performed in Germany, India, the Philippines, and Vietnam showed that controlled and sight-specific N applications increase fertilizer efficiency and reduce greenhouse gases emission. Pesticide application was reduced by more than 90% by using precision agricultural strategies compared to human inspection and recommendation. Likewise, in another study, it was found that weedicide application can also be reduced up to 90% in arable crops by adopting precision agriculture. Sensor-based and site-specific insecticide application can reduce >13% pesticide usage in wheat for aphid control and result in more than 25% savings (Kempenaar et al., 2017). PA technologies also increase irrigation efficiency and potentially save water from 20% to 25%, resulting in less effluent from agricultural fields to water bodies. Though variable rate irrigation, fertilization, pesticide application, and other technologies help increase crop yield and reduce negative impacts on the environment (Finger et al., 2019).

The advent of mechanization has revolutionized industrial and agricultural production. Humans and machines are collaborating to meet the physiological needs of the expanding world population. The green revolution has increased crop production; however, future predictions indicate the need for a sustainable system. This sustainable system should be capable of increasing farm productivity efficiently and respectfully. Remarkable technological advances have given access to reliable agricultural data and its manipulation for obtaining great benefits without compromising climate safety. These technologies imply that farmers must act as supervisors of their crops and livestock rather than workers. In the modern agricultural framework, farm data is handled to address specialized solutions for individual farm issues (Saiz-Rubio and Rovira-Más, 2020).

10.8 PRECISION AGRICULTURE STRATEGIES TO MITIGATE CLIMATIC VARIATION IMPACT ON A FODDER CROP

As described earlier, an increase in temperature, drought conditions, and carbon dioxide concentration are climate change characteristics. These climatic changes pose a severe threat to fodder productivity: reduced forage production, low nutrient content, and decreased planting area. These negative impacts affect the supply of animal feed and hence disrupt human food availability. Fodder plant adaption strategies include increased water content, structural stability of cell membranes, and photosynthetic potential. Fodder crops achieve stability by suppressing stomatal conductance and carbon consumption through respiration. Precision agriculture involving the planting, irrigation, and harvesting of fodder crops is a successful and trusted approach (Harmini and Fanindi, 2020).

A sustainable livestock system requires increased pasture production and modifications in forage species, adequate fertilizer applications, adequate water supply, and the use of supplementary feeds (silage and hay), especially in times of low production (Howden et al., 2007).

10.9 FODDER YIELD AND PRECISION AGRICULTURE

There are various fodder crop adaptations to changing climatic conditions and variability: type of cultivar, direct and indirect selection, genomic selection method, management approach, and their compatibility with other employed techniques. Adaptation costs could be reduced by collecting accurate environment data and predictions. There is a need for an effective seed research system for developing new and diverse fodder varieties that can meet farmers' needs, particularly climatic variability. Along with fodder plant varieties, there is a need to identify and develop advanced and suitable breeding strategies for animals. Tolerant livestock breeds can survive extreme climate changes. For this purpose, local breeds, adapted to local climate conditions and feed sources, are a better option to start with (Haussmann et al., 2012; Koirala and Bhandari, 2019).

During a study on Australian agriculture, it was found that the average productivity and financial performance of farms improved. They classified 64% of farms as 'growing,' adopting the latest precision agricultural strategies and existing technologies. It was concluded that crop farms' productivity and profitability were triple that of livestock farms. Regions facing frost risk and reduced rainfall have higher proportions of less secure farms (reduced farm production) compared to regions with less frost risk and negative impacts of climate change. Not all the farms in the affected region were 'less secure.' Successful farmers focused on maximizing margins rather than maximizing yields. They concentrated on judicious use of inputs, increasing water availability and profit rather than risks. Tramline farming, weed-seeker technology, variable application rates based on soil and crop condition, and yield mapping reduced input use. In comparison, water availability was enhanced by improving soil constraints, tramline farming, and introducing ley year for weed management and soil-water recharge. These precision agriculture strategies and technologies improved production and profitability in climate-affected regions and can be applied where deemed fit (Browne et al., 2013; Plunkett, 2015; Sudmeyer et al., 2016).

For sustainable fodder and forage crop production, various precision agricultural technologies must be incorporated together. Most important among all is the development of temperature-resistant cultivars which can withstand fluctuations in temperature, especially expected high temperature due to climate change. A detailed study in Ghana found that the early rainy season promotes sorghum growth, whereas it impedes maize growth and reduces ensuing production. Along with temperature-resistant varieties, considerable attention must be provided to managing and supplementing available water, particularly in marginal areas. Integrated water conservation/storage and crop-livestock intervention systems have been successfully implemented in Mali, Ghana, and Burkina. Controlled tree integration in the cropping system improves soil fertility; increases fuel, food, and wood supply; reduces wind damage to crops; and serves as fodder requirements in low forage crop yield. Weather forecast data is the most critical factor for successful crop production. Reliable weather forecast data helps in projecting rainfall distribution and temperature variation. If conveyed, this data can help farmers adjust the sowing time of crops for the most use of expected rainfall (Kumar and Tuti, 2016; Magan Singh et al., 2017; Sujakhu et al., 2016).

A mixed crop-livestock system fulfills almost half of the world's food requirements and sustains village dwellers' livelihood in developing countries. Conversely, most studies focus on impact and adaptation strategies for grassland-based livestock production systems rather than mixed cropping systems. So, there is a need to formulate policies and practices for sustainable livestock production while combating climate change (Escarcha et al., 2018).

Certain practical approaches to mitigate the impact of climate change involving precision agricultural technologies are described under.

10.9.1 Cultural Practices

Climatic variations impact the agriculture and livestock of a particular area. Hence, farmers are its first victims. To tackle these environmental changes they, on their own, practice specific approaches to increase crop productivity and better plant adaptation strategies. Various studies have confirmed the effectiveness of farmer adopted strategies such as adjusting sowing and harvesting time, preference for short-duration crops, crop rotation, cropping schemes, and irrigation techniques. All of these abiotic strategies are effective for crop adaptation under climate stress conditions. Drought-resistant varieties, adjustment of planting time, and cultivation of new crops have reduced the danger of climate variation and ensured food safety and security. Precision agriculture strategies related to cultural practices such as judicious irrigation and fertilizer application are critical for improving production and reducing global warming (Henderson et al., 2018).

10.9.2 Conventional Strategies

Plant breeding techniques, under environmental stresses, play a considerable role in crop growth, productivity, and betterment. These resistant cultivars promote plant growth at a crucial stage and help farmers develop stress-resistant varieties that can escape various biotic and abiotic stresses and potentially guarantee food security and safety. Genetic divergence analysis is used to develop new resistant cultivars against biotic and abiotic stresses through polymorphism, inbreeding, and assortment. Genome approaches, such as genome-wide associated studies (GWAS) and marker-assisted selection (MAS), help develop resistant varieties through molecular and related plant breeding strategies (Lopes et al., 2015). Genetic and transcriptomic analysis identify phenotypic characteristics of a particular crop cultivar. Genomic approaches also identify the molecular stress-resistant mechanisms that aid in developing smart fodder crops with high production under various climatic variations.

It is predicted that prices of primary and secondary cereal and oil seed food/feed stuff will increase due to the high demand for animal feeding and human food production. This may force animal production to rely on less digestible feeds, forage co-products, and non-protein nitrogen sources. Increasing demand for agricultural land and food production, along with adverse impacts of climatic changes on crop production and yield, will increase the value of non-arable land for ruminant production to ensure human food security and safety. As discussed, maintaining feed crop yield while considering the negative impacts of climatic variations and greenhouse gas emissions is a considerable challenge for sustainable milk and meat production. It requires integrating climatic sciences, plant sciences, and animal sciences to mitigate the detrimental impacts of climate variations on animal feed supplies (Wheeler and Reynolds, 2013).

10.10 ADAPTATION STRATEGIES FOR GRASSLAND GRAZING SYSTEMS

Temperate grasslands cover almost 70% of the total agricultural land. These grasslands supply 50 % of global livestock needs, which provides 15% of the energy and 25% of the global human population's protein requirements (Herrero et al., 2013). The significant increase in the grassland livestock system's production and profit could be achieved by integrating multiple incremental adaptations, such as an increase in animal reproduction rate, animal genetic improvement, and grassland fertility. The combination of adaptation strategies is particularly successful in dry regions like Australia, as farming systems in these regions, compared to high-rainfall areas, are more vulnerable to climatic variations (Ghahramani and Moore, 2015). Improved nutrition and breeding are the key strategies to improve livestock productivity, especially to cope with high temperatures. Studies have identified that along with developing heat-resistant animals, various nutritional strategies to boost the immune system and livestock productivity are significantly useful, for example, diets with high energy and low rumen degradability protein, increased time and frequency of feeding, and additional supplements. It is possible to reduce greenhouse gases emission from livestock by adopting technical options, such as feed additives, animal selection, and anti-methanogenic vaccines. Emission efficiency could also be improved by adopting specific husbandry and feed management strategies. Effective stocking rate and pest control are cost-effective and practical options to reduce GHG emissions (Ghahramani et al., 2019).

10.11 CONCLUSION

The previous discussion concludes that farmers need to be agile in managing crop systems in the face of extreme climatic conditions for sustainable forage crop production. For sustainable dairy production, farmers and scientists must focus on developing and planting drought-tolerant forage crops and precision agriculture strategies to enhance water retention, such as conservation tillage, cover crops and crop residues, and improvement in adopting weather prediction models. Better use of water resources to produce forage crops, digestible energy, and protein and their ability to elicit milk response will play an important role in future adaptations to climatic changes and economic factors. Precision agricultural practices such as cover cropping and intercropping will improve climatic outcomes and reduce farm production and profitability, particularly in spatially variable rainfall. Perennial legumes such as alfalfa enhance cropping system sustainability and do not require N-fertilizers if nodulated with rhizobia for the conversion of atmospheric nitrogen into amino acids. These leguminous crops provide sufficient nitrogen fertilizer for corn production and a significant reduction in energy and greenhouse gas emissions. Forage crops must be improved for increased fiber digestibility, crude protein, and non-structural carbohydrate content. Forage crops with reduced lignification impacts will serve as a critical dietary forage component for enhanced milk production and dry matter intake. Attempts are being made to develop forage crops such as alfalfa to contain less lignin for increased fiber digestion, condensed tannin in leaves, and water efficiency. Crop rotations and higher forage production that is ecosystem friendly and increases land utilization will result in economic and environmental resilience to meet future milk and meat production demands (Martin et al., 2017).

REFERENCES

Browne N, Kingwell R, Behrendt R, et al. (2013) The relative profitability of dairy, sheep, beef and grain farm enterprises in southeast Australia under selected rainfall and price scenarios. *Agricultural Systems* 117: 35–44.

Dumont B, Andueza Urra D, Niderkorn V, et al. (2015) A meta-analysis of climate change effects on forage quality in grasslands: Perspectives for mountain and Mediterranean areas. *Grass and Forage Science* 70: 239–254.

Elbehri A and Burfisher M (2015) Economic modelling of climate impacts and adaptation in agriculture: A survey of methods, results and gaps. *FAO. Climate Change Food Systems: Global Assessments Implications for Food Security Trade. Viale Delle Terme di Caracalla* 153: 60–89.

Elbehri A, Elliott J and Wheeler T (2015) *Climate Change, Food Security and Trade: An Overview of Global Assessments and Policy Insights.* Rome: FAO.

Escarcha JF, Lassa JA and Zander KK (2018) Livestock under climate change: A systematic review of impacts and adaptation. *Climate* 6: 54.

Finger R, Swinton SM, El Benni N, et al. (2019) Precision farming at the nexus of agricultural production and the environment. *Annual Review of Resource Economics* 11.

Ghahramani A, Howden SM, del Prado A, et al. (2019) Climate change impact, adaptation, and mitigation in temperate grazing systems: A review. *Sustainability* 11: 7224.

Ghahramani A and Moore AD (2015) Systemic adaptations to climate change in southern Australian grasslands and livestock: Production, profitability, methane emission and ecosystem function. *Agricultural Systems* 133: 158–166.

Gitz V, Meybeck A, Lipper L, et al. (2016) Climate change and food security: Risks and responses. *Food Agriculture Organization of the United Nations Report* 110.

Hanna M, Janne K, Perttu V, et al. (2018) Gaps in the capacity of modern forage crops to adapt to the changing climate in northern Europe. *Mitigation Adaptation Strategies for Global Change* 23: 81–100.

Harmini H and Fanindi A (2020) Strategy for adaptation of forage crops to climate change. *WARTAZOA. Indonesian Bulletin of Animal Veterinary Sciences* 30.

Haussmann BI, Fred Rattunde H, Weltzien-Rattunde E, et al. (2012) Breeding strategies for adaptation of pearl millet and sorghum to climate variability and change in West Africa. *Journal of Agronomy Crop Science* 198: 327–339.

Henderson B, Cacho O, Thornton P, et al. (2018) The economic potential of residue management and fertilizer use to address climate change impacts on mixed smallholder farmers in Burkina Faso. *Agricultural Systems* 167: 195–205.

Herrero M, Havlík P, Valin H, et al. (2013) Biomass use, production, feed efficiencies, and greenhouse gas emissions from global livestock systems. *Proceedings of the National Academy of Sciences* 110: 20888–20893.

Howden SM, Soussana J-F, Tubiello FN, et al. (2007) Adapting agriculture to climate change. *Proceedings of the National Academy of Sciences* 104: 19691–19696.

IFAD (2011) Rural Poverty Report 2011. New Realities, New Challenges: New Opportunities for Tomorrow's Génération. London, UK.

Kempenaar C, Been T, Booij J, et al. (2017) Advances in variable rate technology application in potato in The Netherlands. *Potato Research* 60: 295–305.

Kirtman B, Power SB, Adedoyin AJ, et al. (2013) Near-term climate change: Projections and predictability. In: Stocker TF, Qin D, Plattner G-K, Tignor M, Allen SK, Boschung J, Nauels A, Xia Y, Bex V. and Midgley PM (eds) *Climate Change 2013: The Physical Science Basis.* Cambridge, UK. New York, USA: Cambridge University Press.

Koirala A and Bhandari P (2019) Impact of climate change on livestock production. *Nepalese Veterinary Journal* 36: 178–183.

Kumar B and Tuti A (2016) Effect and adaptation of climate change on fodder and livestock management. *Int. J. Sci. Environ. Technol* 5: 1638–1645.

Lobell DB, Schlenker W and Costa-Roberts J (2011) Climate trends and global crop production since 1980. *Science* 333: 616–620.

Lopes MS, El-Basyoni I, Baenziger PS, et al. (2015) Exploiting genetic diversity from landraces in wheat breeding for adaptation to climate change. *Journal of Experimental Botany* 66: 3477–3486.

Magan Singh, Sanjeev Kumar and Marigowda C (2017) Climate change: Impacts and adaptations of fodder and forage crops. In: Shekara BG, Mahadevu P, Lohithasava HC and Chikarugi NM (ed) *Role of Forages as Contingent Crops During Aberrent Weather Conditions.* 2017 ed. U.P. India: University of Agriculture Sciences, Banagaluru.

Martin N, Russelle M, Powell J, et al. (2017) Invited review: Sustainable forage and grain crop production for the US dairy industry. *Journal of Dairy Science* 100: 9479–9494.

Medina-Garcia G, Guadalupe Echavarria-Chairez F, Ariel Ruiz-Corral J, et al. (2020) Global warming effect on alfalfa production in Mexico. *Revista Mexicana De Ciencias Pecuarias* 11: 34–48.

Müller C and Elliott J (2015) The global gridded crop model intercomparison: Approaches, insights and caveats for modelling climate change impacts on agriculture at the global scale. *Climate Change Food Systems: Gglobal Assessments Implications for Food Security Trade.* Rome: FAO.

Plunkett B (2015) Adapting growth for climate change on Bungulla Farm, 2007–14: Shifting focus from yield productivity to profit and minimising the probability of loss by minimising inputs. *Australasian Agribusiness Perspectives* 102: 1–13.

Rahut DB and Ali A (2018) Impact of climate-change risk-coping strategies on livestock productivity and household welfare: Empirical evidence from Pakistan. *Heliyon* 4: e00797.

Raza A, Razzaq A, Mehmood SS, et al. (2019) Impact of climate change on crops adaptation and strategies to tackle its outcome: A review. *Plants* 8: 34.

Riedy C (2016) Climate change. *The Blackwell Encyclopedia of Sociology* 1–8.

Saiz-Rubio V and Rovira-Más F (2020) From smart farming towards agriculture 5.0: A review on crop data management. *Agronomy* 10: 207.

Sirohi S and Michaelowa A (2007) Sufferer and cause: Indian livestock and climate change. *Climatic Change* 85: 285–298.

Sudmeyer RA, Edward A, Fazakerley V, et al. (2016) Climate change: Impacts and adaptation for agriculture in Western Australia.

Sujakhu NM, Ranjitkar S, Niraula RR, et al. (2016) Farmers' perceptions of and adaptations to changing climate in the Melamchi valley of Nepal. *Mountain Research Development and Change* 36: 15–30.

UN (2015) *Millennium Development Goals Report 2015.* New York: United Nations.

Wheeler T and Reynolds C (2013) Predicting the risks from climate change to forage and crop production for animal feed. *Animal Frontiers* 3: 36–41.

Zarco-Tejada P, Hubbard N, Loudjani P, et al. (2014) Precision agriculture: An opportunity for EU farmers—potential support with the Cap 2014–2020 Study. In: Development PDBSACPaaR (ed). European Parliment: Directorate-General for internal Policies.

Zhang N, Wang M and Wang N (2002) Precision agriculture—a worldwide overview. *Computers Electronics in Agriculture* 36: 113–132.

Zude-Sasse M, Fountas S, Gemtos TA, et al. (2016) Applications of precision agriculture in horticultural crops. *European Journal of Horticultural Science* 81: 78–90.

Section III

Challenges of Winter Fodder Crops

11 Winter Fodder
Opportunities and Challenges in Livestock

Saima Naveed, Usman Ali, Muhammad Naveed-ul-Haque, M. Abdullah, Usama Ahmad, Jamshaid Ahmad, and Manzoom Akhtar

CONTENTS

DOI: 10.1201/9781003055365-11

This chapter provides adequate and comprehensive information regarding the challenges and the gaps for winter fodder shortage and prospects to overcome them. We have divided this chapter into two sections for understanding.

Section I: Introduction and Status of Agriculture in Pakistan
Section II: Winter Fodder Challenges to Livestock, Modern Trends, and the Future Prospects

Section I: Introduction and Status of Agriculture in Pakistan

It is a fact that Pakistan is an agriculture-based country where a significant share of the country's economy is provided by agriculture. The country has a semi-industrialized economy with a well-integrated agriculture production system. The agriculture sector is one of the most substantial sectors, with a significant impact on its economic growth. Overall, the country's economic global ranking was 23rd in terms of the Gross Domestic Product, that is, purchasing power parity (FAO Pakistan, 2018). The agriculture sector contributes approximately 19% to the GDP and provides 38.5% of the employment to the national labor force (Anonymous, 2018–2019). This sector has great potential to contribute more, as almost 65% of the country's population is resident in rural capacities where the primary income-generating source of livelihood is agriculture. The agriculture sector is divided into four main sub-sectors. These are:

- Livestock
- Crops
- Fishing
- Forestry

A comparison of agricultural sub-sectors in percentage is given in Figure 11.1(a), and subsequent growth rates of these sub-sectors are represented in Figure 11.1(b).

11.1 LIVESTOCK AND ITS STATUS IN PAKISTAN

Pakistan is blessed with diversified livestock and is the third-largest livestock-rearing country, with a population of around 200 million domestic animals (Anonymous, 2019–2020). Livestock can generate significant earnings via foreign exchange and act as a stimulant for other sectors' development by providing the working power and many valuable animal products (meat, milk, eggs) and by-products (wool, skin, bones, and manure) for different purposes. That is why the backbone of the agricultural sector is livestock. The flourishing of the livestock is usually maligned for different reasons but mainly lack of planning and resources. It is a fact that well-oriented development can decrease hunger and poverty and ensure the flourishing of the food supply and other related businesses that improve the overall lifestyle (Figure 11.1).

11.2 FODDER AND ITS STATUS IN PAKISTAN

Fodder is any agricultural feedstuff fed to domesticated animals rather than that they forage themselves to fulfill their nutritional demands.

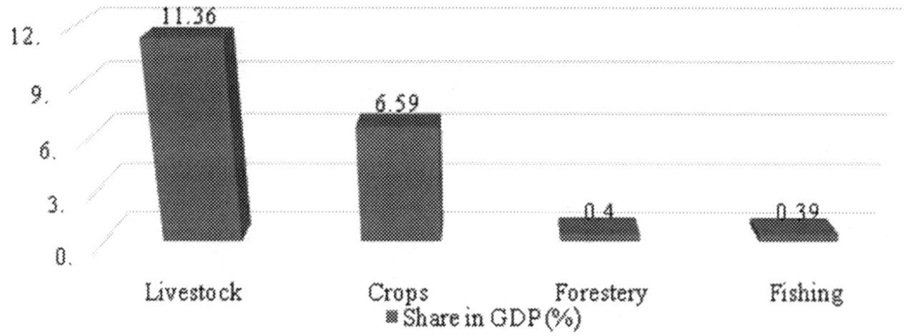

FIGURE 11.1A Share of agricultural sub-sectors (Pakistan Bureau of Statistics, 2018–19).

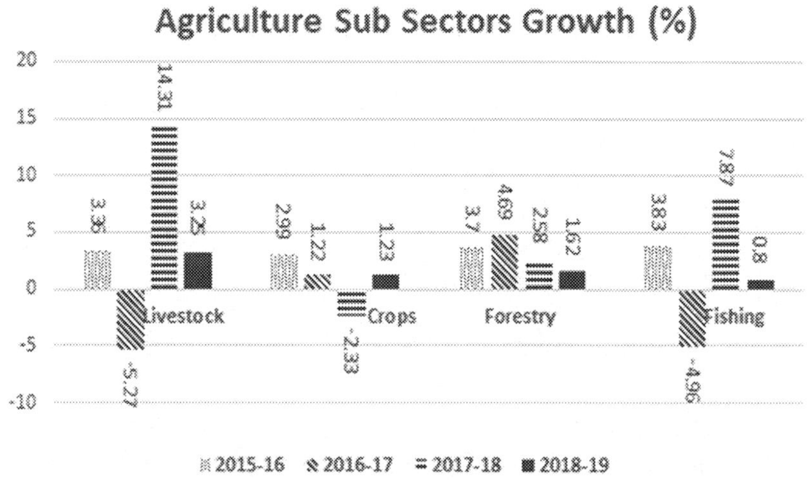

FIGURE 11.1B Agricultural sub-sector growth rates (Pakistan Bureau of Statistics, 2018–19).

Fodder is the most cost-effective and valued feedstuff available for animal feeding. These have a wide range of nutrients, that is, minerals, vitamins, proteins, and carbohydrates required for different animals' bodies (Mishra and Pathak, 2015).

In Pakistan, there are around 79.6 million hectares of land, out of which the cultivation area is slightly over 22 million hectares. The remaining territory is segmented into culturable waste, forests that are densely populated, and rangelands. The documented cropped and forest area covers around 23.3 million hectares and 4.6 million hectares, respectively, out of the total land (FAO Pakistan, 2018).

The crop sector is mainly divided into two categories based on the area of cultivation:

Major Crops: These are defined as 'the crops grown are greater than 300,000 acres area.'

Minor Crops: These are defined as 'the crops are grown on less than 300,000 acres area.'

These crops are further divided into seasonal crops.

11.2.1 Types of Fodder

The domestic fodder (forage) of Pakistan has been divided into two cropping seasons: 'the Kharif,' as the first season of sowing starts in April–June and harvesting months October–December. Sugarcane, cotton, rice, maize, sorghum, rhode grass, and bajra are typical Kharif fodder examples. 'The Rabi,' as the second sowing season, begins in the months of October–December, and harvesting is done in April–May. Wheat, gram, lentil (masoor), rapeseed, barley, lucerne, berseem, oat, and mustard are typical Rabi fodder examples.

The third type is perennial fodder: This is evergreen fodder. These are the multi-cut plants that are usually planted once and can be utilized for many years. These fodders are lucerne, mott-grass, rhode grass, and ryegrass.

11.2.2 Qualities of Good Fodder?

The fodder best for animals' growth and production is referred to as excellent-quality fodder. The development of animals and their production is hugely dependent on the type of fodder they consume. Thus, only good-quality fodder can ensure productive performance of animals. So, the fodders that have the following characteristics should be given to farm animals for profitable earnings.

- The fodder should be palatable.
- The dry matter of the fodder should be in the desired amount.
- The amount of digestible raw proteins should be in a reasonable amount in the fodder to meet the body requirements.
- The digestibility of fodder should be maximum.
- The fodder should have adequate preservation capacity to have an exact idea about the animals' nutritional requirements being fulfilled.

11.3 FODDER PLANTS

Several plants are grown for fodder, but some of the most common plants are:

- Lucerne (alfalfa)
- Maize
- Barley
- Turnip
- Sorghum
- SS hybrid
- Millets
- Oat
- Wheat
- Berseem
- Rhode grass
- Rye grass

Some of the essential fodders used for feeding livestock in Pakistan are tabulated along with their sowing and harvesting times in Table 11.1.

TABLE 11.1
Cultivation and Harvesting Time of Some of the Essential Fodders

Fodder	Cultivation	Harvesting
Lucerne (alfalfa)	October	Yield after 15th January
Berseem	1st to 15th October	10th December to the first week of May
Mott grass	February	November–December
Oat	15th October to15th December	January to10th April
Rhode grass	March to September	First cut after 60 days after cultivation, then 30 days
Ryegrass	October to 15th November	15th December to 15thApril
Sorghum	May to July	July to November

Source: Sattar et al. (2017)

11.3.1 Lucerne (*Medicago sativa* L.)

Lucerne is deep-rooted, perennial fodder, which is well distributed in Asian territories. Lucerne has a very valuable ability to flourish in variable climatic conditions, allowing its cultivation in different environmental regions across the globe.

11.3.1.1 Importance

Lucerne is a vastly cultivated forage throughout the world for livestock. It can be quickly processed into hay or silage but also can be fed as green-chop or grazed. Lucerne has the maximum feeding value among all the forage crops used for haymaking, with high yields when grown in suitable conditions. This is the most common and most important leguminous fodder used in agriculture today, primarily for livestock feeding, due to its high protein and digestible fiber content. Lucerne plants can absorb soil water, thus enhancing soil's ability to reserve the water whenever it is in excess. Its roots contain nitrogen-fixing bacteria that perform nitrogen fixation directly from the air and increase soil fertility. This fodder's nutritive value is more than non-leguminous fodder, such as maize, sorghum, oat, and barley. Lucerne fodder provides a substitute for feeding in the animal production system, especially during scarcity periods of annual crops and pastures. In brief, lucerne:

- Is quality feedstuff for livestock and improves animal health in general.
- Helps maintain soil salinity and alleviate effects due to its water capacity property.
- Is perennial fodder and an excellent choice even in the absence of seasonal fodder.
- Helps maintain soil fertility and structure.

11.3.1.2 Nutritional Aspects

Lucerne produces high-quality green feed for livestock. Its digestibility is around 65–72%, with metabolizable energy of 8–11 MJ/kg and DM around 19% with high protein (18–24%). The nutritional profile of lucerne is tabulated in Table 11.2.

TABLE 11.2
Nutritional Information of Lucerne

Nutrients	Percentage
Dry matter (as fed)	14.1–33.3
Crude protein (DM)	12.0–31.8
Ash (DM)	7.5–19.7
Ether extract (DM)	1.4–4.9
Crude fiber (DM)	15.6–38.2
NDF	25.0–59.6
ADF	18.4–44.8

Source: Heuze et al. (2016c)

TABLE 11.3
Nutrient Requirements of Soil for Lucerne

Element	Recommended levels	Critical level	Action required
Phosphorus (P) (mg/kg)	20–40	<20	9–15 kg/ha P
Potassium (mg/kg)	100–200	<90	20–40 kg/ha K
Sulfur (S) (mg/kg)	>10	<10	5–10 kg/ha S
Aluminum (mg/kg)	<2	>5	Lime can be applied

Source: Dolling (2018)

11.3.1.3 Sowing

If weed germination is not ceased, the seeding can be delayed, as weed control is more critical than sowing time. Usually, sowing is done between15th October and 15th November, but early sowing at the start of the season will have more growth and production than late sowing.

11.3.1.4 Harvesting Stage

Harvesting is done after four months of sowing and then with a gap of one month. In general, the plant is harvested or cut in the late bud to early bloom stage for good-quality fodder with acceptable yields.

11.3.1.5 Irrigation Requirements

The water requirements of lucerne are on the higher side due to the long growing season, dense vegetation, and deep-root system. The irrigation requirements for lucerne are determined by the rainfall and water holding capacity (WHC) of soils. Usually, the plant's normal water requirements are 20–46 inches of water/season but may vary depending on the climate, growing season for the crop, number of cuttings, elevation, and latitude.

11.3.1.6. Fertilizers

Lucerne has a relatively high demand for some nutrients. Each ton of harvested dry matter takes about 14 pounds of phosphate (P_2O_5) and 58 pounds of potash (K_2O).

Similarly, each ton also removes the calcium and magnesium found in about 100 pounds of aglime.

11.3.2 Berseem (*Trifolium alexandrium* L.)

Berseem clover, also known as Egyptian clover, is leguminous fodder that is widely cultivated in the Mediterranean basin, central and western Asia, and now in the United States. It has excellent regrowth ability and is mainly used with rotational grazing from winter to late spring.

Berseem is usually compared with lucerne in nutritional aspects. It is not as efficient as lucerne in drought resistance but has a history of performing well in high-moisture and alkaline soils. Moreover, this crop can also be sown right at the beginning of the autumn season and is available for animal feeding during the colder months. Berseem is also very useful immediately after the winter season when the temperature rises (Suttie, 1999; Hannaway and Larson, 2004).

11.3.2.1 Importance

Berseem is the most palatable multi-cut leguminous fodder of the winter season. It is also called the king of fodders because of its nutritive value and multi-cut ability. In dairy animals, milk production is immensely increased with the feeding of berseem fodder. Berseem also suppresses the growth of herbs in the field. Varieties of berseem are Agati berseem and Pachati berseem, which provides fodder in early and late seasons. Normally, farmers use grass-legume mixtures to increase the herbage yield instead of crop monocultures.

11.3.2.2 Nutritional Aspects

Berseem is characterized by high nutrient concentration, mainly protein (15–20% DM), minerals (11–19%), and also carotene (Heuze et al., 2016b). The nutritional profile of berseem is tabulated in Table 11.4.

TABLE 11.4
Nutritional Information of Berseem

Nutrients	Percentage
Dry matter (as fed)	8.4–25.7
Crude protein (DM)	14.4–26.7
Ash (DM)	11.3–20.2
Ether extract (DM)	1.8–5.1
Crude fiber (DM)	15.7–28.5
NDF	30.6–51.2
ADF	20.6–38.4

Source: Heuze et al. (2016b)

11.3.2.3 Sowing

Sowing is done from mid-September to mid-October. The seeds are usually broadcast in standing water, preferably in the evening hours.

11.3.2.4 Harvesting Stage

Berseem is usually ready for the first cutting within three months (50–60 days) after sowing. After that, cuttings can be done after every 30–40 days (Suttie, 1999). Maximum protein content and relatively low fiber are obtained when the plant is cut at a height of around 40 cm.

11.3.2.5 Irrigation Requirement

Berseem requires a large amount of water to produce high succulent biomass. Adequate and timely supply of water is essential to obtain good production. Usually, 10–12 irrigations are required with intervals.

11.3.2.6 Fertilizers

Berseem is a leguminous crop and hence needs less nitrogen from the fertilizers due to nitrogen-fixing bacteria. In some instances, 20 kg N/ha at the time of sowing is sufficient. The phosphorus and potassium requirements for the crop are usually 80–90 kg P_2O/ha and 30–40 kg K_2O/ha, respectively.

11.3.3 Mott Grass (*Pennisetum benthium* L.)

Mott grass is a multi-cut fodder of great importance. Mott grass yields are almost three times greater than any other traditional crop in a single time and can usually produce fodder for almost eight to ten years. It is also important because of its availability during feed scarcity periods. Mott grass can also be processed into good-quality silage.

11.3.3.1 Importance

Mott grass is evergreen, nutritious, and multi-cut fodder. This fodder looks like sugar cane, so also called 'dhood ka kamad.' It is classical evergreen fodder, as one can get six to seven cuttings per year. Due to the palatability and succulent nature, this fodder is ideal for conservation and silage making (Sattar et al., 2017).

11.3.3.2 Nutritional Aspects

Mott grass is a multi-cut fodder with a rich nutritional profile and can maintain its nutrition over more extended re-growth periods. The nutritional profile of mott grass is tabulated in Table 11.5.

11.3.3.3 Sowing

The fodder is sown between February and March (spring season) and in the months of July–August (monsoon season).

TABLE 11.5
Nutritional Information of Mott Grass

Nutrients	Percentage
Dry matter	24.4–31.6
Crude protein (DM)	11.0–12.8
Ash (DM)	11.6–12.2
Ether extract (EE)	1.66
Crude fiber	32.31
ADF	47.0
NDF	76.8

Source: Altaf-ur-Rehman and Aneela (2004); Touqir et al. (2007)

11.3.3.4 Harvesting Stage

The first cutting of the fodder is achieved after four months of sowing. Then after every two months, fodder is ready to cut.

11.3.3.5 Fertilizers

During land preparation, one trolley of organic fertilizer, such as dung and manure, one bag (50 kg) per acre potash, and two bags (100 kg) per acre nitro-phosphorus should be used after every cutting; one bag (50 kg) per acre urea is used to have maximum production in upcoming cutting.

11.3.4 Oat (*Avena sativa* L.)

Oat is adaptive to a wide range of soil types and usually performs better in acidic ones. These are mostly cultivated in cool and moist climates. In Pakistan, oat forage is grown in various parts of the country and accounts for more than 35% of land under forage use.

11.3.4.1 Importance

Oat was once the predominant fodder for cattle but became marginalized during the latter part of the 20th century (Assefa, 2006). The use of oat for dual purposes such as fodder and feed has made it very valuable. Usually, the crop is harvested in the early stages for fodder and allowed to grow to obtain the grains. Oat forage is used for hay-making, silage making, and grazing, and its straw is usually used for animal bedding (Suttie and Reynolds, 2004).

11.3.4.2 Nutritional Aspects

Oat is a valuable source of feed, especially for ruminants. The high energy content in the oat is offset by the high fiber content, resulting in relatively lower digestibility. Similarly, oat is also somewhat lower in protein content due to the high nitrogen degradability. The nutritional profile of oat fodder is tabulated in Table 11.6.

TABLE 11.6
Nutritional Information of Oat

Nutrients	Percentage
Dry matter	14.1–38.3
Protein	5.8–26.2
Ash	7.0–14.5
Fiber	21.5–34.7
Ether extract	2.0–4.7
NDF	34.2–65.6
ADF	19.6–39.3

Source: Heuze et al. (2016a)

11.3.4.3 Sowing

Usually, oat forage is sown between 15th October and 15th November.

11.3.4.4 Harvesting Stage

The oat can be harvested at the boot stage or milky dough stage, but the yield is highly variable depending on the origin and harvest time. In Pakistan, the reported oat forage yield is around 18 t DM/ha (Bakhsh et al., 2007).

11.3.4.5. Irrigation Requirement

Oat consumes more water than wheat crops. Usually, four to five irrigations can provide a good yield. Irrigation is highly recommended immediately after each cutting.

11.3.4.6 Fertilizers

A ratio of 80:40:0 kg NPK/ha is the recommended dose of fertilizers. One hundred percent P is to be applied as basal. Sixty kg N is to be applied as basal, 10 kg at first irrigation and 10 kg at second irrigation is good for higher yield. Ten kg of N is to be applied after first cutting if sown for fodder-cum-grain (Lafond et al., 2013).

11.3.5 Rhode Grass (*Chloris gayana* K.)

Rhode grass is a prominent perennial tropical grass. This is a leafy grass of tropical and sub-tropical areas.

11.3.5.1 Importance

Rhode grass is a perennial or annual tropical grass. It is one of the improved pasture species available for livestock. Rhode grass is mainly cultivated over the years due to its high dry matter and cost-effective cultivation. It can be used as pasture or cut for hay but is not suitable for silage making. This grass is useful forage of medium to high quality and is drought resistant, mostly when young. It is also very productive as a cover crop and improves soil quality by improving soil fertility and structure and lowering nematode count.

11.3.5.2 Nutritional Aspects

Rhode grass is good-quality forage with variable nutrient composition. The nutrients are at a peak when the forage is relatively young. However, the nutritive value starts to decline with the maturity of forage. Similarly, the digestibility of rhode grass also decreases with plant maturity. The hay of the harvested forage at later stages has decreased protein content and higher fiber content. The nutritional profile of rhode grass is tabulated in Table 11.7.

11.3.5.3 Sowing

Propagate rhode grass seeds by the broadcast method in standing water. The seeding rate should be around 8 kg per acre. For sowing the seeds, the soil should be bedded plain and consistent.

11.3.5.4 Harvesting Stage

The first cut is usually achieved 60 days after sowing, and the remaining is cut in 30 days. Rhode grass is cut above 4 inches from the soil. The maximum observed or reported yield of rhode grass is around 30–40 t DM/ha, and the recorded average yield is in the range of 10–16 t DM/ha (Murphy, 2010).

11.3.5.5 Fertilizer

One bag of DAP and 1 to 1.5 bags urea at the time of sowing are suggested, and then one to two bags of urea after each cut is applied.

11.3.6 Ryegrass (*Secale multiflorum* L.)

Ryegrass belongs to the family Poacea and is a perennial or annual grass and found in many regions across the globe. It is leafy grass and produces high-quality, palatable forage. It is an outstanding annual forage and a very useful cover crop during the winter season. It is equally valuable for pasture, silage, and hay-making.

TABLE 11.7
Nutritional Information of Rhode Grass

Nutrients	Percentage
Dry matter (as fed)	16.7–39.0
Crude protein (DM)	5.1–15.7
Ash (DM)	6.1–13.2
Ether extract (DM)	1.2–3.4
Crude fiber (DM)	28.2–43.4
NDF	69.9–82.1
ADF	36.6–47.1

Source: Heuze et al. (2016d)

11.3.6.1 Importance

Ryegrass is a valuable forage crop and is usually cultivated in those territories with cold winters and hot and dry summers. This ability makes it precious as it can tolerate a wide range of temperatures, that is, 3–31°C, but it has a maximum yield between the temperature ranges of 15–20°C. It is also well adapted to winter hardiness and can trap water (SAREP, 2006). Ryegrass can be cultivated in areas where wheat crops cannot be grown (Ecoport, 2011). These can also be grown in association with leguminous fodders as companion crops and improve soil quality.

11.3.6.2 Nutritional Aspects

Ryegrass is a highly variable nutritive forage depending upon the stage of maturity of the plant. Its digestibility also varies with the age of the plant. The protein content of the forage is highly variable with the stage and origin of the forage. The nutritional profile of ryegrass is tabulated in Table 11.8.

11.3.6.3 Sowing

Ryegrass is sown between October and November. It is essential to select a clean site for sowing that is free of residual herbicides. Ryegrass is sometimes cultivated in companion with clovers to provide earlier grazing and minimize the risks of bloat from clovers. The seeding rates are usually 8–10 kg per acre.

11.3.6.4 Harvesting Stage

Usually, the fodder is harvested between December and April. It is recommended that the ryegrass forage should be harvested before the headings appear or at the early boot stage. At this stage, the forage has maximum nutritive value and good palatability, and it yields around 5 t DM/ha. It is also to be considered that ryegrass is sensitive to too much harvesting and overgrazing, and it should be avoided.

11.3.6.5. Fertilizer

Ryegrass is usually responsive to nitrogen fertilizer. Commonly, 50 kg of potash and phosphorus should be applied in one growing season. Further, nitrogenous fertilizers are applied @ 60–100 kg per acre per cut.

TABLE 11.8
Nutritional Information of Ryegrass

Nutrients (%)	Percentage
Dry matter (as fed)	12.1–22.3
Crude protein (DM)	7.5–21.4
Ash (DM)	4.7–15.2
Ether extract (DM)	4.8
Crude fiber (DM)	24.8–38.7
NDF	88.7

Source: Heuze et al. (2015)

11.3.7. Sorgum-Sudan Grass (Hybrid)

Sorghum-Sudan grass is a hybrid variety of forage for animals. It provides an efficient solution to produce dry forage matter in case of urgency. The yield of Sorghum-Sudan grass is comparatively lower than maize fodder but is better than corn. It is multi-cut; it can be cut twice or thrice during the entire season and preserved as chopped silage or pastured.

11.3.7.1 Importance

In Pakistan, the first time it was introduced to farmers was in 1973. At that time, it was less palatable and tasteless. Later on, the Fodder Research Institute, Sargodha, modified it and prepared a sweet variety that is more palatable. It is a multi-cut fodder of the summer season and gives three to four cuttings a year; that is why it is called berseem of the summer season. As it is a hybrid variety, if seed is used for crop cultivation, then per-acre fodder production will decrease to 15–20% (Sattar et al., 2017).

11.3.7.1.1 Varieties

1. Sweet sorghum-Sudan grass (hybrid)
2. Tasteless sorghum-Sudan grass (hybrid)

11.3.7.2 Nutritional Aspects

A sorghum-Sudan hybrid is more productive for replacement heifers over 12 months old, dry cows, cows for beef purposes, and calves. As the sorghum-Sudan crop matures, there is a significant dip in protein content, while fiber levels increase. This results in decreased feed energy value and rumen digestibility. The nutritional profile of sorghum-Sudan grass is tabulated in Table 11.9.

11.3.7.3 Sowing

In early March, sowing is performed with a single-row cotton drill machine in lines 1½ feet apart.

TABLE 11.9
Nutritional Information of Sorghum-Sudan Grass

Nutrients	Percentage
Dry matter (as fed)	23.40
Crude protein (DM)	7.30
Ash (DM)	27.30
Crude fiber (DM)	30.70
Ether extract (DM)	2.43

Source: Sattar et al. (2017)

11.3.7.4 Harvesting Stage

The first cutting of the crop is achieved after 2 to 2½ months of sowing. The next cuttings of the crop are performed after every 1½–2 months.

11.3.7.5 Fertilizers

One hundred to 150 kg per acre of organic fertilizer such as dung and manure one month before sowing will increase per acre fodder production immensely, or two bags (100 kg) of nitro-phosphorus per acre at the time of sowing. Urea is also necessary for maximum growth and production with first water after every cutting of the fodder.

Section II: Winter Fodder Challenges to Livestock, Modern Trends and the Future Prospects

11.4 FODDER AS NUTRITION AND ITS PHYSIOLOGY

Feeding livestock through fodder is one of the most economical feeding methods and is important in the farming system. However, the quality and quantity of the nutrients in the fodder or feed of the livestock are essential considerations for their selection, voluntary feed intake (VFI), and digestibility. VFI and digestibility are interlinked parameters. The factors that influence any fodder's digestibility are maturity level of the plants, processing of dietary supplements and the nature of the chemical treatment, and dry matter intake (DMI). Usually, the plants' maturity and processing are also essential for digestibility, but the most critical parameter is the dry matter intake, as it ensures the proper release of nutrients required for different physiological functions.

Some of the critical factors that can influence voluntary feed intake are:

- Dry matter
- Crude protein concentration
- Fiber concentration

Voluntary feed intake is directly proportional to the plant species' dry matter. VFI decreases with a decrease in DM percentage. An increase in the DM usually results in a higher absorption rate for the feed and its nutrients. On the other hand, the role of volatile fatty acid indigestion is also significant. Digestion becomes relatively simple when the volatile fatty acid (VFA) concentration of butyric acid and propionic acid is more than the acetic acid.

Voluntary feed intake is also significantly influenced by the crude protein (CP) level in the diet. The CP content in tropical forages is usually on the lower side. It is essential to know that CP is negatively correlated with plant maturity. As the plant matures, CP content decreases and is at a minimum level before flowering.

Another critical aspect of VFI is the fiber content of the diet or roughage. The roughage is:

> The feedstuff has a minimum level of fiber content around 18% and total digestible nutrients (TDNs) around 70%.

Roughage is a source of a wide range of nutrients in the animal diet and plays a significant role in the maintenance and optimization of the gastrointestinal tract (GIT). Similarly, fibrous carbohydrates have a role in maintaining GIT integrity.

The amount of net energy for production purposes also depends on its consumption and utilization efficiency. The utilization of feed also has a correlation with the nutritive value of feed, animal species, and functioning.

11.5 MODES OF FEEDING

All animals' nutritional needs are met by forage, shrubs, and agricultural waste. For this purpose, different modes for feeding animals are used, that is, pasturing and feeding by fodder or grains. Usually, ruminants are fed on fodder crops, and they cover around 70–75% of the nutrient requirements. Another form is grain feeding, which is relatively uncommon in ruminants. In general, 95% of animals are nourished on forage plants.

11.6 FACTORS AFFECTING FODDER QUALITY

Fodder quality and quantity are both important from a production point of view, and in terms of quality, animals' performance is the best indicator. If good-quality fodder is given in a desired amount to animals, they can perform 50% more than other genetic pools. Some factors can influence the yield of the fodder and its quality.

- The age of the plant
- CP and DM

As the plant matures, the crude protein level decreases, while the dry matter content increases. Excellent-quality fodder should have a high CP level and low crude fiber content. Quality of the fodder and the improved breeds are critical if better returns are to be expected.

Continuous supply of quality fodder in an ample amount is the fundamental need for livestock farming. Usually, winter crops have more CP than energy and vice versa in summer.

Different fodder forms and their contribution to livestock in Pakistan are tabulated in Table 11.10.

TABLE 11.10
Contribution of Different Sources to Feed Livestock in Pakistan

Sources	Contribution (%)
Fodder and crop residues	51
Forage/grazing	38
Cereal by-products	06
Post-harvest grazing	03
Oilcake, meals, animal protein	02

Source: NARC (2012)

11.7 FODDERS AND THEIR CHALLENGES TO LIVESTOCK

Fodder production and utilization are highly dependent on the climate, cropping pattern, livestock, and socio-economic conditions. In Pakistan, livestock is the backbone of agriculture, but lack of resources is the primary limiting factor for the production point. There are multiple challenges in the country, such as lack of quality and quantity of fodder and availability throughout the year, especially the scarcity of fodder at the beginning of winter.

The most severe deficit periods are one in winter (November to January) and others in summer (May–June). Fodder scarcity is further worsened when the traditional winter fodders of berseem (the major winter fodder; *Trifolium alexandrinum*), shaftal (*Trifolium resupinatum*), and lucerne (*Medicago sativa*) are dormant.

> The fodders grown at the beginning of the winter season that reach maturity in summer are called the winter fodders, for example, berseem, oat, and ryegrass.

Here, we will discuss some of the significant constraints regarding the shortage of fodder during the winter season and their proposed solutions.

11.7.1 Constraints and Proposed Solutions

In winter, livestock owners usually use some local feed constituents to overcome fodder shortage problems. These constituents include grains, cottonseed cakes, wheat straw, dry bread loaves, and spontaneously growing wild herbs and wild grasses. The supply deficit becomes large if it is expressed in terms of digestible protein. Poor feed ultimately leads to low production and specific health issues along with financial crises. As a result, animals are sold even at the production stage to overcome financial constraints. In some areas, the type of available fodder plays a vital role in deciding to keep the different livestock types, because, sometimes, it becomes tough to ensure the provision of fodder even for the maintenance of livestock

11.7.1.1 Agricultural Land Availability and Utilization

Crop growth and production are fundamental parameters for sustainable livestock production systems worldwide in the agricultural system. However, there is a significant decline in crop production due to soil pollution, environmental factors, and lack of resources, resulting in a decline in refined land production. Agricultural land is:

> The land area is arable under permanent crops and permanent pasture.

In Pakistan, the reported agricultural land area was 47.79% in 2016 (World Data Atlas, 2016). Due to farmers' economic concerns, cash crops and cereals have become more critical than fodders for animals. This is why most of the cultivable land remains occupied by cash crops or cereals, which induce the gap in fodder production. While discussing the availability of cultivable lands for fodder production, farmers' land assets also possess great importance. Most livestock holders have little cultivable land available for fodder production. Up to 65% of livestock holders have no land available for fodder cultivation.

Moreover, 20% of the population has less than 12 acres of land. Simultaneously, only 15% of livestock farmers have more than 12 acres of land assets. Another reason for decreased land availability is the industrialization and urbanization of the agricultural area.

Furthermore, if the land is not appropriate for producing crops, animal husbandry practices can be used to accommodate livestock production. To determine land shortage issues, the humans on these lands play a vital role. The critical fact is that farmers are the key decision-makers in choosing the production systems and patterns for land usage in any community. Land utilization in Pakistan is represented in Figure 11.2.

11.7.1.2 Water Availability for Irrigation

The agricultural outputs and water availability for irrigation are interlinked. In Pakistan, we have the world's best irrigation system, but there is less water in the system due to poor management, and using a tube well is not cost effective due to increasing electricity prices. So, water availability for irrigation is one of the significant constraints in producing good-quality fodder. This is evident from the reports that during fiscal year (FY) 2015–2016, water availability for Kharif crops was around 65.5 million-acre feet (MAF), which is a significant decrease of around 5.5% from the previous year and 2.1% from the usual supplies of 67.1 MAF.

The major fodders grown in winter are berseem barley, oat, lucerne, and mustard. It is a well-known fact that irrigation systems are the backbone for crop production and agricultural productivity depends on water availability. In Pakistan, inefficient and collapsed irrigation systems, overexploitation of the groundwater, inefficient services, limited storage capacities, and water pollution have collectively created water shortage issues in terms of both quality and quantity.

The irrigation system of Pakistan comprises multiple components: canals, rivers, barrages, dams, headworks, and tube wells (on a small scale). However, the

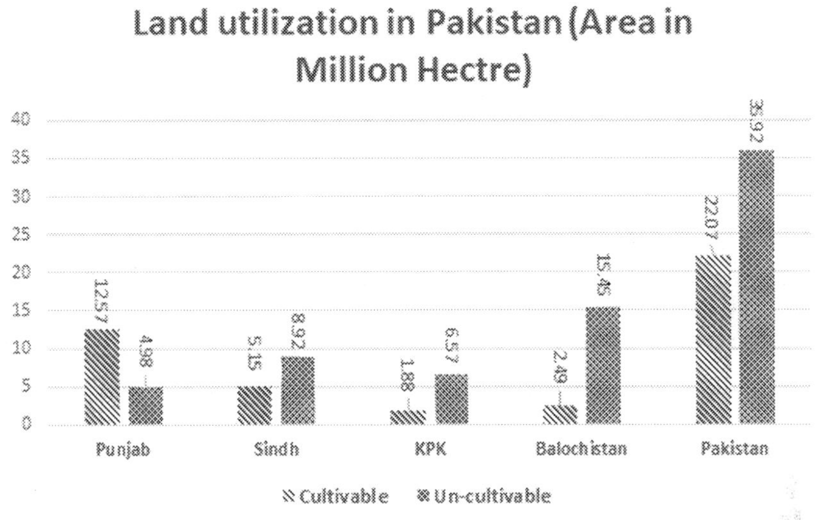

FIGURE 11.2 Comparison of agricultural land in Pakistan (Provincial Agriculture Department, 2017).

prevailing irrigation systems for agriculture in Pakistan are canal water and tube wells. The canal system is sound, but the amount of water is meager in this system, so farmers have to depend on rains and tube wells for irrigation of land. Fodder growth is also dependent on temperature. If water is available, crops will grow faster in summer than in winter. The surface water availability for different seasonal crops is represented in Figure 11.3.

11.7.1.2.1 Solution to Overcome Water Shortage

There are specific proposed solutions to overcome the crises of water shortage and ensure water availability for irrigation. Improved water management at the storage and distribution levels to avoid the pollution and wastage of water will help this cause. Different strategies need to be implemented to maintain the groundwater level and store the excess water during the rainy seasons. The millions of acre-feet of valuable water, which usually remains unused and flows into the sea, could be utilized for irrigation by merely adopting acceptable management practices.

11.7.1.3 Seed Quality and Production

Quality of seed is essential for better fodder production. The availability of good-quality seeds for crops is the limiting factor for fodder production in the country. Generally, seed quality means that it should be resistant to diseases, give the maximum possible per acre production, and be free of weed seeds. There is an estimation that only a small amount (around 5–10%) of fodder is grown with improved seed quality. There are multiple interlinked systems with improved seed production:

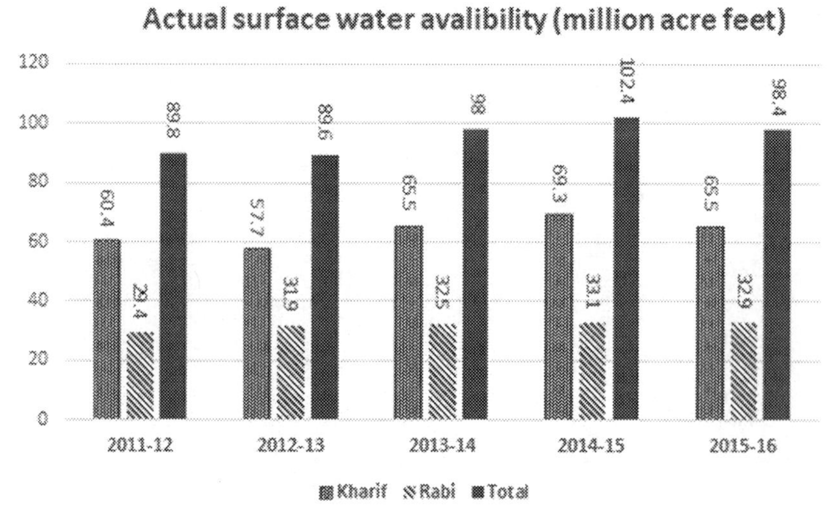

FIGURE 11.3 Comparison of groundwater availability to different seasonal crops (Indus River System Authority, 2016).

- Agro-climate
- Adaptation specificity of crops to the environment
- Political and socio-economic factors
- Management and production

Unfortunately, there is a lack of interest from both the government and private sectors in excellent-quality seed practices. It is evident from the fodder research program reported by NARC (2012) that only 11% of total fodder seed is produced locally with improved quality. The demands are usually achieved by import or using under-quality seeds, that is, that produced by the farmers, without considering seed production guidelines.

Another limitation is that usually, forages are harvested for animal feed and are not kept for seed formation.

11.7.1.3.1 Solution

The development of the dual-purpose varieties of forage might be helpful to ensure the availability of seeds.

11.7.1.4 Decreased Winter Hardiness

This is an essential factor that results in decreased yields during the winter season. It is actually 'the ability of plants to survive all factors influencing survival during winter.'

Generally, winter-season grasses have more tolerance for freezing temperatures than legumes (Ouellet, 1976). Several climatic factors contribute to the risks of damage to many crops. Temperature fluctuating above and below freezing reduces cold hardiness and increases plant susceptibility to winter injury and stand loss (Mckenzie and McLean, 1980). Prolonged exposure to temperatures above 0°C during cold results in gradual loss of winter hardiness.

Ice sheeting, resulting from rainfall and subsequent freezing temperatures, may cause plant death due to anoxia (Guleifsson, 1986). Species such as alfalfa specific to winter needs can be managed to allow hardening and accumulation of adequate energy reserves to survive winter (Belangar et al., 1999).

The ability of plants to survive during freezing weather leads to the scarcity of seasonal fodder.

11.7.1.4.1. Management to Improve Winter Survival

There are some management strategies that can be adopted to improve the winter survival of plants. Harvest management during the fall season of plant species sensitive to winter is critical. It is necessary to provide these plants the required time for proper hardening and reserve the necessary nutrients, that is, energy to survive the winter cold. There should be a fall rest period of around four to six weeks for growth before the first wave of killing frost. The re-growth interval appears to be a better determinant of plants for winter survival. An interval of a minimum of 500 growing degree days (GDDs) where the threshold temperature is 5°C between the fall harvest of plants and the previous one is thought to reduce winter damage risk. However, a simple and effective tool against the winter kill is not to take the plants' full harvest, ensuring productivity for longer durations. This allows the plant species to accumulate the necessary nutrients and harden adequately to tolerate the winter cold.

Another important aspect is the role of soil nutrients, especially potassium (K). Adequate soil potassium (K) is essential for the development of tolerance and resistance against the cold. However, excessive soil-K levels can cause a significant growth fall, hence resulting in decreased winter tolerance, especially in grasses.

11.8 SOME MODERN TRENDS TO OVERCOME WINTER FODDER SHORTAGE

11.8.1 Perennial Fodder

These are herbaceous plants cultivated for feeding livestock, with a life span of more than one year. Another solution to the chronic green fodder shortage during the deficit periods has been recently solved by growing multi-cut fodder crops like SS hybrids, lucerne, mixtures of cereals and legumes, and mott grass.

11.8.2 Companion Crops/Inter-Crops

To obtain excellent and early yields during the winter season, compatible fodder crops on small holdings might be grown in the mix to attain higher yields and improved quality per unit area/season. Leguminous fodders like vetch and berseem may accompany ryegrass, oat, and brassica. Similarly, in the case of lucerne forage, it can be accompanied by oat and barley. The seasonal crop is usually sown between rows of perennial crops. The advantages of inter-crop sowing over pure sowing are:

- In every season, more than one crop/unit area
- Easy to manage weeds
- Yields are usually more than the pure ones
- Fodder quality is improved

The advantages of companion crops are proved and documented in the literature. Mixed sowing of lucerne + oat, red clover + oat, and berseem + oat produced better green and dry matter yields than the sole crops of either legume (Dost, 1997).

11.8.3 Forage Conservation

Forage conservation becomes an essential aspect of managing fodder scarcity problems (both in quality and quantity). The proper procedures for conservation, if adopted, improve not only the bioavailability of nutrients but also maintain the quality of the finished product. Large quantities of surplus Kharif forages are produced, which are frequently more than what is needed. Therefore, it is essential to preserve excess forage during the Kharif and Rabi seasons at the proper stage of maturity to provide nutrients during the lean period. Leaves of leguminous plants can also be sun-dried and stored for feeding as a leaf meal during lean periods. Wheat straw and other grasses should be stored after chaffing them to a 1–2-cm particle size.

Forage can be preserved either as hay (dried fodder) or as silage (wet fodder), depending upon weather conditions and available resources for feeding to livestock during lean periods when the availability of fresh forage is meager or negligible, that is, mid-October to mid-December and mid-April to June.

11.8.3.1 Preservation of Forage by Hay-Making

Hay-making is:

> Conservation of high-quality forage via drying processes.

This is an economical process, especially in warm regions of the globe, as ample sunlight is available. This enables farmers in these regions to dry out green forage in natural sunlight. Drying in this form for hay-making is very economical. The drying process reduces the moisture content and results in the inhibition of enzyme activity in the plants. Crops with thin stems like oats, range grasses, and legumes, particularly all other cultivated leguminous fodders in general, are suitable for hay-making. Legumes have relatively more nitrogen content and possess high buffering capacity, making them convenient for conservation as hay.

During hay-making, perishable green forage is converted into a product which is very easy to conserve and transport without much danger of spoilage, and losses of nutrients and dry matter are minimum. Moisture content is usually reduced from 70–90% to around 20%.

The whole procedure of hay-making can be done mechanically if there is quite enough space. However, the requirement is that the climate or weather remain feasible for this procedure of conservation. In those areas where the climate is uncertain, it is challenging to carry out standard hay-making operations in the open (Suttie, 2000).

11.8.3.1.1 Procedure for Hay-Making

The basic methodology for hay-making is quite simple and is described in the following.

- Generally, fodder is harvested before it reaches its maturity level to have maximum nutritive value. Early harvest of the fodder will yield less, but the nutritive value will be maximum.
- The leaves of crops tend to have more nutrients than stems. So, while cutting the forage, the maximum leaves with the minimum stem are taken.
- The forage should be conserved in a dry environment because any moisture will encourage mold growth, which can be very harmful.
- The forage should be layered as thinly as possible in the sun and should be raked and turned regularly to hasten drying.
- Chopping the forage into small pieces will be useful for drying quickly.
- The drying of the forage may take around three days under normal circumstances.
- Hay should be kept from over-drying, as it can become a fire hazard.
- The dried forage in hay should be baled to storage when the moisture level is below 15%. Baled storage requires less space.

11.8.3.1.2 Types of Hay

There are different forms of hay depending upon the conditions and their intended usage:

Long hay: This is the traditional and ancient form in which herbage is mowed and then carted.

Chopped hay: The form in which forage is chopped and dried is called chopped hay. This is usually an option where the drying conditions are right; highly mechanized systems are available. This form is less bulky and easier to handle. This form of hay should be conditioned and windrowed before collection through a forage harvester.

Baled hay: This is also a widespread type. Initially, this was performed by the hands, but now machines are available. Automated baling has been used since the 1950s. The most simple and popular shape is the round bale. This shape helps to resist water and shed rain more effectively.

Hand trussed: This form of hay is widespread in hay-making manually, often as a means of reducing shattering.

Wafered/pelleted hay: It is a dense form, easy to store, transport, and handle. Field units are available but expensive; they are used for high-quality legume hay in climates, which allow rapid drying. Losses are lower than with baling.

Dried-grass: This is artificially dried herbage at very high temperatures. The conservation through this process is a bit expensive but allows young and good-quality material to be stored effectively.

Barn-dried hay: In this form, the herbage is dried with fan-equipped machinery with or without a heat source. This procedure is not very common.

Haymaking is a beneficial process for preserving high-quality forage but also has some limitations.

11.8.3.2 Limitations

There are specific hay-making challenges, which differ according to the weather and climatic conditions and the type of forage.

- Under sub-humid and humid temperate conditions, the main problems are related to gradual drying, intending to avoid loss by spoilage and to dry the crop as quickly as conditions will allow.
- In contrast, under hot, dry conditions, the problems are more likely to be either shattering of the more delicate parts of the plant through too rapid drying or bleaching, with consequent loss of carotene and vitamins.

11.8.4 Preservation of Forage by Silage-Making

There is another conservation technique of forage known as ensilage. Grown fodder is harvested when it is still green and nutritious and is preserved through a natural process known as pickling. Lactic acid is produced as a result of the fermentation of sugars present in the forage plants by bacteria in an air-sealed container, the silo, where the anaerobic environment is ensured. In this way, the conservation of forage is known as silage or ensiled forage. After preservation, the silage can be kept for around three years without much deterioration in the quality. Silage has good palatability and can be fed to livestock at any time. In essence, silage is

> Any material as a result of controlled fermentation of forages or crop residues, with high moisture levels and an anaerobic environment.

Silage is wet preservation of fodder with a moisture content of around 60%, while hay is the dried form of forage with a moisture content of around 15–20%.

Silage vs. Hay

A brief comparison of silage and hay is given in the following.

Forages are usually processed into hay to conserve the plants' nutrients, especially the proteins, before it declines in the plant. However, the plants are often too wet and take longer to dry successfully in normal circumstances unless special machinery is used. Furthermore, sometimes forage crops are too thick-stemmed to dry successfully as hay, like maize.

In contrast, silage is a much better way to preserve forage crops. Forage crops can be harvested in the early stage, and they need to have only 30% dry matter for successful ensilage. So, there is no compelling reason to dry forage plants for conservation, and wet weather is not a constraint in making silage, which is the case in hay-making. Silage making is extensively practiced in the agriculture sector, but it requires heavy equipment to dig or build the conservation units and pits for large-scale production; the production method relies on heavy equipment and large production, which is a constraint for smallholder farmers (USAID and FAO, not dated).

11.8.4.1 Principles of Silage Making

During the harvesting stage, the cells of the plant do not die immediately. They manage to remain functional until there is enough water and oxygen to respire. Oxygen is required for the respiration process, which results in supplementing the energy to the cells for proper cellular functioning. During this process, the plant sugars are oxidized by the cells in oxygen and yield metabolic water, heat, and CO_2.

The heat generated during respiration is not dissipated readily, which results in the rise of silage temperature. Some of the naturally occurring molds, bacteria, and yeast on plants also aid respiration. The temperature rise indicates the respiration process due to the production of heat that is unable to dissipate. Although a slight rise in temperature of about 10°F is acceptable, the objective is to eliminate all the trapped air in the forage to inhibit respiration and make the environment oxygen free. The temperature rise can be controlled by harvesting the crops at the proper time, with desirable moisture levels.

To check and limit respiration during the fermentation process, some standard techniques can be considered:

- Careful inspection of the walls of silos before filling
- Harvesting of fodder with adequate moisture levels
- Proper chop size
- Rapid filling
- Complete packaging and sealing

11.8.4.2 Advantages of Silage

Some of the advantages of ensiling the fodder are mentioned in the following.

- The stable nutrient composition of the silage for longer durations. Silage has more nutritive values than hay.
- The nutrient losses are reduced in silage (usually less than 10% of DM) when compared with hay (usually around 30% of DM).
- Forage plants can be cut in the early stages of development.
- The plants are used effectively and economically during silage making, which results in high yield.
- Utilization of land is proper by rotation of the crops.
- Fodder can be ensiled in both cloudy and cold weather.
- The fermentation process during silage-making helps reduce nitrates, which are usually accumulated in plants during overfertilized crops and dry periods.
- Silage requires significantly less storage space (approx. ten times) as compared to hay.

11.8.4.3 Disadvantages of Silage

There are some disadvantages of silage, which are as follows.

- Silage is tricky to market because it is difficult to determine the cost.
- It is challenging to transport silage to far places or long distances.
- Silage usually lacks vitamin D content, which is a critical nutrient for farm animals.
- The weight of the silage results in manipulation of costs.

11.8.4.4 Silage Bags

There is a proposed solution using the silage bag, which can eliminate some of the silage disadvantages. This is one of the modern forms used for the preservation of silage instead of pits and silos. These are considered a significant and cheap replacement for traditional conservation systems, especially in storage losses. Some of the advantages of using silage bags are:

- Silage bags are useful in minimizing nutrient loss while conserving the feed. The anaerobic environment helps eliminate mold spoilage growth, yeast, and bacteria without compromising the essential nutrients.
- Silage bags provide a useful cushion to farmers to conserve silage at the place of their choice. The only condition is needing a dry spot.
- Silage bags have a proper seal, which ensures a proper anaerobic environment. As a result, all the acid produced during the fermentation process will remain in the bags, which will compensate for irregular chop size and compaction issues. So, the quality of silage remains uncompromised.
- The preservation of forage in silage bags minimizes the farmer's physical workload compared to pits and silos.
- Usually, while feeding silage to animals, the whole bag is utilized to avoid aerobic contact and contamination and increase shelf life. This is difficult in the case of pits and silos.
- The transportation of silage bags is easy, while it is difficult in pits and silos.

11.8.4.5 Chopping and Compaction of Silage

Green fodder should ideally be chopped into small pieces of regular size, up to 3 cm long, before filling the silage bags. Then the bag is compressed so that any entrapped air can escape. Then the silage bags are sealed. In this way, the feed can be preserved for longer durations without compromising its nutrient profile. This also allows farmers to maintain animals' feed levels during the whole year, especially for scarce periods.

11.8.4.5.1 Objectives of Chopping and Compression

Some of the objectives of chopping and compaction are:

- To cause the maximum possible release of the sugar content for the fermentation process.
- To make sure that all the entrapped air in the bag escapes so that lactic acid can be adequately produced as a result of a smooth fermentation process.

11.8.4.6 Disadvantages of Silage Bags

There are also some issues in using silage bags, which are listed in the following.

- The use of pesticides for the preservation of bags can contaminate feed and cause harm to animals.
- Proper disposal of the silage bags once they are emptied because of plastic in their manufacturing.

- Forage needs to be chopped, as proper sizing is required to eliminate the chances of air entrapping during the packaging and puncturing of the bags, which requires some extra effort.
- These hindrances can be overcome by adopting the proper approach. The possible causes of losses during the processing of silage are:
- When the dry matter is less than 30–32% in the forage, resulting in increased chances of seepage.
- Dry matter should not be more than recommended levels.
- The approach of birds and rodents should be monitored and controlled, as they can cause harm to the silage bags, and chances of spoilage increase.

TERMINOLOGY

Pasture is used for grass or other plants grown for feeding grazing animals.

Fodder is defined as any plant that is cut before being fed to animals in the green stage or after converting to hay and silage.

Forage is used broadly to means all the plant material that is eaten by herbivores or the edible parts of the plants other than grains that provide feed for animals or can be harvested for feeding.

Grazing is when animals eat or partially defoliate any standing vegetation.

Silage is forage preserved by field crops and dried to a moisture level to prevent microbial activity that leads to spoilage.

Hay is an animal feed produced by drying green fodder to a moisture content of about less than 15% so that biological processes do not proceed rapidly.

Haylage, also called no-moisture silage, hay crop silage, or drylage, is the combination of hay and silage in which moisture in the grass or forage is reduced 40–60% by cutting and wilting in the field before it is chopped and stored in silage.

Herbage is a collective term for the above-ground succulent biomass of forage crops fed to livestock.

Straw is dry stocks of cereals after threshing and removing the seeds.

Stower is a term for cereal stubbles and broken pieces of threshing.

REFERENCES

Anonymous (2018–2019) *Economic Survey of Pakistan*, Finance Division, Government of Pakistan.

Anonymous (2019–2020) *Economic Survey of Pakistan*, Finance Division, Government of Pakistan.

Assefa G (2006) Avena Sativa L. In: Brink M and Belay G (eds) *Cereals and Pulses/Céréales et Légumes*. Wageningen, Netherlands: PROTA, pp. 28–32.

Bakhsh A, Hussain A, Khan S, Zulfiqar AZ and Imran M (2007) Variability in forage yield of oats under medium rainfall of Pothowar tract, *Sarhad Journal of Agriculture* 23(4): 867–870.

Belanger G, Kunelius T, McKenzie D, Papadopoulos Y, Thomas B, McRae K, Fillmore S and Christie B (1999) Fall cutting management affects yield and persistence of alfalfa in Atlantic Canada, *Canadian Journal of Plant Science* 70: 57–63.

Dolling P (2018) *Lucerne: The Plant and Its Establishment*, Available at https://agric.wa.gov.au/n/3549 (Accessed 3rd September 2020).

Dost M (1997) Fodder oats in Pakistan, In: Suttie JM and Reynolds SG (eds) *Fodder Oats: A World Overview*. Rome, Italy: FAO, pp. 28–32.

Ecoport (2011) *Ecoport Database*, Available at www.ecoport.org (Accessed 3rd September 2020).

FAO in Pakistan (2018) *Pakistan at a Glance*, Available at www.fao.org/pakistan/our-office/pakistan-at-a-glance/en/ (Accessed 3rd September 2020).

Guleifsson BE (1986) Ice encasement damages in grasses and winter cereals. In: *84th Nordiske Jordbrugsforskeres Forening Seminar on Overwintering of Agricultural Crops*, Jokioinen, Finland, June 1986 68(4), pp. 578.

Hannaway DB and Larson C (2004) *Berseem Clover (Trifolium alexandrinum L.). Oregon State University, Species Selection Information System*, Available at http://forages.oregonstate.edu/php/fact_sheet_print_legume.php (Accessed 3rd September 2020).

Heuze V, Tran G, Boudon A and Lebas F (2016a) *Oat Forage*, Available at www.feedipedia.org/node/500 (Accessed 3rd September 2020).

Heuze V, Tran G, Boudon A, Bastianelli D and Lebas F (2016b) *Berseem (Trifolium alexandrinum)*, Available at www.feedipedia.org/node/275 (Accessed 3rd September 2020).

Heuze V, Tran G, Boval M, Noblet J, Renaudeau D and Lebas F (2016c) *Alfalfa (Medicago sativa)*, Available at www.feedipedia.org/node/275 (Accessed 3rd September 2020).

Heuze V, Tran G, Boudon A and Lebas F (2016d) *Rhode Grass (Chloris gayana)*, Available at www.feedipedia.org/node/480 (Accessed 3rd September 2020).

Heuze V, Tran G and Noziere P (2015) *Rye forage (Secale cereale L.*), Available at www.feedipedia.org/node/385 (Accessed 3rd September 2020).

Indus River System Authority (2016) Actual Surface Water Availability (Million Acre Feet). Available at www.indusbasin.org (Accessed at 3rd September 2020).

Lafond GP, May WE and Holzapfel CB (2013) Row spacing and nitrogen fertilizer effect on no-till oat production. *Agronomy Journal* 105(1): 1–10.

McKenzie JS and McLean GE (1980) Some factors associated with injury to alfalfa during the 1977–78 winter at Beaver-lodge, Alberta. *Canadian Journal of Plant Science* 60: 103–112.

Mishra S and Pathak P (2015) Fodder production and conservation: A potential source of livelihood for women, In: Vasudevan P, Sharma S, et al. (eds) *Women, Technology and Development*. New Delhi, India: Narosa Publishing House, pp. 136–150.

Murphy SR (2010) Tropical perennial grasses—root depths, growth and water use efficiency. *NSW Industry and Investment, Prime facts 1027*, June 2010.

NARC (2012) *Fodder Research Program*, Available at www.parc.gov.pk/index.php/en/research-institutes/137-narc/crop-sciences-institue/714-fodder-program (Accessed 3rd September 2020).

Ouellet CE (1976) Winter hardiness and survival of forage crops in Canada. *Canadian Journal of Plant Science* 56(3): 679–689.

Pakistan Bureau of Statistics (2018–19) *Agriculture Statistics*. Available at www.pbs.gov.pk/ (Accessed 3rd September 2020).

Provincial Agriculture Department Government of Punjab (2017) Agriculture, an Overview. Available at www.agripunjab.gov.pk/overview (Accessed 3rd September 2020).

Rehman A and Aneela K (2004) Effect of ensilage on chemical composition of whole crop maize, maize stover and mott grass. *Pakistan Veterinary Journal* 23(4): 157–158.

Sarep UC (2006) *Cereal Rye: Cover Crop Database*. Available at http://sarep.ucdavis.edu/cgi-bin/ccrop.EXE/show_crop_12 (Accessed 3rd September 2020).

Sattar MMK, Ahmad H and Shahzad F (2017) *Fodder Production: A Module for Extension Workers*. Pakistan: Createspace Independent Publishers.

Suttie JM (1999) *Trifolium alexandrinum* L. Grassland Index. A searchable catalogue of grass and forage legumes. *FAO.*

Suttie JM (2000) Hay Making. In: Suttie JM (eds) *Hay and Straw Conservation—For Small Scale Farming and Pastoral Conditions.* UN: FAO.

Suttie JM and Reynolds SG (2004) *Fodder Oats: A World Overview.* Rome: FAO.

Touqir NA, Khan MA, Sarwar M, Mahr-un-Nisa, Ali CS, Lee WS, Lee HJ and Kim HS (2007) Feeding value of jambo grass silage and mott grass silage for lactating nilibuffaloes. *Asian Australian Journal of Animal Science* 29(4): 523–528.

USAID and FAO (not dated) *Silage Making for Small Scale Farmers.* Available online at http://pdf.usaid.gov/pdf_docs/PNADQ897.pdf (Accessed 3rd September 2020).

World Data Atlas (2016) Land Use: Agriculture Land (% of land area). Available at https://knoema.com/atlas/Pakistan/Agricultural-land area (Accessed 3rd September 2020).

12 Microbial Determinants in Silage Rotting

A Challenge in Winter Fodders

Imran Ul Haq, M. Kaleem Sarwar, and Zia Mohyuddin

CONTENTS

DOI: 10.1201/9781003055365-12

12.1 ANCIENT AND MODERN HISTORY OF SILAGE

The ensilage system at the end of the 19th century was summed up by Prof. Wrightson as 'burying grass in trenches.' This idea of burying and preserving in trenches was much older than the 19th century, as depicted by ancient history. This trench or pit used for burying fodder crops for silage making is called a silo. Silo is derived from the Greek word 'siros,' for a pit used to store corn.

There is some pictorial evidence from Egypt, dating 1000–1500 BCE, depicting the filling of silos by farmers with sorghum-like green fodder. Some other pieces of evidence—Carthage ruins, dating around 1200 BCE, and the writings of Roman historian Cato about green fodder storage and covering with dung to provide anaerobic conditions from the 1st century CE—support the argument of ensilage from ancient times (Mannetje, 2016; Bernardes et al., 2018).

From modern agricultural history, reports of ensiling wilted grass in Italy from the late medieval period and leaves and beet tops in the 18th and early 19th century from Sweden, northern Russia, and Germany are available. In modern agricultural literature, ensilage in its modern connotation was mentioned and practiced in the second half of the 19th century. It started with the French-translated letters of a German, Herr Reihlen, who in 1861 successfully preserved 400 acres of sugar beet leaves and tops in silos. This process took a big leap and attracted the agricultural community and scientists when Herr decided to take maize for ensiling. Maize, being hardly a new crop in other parts of Europe, was facing ripening issues due to climatic conditions at Herr's farm in Stuttgart—an area at the northern limit for grain maize production. He successfully preserved it, alone and with a combination of beet pulp, in his 10 feet deep and 15 feet wide silos. His letters were officially published in 1870. At the same time, other scientists performed successful ensilage of green maize and oat. When Goffart in 1877 published a detailed ensilage process with this background, it influenced agricultural society in France and the United States. This success story of silage making was taken to Europe by Chezelles in 1882, where he described lucerne's ensiling and red clover, meadow grass, and maize (Alonso et al., 2013).

Contrary to the fact that the principle of ensiling was known from ancient times, it does not significantly impact farming until the 1870s. The farming community in the 1880s was curious about an innovative and effective technique, which was extolled in the British farming press, to cope with fodder shortage problems, especially in winter months of no sunshine. This technique, termed 'ensilage,' attracted innovators and opinion formers of the British agricultural industry. Scientists started experimentation, the Royal Agricultural Society published reports confirming its usefulness, and an ensilage society was formed for further studies. At the time when the ensilage process was being introduced in Britain, it was rapidly being established and getting fame among the livestock community of the United States—where the first silo was built in 1873—for its better and striking results (Brassley, 1996).

It took almost a century for this technique to surpass hay as the most popular and effective fodder preservation method in Britain. Despite the attempts of scientists and official campaigns to popularize it, developing countries face different issues and challenges regarding the adoption and commercial use of this technique.

12.2 PRINCIPLES OF SILAGE PRODUCTION

In respiration, plants consume carbohydrates to produce energy for normal cell functioning—and carbon dioxide and water as by-products. Oxygen is inevitable for these energy-producing physiological processes. This process of respiration continues in plants even after harvesting, provided adequate hydration and aerobic conditions. Certain naturally occurring microorganisms on fodder crops, including bacteria and fungi, also become significant respiration sources during ensilage. This increased respiration will eliminate the oxygen trapped in ensilage fodder. Increased respiration can slightly raise ensilage temperature, which can be managed by harvesting fodder at a high moisture content and increasing its bulk density. Limiting respiration during fermentation is the most desirable and critical feature of silage-making techniques, which is achieved by proper silo inspection, harvesting fodder crops at adequate moisture content, finely chopped fodder, rapid and thorough filling, and close inspection of plastic sealed ensilage for any possible damage (Piltz and Kaiser, 2004).

Silage can be defined as 'concentrated and preserved green fodder pickled under anaerobic conditions by fermentation of high moisture green fodder crop, using microorganisms in an acidic environment to keep the nutrients high.' When bacteria ferment sugars of ensiled fodder in airtight sealed containers, they produce lactic acid (LA). This acid serves as a natural pickling agent and can preserve this green, palatable, and nutritious fodder for three to five years without deterioration (Kaiser et al., 2004; Adesogan and Newman, 2010).

12.3 LIVESTOCK IMPORTANCE AND ITS FEED REQUIREMENTS, ESPECIALLY DURING LEAN PERIODS

12.3.1 Importance of Livestock

At the dawn of the 21st century, the human population was alarmingly double that of the 1960 world population. Some domestic animal populations, such as pigs and goats, kept up with the human population growth rate, and others, like cattle and sheep, have decreased in number compared to humans. In contrast, the chicken population is the most successful among all domestic animal species, increasing by almost fivefold since 1960. According to careful predictions, it is estimated that by the mid of the 21st century, the world human population will be roughly 9 billion (Von Braun, 2010).

These figures of domestic animal populations compared to 1960 reveal that animal herd composition is changing. This change in herd structure indicates improved wealth conditions of humans, especially in the developing world. These figures and predictions of computer models indicate the high purchasing power of future people, especially land-intensive foods, such as meat and milk. Livestock input is a growing sector, and it has direct impacts on the livestock economy. More than 60% of rural people, especially women, keep livestock as their primary income source (Sugiyama et al., 2003).

12.3.2 Livestock and Quality Feed

The world domestic animal population is increasing, but the number of farm households has drastically decreased, especially in developing Asiatic countries, due to industrialization and technological advances in farming. Though the world animal population increased, large-scale farming replaced small-scale farming of rural areas in developing countries. So, livestock sector development program success is to improve productivity by utilizing available resources, primarily feed, forage, and pasture. Most rural farmers of Asiatic countries have small landholdings, which is a significant constraint in large-scale livestock production. This problem can be solved by gaining the maximum potential from the field by utilizing the latest technologies of fodder preservation (Sugiyama et al., 2003).

Livestock management demands an efficient and uninterrupted supply of feed to meet animal requirements for a profitable business throughout the year. Although fodder crops are available in most months of the year, seasonal shortages, not just quantity but also quality, during summer and winter, are common in most European regions. Grazing is a low-cost livestock production system in Australia but is not the most profitable; hence, it requires feed supplements to meet production targets for a sustainable system. These feed gaps require a readily available product with high nutritive value to meet market specifications. This crucial gap to meet livestock production targets is fulfilled by the ideal fodder conservation technique of ensilage during surplus growth periods in times of deficit (Rotz et al., 2003).

12.3.3 Livestock and Energy Requirement during Lean Periods

Green forages play a vital role in improving livestock productivity. Livestock, as the subsidiary occupation of the rural population of developing countries, plays a critical part in farmers' wellbeing. The profitability of livestock is directly linked with the supply of nutrients, especially during scarce periods. In the Indian sub-continent, green fodder deficiency is a primary concern compared to dry fodder—a sufficient quantity of wheat and rice straw meets its requirements. During summer (May—June) and winter (November—December) lean periods, farmers have to feed nutrient-deficient, inadequate, and economically impractical concentrates to meet livestock requirements. Countries like Pakistan and India, being in the tropics, have abundant and surplus winter and summer fodder crops during the flush period, preserved at the optimum time, and with better techniques can provide the nutrient requirements of livestock during lean periods. Two preservation techniques for green fodder crops are being used extensively, hay and silage. Nevertheless, during the last two to three decades, silage has surpassed hay in most countries of the world. A brief comparison is made in the following to understand their principles and advantages (Taran, 2019).

12.4 SILAGE AND HAY

As described in the principles of silage preparation, when fodder crops are cut down, they continue to respire and, consequently, keep converting sugars into

CO_2. This process continues until the dry matter (DM) content of the product reaches up to 85%. Reaching this high DM content requires fresh fodder crops with just 25% DM, a considerable amount of water to be lost. This is achieved by sun and wind drying during the summer months. This process of drying green fodder crops, increasing their DM content, and preserving them for months with less pasture growth is popularly known as haymaking, whereas during ensilage, ensiled material is arrested from respirational degradation on an altogether different principle of increasing acidity in the absence of oxygen. Ensiled material is preserved using crop bacteria by converting sugars to lactic and other acids under anaerobic conditions. The main advantage and goals of silage production are to conserve digestible nutrients and minimize biological degradation losses (Pitt, 1990; Collins et al., 2017).

Conservation of crop by hay requires successful quick-drying, which is difficult to achieve, especially during winter and for thick-stemmed crops, such as maize and sorghum. For ensilage, fodder crops are harvested early at their optimal development phase—more lush green leaves to ensure more nutrients—with as low as 30% DM content. Silage production reduces the nutrient losses from 30% to 10% compared to hay and requires 10% less storage space. Silage production also does not require dry weather conditions to continue, as it can be produced in dry, cold, and cloudy weather. It was also proved during experiments that when hay and silage from the same crop on the same day were processed, cattle feeding on silage have consistently higher milk production compared to hay feeding (Jones et al., 2004). The main advantages of silage over other preservation techniques are;

- Fodder crops provide more proteins and other nutrients when harvested and preserved for silage than hay due to different principles.
- Field losses to fodder crops harvested for silage preparation are less than other preservation techniques by avoiding extended bad weather, other physical losses of grain and leaves shattering, and premature ear declination.
- Silage-making provides flexible harvesting dates of fodder crops to farmers. This can be of great importance, mostly when decisions are made late in the growing season.

12.4.1 Potential Role of Silage on the Farm

The potential role of silage on a farm can be assessed by developing a feed budget that indicates pasture surpluses and deficits and outlines feed demand and supply over the year. The farm's historical data can be utilized to estimate feed demand, production, low season for fodder crops, and drought. This data will help to compare the pasture growth rate and animal requirements. This feed budget must account for fodder crop growth from the vegetative phase to the reproductive phase, impacting the forage's quantity and quality. So, combining all that data, crucial months for supplementation can be determined to meet animal health and production needs (Rotz et al., 2003).

12.5 CRITERIA FOR SELECTION OF FODDER CROP FOR SILAGE

Fodder crop selection is a pivotal point in livestock management, production, and profitability. Fodder characteristics such as nutritional value, production/acre, cuts per season, digestibility, and preservation ability should be considered, along with better and innovative crop production technologies while selecting fodder crops.

Countries with temperate climates, with plenty of persistent rains and fertile soils, produce quality grasses and legumes with multiple fodder cuts. These climatic conditions, such as in Europe, make both the silage and haymaking process successful. Countries like the United States, with an abundant supply of irrigation water, prefer legumes fodder crops, such as high-quality alfalfa crops, for silage and hay. Countries with winter rainfall, like Israel, Australia, and South Africa, prefer lucerne (alfalfa) and winter wheat growth. Before harvesting, this heavy rainfall poses a threat to haymaking due to potential losses and diseases but is ideal for ensilage. Similarly, ensilage is the best option in tropical countries with irrigation systems and rainfall, as it preserves nutrients in its original form and has other characteristics of green fodder (Titterton and Bareeba, 2000).

Harvesting time for fodder crops selected for preservation to meet requirements of lean periods is crucial. It should be at the maximum limit of its nutritional value while staying green. Crops harvested at any time of their growth can be used for silage making alone and sometimes with other crops to make the ingredients/conditions right.

The answer to the best fodder crop for silage lies in the climatic conditions of the plant's particular area and physiology. Broad recommendations for fodder crop species or varieties are challenging to propose due to a significant harvest index (Kaiser et al., 2004). Winter fodder crops with considerably high nutrition and protein content—multi-cut berseem and alfalfa have high protein percentages of 20% and 24%, respectively—as compared to summer fodder crops are preferred fodder and first choice for ensilage. In this chapter, our focus is on 'ensiling' issues and challenges of winter fodder crops, emphasizing harmful microorganisms.

12.6 PHASES/STEPS IN SILAGE MAKING

Biological degradation is a continuous process that starts right after fodder harvesting. As discussed in detail under the section 'Principles of Silage Making,' the purpose of silage preparation is to preserve the maximum amount of energy and digestible proteins. Beforehand, phases or steps in silage making are discussed; a few preliminary precautions are very critical.

1. Harvesting of fodder crop with maximum nutrient contents at optimum moisture concentration.
2. Cutting the crop at the right height from soil to avoid unwanted microorganisms in the final product.
3. Chopping of fodder crop by balancing between preferred large and must-have small length particles (Bernardes and Do Rêgo, 2014), so that ensiled material without compromising quality can be effectively compacted in silos.

After these preliminary actions, the ensiling process progresses through the following four fermentation phases/steps;

12.6.1 Pre-Seal/Aerobic/Respiration Phase

This phase starts just after harvesting. Even though the fodder crop is harvested and chopped into small particles, many cell walls remain intact, as well as protease enzyme activity and oxygen consumption. This also encourages aerobic bacteria, present on crop surface, to proliferate and keep feasting on available carbohydrates. Plant respiration and aerobic bacteria fermentation will continue unless all oxygen is consumed from the ensiled material.

From a management standpoint, the elimination of aerobic conditions throughout the storage period to shorten the aerobic phase is the primary goal. Short covering, sealing of storage structure, and good compaction of ensilage deplete oxygen and reduce air spaces between particles (Jones et al., 2004). This is a race against time to stop unfavorable fermentation, as this phase can last from three hours to one day, depending on available oxygen.

12.6.2 Fermentation Phase

The fermentation phase begins when anaerobic conditions are achieved, and the growth of anaerobic bacteria is encouraged. The first short period of the fermentation phase, which lasts for 24 hours, is also known as the lag phase or acetic acid production phase as cells are broken down, releasing fermentation substrates. In this part, bacterial multiplication is exponential, 1 billion/g of fresh fodder. During this first part of the fermentation phase, as the term indicates, LA is produced, and, resultantly, pH is lowered to 5–5.5 from 6–6.5 compared to the start of the ensiling process.

The condition produced by the first part of this phase favors lactic acid bacteria (LAB) to grow and multiply, which causes a rapid reduction in the pH of silage. Homo-fermentative/homo-lactic bacteria (produce LA only) are preferred compared to hetero-fermentative bacteria (produce LA, acetic acid, and ethanol) because they are more effective in rapid pH reduction, which is necessary for better final silage quality. This phase continues from the second day of ensiling to the third week provided the correct anaerobic conditions, causing ensilage pH to drop from 5–5.5 to 3.5–4.5 depending upon factors like fodder crop species and moisture content (Weinberg and Muck, 1996; Piltz and Kaiser, 2004).

12.6.3 Stable Phase

The fermentation phase is the primary preservation phase of silage making, ensuring the quality of preserved fodder. After lactate silage has been attained, the stable phase continues, causing all biological activity, including LA bacterial growth, to stop, as low pH levels have been attained. During this phase, further pH reduction, in some cases, may also occur as a result of hemicellulose breakdown (Bolsen et al., 1996; Moran, 2005).

During the fermentation and stable phases, special precautionary measures to prevent oxygen entry to packaging/silo are crucial; otherwise, harmful bacteria and fungi can proliferate and damage its quality, which will be discussed later.

12.6.4 Feed-Out/Aerobic Deterioration Phase

When the silo/packaging of ensiling material is exposed to air either due to physical damage by mice, birds, other rodents, insects, or during feeding purposes, it begins to spoil. It happens because oxygen exposure activates dormant aerobic microorganisms that unable to grow under anaerobic conditions. Deterioration of preserved fodder causes pH to rise as degradation of organic acids starts. This rise in pH will cause temperature increase and subsequent spoilage by microorganisms, including fungi, *bacilli*, and *enterobacteria*.

If the spoilage phase is initiated by physical damage to packaging/silo, it can be prevented by reducing air penetration, whereas, if opened for feeding purposes, a significant increase in feed-out rate and measures to keep silage face disturbance as low as possible are strictly recommended because, at this point, microorganisms have an unlimited supply of oxygen as compared to the minimum possible supply in stable phase (Moran, 2005).

12.7 ISSUES AND CHALLENGES IN SILAGE MAKING

The best-quality silage making involves several factors starting from selecting fodder species, harvesting at a proper height, and time for maintenance of ideal conditions during each phase of ensiling, as minute variations at each step affect the final product. Pre-harvest factors, such as moisture and nutrient contents of fodder, and post-harvest factors before ensiling, such as drying and in some cases chopping at the proper length, are discussed briefly in the early part of this chapter. Pivotal points ensuring good-quality silage are those that involve the process of ensilage.

Ensilage is a potentially dynamic preservation technique, and its product is prone to rotting/deterioration during the entire process, provided appropriate aerobic conditions at any stage. We will discuss the cause and effect of several undesirable microorganisms on silage quality, animal performance, and their repercussions on human and animal health.

12.8 ISSUES OF BACTERIAL AND FUNGAL ROTTING OF SILAGE

Microbiological studies of fodder before ensiling reveal that different bacterial and fungal species are present on the plant surface, mostly on lower parts—where microorganisms are protected from the harsh climate and UV radiation. These microbial populations differ considerably from the microbiology of ensiling fodder and silage.

Typical bacterial and fungal populations from fresh fodder before putting into silos are enumerated: aerobic bacteria; LAB; acetic acid bacteria; propionic acid bacteria; *Enterobacteria*, *Bacilli*, and clostridial endospores (as these are scarce on standing crops); *Aspergillus* and *Penicillium* species; and other yeast and yeast-like fungal species.

Many of these mentioned microbial species are obligate epiphytic and aerobic bacterial/fungal populations and do not contribute to fodder preservation, as their growth is inhibited just after compaction in silos. Our focus in this chapter is on undesirable/deteriorating bacterial and fungal species during ensiling.

12.8.1 Bacterial Rotting of Silage

Harmful bacteria in silage are associated with the rotting and degradation of silage. However, they are also responsible for human and animal illnesses upon direct interaction, such as *Listeria* species' contamination, production of toxic compounds, and biogenic amines, in the case of certain species.

12.8.1.1 Enterobacterial Fermentation and Silage Rotting

Enterobacteria species are facultative, anaerobic, rod-shaped, often motile Gram-negative bacteria that reduce nitrate and are found in poorly preserved acetic ensiled fodder. Usually, these enterobacterial species are non-pathogenic and are indigenous to animals' intestinal tract. These bacteria contain endotoxin in their outer cell membrane, responsible for feeding problems in animals, and have a possible mastitis association. *Enterobacteria* have a better ability to survive under unfavorable conditions of cold, dry, and low-temperature winters. They are less fastidious and have higher populations, often 100-fold higher than LAB. Due to their epiphytic fitness, they are first detected in sizeable numbers from post-winter samples of winter fodder crops when other bacterial species are even difficult to detect (Ávila and Carvalho, 2020).

Enterobacteria populations are higher throughout the vegetative period of fodder crops. Dramatically, if the crop to be ensiled undergoes prolonged drought conditions before harvesting, their population increases exponentially. Irrespective of initial *Enterobacteria* populations, before ensiling, their numbers could reach up to 10^{10} CFU/g during the first day of ensiling in the aerobic phase. During this early stage of silage making, crops supporting rapid acidification, such as readily fermentable maize crops, also promote *Enterobacteria* proliferation until a pH below 4.5 precludes its multiplication.

Therefore, just large populations of *Enterobacteria* during ensilage are not directly correlated with the rotting of silage. Their detrimental impact (ethanol and butanediol production and nitrate reduction) starts as soon as LAB fails to proliferate extensively due to undesirable ensilage conditions: wilting of fodder before ensiling, improper harvesting time, or fodder species (Nishino, 2011).

In the anaerobic environment of ensiled mass, *Enterobacteria* are undesirable, as they compete with LAB for nutrients, especially fermentable carbohydrates, hexose, and pentose. Apart from hampering LAB growth, *Enterobacteria* are responsible for producing ammonia in silage by protein degradation and nitrate reduction. In the enclosed environment of ensiled mass, this activity will increase the buffering capacity of ensiled fodder, which will prevent the rapid reduction of pH, a critical step for preservation.

The decline in their population during ensilage is an important indicator and a better single parameter of good-quality silage: it ensures that ensiling conditions are right, nutrients and water are abundant, desirable fermentation by LAB is in progress, pH is low, and the temperature is moderate.

Under reducing ensilage conditions, the nitrate degradation process positively impacts the hygienic and chemical quality of silage; intermediates of this nitrate degradation, namely nitrite and nitric oxide, selectively inhibit other undesirable microorganisms, discussed later in this chapter, like clostridia. This process is of great importance during the early aerobic phase of ensilage when pH is still high and favorable for clostridial endospores.

12.8.1.1.1 Biogenic Amines

Enterobacteria species are facultative anaerobes. They survive by shifting their electron acceptor source from O_2 to nitrate and producing various mixtures of NO-based silo gases (Muck, 2010). Though compounds produced by *Enterobacteria* species through fermentation such as LA, acetic acid, succinic acid, carbon dioxide, and water are non-toxic, amino acid degradation produces undesirable biogenic amines silage. Production of biogenic amines and ammonia during ensilage have a detrimental impact on silage quality, palatability of feed, and livestock health.

These are formed during the hydrolysis of peptide bonds involving amino acid decarboxylation by plant and bacterial enzymes. These amines formed during the early phase of ensiling, the aerobic phase, by *Enterobacter* species are risk factors for slow acidification. Earlier biogenic amines are produced from proteins/AA decarboxylation, whereas later amines result from poor ensilage conditions that promote clostridial species' proteolytic activity.

The low fermentation process, amino acid deamination, nitrate, and nitrite reduction elevate biogenic amines and ammonia concentrations, which predominately reduces silage intake in livestock and loss of dry matter and cattle performance. When feeding on this poorly fermented silage with high biogenic amine concentrations, livestock suffers from health issues, even as severe as impaired fertility, mainly due to ammonia absorption to the liver and blood. Biogenic amine formation could be suppressed by increasing LAB growth in ensiled mass to counter undesirable bacterial growth (Driehuis et al., 2018; Wang et al., 2019).

12.8.1.1.2 Verocytotoxins/Shiga Toxins

Enterobacterial cells, upon breakdown, release endotoxins from their cell membranes into the ensiled mass. Livestock has different susceptibility to these toxins in the diet; however, some can induce resistance upon exposure, as described earlier. Most members of the genus *Enterobacter* are non-pathogens the same as *E. coli*, which are part of the normal microbiota of the human intestinal tract and livestock. Whereas few *Escherichia coli* are pathogenic to humans, the Shiga toxin-producing *Escherichia coli* (STEC) strain, also known as verocytotoxigenic EC (VTEC), recovered from silage can use cattle as its reservoir. According to careful estimates10–30% of cattle in Europe and the United States are carriers. Like other *Enterobacter* species, this particular strain can survive the extreme temperatures and pH of ensiled fodder for at least 100 days (Callaway et al., 2006).

Pathogenic strains of *E. coli* are a source of different human disorders ranging from diarrhea and hemorrhagic colitis to renal failure. Renal failure, particularly in young children, is due to hemolytic uremia syndrome (HUS), which is mediated by the virulence factor of these strains: verocytotoxins (Vtx) or Shiga toxins (Stx). The shedding rate of VTEC/STEC through livestock feces is almost 21%, without causing any disease or complexity. When used as fertilizer, these feces are the primary source of *E. coli* contamination through contaminated fodder crops to be ensiled. In an experiment by (Ogunade et al., 2016), it was shown that VTEC survived the alfalfa ensilage process. This can be attributed to the high buffering capacity of the ensiled alfalfa crop, as it hampers the rapid reduction of pH during ensilage.

(Nyambe et al., 2017) performed a detailed experiment to determine the vtx gene transfer from bacteriophage to *E. coli* and its survival during ensilage. This comprehensive experiment demonstrated and proved the survival ability of VTEC during unfavorable conditions of ensilage. This experiment also proved, albeit in a single treatment, transduction and after lysis infection of new recipient cells. Bacteriophages carrying the Vtx gene, responsible for endotoxin production, can induce their DNA in the chromosome of *E. coli* prophage under unfavorable environmental conditions of low O_2 levels and acidic pH; both factors are favorable during ensilage. This further will cause lysis of *E. coli* cells releasing abundant phage particles ready to infect other bacterial cells (Cornick et al., 2006).

Hence it was proved that silage is the cause of VTEC/STEC spread to human-provided high initial contamination and suitable ensilage environment for new strain emergence. Air penetration through physical damage to silos or during the feed-out stage (Pedroso et al., 2010; Dunière et al., 2011) allows VTEC survival. STEC contamination chances during the last phase of the ensilage, the feed-out phase, increase exponentially and are a significant cause of pathogen cycling on dairy farms (Driehuis et al., 2018).

12.8.1.1.3 Conclusion

Although *Enterobacteria* play an indirect positive role in the early stages of the ensiling, their population's rapid suppression is desirable: due to the factors mentioned previously. This will reduce their undesirable fermentation of amino acids to ammonia and carbohydrates to acetic acids and other compounds that do not decrease the ensilage mass's pH. Resultant silage dominated by enterobacterial species is acetic, less palatable to livestock, and has significant energy losses. Anecdotally, it is attributed by some scientists that acetic acid is not the sole compound responsible for less intake because *Lactobacillus* acetic silage has a high intake as compared to *Enterobacter* acetic silage. So, it is supposed that *Enterobacteria* during ensilage produce unidentified and unproved compounds responsible for its reduced palatability.

12.8.1.2 *Bacillus* Rotting of Silage

The genus *Bacillus* encompasses spore-forming bacterial species, including *Bacillus cereus, B. licheniformis, B. coagulans, B. polymyxa,* and *B. pumilus*, which, like *Enterobacter* spp., belong to obligate and facultative anaerobes. Their population at feed-out time differs considerably in silage. Soil is the primary habitat and source of *Bacillus* species during ensiling fodder crops. Fodder crops harvested at lower stem heights with soil or manure contamination have many chances of *Bacillus* population growth during ensilage. Due to possible air infiltration, the top layer of ensiled matter has higher *Bacillus* spp. populations compared to the deeper/bottom layer of the silo (Muck, 2010; Queiroz et al., 2018).

From different earlier experiments, it was evident that aerobic exposure marked an increase in *Bacillus* spp. over LAB spp. is common, but their role as initiators of deterioration under aerobic conditions was arguable. Some species of this genera cause fermentation of carbohydrates and organic acid during ensiling. They play a significant role in impairing the aerobic stability of silage by advancing the rotting activity. When other undesirable microorganisms increase silage pH and temperature, providing conditions favorable for *Bacillus* growth further increases temperature, in some cases up to 50°C. Though aerobic spoilage by other silage microbiota is a prerequisite for *Bacillus* contamination, in some instances with scarce supporting studies, they can also start spoilage under high temperatures, common in big baled silages or after inoculating formaldehyde (Nishino, 2011).

Resultantly, this secondary aerobic deterioration of silage produces low-quality silage with a slimy feel. However, under anaerobic conditions, their fermentation is of relatively little importance.

12.8.1.2.1 Milk Spoilage by Bacillus *spp.*

Their indirect impact on livestock produce, milk, is of great importance for health reasons. As we know, silage is a significant source of spore-forming bacteria, both aerobic and anaerobic, in milk. It was shown that spore-forming species of ubiquitous *Bacillus* spp. in silage ultimately contaminate milk and cause food spoilage. Due to this indirect detrimental impact on human health, proliferation and spore production of *Bacillus* spp. during ensiling and exposure to aerobic conditions of the feed-out phase must be inhibited. These aerobic spore-forming bacteria, ASFB, like the mentioned *Enterobacter* species, compete with LAB when aerobic conditions prevail in silage with comparatively high pH. Common ASFB species identified from milk in various countries include *Bacillus sporothermodurans, B. polymyxa, B. subtilis, B. licheniformis, B. pumilus, B. megaterium*, and *GeoBacillus stearothermophilus*.

Milk contaminated with these species is given heat treatment to mitigate their harmful impact on human health. *B. cereus* is the most concerning and problematic species for the food industry whose few strains can produce enterotoxins: hemolytic and non-hemolytic toxins (Ávila and Carvalho, 2020).

Enterotoxins cause foodborne illnesses such as diarrhea and vomiting. A significant positive relationship was found between spore count in feces and raw milk. It was concluded that an spore count exceeding 10^3–10^4 in feces of livestock feeding on a particular diet increased raw milk spore count and spoilage (Te Giffel et al., 2002; Bennett et al., 2013).

Favorable conditions for toxins producing *B. cereus* strains: temperature from 30–32°C, aerobic conditions, sugars as substrate, and alkaline pH are available during the feed-out/aerobic deterioration phase. However, it was observed that no toxins are produced during ensilage, attributed to the low viability of these toxins producing strains during ensilage. Soil, as with other *Bacillus* spp., is a significant source of *B. cereus* spore contamination along with silage and other feed sources.

According to different studies, silage was found as the primary source of raw milk contamination with *B. cereus* of up to 3×10^2 spores/g, whereas considerably higher concentrations of spores, 10^8–10^9, were observed in upper and moldy silage layers, causing impaired aerobic stability.

The maximum spore count of *B. cereus* in silage, 10^4/g, compared to other members of ASFB, is considerably less. Its spores survive the livestock's gastrointestinal tract and shed in excrement as of *E. coli* toxic strains. Teat surface soiled with feces containing *Bacillus* spores serve as a source of contaminating raw milk during milking operations. *B. cereus* is notorious and hardy spoilage bacteria and can germinate and proliferate even in pasteurized milk and milk products as low as 5°C. These psychrotrophic bacterial spores significantly reduce the shelf life of milk and its products. This sturdy pathogen even survives in dry milk products and has been isolated from milk powder, whey powder, and whey protein (Magnusson et al., 2007; Pinto et al., 2018).

Manure fertilization in crops can increase *Bacillus* spore contamination and serve as abundant initial inoculum for silage proliferation. According to an experiment by (Rammer et al., 1994), it was shown that solid farmyard manure fertilizer had almost a hundred times more spore concentration than inorganic fertilizer.

12.8.1.2.2 Conclusion

Few studies suggest that *Bacillus* spp. competes with undesirable microorganisms and might have a positive role in reducing corn silage aerobic rotting. It was also suggested that few of its species produce bacteriocin; zymocin, a potential inhibitor of spoilage yeasts; and fungi. Despite all possible positive effects of *Bacillus* spp. from the various latest studies, it is concluded that *Bacillus* species' proliferation, especially facultative anaerobes, in silage should be discouraged, because they are not just less efficient LA producers than LAB but also deteriorate silage quality during aerobic deterioration and, in the case of *B. cereus*, spoil milk.

Standard mitigation techniques/methods used to control food spoilage pathogens, heat treatment, UV radiation, and high pressure, are not possible or adequate to treat large-scale ensiling mass to prevent milk spoilage and consequent food spoilage illnesses of ASFB spp. It is essential to prevent the initial contamination of raw milk. As milk contamination has been studied, preventive measures should be taken to avoid spore entry in milk: control silage fermentation process and improve aerobic stability by inoculating LAB cultures.

Both for *Bacillus* and *Enterobacter* spp., successful control chain management approaches start from raw material; fodder harvesting to final processing should be considered during product packaging. It will help prevent *B. cereus* and *E. coli* (few detrimental strains) species' entries in the human food chain and food spoilage.

12.8.1.3 Listerial Rotting of Silage

Listeria is an opportunistic facultative anaerobic bacteria that cause a wide range of diseases; meningitis, gastroenteritis, abortions, mastitis, and septicemia, especially in immunocompromised animals and humans. Listeriosis, a systemic infection, has a fatality rate of 20% in humans, whereas livestock's is dangerously as high as 100%. *Listeria monocytogenes* are the leading causative agent for listeriosis in both animals and humans.

12.8.1.3.1 Listeria monocytogenes

During ensilage, *listeria* genus species require micro-aerobic conditions and, by glucose fermentation, produce different acids without silo gas production. Different *Listeria* spp. have been isolated from ensilage mass: *L. monocytogenes, L. welshimeri, L. ivanovii, L. innocua*, and *L. seeligeri*. Isolated species' percent incidence depends on the stage and quality of silage; however, among the identified *Listeria* spp., *L. monocytogenes* is most worrisome. *Listeria* is part of fodder crop microbiota before harvesting and survives silage fermentation phases due to its ability to withstand low pH and high-temperature variations. Some species can survive anaerobiosis and acidic environments during ensilage and proliferate when conditions are favorable.

Notwithstanding the higher fatality rate and pathogenicity of *L. monocytogenes*, it is not always most frequent in ensiled mass. Listeriosis is a foodborne ruminant disease when including silage as a primary feed source. It can cause uterine infection in livestock, causing abortions and eye infection upon direct contact with contaminated silage (Driehuis, 2013).

12.8.1.3.2 Listeriosis Outbreaks

Listeria spp. has been associated with food spoilage disease outbreaks: a 1983 outbreak of listeriosis in Massachusetts involved 49 patients, and milk was implicated as a carrier; likewise, a 1985 outbreak in California involved 314 people, where, a milk product, cheese, was identified as a carrier (Perry and Donnelly, 1990). Yearly reported cases of human listeriosis, as described earlier, are rare: 7.5/million in Europe, 4.4/million in the United States, and 3/million in Australia.

12.8.1.3.3 Types of Listeriosis in Humans

Listeriosis is divided into two forms depending upon the host's health conditions: non-invasive and invasive gastrointestinal disease. Healthy adults with any immunocompromised complexities suffer from non-invasive types with typical gastroenteritis symptoms such as diarrhea, emesis, and fever. In comparison, more severe and fatal disease form occurs in neonates and adults with compromised immune systems (especially elderly pregnant women, patients of AIDS and cirrhosis). Invasive listeriosis is significant with a high mortality rate, though these outbreaks and sporadic cases are not expected (Organization, 2004).

12.8.1.3.4 Silage as a Source of Listeria in the Human Food Chain

Feeding of silage, particularly of low quality, has had a strong association with listeriosis in livestock since the beginning of silage feeding in the 20th century: in 1922, Iceland farmers named this disease 'Votheysveiki,' meaning 'silage disease,' Proper experimentation to determine *Listeria* and livestock disease's association started in the second half of the 20th century (Woolford, 1990).

During different contaminated silage feeding experiments, it was concluded that listeriosis is confined to sheep and goats, and its risk recedes when silage is removed from the feed. It was suggested that few fungi could use LA as a substrate, reducing silage pH and making conditions conducive to *Listeria* growth. First, it was concluded that low-quality silage is responsible for listeriosis, and if anaerobic conditions are provided during ensilage, *Listeria* proliferation will be hindered. Although, contrarily, it is possible to obtain good-quality silage with relatively basic pH: when heavily dried fodder is ensiled, it produces good silage but with high pH and also in case of big baled silage. Recent research has suggested and explained that *Listeria* is commonly found on pre-harvested fodder, and when ensiled under compromised aerobic conditions, it can proliferate exponentially (Ryser et al., 1997; Coblentz and Akins, 2018).

This pathogen in the human food chain has been reported from raw milk, a big concern for the soft cheese industry. This pathogen, like *B. cereus*, can gain entry in raw milk through fecal-contaminated teat ends. *L. monocytogenes* is ubiquitous and more prevalent in livestock operations. It is suggested that contaminated silage is the primary source of its entry in ruminants. It can tolerate adverse environmental conditions of low water activity and refrigeration temperature along with a wide pH range: from 4 in the case of a few strains to optimal above 5 (Sanaa et al., 1993).

The association of poorly fermented silage with listeriosis observed in the second half of the 20th century in Europe has been verified statistically by a detailed raw milk study from various dairies in Spain (Vilar et al., 2007). In other tank milk observations from dairy farms, made by De Reu et al. (2004) and Marshall et al. (2016), of the United States, New Zealand, and Belgium, up to 6.3% of samples were found to be contaminated by *L. monocytogenes*.

It was observed that deteriorated silage with pH 4.5 or above has more *L. monocytogenes* contamination than more acidic silages, 29.5% and 6.2%, respectively. The pathogen can thrive in silage or parts of ensiled mass with aerobic conditions; either oxygen is not excluded properly or infiltration has occurred due to holes. A surge in the thriving *Listeria* population is observed after silo opening in the feed-out phase because conditions are favorable for those pathogens that survived ensiling. Sheep are more susceptible to listeriosis outbreaks than cattle, but cattle serve as the pathogen's asymptomatic carrier. Livestock is also fed more with baled silages, which also have relatively more *Listeria* contamination chances than other silages (Ávila and Carvalho, 2020).

12.8.1.3.5 Conclusion

It is concluded that *Listeria* spp., especially *Listeria monocytogenes*, is a significant threat to human and animal health. It gains entry into the human food chain through contaminated feed, most probably silage.

As was suggested under the feed-out phase of ensilage earlier in this chapter, there should be no holes/punctures in silos/packaging film; the same is advised for *Listeria* management.

Different combinations of antagonistic microorganisms, such as *Lactococcus* and *Pediococcus*, have proven effective in controlling *L. monocytogeneses* in ryegrass silage with one day of fermentation. During another experiment, bacteriocin, pediocin alone, eradicated *L. monocytogenes* from ensilage mass within eight days, whereas in control, it persisted until the last silo opening (Amado et al., 2016).

12.8.1.4 Clostridial Rotting of Silage

Clostridium genus belongs to the Gram-positive, endospore-forming obligate anaerobic bacteria and usually proliferates long after LAB has stopped active growth.

Clostridial contamination and subsequent spoilage produce stinky silage. The most problematic and commonly isolated clostridial species of silage include *C. botulinum, C.saccharolyticum, C. disporicum, C. butyricum, C. tyrobutyricum*, and *C. sporogenes*. The main signs of clostridial fermentation in silage are a higher concentration of butyrate, acetate, and soluble proteins, with lower lactate levels (Julien et al., 2008).

Clostridial sporulation and fermentation are higher in the outer layer of baled silos and the top layer of bunker silos. During an experiment in 2013, it was confirmed by Borreani et al. that bunker silos have higher spore concentration (10^6 spores/g) in peripheral areas as compared to (10^3 spores/g) central parts of the silo (2013). During the proteolysis phase of clostridial fermentation, biogenic amines such as histamine, and cadaverine, though in minute concentrations, are produced, which are potentially toxic to ruminants as described previously. Many of these putrefaction compounds in silage adversely affect the digestive system, milk production, and ruminants' growth rate. *C. perfringens* is responsible for hemorrhagic bowel syndrome in cows.

For optimal growth, clostridia species require >4.5 pH value, >70% of moisture concentration, and water activity between 0.95–0.97. As these conditions are not available during ideal rapidly acidified silage, their growth is inhibited. So, the factors that predispose the ensiled mass to delayed pH decline are favorable for clostridial growth: high moisture content at ensiling and fodder crop with high buffering capacity, requiring more LA to achieve critical pH <4.5.

A combination of different factors allows *Clostridia* spp. to proliferate near aerobic spoiling layers, such as increased pH and temperature, possibly due to *Enterobacter* and *Bacillus* species' activities. These aerobic spoilers favor clostridial growth by exhausting O_2 supply near the source and utilizing NO_3, which retards clostridial proliferation. This reduced nitrate concentration before clostridial growth supports the argument of *Enterobacter* activity: as mentioned earlie, *Enterobacter* reduces nitrates in aerobic silage (Muck, 2013).

Their obligate anaerobic nature makes them an active group of spoilage bacteria in later stages of fermentation rather than in the early aerobic instability phase. Hence, their population, especially proteolytic and LA assimilating spp., increases during the feed-out stage, ultimately reducing silage's nutrition value. The importance of pH for *Clostridium* species' growth during ensiling is crucial, as maintaining pH lower than favorable for clostridial activity plays a significant role in obtaining good-quality silage. Different experiments were performed to cease clostridial activity in ensiled mass by lowering pH. It was found that water activity, which is directly dependent upon the dry matter content of the fodder crop, significantly affects the pH of legumes and ryegrass silage. *Clostridium tyrobutyricum*, an important silage species, ceases to proliferate as a function of dry matter content according to DM content of the ensiled crop (Muck, 1988).

12.8.1.4.1 Groups of Clostridia

Usually, *Clostridia* species from mature silage are isolated in endospores, so there is obscurity regarding their ecological niche and proliferation. The ecological niche is determined by studying the characteristics of isolated species as they are substrate-specific. Based on their substrate fermentation, clostridial species can be divided into two groups: saccharolytic and proteolytic. The former group, including *C. butyricum* and *C. tyrobutyricum*, ferments residual carbohydrates and LA into butyric acid, increasing pH. The latter group, including *C. bifermentans* and *C. sporogenes*, causes protein fermentation into various products: including ammonia and amines (D Mello, 2004).

Clostridia species' activity produces CO_2 and H_2, which indicate loss of digestible dry matter and energy. Replacement of LA with butyric acid is lost to silage quality, as it raises the pH of the ensiled mass, making it susceptible to undesirable microorganisms.

12.8.1.4.2 *Clostridial Fermentation and Quality of Ensiled Fodder*

Clostridial fermentation is directly related to the high moisture content of the harvested fodder crop to be ensiled. Fodder crops used for silage could have higher moisture content than recommended due to several conditions: crop was not dried, heavy rain before harvesting, and on-farm management. A detailed case study of clostridial fermentation of second crop alfalfa silage in Wisconsin was done when dairy cows suffered from severe acidosis and abomasum upon feeding. The first cut of alfalfa crop was at the proper time and ensiled after respected drying, whereas during the second crop ensiling, rainy weather forced early ensiling. When the farmer had a few dry days in the rainy season, he decided to ensile alfalfa as quickly as possible before the next rainy spell. Unfortunately, within a few hours of cutting, a rainstorm hit the field, and farmworkers hastened in ensiling freshly harvested fodder crops. Samples of second alfalfa ensiled crops were analyzed in the laboratory and compared with the first crop samples. Analysis of ensiled alfalfa crop, fed to dairy cows, proved clostridial fermentation: it had a high moisture content of 75%, pH of >5.5, butyric acid 7%, ammonia nitrogen 50%, and astonishingly no LA. When this heavily infested silage was fed to fresh cows for confirmation, a rapid decline in health was observed. This contaminated silage was fed to heifers and steers with no complex problems.

Clostridial silages act as a preservative for yeast and mold activity due to their high butyric acid contents. Even when feeding to other livestock classes, a daily dose of contaminated silage should be reduced for minimal diet inclusion levels. It is also recommended that one day before feeding, contaminated alfalfa silage should be removed from the silo and exposed to air so that undesirable odors can be removed by evaporating volatile compounds: adding molasses to improve nitrogenous products' palatability such as amines and amides.

12.8.1.4.3 *Impact on Clostridia Dairy Products*

Semi-hard cheese types, such as cheddar and gouda, suffer from flavor and texture defects when spore-forming butyric acid bacteria develop. These bacteria damage cheese quality by producing off-flavor, bad smell, and gas formation, termed 'late blowing.' These butyric acid bacteria belong to the genus *Clostridium* and convert LA to butyric acid and carbon dioxide, ammonia, and hydrogen under low pH. This BAB originates in the farm environment, gains entry in raw milk, and subsequently ends in cheese deterioration (Vissers et al., 2007).

Clostridia species usually inhabit manure and soil. These bacteria can grow and deteriorate silage under anaerobic conditions, typically after LAB stop proliferating.

They utilize proteins, carbohydrates, acetic acid, LA, and other acids as substrate, formed during previous fermentation phases of ensilage. As a result, these bacteria, befitting their name, produce butyric acid, a foul-smelling and low-energy acid. Resultantly, it reduces palatability, nutrients, and digestible DM content of silage and lessens its intake by livestock. Foul unpleasant smell in deteriorated silage is mainly the result of unwanted putrescine and cadaverine compounds. These unpleasant compounds are nitrogenous apart from proteins produced as plant proteins undergo undesirable fermentation by *Clostridia* spp. during ensilage (Jones et al., 2004).

12.8.1.4.4 Hazards of Butyric Acid in Silage

As described earlier, butyric acid is produced due to undesirable fermentation of clostridial species during the ensilage period and in the feed-out phase and other hazardous compounds. Livestock health suffers upon consuming butyric acid silage mainly due to ketosis, directly linked with butyric acid and partially with less energy intake from low nutrient silage. During the study of all three types of ketosis in dairy cows, β-hydroxybutyric acid is found in elevated concentrations. Although butyric acid is produced during rumen fermentation and rumens frequently taste it when chewing their cud, observed concentrations in cows suffering from ketosis were significantly higher. During other experiments, it was suggested that if higher concentrations of β-hydroxybutyric acid are found in blood, silage should be examined for butyric acid concentrations, and it should not be more than 50g/d. It was highly recommended that butyric acid silage should be avoided as part of the feed, especially for newly calved and close-up dry cows. If concentrations of butyric acid in silage are extreme, it should not be a part of any livestock class; instead, it should be used for crop fertilization (Oetzel, 2007).

12.8.1.4.5 Clostridium Tyrobutyricum

C. tyrobutyricum is the most studied member of the *Clostridium* genus during ensilage. It plays a significant role in butyric acid production due to its ability to use LA as an energy source and survive under low pH conditions, which are not conducive for other *Clostridium* spp. Another primary reason is its ability to transfer from contaminated silage to raw milk and, ultimately, in milk products, where it causes late blowing in semi-hard cheeses, as explained earlier in detail. Hence, *C. tyrobutyricum* harms the preservation, nutrition, and quality of ensiled mass and poses a health risk for being a possible foodborne pathogen. During different experiments on gouda cheese, *C. tyrobutyricum* was detected in commercial cheese samples and was exclusively found responsible for late blowing. Unlike *C. botulinum*, it is not harmful to animals and humans, so its losses are of just economic importance (Driehuis, 2013).

12.8.1.4.6 Clostridium botulinum

Among pathogenic species associated with silage, *C. botulinum* is most notorious due to its lethal and extremely toxic neurotoxin producing ability: botulinum. Botulinum is one of the few most potent neurotoxins in nature and causes botulism in animals and humans. Botulinum toxin has different types, from A–G; type A, B, E, F, and rarely G are associated with human botulism, whereas type C, D, and occasionally B are associated with other mammals, birds, and fish. Botulism is a severe disease, with a fatality rate of more than 10% in cattle, resulting from botulinum toxin's intoxication through contaminated silage (Driehuis et al., 2018).

During the outbreak of cattle botulism in the 1980s, in various farms of the Netherland, spores and toxin of *C. botulinum* were detected in ensiled grass silage. Its spores were also detected in feces and manure used for fertilization on the farm, from cattle fed with contaminated silage. In both these studies, clostridial spores were associated with weak silage preservation techniques with higher pH values. A report of botulism from a cattle herd in Tennessee by Kelch et al. (2000) was also published, which proved its association with feeding contaminated barley silage (Kelch et al., 2000).

Spores of *C. botulinum* are commonly distributed in soil and consequently are found in farm crops and the intestinal tract of livestock, though in low numbers, which is not harmful. Despite their low numbers, botulinum toxin's extreme toxicity urges extra care during ensilage to eliminate any chances of their potential growth and proliferation. It is estimated that spore concentration of about 1/liter of raw milk is unaffected; that is why botulism outbreak chances in dairy products are generally not common. During detailed studies in the UK by (Brett et al., 2005), it was traced that infant formula milk products have 1–4 spores/kg of *C. botulinum*. As these spores could be a potential source of infant botulism, special measures should be taken to stop their raw milk entry. Clostridial spores' transmission pathway is the same as other bacterial spores' entry in the human food chain: selection of contaminated fodder crop for ensiling, flawed ensilage methodology, surviving gastrointestinal tract of cattle, and finally access during milking operation.

As under standard farming practices, it is inevitable for udder and teats to be soiled with bacteria from bedding and feces. Although teat cleaning is commonly practiced before milking but is ineffective from a microbiological perspective, many spores and bacteria contaminate raw milk.

12.8.1.4.7 Conclusion

Clostridium species are more widespread and common in soil and fresh and ensiled fodder. This genus includes one of the dangerous neurotoxin producing species, *C. botulinum*, which has high chances of proliferation during ensilage, transfers to raw milk, and ultimately affects human health. The problems regarding extremely toxic clostridia species, including botulinum, can be avoided by providing optimum fermentation conditions during ensilage.

Fodder should be harvested at optimal growth and dried before ensilage for inhibiting clostridial fermentation. LAB additives could also be useful in the rapid acidification of ensiled mass to cease *Clostridia* species sporulation. Fodder crops with low buffering capacity should be preferred for ensilage; otherwise, different fodder could be mixed to improve it. Different chemicals, sodium nitrite, sodium propionate, and hexamine, are also useful in reducing clostridium fermentation.

12.8.2 Fungal Rotting of Silage

Fungi isolated from ensiled fodder are generally differentiated into endophytic symbionts and those which develop in poorly managed silage. Fungi are sub-classed into yeasts and molds for studying their effect during the ensilage process. Most fungal species require aerobic conditions to propagate, although several species can proliferate in an anaerobic environment. Fungal species have no role in the feed preservation process. Their presence confirms and evokes silage spoilage, especially aerobic deterioration, during the initial and last ensilage phase. Aerobic fungi, usually restricted to surface layers, can grow in different oxygen exposed parts of ensiled fodder. These fungi reduce feed intake value and palatability of silage and have a detrimental effect on animal and human health due to their toxins producing capability. Anaerobic fungi compete with LAB for water-soluble carbohydrates and lactate when anaerobic conditions are achieved. Their growth continues to decline as concentrations of silage acids, LA, and acetic acid increase (Mickan et al., 2003).

12.8.2.1 Yeast Species during Ensilage

Yeasts, usually single-celled but occasionally with pseudo-mycelium, originated millions of years ago and have more than 1500 discovered species widely distributed in various climatic zones. Yeasts have a striking ability to survive varied osmotic concentrations, pH, and temperature of the substrate. As discussed previously in the chapter, each phase of the ensilage has its unique requirements and characteristics and slight deviations from standard compromise silage quality. Yeasts are facultative and heterotrophic eukaryotic microorganisms which commonly propagate through budding. They include various acid-tolerant ensilage species of different genera: *Candida, Hansenula, Debaryomyces, Pichia, Saccharomyces, Geotrichum*, and *Torulopsis*. Yeast, with the ability to tolerate these unfavorable conditions and utilize LA as an efficient C source, rapidly propagates when exposed to favorable aerobic conditions, that is, feed-out and aerobic phases.

Silage susceptibility to contamination under aerobic conditions is determined by the population and lactate diversity utilizing yeasts. Yeasts can efficiently exploit carbohydrates and LA, which are abundant in ensiled fodder crops. They are the most significant undesirable microorganism in deteriorating quality of silage during the aerobic phase—though few species can develop anaerobically—and are usually first to grow in ensiled mass when conditions become aerobic. They can thrive at a highly acidic pH of 3.5, which slows or inhibits other microbes, filamentous fungi, acetic acid, and butyric acid bacteria.

The fermentation pathway dominantly followed by yeasts is pyruvate decarboxylation to acetaldehyde, which produces ethanol upon subsequent reduction. This pathway of ethanol production is highly undesirable during ensilage, as it not only loses dry matter in the form of carbon dioxide but also increases pH, promoting the growth of other undesirable bacterial and fungal microbiomes. Yeasts possess an electron transport chain and consequently produce substantial ATPs compared to bacterial species, as discussed earlier. Yeasts completely oxidize carbohydrates and LA in the presence of oxygen to CO_2 and H_2O by following the glycolytic and tricarboxylic acid cycle, respectively. These biochemical changes in ensiled fodder, under aerobic conditions, promote the propagation and other secondary fermenters (Rooke and Hatfield, 2003).

Yeast is responsible for dry matter losses, especially when water-soluble carbohydrates are comparatively high in the ensiled fodder crop. They can follow both aerobic and anaerobic routes for their growth and fermentation, resulting in different metabolites. Under aerobic conditions, they cause LA oxidation, whereas under anaerobic conditions promote sugar fermentation—such as glucose, maltose, and sucrose—resulting in DM and available sugar losses. In anaerobic conditions, yeasts also produce alcohol like propanol, pentanol, and volatile fatty acids. Consequently, yeast growth increases soil pH, making the ensilage environment suitable for other aerobic deteriorating bacteria. When oxygen is depleted, anaerobic species of yeasts grow by fermenting carbohydrates into ethanol and carbon dioxide. This characteristic makes them the presumed cause of ethanol silage, especially in silage, which has significant sugar content even after LAB growth (Driehuis and Elferink, 2000).

12.8.2.1.1 Yeast Flora Composition during Ensilage

Generally, acidic pH during ensilage is hostile to most yeast cells, usually due to propionic and acetic acids. These un-dissociated fatty acids are significantly inhibitory as compared to dissociated acids under the acidic environment of silage. These un-dissociated molecules can diffuse into cells and lower intracellular pH, resulting in yeast cell death. It can be counteracted by pushing H^+ ions out of the cell by any energy-consuming process. Under aerobic conditions, this energy is obtained by oxidation of the substrate; however, it is obtained by fermentation of carbohydrates under the ensilage anaerobic environment. The majority of yeast species require oxygen to trigger alcoholic fermentation, but, contrarily, *Saccharomyces cerevisiae* can produce ethanol in the presence of O_2 and can propagate in its absence. Except for *Candida famata* and *Geotrichum candidum*, several yeast species vigorously cause glucose fermentation compared to non-fermentative fresh fodder fungal species (Jonsson, 1989).

Yeast flora composition during ensilage depends on several factors, air ingress being the most important. Aerobic conditions will support lactate assimilating genus *Candida* species, whereas anaerobic ensilage mass will favor fermentative non-lactate assimilating *S. cerevisiae*. *Candida* and *Hansenula* species predominate aerobically compromised ensilage due to their lactate assimilation and higher sugar affinity. Yeasts displaying the crabtree effect, such as *S. cerevisiae*, have a lower affinity for sugar than yeast, such as *Candida*, with a negative crabtree effect. Fodder crop to be ensiled significantly impact the yeast flora composition during ensilage. Fodder crops with menthol, such as a few *Brassica* species and spearmint, favor non-fermentative yeast species. A few other factors include silage additives, time of harvest, and ensilage conditions (Pahlow et al., 2003).

12.8.2.2 Filamentous Fungi during Ensilage

As a preservation system for fodder crops over an extended time, ensilage retains its final product's highest nutritive value. Various undesirable bacteria threaten the silage quality described earlier in the first part of this chapter. Fungal contamination is the most significant risk to nutrients, DM, palatability, consumption, and, ultimately, animal performance. Although yeasts and filamentous fungi naturally contaminate fodder crops in the field, it can also occur during harvest, transport, and storage. This becomes a significant silage rotting factor when poor post-harvest handling and management conditions are present. When water activity during ensilage exceeds its critical level of 0.8, it supports the proliferation of degrading fungi of genus *Penicillium, Eurotium, Fusarium, Trichoderma aspergillus*, already present on fodder crop.

The ensilage process is characterized by anaerobic and acidic conditions unfavorable for most fungal species to grow. However, compromised storage conditions triggered by inadequate drying, insect infestation, insufficient packaging could promote undesirable fungal communities to proliferate (Garon et al., 2006).

During fermentation, diversity and propagation of these filamentous fungi cause energy losses and mycotoxins contamination of animal feed with possible colonization of animal tissues as in aspergillosis. Increased respiratory activity of these fungi favors other thermophilic fungi, subsequently worsening the deterioration process. Fungal concentrations beyond 1×10^4 colony-forming unit/gram of feed are responsible for various livestock complexities of the respiratory, digestive, and reproductive systems; kidney damage; and skin and eye allergies (Magan and Aldred, 2007; O'brien et al., 2008).

Among different mold species isolated from ensiled fodder, *Penicillium roquesforti* and *P. paneum* are potentially toxic fungi that can tolerate acidic pH under a limited supply of O_2. *Aspergillus fumigatus*—linked with hemorrhagic bowel syndrome of dairy cows—is also a widely disseminated species of *Aspergillus* genus in silage producing farms of tropical climate high temperatures. HBS is a severe disease of lactating cows characterized by a rapid decrease in milk production, anemia, and abdominal pain. *A. fumigatus* is a possible causative agent of HBS disease because symptoms in dairy cows are the same as human enteral hemorrhages caused by this pathogen, and it is also isolated from infected HBS cattle. In addition to *Monascus* and *Byssochlamys* genus, these fungal species are well adapted to ensilage conditions of high organic acids and CO_2 concentrations, limited supply of O_2, and acidic pH (Elhanafy et al., 2013; Wambacq et al., 2016).

12.8.2.3 Fungal Diversity during Ensiling

Fungal diversity changes rapidly during ensilage, even at the phylum level, when silos are exposed to oxygen accidentally or during the feed-out phase. During different observations, *Basidiomycota* dominated the fungal community (>75%) in the early stages of ensiling. At later stages of ensilage, *Ascomycota* species predominate fungal core microbiota, especially in untreated silages. *Ascomycota* members grow exponentially after aerobic exposure and can account for 99% of the fungal community. The order of fungal community abundance observed during the study of different ensilage fodder crops also suggests that phylum Ascomycota is the leading contaminant, followed by *Basidiomycota zygomycota* (Liu et al., 2019).

During ensilage, fungal diversity tends to decline, like microbial diversity. The population of fungal orders on fodder crops in the field alters. In some cases, the fungal microbiome's relative abundance in the ensilage process accounts for <1% variation. Different forage crops have different impacts on the fungal population, as barley during ensilage increases the relative abundance of operational taxonomic units of order *Pucciniales, Sporidiobolales*, and *Cystofilobasidiales*. The number of OTUs during barley and oat ensilage in *Tremellales* and *Agaricales* increases, whereas *Cladosporium* decreases, sometimes to a low level of 8% of the total fungal microbiome. Terminal stages of ensilage represent *Saccharomycetales* as the most abundant fungal core microbiome representing most OTUs of *Kazachtania* and *Pichia*. The relative abundance of *Saccharomycetales* increases by 1.5% to 96% of the total fungal microbiome—depending on the ensiled crop—during terminal ensilage stages, whereas *Pleosporales* decreases. After aerobic exposure, the fungal microbiome of silage changes during the feed-out stage, but *Saccharomycetales* still represent up to 70% of fungal populations.

Yeast species, *Saccharomycetales*, dominate silage fungal core microbiome before and after oxygen exposure. Most of these operational taxonomic units belong to the spoilage genera of *Candida* and *Pichia*. When the aerobic stability of silage is stable, it inhibits fungal growth. During some ensilage fodder crops, aerobic exposure of silage decreases the diversity of *Saccharomycetales* and promotes the proliferation of less acid-tolerant species such as *Hypocreales*. Filamentous fungi such as *Aspergillus* and *Penicillium* genera, though present relatively in less abundance, are dangerous and undesirable during ensilage (Duniere et al., 2017).

12.8.2.4 Mycotoxin Production during Ensilage

Mycotoxins are secondary metabolites produced by toxigenic fungal pathogens under specific conditions. Mycotoxins commonly associated with poorly ensiled forages include aflatoxins, ochratoxins, trichothecenes, fumonisins, zearalenone, mycophenolic acids, and roquefortine C. As mentioned in silage, the main prevalent genera producing mycotoxins are *Aspergillus, Penicillium*, and *Fusarium*, with *A. flavus, A. fumigatus A. parasiticus, F. verticilloides, F. graminearum*, and *P. roqueforti* as primary frequently isolated species (Vila-Donat et al., 2018).

During a global survey of feed and feed-related commodities for different mycotoxins, aflatoxins, ochratoxin A, fumonisins, zearalenone, and deoxynivalenol (DON), it was observed that 72% of samples have one or more detectable level of these toxic secondary metabolites. Mycotoxins' presence in feed is of dual concern as it causes economic losses by affecting animal health and production and, as in aflatoxins, a significant food safety issue in animal products, such as milk.

Another health hazard from contaminated silage to both humans and animals, apart from mycotoxins, is associated with spore inhalation of *Aspergillus fumigatus*, cause of aspergillosis. *A. fumigatus* cause a hypersensitive response in the lungs, a disease given eponym of 'farmer's lung' after its high prevalence in farmers (Dagenais and Keller, 2009).

Apart from quantitative and qualitative losses, and reduced feed intake, fungal contaminated silage cause severe respiratory allergies and mycoses. Both animals and humans are equally susceptible to toxigenic mold spores, especially in poorly ventilated sheds/buildings. Fungi frequently produce mycotoxins during the aerobic deterioration of ensiled fodder. These secondary metabolites account for the primary cause (15%) of abortions in farm cattle and gain entry through the respiratory tract or digestive system. During an experiment performed in Italy to compare two famously known preservation methods, hay and silage, alfalfa preservation was shown using ensilage produced more mycotoxins than hay (Tomasi et al., 1999).

During a study of baled silage mycobiota in Norway, 76 fungal species were isolated. Among these mold species, *Aspergillus fumigatus* and *Penicillium roqueforti* were most common and frequently isolated. Most of the other recorded species were of the genus *Penicillium*, *Rhizopus*, *Mucor*, and *Geotrichum*. The most common silage contaminating fungal species in various studies is *P. roqueforti*, which is also associated with mycotoxins, roquefortine, and mycophenolic acid. Apart from mycotoxins, studies have been conducted to observe visible fungal growth, their occurrence, and risk assessment for livestock and human. Baled silage in Ireland was observed on various farms, and it was noted that the infestation rate was 90% with 1–12 colonies/bale. *P. roqueforti* was the most prevalent and dominating isolate (O'Brien et al., 2005).

Many isolated fungal species from silage can produce different mycotoxins, resulting in complex animal disorders, including abortions, hormonal imbalance, and compromised immune systems. *P. roqueforti*, the predominant fungal species during ensilage, as mentioned earlier, can produce a range of toxins: roquefortin C, patulin, mycophenolic acid, PR-toxin, and penicillic acid. Characteristics that enable *P. roqueforti* to grow under unfavorable ensilage environments tolerate higher carbon dioxide and acetic acid concentration and survive low oxygen levels. Among toxins produced by this fungus, PR-toxin and roquefortin C are most commonly detected, especially in molded samples.

Molds grow when all components of the disease triangle are favorable. These components, except O_2, are conducive for fungal proliferation during ensilage. Oxygen ingression into silage during poor ensiling propagates hibernating mold species to grow. However, all mold species inhabiting silage are not mycotoxigenic, save a few, producing these secondary metabolites basically in the field to gain competitive advantage and compromise the host's immune system. During ensilage, visible observation of these molds does not confirm mycotoxin production and vice versa. Silage stored for an extended period is more prone to fungal spoilage and subsequent mycotoxin production. It is also worth mentioning that pre-harvest mycotoxin contamination of fodder crops is unaffected instead of increasing during the ensiling process and adding to overall contamination (Fink-Gremmels, 2008).

12.8.2.4.1 Mycotoxin Classes with Relevance to Ensiling

Mycotoxin contamination of silage can be divided into two major classes: pre-harvest contamination of forage crop with mycotoxins—typical in the field—and during the ensilage and storage/feed process phase. Different fungal species, toxins, and environmental factors are required to cause different mycotoxin contamination levels at each stage.

12.8.2.4.2 Mycotoxin Contamination in Field/Pre-Harvest

In the field, before harvesting, fodder crops are attacked by toxigenic fungal species of genera *Fusarium*, *Aspergillus*, *Claviceps*, and *Neotyphodium*. These toxigenic fungal species gain entry through disrupted cell walls due to mechanical or insect damage—resulting in infestation and mycotoxins production. Mycotoxin production is influenced by different factors, such as host nutrients, water activity, temperature, and other environmental factors.

12.8.2.4.3 Fusarium Mycotoxins

Though *Fusarium* species are ubiquitous, these are most prevalent in a temperate climate with a broad host range. Primary sources of *Fusarium* in the field are soil and plant debris, whereas insect and mechanical damages of plant surface accompanied by optimal environmental conditions result in high mycotoxin production. *Fusarium* species produce fumonisins, zearalenone, and trichothecene toxins at different levels by different species. This genus has a low prevalence in acidic and anaerobic ensilage environments as compared to *Aspergillus* and *Penicillium*.

12.8.2.4.4 Aspergillus Mycotoxins

Aspergillus flavus and *A. parasiticus* produce highly toxic aflatoxins, AFB1, which is more prevalent, which transforms to AFM1 in cattle liver and then carries over in milk. These species are mostly associated with maize and peanut.

12.8.2.4.5 Claviceps and Neotyphodium Mycotoxins

Claviceps purpurea and *Neotyphodium* produce alkaloid (higher concentration of clavines and lysergic acid) mycotoxins in rye, barley, and several kinds of grass, especially at the flowering stage. The resting spore of *Claviceps purpurea*, sclerotia, allows it to survive extreme conditions. *Neotyphodium* is highly prevalent in the wild and managed grass pastures of Australia, Europe, New Zealand, and the United States.

12.8.2.4.6 Mycotoxin Production during Ensilage

During the ensilage period, mycotoxin contamination depends on the duration and extent of oxygen ingression in a silo and the ensiled mass's porosity and density. Ideally, it is not possible practically to keep ensiled mass completely anaerobic, first because covering/packing material is not 100% airtight, and second, rodents, insects, and birds physically damage the silo. Moreover, the aerobic condition is inevitable during the feed-out phase, which favors mold growth and is a reason for high infestation in the surface layer. *P. roqueforti* and *A. fumigatus* are prevalent and health risk species at this stage. *Monascus ruber* and *Byssochlamys nivea* also produce mycotoxins monacolin K, citrinin, and patulin, respectively (Driehuis, 2011).

12.8.2.4.7 Mycotoxin Distribution within the Silo

Mycotoxin distribution within a silo depends on the environmental conditions and physical properties of the ensiled fodder crop. During different experiments regarding *Fusarium* toxins, it was observed that in some cases, DON concentration was higher in the surface layer, whereas in others, it was significantly higher in the bottom layer (Richard et al., 2009). In another study about aflatoxin distribution within the silo, it was noted that AFB1, citrinin, and gliotoxin concentrations were more in the bottom layer of ensiled maize. The distribution of *A. fumigatus* was relatively higher in the middle and surface layer of silage, whereas the bottom layer was absent. This could be attributed to the distribution of favorable microaerophilic conditions, humidity, and pH within the silo. As mentioned earlier, fungal aerobic deterioration of silage is highest during the feed-out stage due to the recolonization of pre-harvest infestation of undesirable toxigenic fungi upon oxygen availability at the silo opening (Cheli et al., 2013).

12.8.2.4.8 Conclusion

Mycotoxin production during ensilage can be discouraged by following proper plant production practices and fermentation. The former is done to reduce inoculum during the ensiling fodder crop. Soil fertility, resistant fodder crop varieties (if available), crop rotation (to reduce carry-over risk), and proper agronomic practices are recommended to limit toxigenic fungal species infection. After harvesting, proper drying, chopping, and quick removal of oxygen before ensiling effectively maintain silage quality free of mold growth. Silo size is crucial, especially when dealing with fungal contamination. Silo size should be according to the feeding rate to keep its surface fresh and free of fungal proliferation, which is more likely after air exposure. Though silage additives do not affect mycotoxin production, they can be useful by reducing mold growth in the first place. Spoiled silage should not be fed to young livestock classes, as they are more susceptible to developing complexities. If mycotoxin presence is suspected in feed, silage testing at different locations is recommended, as it varies depending upon favorable aerobic micro-pockets.

REFERENCES

Adesogan AT and Newman YC (2010) *Silage Harvesting, Storing, and Feeding*. EDIS 2010.

Alonso VA, Pereyra CM, Keller LAM, et al. (2013) Fungi and mycotoxins in silage: An overview. *Journal of Applied Microbiology* 115: 637–643.

Amado IR, Fuciños C, Fajardo P, et al. (2016) Pediocin SA-1: A selective bacteriocin for controlling *Listeria monocytogenes* in maize silages. *Journal of Dairy Science* 99: 8070–8080.

Ávila C and Carvalho B (2020) Silage fermentation—updates focusing on the performance of microorganisms. *Journal of Applied Microbiology* 128: 966–984.

Bennett SD, Walsh KA and Gould LH (2013) Foodborne disease outbreaks caused by *Bacillus cereus, Clostridium perfringens*, and *Staphylococcus aureus*—United States, 1998–2008. *Clinical Infectious Diseases* 57: 425–433.

Bernardes T, Daniel J, Adesogan A, et al. (2018) Silage review: Unique challenges of silages made in hot and cold regions. *Journal of Dairy Science* 101: 4001–4019.

Bernardes T and Do Rêgo A (2014) Study on the practices of silage production and utilization on Brazilian dairy farms. *Journal of Dairy Science* 97: 1852–1861.

Bolsen K, Ashbell G and Weinberg Z (1996) Silage fermentation and silage additives—review. *Asian-Australasian Journal of Animal Sciences* 9: 483–494.

Borreani G, Dolci P, Tabacco E, et al. (2013) Aerobic deterioration stimulates outgrowth of spore-forming *Paenibacillus* in corn silage stored under oxygen-barrier or polyethylene films. *Journal of Dairy Science* 96: 5206–5216.

Brassley P (1996) Silage in Britain, 1880–1990: The delayed adoption of an innovation. *The Agricultural History Review*: 63–87.

Brett M, McLauchlin J, Harris A, et al. (2005) A case of infant botulism with a possible link to infant formula milk powder: Evidence for the presence of more than one strain of *Clostridium botulinum* in clinical specimens and food. *Journal of Medical Microbiology* 54: 769–776.

Callaway T, Edrington T, Brabban A, et al. (2006) Fecal prevalence of *Escherichia coli* O157, Salmonella, Listeria, and bacteriophage infecting *E. coli* O157: H7 in feedlot cattle in the southern plains region of the United States. *Foodbourne Pathogens Disease* 3: 234–244.

Cheli F, Campagnoli A and Dell'Orto V (2013) Fungal populations and mycotoxins in silages: From occurrence to analysis. *Animal Feed Science Technology* 183: 1–16.

Coblentz W and Akins M (2018) Silage review: Recent advances and future technologies for baled silages. *Journal of Dairy Science* 101: 4075–4092.

Collins M, Nelson CJ, Moore KJ, et al. (2017) *Forages, Volume 1: An Introduction to Grassland Agriculture*. John Wiley & Sons.

Cornick NA, Helgerson AF, Mai V, et al. (2006) *In vivo* transduction of an Stx-encoding phage in ruminants. *Applied Environmental Microbiology* 72: 5086–5088.

D Mello J (2004) Microbiology of animal feeds. *FAO Animal Production Health Paper* 89–106.

Dagenais TR and Keller NP (2009) Pathogenesis of *Aspergillus fumigatus* in invasive aspergillosis. *Clinical Microbiology Reviews* 22: 447–465.

De Reu K, Grijspeerdt K and Herman L (2004) A Belgian survey of hygiene indicator bacteria and pathogenic bacteria in raw milk and direct marketing of raw milk farm products. *Journal of Food Safety* 24: 17–36.

Driehuis F (2011) Occurrence of mycotoxins in silage. *Proceedings of the II International Symposium*.

Driehuis F (2013) Silage and the safety and quality of dairy foods: A review. *Agricultural Food Science* 22: 16–34.

Driehuis F and Elferink SO (2000) The impact of the quality of silage on animal health and food safety: A review. *Veterinary Quarterly* 22: 212–216.

Driehuis F, Wilkinson J, Jiang Y, et al. (2018) Silage review: Animal and human health risks from silage. *Journal of Dairy Science* 101: 4093–4110.

Dunière L, Gleizal A, Chaucheyras-Durand F, et al. (2011) Fate of *Escherichia coli* O26 in corn silage experimentally contaminated at ensiling, at silo opening, or after aerobic exposure, and protective effect of various bacterial inoculants. *Applied Environmental Microbiology* 77: 8696–8704.

Duniere L, Xu S, Long J, et al. (2017) Bacterial and fungal core microbiomes associated with small grain silages during ensiling and aerobic spoilage. *BMC Microbiology* 17: 1–16.

Elhanafy MM, French DD and Braun U (2013) Understanding jejunal hemorrhage syndrome. *Journal of the American Veterinary Medical Association* 243: 352–358.

Fink-Gremmels J (2008) Mycotoxins in cattle feeds and carry-over to dairy milk: A review. *Food Additives Contaminants* 25: 172–180.

Garon D, Richard E, Sage L, et al. (2006) Mycoflora and multimycotoxin detection in corn silage: Experimental study. *Journal of Agricultural Food Chemistry* 54: 3479–3484.

Jones C, Heinrichs A, Roth G, et al. (2004) From harvest to feed: Understanding silage management. *Pennsylvania State University. College of Agricultural Science*: 2–11.

Jonsson A (1989) *The Role of Yeasts and Clostridia in Silage Deterioration. Identification and Ecology*.

Julien M-C, Dion P, Lafreniere C, et al. (2008) Sources of clostridia in raw milk on farms. *Applied Environmental Microbiology* 74: 6348–6357.

Kaiser AG, Piltz J, Burns HM, et al. (2004) *Top fodder successful silage*. NSW Department of Primary Industries.

Kelch WJ, Kerr LA, Pringle JK, et al. (2000) Fatal *Clostridium botulinum* toxicosis in eleven Holstein cattle fed round bale barley haylage. *Journal of Veterinary Diagnostic Investigation* 12: 453–455.

Liu B, Huan H, Gu H, et al. (2019) Dynamics of a microbial community during ensiling and upon aerobic exposure in lactic acid bacteria inoculation-treated and untreated barley silages. *Bioresource Technology* 273: 212–219.

Magan N and Aldred D (2007) Post-harvest control strategies: Minimizing mycotoxins in the food chain. *International Journal of Food Microbiology* 119: 131–139.

Magnusson M, Christiansson A and Svensson B (2007) *Bacillus cereus* spores during housing of dairy cows: Factors affecting contamination of raw milk. *Journal of Dairy Science* 90: 2745–2754.

Mannetje L (2016) Silage for animal feed. *Encyclopedia of Life Support Systems*: 123–135.
Marshall J, Soboleva T, Jamieson P, et al. (2016) Estimating bacterial pathogen levels in New Zealand bulk tank milk. *Journal of Food Protection* 79: 771–780.
Mickan F, Martin M and Piltz J (2003) Silage storage. *Successful silage* 2: 217–252.
Moran J (2005) *Tropical Dairy Farming: Feeding Management for Small Holder Dairy Farmers in the Humid Tropics*. Csiro publishing.
Muck R (1988) Factors influencing silage quality and their implications for management. *Journal of Dairy Science* 71: 2992–3002.
Muck R (2013) Recent advances in silage microbiology. *Agricultural Food Science* 22: 3–15.
Muck RE (2010) Silage microbiology and its control through additives. *Revista Brasileira de Zootecnia* 39: 183–191.
Nishino N (2011) Aerobic stability and instability of silages caused by bacteria. *Simpósio Internacional Sobre Qualidade E Conservação De Forragens* 2.
Nyambe S, Burgess C, Whyte P, et al. (2017) The fate of verocytotoxigenic *Escherichia coli* C600φ3538 (Δvtx2: cat) and its vtx2 prophage during grass silage preparation. *Journal of Applied Microbiology* 122: 1197–1206.
O'Brien M, Egan D, O'kiely P, et al. (2008) Morphological and molecular characterisation of *Penicillium roqueforti* and *P. paneum* isolated from baled grass silage. *Mycological Research* 112: 921–932.
O'Brien M, O'kiely P, Forristal PD, et al. (2005) Fungi isolated from contaminated baled grass silage on farms in the Irish Midlands. *FEMS Microbiology Letters* 247: 131–135.
Oetzel GR (2007) Herd-level ketosis—diagnosis and risk factors. *Preconference Seminar C*. Citeseer 67–91.
Ogunade I, Kim D, Jiang Y, et al. (2016) Control of *Escherichia coli* O157: H7 in contaminated alfalfa silage: Effects of silage additives. *Journal of Dairy Science* 99: 4427–4436.
Organization WH (2004) *Risk Assessment of Listeria Monocytogenes in Ready-To-Eat Foods: Technical Report*. Food & Agriculture Org.
Pahlow G, Muck RE, Driehuis F, et al. (2003) Microbiology of ensiling. *Silage Science Technology* 42: 31–93.
Pedroso A, Adesogan A, Queiroz O, et al. (2010) Control of *Escherichia coli* O157: H7 in corn silage with or without various inoculants: Efficacy and mode of action. *Journal of Dairy Science* 93: 1098–1104.
Perry CM and Donnelly CW (1990) Incidence of *Listeria monocytogenes* in silage and its subsequent control by specific and nonspecific antagonism. *Journal of Food Protection* 53: 642–647.
Piltz J and Kaiser A (2004) Principles of silage preservation. *Successful Silage*: 25–56.
Pinto CL, Souza LV, Meloni VA, et al. (2018) Microbiological quality of Brazilian UHT milk: Identification and spoilage potential of spore-forming bacteria. *International Journal of Dairy Technology* 71: 20–26.
Pitt RE (1990) *Silage and Hay Preservation*. CIRAD.
Queiroz O, Ogunade I, Weinberg Z, et al. (2018) Silage review: Foodborne pathogens in silage and their mitigation by silage additives. *Journal of Dairy Science* 101: 4132–4142.
Rammer C, Östling C, Lingvall P, et al. (1994) Ensiling of manured crops—Effects on fermentation. *Grass Forage Science* 49: 343–351.
Richard E, Heutte N, Bouchart V, et al. (2009) Evaluation of fungal contamination and mycotoxin production in maize silage. *Animal Feed Science Technology* 148: 309–320.
Rooke JA and Hatfield RD (2003) Biochemistry of ensiling. *Silage Science Technology* 42: 95–139.
Rotz CA, Ford SA and Buckmaster DR (2003) Silages in farming systems. *Silage Science Technology* 42: 505–546.
Ryser ET, Arimi SM and Donnelly CW (1997) Effects of pH on distribution of *Listeria* ribotypes in corn, hay, and grass silage. *Applied Environmental Microbiology* 63: 3695–3697.

Sanaa M, Poutrel B, Menard J, et al. (1993) Risk factors associated with contamination of raw milk by *Listeria monocytogenes* in dairy farms. *Journal of Dairy Science* 76: 2891–2898.

Sugiyama M, Iddamalgoda A, Oguri K, et al. (2003) Development of Livestock Sector in Asia: An Analysis of Present Situation of Livestock Sector and Its Importance for Future Development. Gifu, Japan: Gifu City Women's College, 52.

Taran T (2019) Nutritious Feed for Farm Animals During Lean Period: Silage and Hay—A Review.

Te Giffel Mt, Wagendorp A, Herrewegh A, et al. (2002) Bacterial spores in silage and raw milk. *Antonie van Leeuwenhoek* 81: 625–630.

Titterton M and Bareeba F (2000) Grass and legume silages in the tropics. *FAO Plant Production Protection Papers* 43–50.

Tomasi L, Horn W, Roncada P, et al. (1999) Preliminary studies on the presence of B1 aflatoxin in cured forages in the province of Reggio Emilia (Italy). *Fourrages.*

Vila-Donat P, Marín S, Sanchis V, et al. (2018) A review of the mycotoxin adsorbing agents, with an emphasis on their multi-binding capacity, for animal feed decontamination. *Food Chemical Toxicology* 114: 246–259.

Vilar MJ, Yus E, Sanjuan ML, et al. (2007) Prevalence of and risk factors for *Listeria* species on dairy farms. *Journal of Dairy Science* 90: 5083–5088.

Vissers M, Driehuis F, Te Giffel M, et al. (2007) Concentrations of butyric acid bacteria spores in silage and relationships with aerobic deterioration. *Journal of Dairy Science* 90: 928–936.

Von Braun J (2010) The role of livestock production for a growing world population. *Lohmann Information*. 3–9.

Wambacq E, Vanhoutte I, Audenaert K, et al. (2016) Occurrence, prevention and remediation of toxigenic fungi and mycotoxins in silage: A review. *Journal of the Science of Food* 96: 2284–2302.

Wang C, He L, Xing Y, et al. (2019) Effects of mixing *Neolamarckia cadamba* leaves on fermentation quality, microbial community of high moisture alfalfa and stylo silage. *Microbial Biotechnology* 12: 869–878.

Weinberg ZG and Muck R (1996) New trends and opportunities in the development and use of inoculants for silage. *FEMS Microbiology Reviews* 19: 53–68.

Woolford M (1990) The detrimental effects of air on silage. *Journal of Applied Bacteriology* 68: 101–116.

13 Impact of Pests on Winter Fodder Preservation

Asadullah Azam and Sakhidad Saleem

CONTENTS

DOI: 10.1201/9781003055365-13

13.1 INTRODUCTION

Fodder crops are the primary and cheapest feed for livestock and are produced annually for better production of meat, milk, and other dairy products that significantly affect most developing and developed countries' economies. According to the International Feed Industry Federation, the worldwide animal feed industry produces 873 million tons of feed in 2011 (Dick, 2012). The production of fodder crops is always associated with seed quality, pests, diseases, preservation, climate change, and so on. These problems lead to fodder shortage in the growing season and during the lean periods of feed to the dairy farms, resulting in decreasing the livestock products. Low-quality feed for livestock results in non-infectious diseases such as grass tetany (Martens and Schweigel, 2000), ketosis (Lean et al., 1994), milk fever (Oetzel and Goff, 2008), pregnancy toxemia, and so on. Insect pests and diseases significantly affect the establishment, yield, and longevity of grass and forage crops (Koli and Bhardwaj, 2018). Locusts severely affect green pasture and fodder crops. Shorter pasture species and newly sown grass pastures are particularly vulnerable to locust infestation (FAO, 2015). Pasture and fodder crops are at risk during massive numbers of locusts, and they may compete with livestock for herbage.

As lucerne is green for a longer period than other fodder crops, it may be the prime feeding target for the locust in fodder crop areas (FAO, 2015). For a sustainable fodder production system, biotic and abiotic stresses should be kept under the economic threshold level. Insect pests also have an indirect effect by reducing nodule formation in legumes, resulting in reduced nitrogen fixation, while plant pathogens may produce toxins such as mycotoxins, which affect animal health after consumption (Gashaw, 2016). Using pesticides for pest and disease management is also of concern because of pesticide residue on fodder and the risk of animal feed contamination. Organophosphate, organochlorine, and pyrethroid are effective pesticides that contaminate feed (Van Barneveld, 1999). Pesticides can enter the livestock through their application to animals or direct application to fodder crops. The rearing of livestock on pesticide-contaminated crops accumulates pesticide residue in edible tissue, milk, and other fat-rich substances (John et al., 2001). Dikshit et al. (1989) reported a pesticide residue up to 6 ppm in agriculture to control crop pests in India. Pesticide residue sometimes may be higher in dry fodder during the preservation period than green fodder, as Sharma et al. (2012) stated.

Pest resistance against insecticide is also another big issue in terms of fodder crop production. During the application of insecticides, the resistant insects pass their genes to the next generation, and the future generation becomes resistant to the same insecticide. Due to pest resistance development, farmers are using extensive doses of insecticide, which is a worldwide concern in terms of pesticide residue. The increasing pest resistance and the impact of pesticides on the environment have led to the need for an effective and environmentally friendly method for managing fodder crops, pests and field crops, fruit trees, and so on. The integrated pest management (IPM) approach for the management of insect pests is a significant trend that applies multiple methods for suppressing the pest population. Host plant resistance, biopesticides, pheromones, biological control, predators and parasitoids, *Bacillus thuringiences*, baculoviruses, entomopathogenic nematodes, and entomopathogenic fungi are among the IPM methods that recently been used as a safe alternative to synthetic insecticides.

13.2 FODDER CROPS

Fodder is a type of animal feed specifically used to feed domesticated livestock in plant cut forms. The worldwide animal feed industry produced 873 million tons of feed in 2011 (Dick, 2012). Fodder crops are the cheapest resources for livestock feeding and provide various important nutrition throughout the year (Anthony, 2005). Recently, modern trends for fodder production are advised to increase the amount of fodder per acre. These trends include fodder selection with higher nutritional value, high production per acre, evergreen fodder use, and preference for fodder with easy preservation (Anthony, 2005). There are three types of fodder crops: multi-cut/evergreen, winter fodder, and summer fodder.

13.2.1 Evergreen Fodder

This type of fodder is planted once and utilized for many years and includes lucerne, mott grass, Sudan grass, and so on (Anthony, 2005).

13.2.2 Winter Fodder Crops

This type of fodder is planted at the beginning of winter, provides feeding during the winter, and reaches maturity in summer. Berseem, oats, ryegrass, barley, and mustard are winter season fodder crops (Anthony, 2005). In some countries like Afghanistan, wheat and barley straw and other legumes and pulses comprise winter fodder's primary resource (Anthony, 2005).

13.2.3 Summer Fodder Crops

They are planted at the beginning of summer and provide feeding until the beginning of the winter season. Maize, sorghum, millet, and so on are summer season fodder crops (Anthony, 2005).

13.2.4 Lucerne

It is a leguminous multi-cut fodder and directly fixes nitrogen through nitrogen-fixing bacteria from the air. Its nutritive value is higher than the non-leguminous fodder, it contains more protein, and farmers store it as hay (Chaudhry and Mian, 2017). It is not suitable for ensiling because of high moisture and protein concentration: 85–95% and 14–18% (based on the dry matter), respectively (Tripathi et al., 1995).

13.2.5 Berseem

It is the most palatable multi-cut leguminous fodder of the winter season and is called the king of fodders due to its nutritive value. Berseem increases milk production and suppresses herb growth in the field (Chaudhry and Mian, 2017).

13.2.6 Pastures

They are lands that were cultivated once and used for domesticated grazing livestock. They are divided into summer and winter rainfall pastures. Summer rainfall pastures in Australia consist of native grasses and annual forbs (Elder, 2007). Grasses, legumes, and broadleaf plants are among winter rainfall pastures. They can be grown for seed and to make hay (Pavri, 2007).

13.2.7 Legume Pastures

They may serve a dual purpose, such as fodder for grazing stock and nitrogen provision for the upcoming season crop when integrated into a cropping rotation. Medic pastures and clover are among legume pastures and are susceptible to insect damage (Pavri, 2007).

13.2.8 Pulses

Pulse crops are grown as a cash crop, cereal disease control break, weed control, and soil fertility (Miles et al., 2007). They are grown in the summer and winter seasons. Summer season pulses include cowpeas, lima beans, mung beans, peanuts, pigeon peas, soybeans, and so on and are presently grown in smaller quantities than winter season pulses (Brier, 2007). Chickpeas, faba beans, field peas, lentils, lupins, and vetch are winter season pulses (Miles et al., 2007).

13.3 FODDER PRESERVATION

Fodder crops supply fiber, energy, proteins, and minerals to livestock that are continuously required for feeding throughout the year (Weinberg, 2008). The conservation of fodder helps avoid the loss of nutrients from green fodder when they are abundant. Fodder crops are preserved as either hay or silage. Fodder conservation is an essential element for productive and efficient ruminant livestock farms. However, the availability of fodder crops is seasonal. For example, in a temperate climate, they are not available during the winter; in semi-arid areas in the dry season; and in other areas, rain interferes with harvesting operations (Weinberg, 2008). Therefore, it is needed to produce fodder crops during the growing season and preserve them for scarce periods. If crops are not preserved, they will eventually spoil and rot. Preservation of fodder crops is used to inhibit spoilage factors such as microorganisms, insects, enzymes, and chemical reactions such as oxidation and browning (Weinberg, 2008). In most countries, fodder and grasses are preserved as haymaking and to overcome the feeding shortage in scarce periods. Haymaking is based on sun drying, while anaerobic conditions are involved in the ensiling process (Weinberg, 2008).

13.3.1 Haymaking

Haymaking is a suitable method for preserving fodder and grasses due to the reduction of efforts in transportation and handling (Tripathi et al., 1995). In haymaking, the crop is dried, resulting in inactivity of the crops concerning plant enzyme activity and microbial spoilage (Muck and Kung, 1997). It is mainly used in those areas of the world which have good dry conditions. However, it is also practiced in humid climates where the ensiling process is difficult due to forage characteristics, high temperatures, or tradition (Muck and Kung, 1997). It is suitable for thin crops dried quickly in the sunny season (Weinberg, 2008). Losses during the haymaking process may occur due to continuous respiration especially forage legumes such as alfalfa, clover, and vetch, because the protein is present in their leaves and susceptible to nutritional losses (Weinberg, 2008). Sometimes improper haymaking processes may lead to spoilage of preserved fodder and the occurrence of non-infectious diseases in animals. Respiration, shattering, dropping of leaves, leaching, bleaching, fermentation, and molding are among the significant losses during the haymaking process (Muck and Kung, 1997). Therefore, important points such as timely harvesting on rain-free days, proper drying time, proper racking and baling, and proper storage moisture should be considered during haymaking.

13.3.2 Silage Making

In silage making, the ensiled fodder's sugar contents are converted to the lactic acid that reduces the pH of the silage and results in the inhibition of the biological activities responsible for spoilage (Tripathi et al., 1995). Silage is less weather dependent and can be made of almost any crop (Weinberg, 2008). It is consistent with hay in nutrient content but requires high capital investment (Weinberg, 2008). The crop is fermented anaerobically by lactic acid bacteria during the ensiling process and depends on low pH to prevent anaerobic microorganisms and anaerobic conditions to prevent aerobic organisms' growth (Muck and Kung, 1997). The silage-making process includes harvesting, wilting, chopping, enrichment, transportation, compaction, sealing, and feed-out (Weinberg, 2008).

During the ensiling process, losses may occur in dry matter and energy. Dry matter losses may occur due to the conversion of organic matter to carbon dioxide. Energy is retained in the fermentation process, and thus its losses are usually smaller than dry matter losses (Weinberg, 2008). To avoid losses, silage additives are developed to improve silage fermentation and preserve its nutritive value. These additives include fermentation inhibitors, fermentation enhancers, aerobic deterioration inhibitors, nutritional additives, and moisture absorbents (Weinberg, 2008). The additives will upgrade the silage quality and reduce preservation failure (Savoie and Jofriet, 2003; Tripathi et al., 1995).

The silage must be safe for use because it is the first step in the food production chain for the people, and its quality will affect animal health. Detrimental organisms, toxic chemicals, and excess acidity are health hazards associated with silage (Weinberg, 2008).

13.4 MAIN PEST OF PULSES

13.4.1 Blue Oat Mite (*Penthaleus* spp.)

13.4.1.1 Introduction

Blue oat mites (BOMs) are major agricultural pests that attack various pasture, vegetable, and crop plants. The BOM was first recorded in New South Wales in 1921 and was introduced from Europe (Chen et al., 1967). *Penthaleus major*, *Penthaleus falcatus*, and *Penthaleus tectus* are the three main species of blue oat mite and differ in their distributions (Capinera, 2001).

13.4.1.2 Description

Penthaleus major, *Penthaleus falcatus*, and *Penthaleus tectus*. *P. major* has long setae arranged in four to five longitudinal rows, while *P. falcatus* has a higher number of short setae scattered irregularly. *P. tectus* has setae of medium length and number. Adults have a blue-black body and are approximately 1 mm long, with a distinctive red mark on their back (Capinera, 2001). Larvae are oval, approximately 0.3 mm long, with three pairs of legs. After hatching, the color of the BOM is pink-orange and soon becomes brownish and then green. The larvae are active during the day's cooler time and mostly feed in the mornings and cloudy weather. In the warmer parts of the day, they hide under moist soil surfaces or foliage and sometimes dig into the soil under extreme conditions (Chen et al., 1967).

13.4.1.3 Distribution

Environmental conditions limit the distribution of BOM, and their emergence needs to coincide with the growth of winter pastures and crops (Chen et al., 1967). In southern Australia, BOM is considered an important crop and pasture pest that is found in the Mediterranean climates of Victoria, New South Wales, South Australia, Western Australia, and eastern Tasmania. They can also be found in some regions of southern Queensland (Child, 1991). The adult individuals of BOM move between plants over distances of several meters during the winter. Sometimes long-range dispersal occurs via the movement of eggs carried on soil adhering to livestock and farm machinery and through the transportation of plant material. Winds also play an essential role in the movement of diapause eggs during heavy winds (Chen et al., 1967).

13.4.1.4 Life cycle

Between April and late October, the BOM is active and goes through two or three generations, and each generation lasts eight to ten weeks. They protect themselves as diapause or over-summering eggs that are resistant to the heat and desiccation of summer. Following cool temperatures and adequate rainfall, the eggs hatch in autumn when conditions are optimal for mite survival (Capinera, 2001). The female chooses the leaves, stems, and roots of food plants or the soil surface for oviposition laid either singly or in clusters of three to six. The eggs deposited on the leaves are fastened by a slimy substance secreted next to and on the stem of plants.

13.4.1.5 Host Plants

The BOM has a broad host range on various agriculturally important plants, such as cereals, grasses, canola, field peas, legumes, and numerous weeds. Host plants and pastures are vulnerable to attack, particularly susceptible at the seedling stage. *P. major* prefers pastures, cereals, and pulses; *P. falcatus* is found on canola and broad-leaved weeds, while *P. tectus* prefers cereals. These preferences can help us in the indication of the species present in a particular field. However, confirmation by an expert is suggested.

13.4.1.6 Damage

The damage of BOM is visible as silvering of the leaves. In the autumn season, the seedling dies, and both grasses and legumes' growth is reduced. In the growing season, the production and quality of older green plants decrease, while reduced seed yield in legumes can occur in spring. The majority of damage and production loss is to grasses in the pasture and can occur in winter pastures (Child, 1991). Sometimes adverse conditions such as prolonged dry weather or waterlogged soils impact the plants and increase mite damage. Ideal growing conditions for seedlings enable plants to tolerate higher numbers of mites (Capinera, 2001).

13.4.1.7 Monitoring

BOM monitoring can be started from seedling emergence that coincides with the hatching of the over-summering eggs. As the new larvae are tiny and difficult to see with the naked eye, close inspection is essential (Child, 1991). Plant pastures and crops are inspected from autumn to spring for mites and evidence of damage. The inspection is essential in the first three to five weeks after sowing. Mites are best detected feeding on the leaves in the morning or on overcast days. Sometimes the mites cannot be found on plant material, so inspecting the soil is recommended (Capinera, 2001).

13.4.1.8 Biological

Several predator species are known to attack earth mites, among which small beetles, spiders, and ants may also play an essential role in decreasing their number (Chen et al., 1967). In Australia, *Anystis wallacei* was introduced from France for biological control of BOM, red-legged earth mite, and lucerne flea. The predatory mites of *Balaustium* sp. have also been observed preying on blue oat mites. During the growing feeding time of BOM, a generalist predatory mite named *Bdellodes lapidaria* occurs in pastures and may feed on it. However, all of the predators mentioned have a limited impact on BOM populations (Child, 1991). Other predatory mites, such as snout mites, are also effective natural enemies of BOM. For conservation and maintaining of the natural enemy populations, leaving shelterbelts or refuges between paddocks is helpful (Capinera, 2001).

13.4.1.9 Cultural

Crop rotation will significantly help to reduce the population of BOM. Rotating the crops or pastures with non-host crops can reduce pest colonization, reproduction, and survival, which will decrease the need for chemical control (Chen et al., 1967). Rotate paddocks with non-preferred crops (*P. major*—canola; *P. tectus*—chickpeas; *P. falcatus*—wheat, barley). Weed management at pre-and post-sowing is important. Many broad-leaved weeds are alternative food sources for BOM, especially the juvenile stages (Chen et al., 1967), so heavy grazing of pastures in winter and spring significantly reduces BOM populations (Child, 1991).

13.4.1.10 Chemical

Chemicals are the last weapon against any pest problem and the most common method of control. All the recent pesticides that are registered for mite control are only effective against the mites' larval stage and do not kill the mite eggs (Capinera, 2001). An excellent time for controlling BOM is two to three weeks from the season's break. Due to their differing diapause strategies, spring spraying is not practical. There is also evidence of tolerance to some insecticides by *P. falcatus* (Child, 1991; Robinson and Hoffmann, 2001). Insecticide seed treatment is an effective method for low-moderate mite populations. Apply insecticides only if control is warranted and if you are sure of the mite identity (Capinera, 2001). Bare-earth treatments can be applied before or at sowing to kill emerging mites and protect the plants throughout their seedling stage (Chen et al., 1967). Within three weeks of mites' first appearance, pesticides should be used only when necessary, and rotate pesticides from chemical classes with different modes of action for preventing the build-up of resistant populations. Also, chemical classes are rotated to avoid developing multiple pesticide resistance across generations (Chen et al., 1967).

13.4.2 Blue-Green Aphid

13.4.2.1 *Acyrthosiphon kondoi* (Aphididae: Hemiptera)

13.4.2.1.1 Introduction

The blue alfalfa aphid (*Acyrthosiphon kondoi*), also known as blue-green aphid, is a true bug and sucks sap from leguminous plants, particularly alfalfa, and belongs to superfamily Aphidoidea in the order Hemiptera (Capinera, 2001). They are gray-green to blue-green colored, attacking lupins, lucerne, medics, and clovers (Bailey, 2007).

13.4.2.1.2 Description

The blue-green aphid has a waxy appearance with a body size of about 2 to 3 mm long and may be winged or wingless, with six long and pale with black tips legs. A pair of long cornicles extend beyond the abdomen, which ends in a short tail. Two slender, uniformly dark-colored antennae are located at the heed and angled back over the body, at least as long as the body. Immature aphids are smaller and paler than adults (Robinson and Hoffmann, 2001). They resemble the closely related pea aphid (*Acyrthosiphon pisum*) but are often a more bluish shade of green (Capinera, 2001).

13.4.2.1.3 Distribution

The aphid is native to Asia, and later it spread to North America, South America, Australia, and New Zealand. It was first detected in the United States in California in 1974 and had spread to Nebraska by 1979, Georgia and Kentucky by 1983, and Maryland by 1992. This pest's main hosts are plants in the family Leguminosae including alfalfa, pea, lentil, and cowpea (Capinera, 2001). The aphids occur in spring but are also active in autumn and winter.

13.4.2.1.4 Life Cycle

After spending the winter season on broadleaf weeds, pasture legumes, and medics, colonies of winged aphids fly to the host plants and start to reproduce asexually or sexually (Bailey, 2007). Aphids reproduce asexually, whereby females give birth to live young, often referred to as clones (Bailey, 2007). As the aphids grow in mild weather, they are most likely to be found from autumn to spring. The aphids survive in very low numbers on host plants during the dry months. In autumn, they migrate to germinating annual legumes or lucerne, and large colonies can rapidly develop with optimal conditions (Robinson and Hoffmann, 2001).

Nymphs are wingless and go through several growth stages. Many generations are produced during the growing season. Development rates are dependent on temperature, and either cooler or extremely hot temperatures slow development (Bailey, 2007).

13.4.2.1.5 Host Plants

This aphid's host range includes lucerne, lupins, faba beans, canola, and a wide range of pasture legumes, including subterranean clover, white clover, red clover, and annual medics (Robinson and Hoffmann, 2001).

13.4.2.2 Damage

13.4.2.2.1 Direct Feeding Damage

Blue-green aphids severely damage the upper leaves, stems, and terminal buds of host plants. Heavy infestations directly remove plant nutrients, causing leaf deformation and wilting. Stunted growth, leaf curling, and leaf drop are damage associated with lucerne and medics, resulting in reduction of dry matter production. Honeydew is an aphid secretion that can cause secondary fungal growth, inhibit photosynthesis, and decrease plant growth (Bailey, 2007).

Capinera (2001) reported that in 2010, there was a more virulent blue-green biotype that caused damage to some varieties of lentils, annual medic, and sub-clover that were previously thought to be tolerant to blue-green aphids.

13.4.2.2.2 Indirect Damage (Virus Transmission)

Aphids are among the most significant numbers of insects that spread viral diseases and are considered indirect damage. Aphids spread viruses between plants by feeding and probing as they move between plants and paddocks. Cucumber mosaic virus (CMV) and bean yellow mosaic virus (BYMV) are among the important viruses that are transmitted by blue-green aphids (Bailey, 2007).

13.4.2.2.3 Monitoring

Regular monitoring of the crops during bud formation to late flowering is essential for good management of this aphid. At least five sampling points over the paddock are necessary, and inspect at least 20 plants at each sampling point (Bailey, 2007). Each plant's youngest inflorescence is visually searched for aphids and for clusters of aphids or symptoms of leaf-curling. Heavy rain events or sustained frosts can reduce aphid infestations (Bailey, 2007).

13.4.2.2.4 Biological

Aphids have many effective natural enemies (Bailey, 2007). Parasitic wasps, hoverflies, ladybirds, damsel bugs, lacewings, and fungal disease are biocontrol agents and are only effective when aphid numbers are low (Robinson and Hoffmann, 2001).

Aphid parasitic wasps lay eggs inside bodies of aphids, and the larvae develop inside the aphids and finally leave the parasitized aphids as bronze-colored enlarged 'mummies.' The larvae go through different stages and finally emerge as adults from the mummified aphids. *Pandora neoaphidis* and *Conidiobolyus obscurus* are naturally occurring fungal diseases of aphids and reduce their populations (Bailey, 2007).

13.4.2.2.5 Cultural

Using resistant cultivars is the best cultural practice to reduce and avoid aphid feeding damage. The aphid numbers peak in the spring, so using early sowing varieties will enable plants to begin flowering before the aphid number increases. Controlling summer and autumn weeds will reduce the availability of alternate hosts between growing seasons (Bailey, 2007). Sowing resistant or tolerant cultivars of lucerne, medic, clovers, or lupins can play a significant role in overcoming economic losses. Use a high sowing rate to achieve a dense crop canopy, which will deter aphid landings (Bailey, 2007).

13.4.2.2.6 Chemical

Insecticide applications generally harm beneficial species, many of which are more susceptible to broad-spectrum chemicals than pest insects. Broad-spectrum chemicals have a negative impact on beneficial insects, so it is recommended to use selective insecticides. Pirimicarb provides effective control and has little impact on non-target insects. However, before making spray decisions, monitoring for beneficial insects is essential (Bailey, 2007).

13.4.2.3 Economic Thresholds

13.4.2.3.1 Canola

NSW: >50% of plants with clusters 25 mm long on stem or four to five stems per m2 with cluster 50 mm long on stems (Bailey, 2007).

- WA: 20% of plants infested OR >10% of plants with >25 mm of stem infested.

13.4.2.3.2 Lupins

- NSW: Treat at the first indication of virus-infected plants or first appearance of aphid clusters on flowering spikes.
- WA: >30% of inflorescences infested with 30 or more aphids; however, also take into account the aphid tolerance of the variety. Lucerne—100 aphids per 5 m sweep of an insect net for fresh regrowth and increasing to 400 aphids for older stands. For seedlings—one to three aphids per plant are usually worth spraying (Robinson and Hoffmann, 2001).

13.4.3 Lucerne Flea

13.4.3.1 *Sminthurus viridis* (Sminthuridae: Collembola)

13.4.3.1.1 Introduction

Lucerne flea *Sminthurus viridis* is a pest of winter seedling crops and pastures.

13.4.3.1.2 Description

Flea color is variable and often green or yellow with a white stripe down their back. The body is stout and globular, and soft, 2–4 mm long and 2–4 mm wide, with no wings. Mouthparts are chewing, and antennae have four segments and legs with five to six segments. Head: primitive eyes with eight ocelli and thorax with three segments. The abdomen is globular and has six segments. Has a springtail for jumping on segment 4. Eggs are spherical, pale, and smooth, and the nymph passes through 5–6 instars 0.5 mm long when born. They look like a small version of the adult. As habits, they jump up to 30 cm high when approached (Bailey, 2007).

13.4.3.1.3 Distribution

They are a significant pest in southern New South Wales, South Australia, Victoria, Tasmania, and Western Australia and are most prevalent in areas with a

Mediterranean climate and winter rainfall. They only occur on soils with significant clay content.

13.4.3.1.4 Life Cycle

The insects over-summer in the egg stage, hatch with the winter season break, and pass through three to nine nymphal stages. The adults emerge within three to five weeks. Females lay several clusters of eggs in the soil and hatch in a few days. There are three to four generations from autumn to late spring. Hot weather in early summer often kills the adults, while their population peaks in autumn, falls in winter, and peaks again in spring.

13.4.3.1.5 Host Plants

Lucerne flea is a severe pest of lucerne, legume pastures, capeweed, legume crops, cereals, and ryegrass. Heavy infestations will result in seedling deaths. Broadleaf seedlings are most susceptible (canola, lupins, faba beans, field peas, clovers, lucerne). Grass species (wheat, oats, barley, and pasture grasses) are less preferred hosts, but they will damage seedlings (Bell and Willoughby, 2010).

13.4.3.1.6 Damage

Soft tissue on the leaves' underside is usually the young nymph's feeding site, which results in transparent windows. Adults and older nymphs usually chew the leaves and make irregular holes in them, which can completely defoliate plants (Bell and Willoughby, 2010). Mature plants are retarded as the leaf material is removed, leaving the leaf as a white film or leaving holes or windows in the leaf. Damage is often patchy and usually on the heavier soil types (Robinson and Hoffmann, 2001).

13.4.3.1.7 Monitoring

Monitoring is essential for damage levels and the number of insects at least once weekly in crops, every two weeks in the pasture, for the first three to five weeks after emergence. Because lucerne flea distribution is patchy, check more sites for other species and examine foliage for damage and soil surface where insects may shelter (Bell and Willoughby, 2010).

13.4.3.1.8 Biological

Predatory mites are among the beneficial agents and provide some control. *Bdellodes lapidaria* was introduced as a biocontrol agent. Pasture snout mite (*Bdellodes lapidaria*), spiny snout mite (*Neomulgus capillatus*), and French anystis mite *Anystis wallacei* are species of predatory mites that feed on lucerne flea and provide effective suppression. Some field experiments show 70–90% control of lucerne fleas by predatory mites. They are slow to spread and can only do so by crawling (Bell and Willoughby, 2010).

13.4.3.1.9 Cultural

Check the lucerne flea risk to establishing crops and pasture and their severity in the paddock or on the farm (Bell and Willoughby, 2010; Taverner et al., 1996). Lucerne flea feeds on a range of broad-leaved weeds, so control of alternative hosts is essential for managing the lucerne flea population. Minimize weeds of fallows and pastures and control them around the crop edges. Shorter pastures and lower relative humidity will increase insect mortality and limit food resources (Taverner et al., 1996). Lucerne flea prefers lucerne and clover, so strip cropping with lucerne/clover may lower lucerne flea risk in other crops.

13.4.3.1.10 Chemical

Most of the registered insecticides are not effective against lucerne flea, so their impact on the predatory mite species should be checked in the pasture (Bell and Willoughby, 2010; Taverner et al., 1996). Systemic and contact insecticides are used for control. On sandy soils, this pest rarely occurs, while control is often practiced on heavier soils. Lucerne flea has developed resistance against synthetic pyrethroid (Robinson and Hoffmann, 2001). Spray timing for controlling the first generation in a crop or pasture will reduce the paddock's subsequent population growth. A border spray will control invasion from adjacent fields, while extreme infestations can be controlled with spot spraying, preventing population build-up and spread through the field. Spraying of chemicals in the late season can reduce the number of insects following autumn (Bell and Willoughby, 2010; Taverner et al., 1996).

13.4.3.1.11 Economic Threshold

If more than 50 fleas per 5 m sweep on an insect net are present, chemical control is recommended. The economic threshold for seedling stands is one to two fleas on each seedling in young canola spray when more than ten holes per leaf are found. Damage by the lucerne flea appears as window-like holes in the leaf (Robinson and Hoffmann, 2001).

13.5 MAIN PESTS OF PASTURES AND TURF

13.5.1 Red-Legged Earth Mite

13.5.1.1 *Halotydeus destructor* (Penthaleidae: Acarina)

13.5.1.1.1 Introduction

The red-legged earth mite (RLEM) (*Halotydeus destructor*) originates from South Africa and is a significant pest of pastures, crops, and vegetables that occur during cool, wet winters and hot, dry summers (Child, 1991).

13.5.1.1.2 Description

An adult red-legged earth mite is 1 mm long and 0.6 mm wide with eight red-orange legs and a completely black velvety body. When the eggs hatch, the mites have a pinkish-orange body with six legs, and they are only 0.2 mm long, which cannot be seen with the naked eye. The larva has three nymphal stages in which the mites have eight legs and resemble the adult mite but are smaller and sexually undeveloped (Child, 1991).

RLEM has different feeding habits, generally feeding in large groups of up to 30 individuals, different from other species that tend to feed singularly (Child, 1991).

13.5.1.1.3 Distribution

The RLEM is widespread throughout most agricultural regions of different countries. They are found in southern NSW, on the east coast of Tasmania, the south-east of SA, the south-west of WA, and throughout Victoria. The adult RLEM only moves short distances between plants in winter. However, recent surveys have shown that they can expand their movement over long distances. The movement of eggs in the soil through livestock and farm machinery or transportation of plant material is among the principal means of dispersal. Sometimes the wind also plays a vital role in disseminating the over-summering eggs (Child, 1991).

13.5.1.1.4 Life Cycle

The earth mites are active from autumn to spring and can produce drought-resistant over-summering eggs in spring that remain dormant following autumn (Giachino, 2005). The mites have three to four generations per year which survive six to eight weeks. The over-summering eggs are normally oviposited on the leaves' lover side or the soil surface around the host. A combination of low temperatures and moisture breaks the dormancy of the eggs. The mites grow for three to eight weeks and produce eggs with no dormancy that hatch almost immediately and pass through four nymphal stages. Depending on the conditions, each nymph lasts for one to two weeks. The first generation takes much longer to mature than the second (Liu et al., 1996). The female can produce up to 100 winter eggs that usually hatch in about eight to ten days. The RLEM reproduces sexually, which occurs when the male RLEM produces webbing, usually on the soil's surface. It then deposits spermatophores on the threads of this webbing, which the female mite picks up to fertilize her eggs (Child, 1991).

13.5.1.1.5 Host Plants

Pasture legumes, subterranean and other clovers, medics, and lucerne are the primary hosts of red-legged earth mites. They feed on ryegrass and young cereal crops, especially oats. They also feed on several weed species, including Patersons' curse, skeleton weed, variegated thistle, ox-tongue, smooth cats' ear, and capeweed (Child, 1991).

13.5.1.1.6 Damage

The RLEM spends 90% of its time on the soil surface; that is why they are called earth mites and rarely found on plants' foliage. The mites feed on the foliage for a shorter time and then move to another feeding site (Child, 1991). It feeds on the upper surface of leaves using its sharp chelicerae to pierce the surface and feed on the exuded plant sap (Liu et al., 1996). It causes more damage than any other pasture pest, and as a result, the cell and cuticle are desiccated, and photosynthesis is retarded (Child, 1991). If the pest is established, the losses of dry matter or seed yield can reach 10 to 80% in pastures. They also can be found on a wide range of vegetables and broadleaf crops such as canola, peas, and lupins (Liu et al., 1996). RLEM are most damaging to newly establishing pastures and emerging crops, significantly reducing seedling survival and development (Arthur et al., 2014).

13.5.1.1.7 Monitoring

Inspection of the plants is critical, especially susceptible pastures and crops from autumn to spring, for the presence of mites and damage symptoms. The crops must be regularly checked in the first to fifth weeks after sowing. The mites will be found easily at the base of plants during the warmer part of the day, sheltering under debris or leaf sheaths. To avoid the heat and cold, they crawl into cracks in the ground and seek shelter. Commonly, using pesticides controls the mites; however, because of evidence of resistance and concern, long-term sustainability non-chemical options are becoming increasingly important (Child, 1991).

13.5.1.1.8 Biological

Currently, 29 species of predatory mites feed on red-legged earth mites in Australia; scientists are looking for more effective bio-control agents (Liu et al., 1996). Although small beetles, spiders, and ants play a role in reducing the mite population, other mites are recognized as the most important predator of red-legged mites. A predatory mite (*Anystis wallacei*) with slow dispersal and establishment rates has been introduced as a means of biological control. Natural enemies residing in roadside vegetation and windbreaks demonstrated that they have an important role in suppressing red-legged earth mites in adjacent pasture paddocks (Child, 1991).

13.5.1.1.9 Cultural

To decrease the need for chemical control, we have to apply cultural control. Rotating crops can reduce pest colonization, reproduction, and survival. Use non-preferred crops such as weed-free lentils for crop rotation to reduce the risk for subsequent years. Increase seeding rates to compensate in case of seedling losses (Child, 1991). Plant a less susceptible crop like wheat/chickpeas. Choose selective and appropriate insecticide for seed treatment based on the resistance level. Avoid applying post-sowing pre-emergent insecticide to control red-legged earth mite (Chen et al., 1967; Kibedi, 2016). It is possible to reduce the red-legged earth mite population below damaging thresholds level with appropriate grazing management. Shorter pasture results in lower relative humidity will increase mite mortality and limit food resources (Child, 1991).

13.5.1.1.10 Chemical

Protect seedlings with a systemic insecticide, but in severe attacks, contact insecticide can be used. Systemic insecticides are preferred when the plants have emerged more than 60%. Timing is critical for long-term control, as insecticides have a more negligible effect on eggs (Liu et al., 1996).

13.5.1.1.11 Autumn Sprays

The first generation of mites must be controlled before it has the chance to lay eggs because it is the only effective way to avoid the need for a second spay. Use pesticides after sowing and apply within three weeks of mites' first appearance before female adults begin to lay eggs. Timing of chemical application is critical. Foliage sprays are applied once the crop has emerged, which is an effective method of control for minimizing the damage to plants during the sensitive establishment phase (Liu et al., 1996).

13.5.1.1.12 Spring Sprays

Dense pastures require higher rates of application during springtime than autumn to emerging pasture. The mite number and effect will reduce by heavy grazing (Liu et al., 1996). Applying spray is recommended when necessary; otherwise, it will create population resistance to the strategy. Using selective rotation of products with different modes of action will prevent resistance development (Liu et al., 1996).

13.5.1.1.13 Economic Thresholds

13.5.1.1.13.1 Canola Cotyledon: mite damage (silvering) affects 20%+ of plants and mites are present one per true leaf: >10 mites/plant two true leaves: when plant numbers are low <30/m^2 and mites are present on more than three true leaves: only when plants are under severe stress and >2000 mites/m^2 (Chen et al., 1967; Kibedi, 2016).

13.5.1.1.13.2 Other Crops The economic threshold for other crops is reported as follows:

Wheat/barley: 50 mites/m^2
Linseed: 100 mites/m^2
Pulses: 50 mites/m^2
Establishing medic pasture: 20 to 30 mites/m

The sampling time is significant; in this case, try to sample in the early morning or late afternoon and avoid sampling during bright light or after rain. Take samples around the plants' base and multiply by 10 to get the mites/m^2 (Liu et al., 1996).

13.5.2 Black Field Cricket

13.5.2.1 *Teleogryllus commodus*

(Gryllidae: Orthoptera)

13.5.2.1.1 Introduction

Telegryllus commodus is a cricket species native to Australia. They are significant pests to most plants in Australia and New Zealand (Chen et al., 1967). It is commonly known as the black field cricket, and they feed on the leaves and stems of seedlings (Pavri, 2007).

13.5.2.1.2 Description

An adult black field cricket is 2.5 cm long and 1 cm wide. The body and wings of adults are chestnut-brown and 30 mm long. The cricket has a large and black head with very long antennae. The wings, when folded back, form a point that extends beyond the body (Pavri, 2007). Eggs are laid singly, white, ovoid-shaped, and about 0.3 cm long but loosely clustered, about 1 cm deep in damp soil. The adults' wings are black or brown, and the head and mouthparts are inclined downwards with large hind legs modified for jumping. The nymphs are similar in shape but are smaller, paler, and wingless. Small nymphs have a white band across their back (Pavri, 2007). More than 30 segments are present in the antennae. When they are approaching, they stop singing or clicking sounds. They have strong rear legs for jumping. The pronotum of the thorax has a strong shield. The abdomen has 11 segments, and the spiracles are located on segments 1 to 8 (Child, 1991).

13.5.2.1.3 Distribution

The black field cricket originated in Australia and was then introduced to New Zealand, and they are known to cause significant damage to pasture and gardens and are common in cracking soils (Chen et al., 1967). It is also widespread in Australia. It is found throughout the North Island and milder coastal regions of the South Island, but usually the species is only economically important in Northland, Auckland, Waikato, Bay of Plenty, Hawkes Bay, and Taranki Manawatu.

13.5.2.1.4 Host Plants

The black field cricket damages different field crops, including pulses, and when they are present in huge numbers, they will cause damage to cotton as well (Pavri, 2007).

13.5.2.1.5 Life Cycle

The female lays eggs from February through to May. Eggs overwinter and hatch in spring. Nymphs grow quickly, passing through 12 instar periods to become adults. Individual females lay 500 to 1500 eggs within two to three months. Adults appear from February to May (Pavri, 2007). The black field cricket feeds on dry pasture or early germinated plants. They will not survive during the cold weather of winter. Eggs are inserted at 1–2 cm depth in the soil in grassy areas during autumn (Child, 1991). Adults and nymphs hide during the day in cracks in the soil or under trash and come out during the night and feed on grasses, weeds, or crops (Pavri, 2007).

13.5.2.1.6 Damage

Attacking black field crickets is dependent on plant and climate conditions; in the pasture, attacking the growing crowns of grasses in dry periods often kills the plants, leaving the soil open to weed invasion. However, later in the season, seedlings, whether self-sown or in the reseeded pasture, might also be destroyed. However, the black field cricket is a selective feeder that prefers grasses and legumes, weeds, and seeds. Pasture is very susceptible to cricket damage in their first year (Pavri, 2007).

- Adults and nymphs will damage leaves, stems, and pods.
- All stages of the plant will be attacked by crickets.
- Heavier soil crops have higher risk.
- At podding, adults chew into pods to reach the seeds.

Black field crickets will feed on seedlings, the back of flower heads, and the sunflower's maturing seeds.

13.5.2.1.7 Economic Threshold

The threshold level of black field cricket is 5–10/m^2 to control them, and the sampling time is early summer when nymphs are present (Pavri, 2007).

13.5.2.1.8 Monitoring

Inspect crops at sunset when black field crickets are active. To monitor activity, use light traps and take action when a high population is present (Pavri, 2007). Control is not economic, but in drought years, when feed is sparse and numbers tend to build up, insecticide application is possible but usually requires destocking of the pasture for a period. We must remember that early treatment will prevent damage and the laying of eggs for next season's infestation (Child, 1991).

13.5.2.1.9 Biological

Natural enemies (predators, pathogens, and parasitoids) have little effect on black field crickets (Pavri, 2007). The natural enemies of *T. commodus* include bats and ground beetles. The black field cricket will survive if it is targeted by predators but avoids detection. *T. commodus* mainly relies on hearing when avoiding predators, which is affected by their surroundings (Chen et al., 1967; Kibedi, 2016).

13.5.2.1.10 Cultural

Sanitation or good farm hygiene will help control crickets, as weedy fields before planting may encourage them.

13.5.2.1.11 Economic Threshold

The economic threshold level for black field crickets is ten crickets per m^2; a population of more than 10/m^2 will cause economic damage.

13.5.2.1.12 Chemical

Applying pesticides such as Malathion 50 EC and Yates Madison 50 combined with bran or wheat bait will control effectively. Crickets will cannibalize dead crickets, which may increase the uptake of insecticide from the dead insect. When cricket populations are low, we can prevent more egg laying by poisoning in early summer (Pavri, 2007). Use insecticide-treated cracked-grain baits.

13.6 MAIN PESTS OF LEGUME

13.6.1 Redheaded Cockchafer

13.6.1.1 *Adoryphorus coulonii* (Scarabaeidae: Coleoptera)

13.6.1.1.1 Introduction

Pastures and some cereal crops are attacked by redheaded cockchafers, which are important pests. They belong to the scarab family and are the larvae of native cockchafer beetles. The redheaded cockchafer (*Adoryphorus coulonii*) is considered an Australian scarab beetle in the genus *Adoryphorus*. The species is regarded as a pasture pest in Victoria, New South Wales, South Australia, and Tasmania. The larvae feed on the roots of plants and other decaying soil material (Pavri, 2007).

13.6.1.1.2 Description

The adults are about 13–15 mm long and 8 mm wide, and typically brown-black. Larvae are soft, whitish c-shaped grubs with three yellowish legs and a hard, reddish-brown head capsule. Newly hatched larvae are only 5 mm long, but mature grubs reach up to 30 mm in length. The larva is white-gray in the early stages of growth and becomes white when mature. The redheaded cockchafer has a life cycle of two years, with the larval stage completed mostly underground. The older larva has yellowish legs and a hard red-brown head. The adult beetles emerge from the soil from late winter to early spring and fly at dusk to mate before the females then lay their eggs into the soil (Pavri, 2007).

13.6.1.1.3 Distribution

Redheaded pasture cockchafers are a significant agricultural pest that is native to south-eastern Australia. They can be found in south-west and central Victoria, northern Tasmania, south-eastern South Australia, and the southern tablelands of New South Wales, appearing to be problematic where the annual rainfall exceeds about 500 mm. Although typically found in higher-rainfall areas, they tend to occur in higher numbers and are more of a problem in drier years (Pavri, 2007).

13.6.1.1.4 Life Cycle

The redheaded pasture cockchafer completes its life cycle in two years, and the entire life cycle occurs below the soil surface except for limited crawling on the ground and flight activity by the adults. Female beetles lay eggs at night at a depth of up to 80 mm in the soil. Each female may lay up to 25 eggs in her lifetime. The eggs hatch in late spring, six to eight weeks after oviposition, and the young grubs have three stages. The first two instars pass quickly and are not so dangerous. However, the final (third) stage is the most damaging in autumn and feeds for almost ten months. Feeding is intense during the autumn but is interrupted by cold weather in June (Pavri, 2007). During late summer to early autumn, the adults emerge from pupae in the soil but remain deep in the soil until late winter or early spring. They remain at this stage until early the following summer. When they are about a year old, larvae move deeper into the soil and pupate around December. Next-generation adults emerge from the pupae around the end of January, remaining in the soil until early next spring. After a brief flight period, they return to the pasture and burrow into the soil to mate and lay eggs (Pavri, 2007).

13.6.1.1.5 Host Plants

Pastures, wheat, ryegrass, and pastures with high clover content are among the important hosts of this pest, which feeds on roots at the top 10 cm of soil. Deep-rooted perennial plants such as lucerne, cocksfoot, and phalaris are less susceptible to damage (Pavri, 2007).

13.6.1.1.6 Damage

Most of the damage is caused by the grub that feeds on roots and humus in the root zone, usually within 50 mm of the soil surface. Increased soil moisture in autumn stimulates larvae to move closer to the soil surface to feed on plant roots. Larvae prune or completely sever roots, with damaged plants sometimes dying or showing signs of reduced growth. Damage is typically most serious from March to June. If the soil temperature goes down, their feeding activity slows down. This pest's economic threshold level is the existence of approximately 70 per m^2 in March, and populations have been known to reach 1000 per m^2 (Pavri, 2007).

13.6.1.1.7 Monitoring

Pastures are monitored from late March until June. Inspect susceptible paddocks before sowing by digging to a depth of 10–20 cm with a spade and counting the number of larvae present. This should be repeated 10–20 times to get an estimate of larval numbers. In existing pastures, management practices must be integrated and aimed at limiting damage as much as possible (Pavri, 2007).

13.6.1.1.8 Biological

Several natural enemies such as birds, parasitic wasps, and flies are the most effective. Birds prey on larvae and are most valuable after cultivation. *Metarhizum* spp. is pathogenic fungi that attacks and reduces cockchafer populations. *Heterorhabitis zealandica* is an entomopathogenic nematode used to control this pest in turf and nurseries (Pavri, 2007).

13.6.1.1.9 Cultural

Increasing the grazing in spring will make paddocks less favorable for adult females to lay eggs. Cultivation of tolerant pasture species such as phalaris, cocksfoot, tall fescue, lucerne, or less palatable crops such as oats will significantly reduce the pest problem. Resowing the affected areas with a higher seeding rate will assist plant establishment. Delay re-sowing until cockchafer activity ceases. Rolling damp but not too wet pastures can be used to reestablish contact of the roots with the soil, and larvae close to the soil surface can be killed (Pavri, 2007).

13.6.1.1.10 Chemical

As the most damaging third instar larva will live underground and will not be affected by foliar applications of insecticides, there are currently no synthetic insecticides registered for control of redheaded pasture cockchafers (Pavri, 2007).

13.6.1.1.11 Economic Threshold

If four larvae per spade square are equivalent to 100 larvae per m^2, there is a need for the control measures. In older pastures, 250–500 grubs per square meter (more than 15 per spade square) can sever most roots (Pavri, 2007).

13.7 MAIN PESTS OF LUCERNE

13.7.1 Pointed Snails

13.7.1.1 *Prietocella acuta*

Cochlicella acuta

(Helicidae: Eupulmonata)

13.7.1.1.1 Introduction

Pointed snails prefer dead organic matter as a food source and have rarely been recorded feeding on crops or pasture. They are a contaminant of grain, particularly barley. They will be found under stones and stumps, on posts and vegetation, in cereal stubble, and at the plant's base in the ground (Bailey, 2007).

13.7.1.1.2 Description

The body length of the pointed snail is 12–18 mm with a shaper cone. The body color is grayish brown with brown bands of varying widths. A pointed snail body is a dark, soft, and slimy body enclosed within a hard conical shell and no wings (Micic et al., 2007). The mature snail shell has an 18 mm length, and the ratio of shell length to its diameter is always >2 (Bailey, 2007).

(Provided by Welter Schultes, Francisco.)

13.7.1.1.3 Distribution

The highest numbers of pointed snail are found on the Yorke Peninsula in South Australia, and isolated populations are also found in coastal areas of SA, New South Wales, and Western Australia and inland areas of Victoria. Calcareous and highly alkaline soils retain stubble, and no tillage is particularly favorable for *Cochlicella acuta*. Moist seasonal conditions favor conical snails (Bailey, 2007).

13.7.1.1.4 Life Cycle

Pointed snails avoid high temperatures; for this reason, they seek shelter under plants or debris during the day, and over summer, they aestivate to debris, crops, up to 50 mm under the soil, under stumps or stones, or they climb vegetation or posts. The pointed snail becomes active after the autumn rains, and 1–2 mm of rain activates feeding. Mating will occur two to three weeks after good autumn rains and lower temperatures, and egg-laying begins soon after mating. The pointed snail female will lay eggs in clusters in the topsoil from autumn to spring, and eggs hatch about two weeks after laying. The juveniles feed in winter and spring and aestivate over summer to become sexually mature at one year old (Micic et al., 2007).

13.7.1.1.5 Damage

The pointed snail mainly feeds on dead organic matter; they will not directly damage crops. They are known a pest typically at harvest as a contaminant of grain, mainly barley (Bailey, 2007).

13.7.1.1.6 Monitor

Pointed snail feeding activity must be monitored early in the season. Focus on areas where the pointed snail population occurs in the greatest densities, such as limestone ridges close to the soil surface (Bailey, 2007). Season-wise monitoring is necessary for different purposes, for example, in summer for stubble management options; autumn for burning, cultivating, and baiting; winter for re-baiting and refuge treatment; and spring for grain contamination. To remove refuges, graze or burn stubble (Micic et al., 2007).

13.7.1.1.7 Biological

A parasitic fly, *Sarcophaga pencillata*, was introduced into South Australia from Europe during 2001–3, which targets conical or pointed snails. However, to date, no evidence report suggests this has impacted the snail population (Bailey, 2007).

13.7.1.1.8 Cultural

As lucerne is a very uncompetitive crop during early growth stages, light infestations will be removed in the first cut and smothered by the regrowth, but it depends upon successful initial establishment of the population. Deep cultivation of at least 5 cm to bury pointed snails will reduce the snail population 40–60%. Early harvest in swathed crops, as snails will migrate to the swath with time and increase grain contamination likelihood (Micic et al., 2007). For crop hygiene purposes, a rotation of five years is directed between lucerne crops. Upon stability of the stand, the crops may be down for three to five years.

13.7.1.1.9 Chemical

Apply chemicals to control the snail that minimize damage to biological control agents such as ground beetles. Using liming paddocks should be avoided because that helps snail survival. As many snail populations are relatively immobile juveniles and sample alternative feed, spring baiting is often ineffective (Micic et al., 2007).

13.7.1.1.10 Economic Threshold

According to Micic et al., 2007 for finding the snail number per square meter, we have to count the number of snails in 10 quadrants that are 32 × 32 cm in the field, and control is worthwhile if the snails are present more than $20/m^{-2}$ in oilseeds, $40/m^{-2}$ in cereals, or $100/m^{-2}$ in pastures (Micic et al., 2007).

13.7.2 Spotted Alfalfa Aphid (*Therioaphis trifolii Therioaphis maculate*)

13.7.2.1 Introduction

The spotted alfalfa aphid, *Therioaphis maculate*, was found in the United States in 1953 and Alberta in 1979. *Therioaphis maculate* is more severe on alfalfa than the pea aphid because, besides sucking out the juice, it injects a toxin into the plant, too.

13.7.2.2 Description

The aphid adults are about 2 mm long, with a pale yellow-green color and have four to six rows of tiny black spots running lengthwise on their backs, which are just visible to the naked eye. The adults will be winged or wingless, but the winged form has smoky-colored areas along the wing veins (Bailey, 2007). The nymphs are similar to adults but are smaller in size and have no wings. Two biotypes of spotted alfalfa aphid are present in Australia. These two biotypes are morphologically indistinguishable but genetically distinct. The biotype predominantly attacks lucerne and medics, and the other predominately attacks clovers (Bailey, 2007).

13.7.2.3 Distribution

Therioaphis trifolii was introduced to Australia in 1977, and *Therioaphis maculata* (Buckton) was introduced into central New Mexico in 1954 and spread throughout the alfalfa-growing areas of the United States. Spotted alfalfa aphids are recorded as a minor pest of lucerne and clover pastures in Queensland, New South Wales, Victoria, South Australia, Tasmania, and Western Australia. The aphids are active during spring and summer but are most abundant in autumn (Bailey, 2007).

13.7.2.4 Life Cycle

Most individuals are wingless females that can produce living young without mating. Each female gives birth to more than 100 offspring, and the aphid goes through four instars in five days. The entire life cycle of the spotted alfalfa aphid will complete in about seven days. In the fall, males can produce that mate with oviparous females, and the eggs laid from this sexual cycle are the overwintering stage.

Spotted alfalfa aphids thrive in warm, dry conditions and are most active during late spring, summer, and autumn, but they can survive cooler weather. Spotted alfalfa aphids have 20 to 40 generations per year. A temperature range of 26–29°C is optimum for spotted alfalfa aphid development. An early autumn break tends to be a favorite for these aphids (Bailey, 2007).

13.7.2.5 Host Plants

Spotted alfalfa aphids are restricted to pasture legumes but predominantly attack lucerne, annual medics, and subterranean white and red clovers and other legumes as well (Bailey, 2007).

13.7.2.6 Damage

Spotted alfalfa aphids prefer to feed on alfalfa but also infest some clovers. An early symptom of aphid damage to alfalfa appears as vein clearing of newly emerging leaves. With continuing aphid feeding, veins may whiten, and the remaining areas of the leaf will become yellow, and finally, leaves die and fall from the plant. Overall, they are found on lower leaves' undersides and move up the plant as these leaves are killed.

13.7.2.7 Direct Feeding Damage

Usually, aphids suck the sap from the phloem of leaves and stems and inject a salivary toxin which causes damage to the plant and yellowing and stunting of entire mature plants. The aphid damage will proceed from the plant's base upwards until only stems remain standing, and the plant will die. In high infestations, excretion of honeydew will cause secondary fungal growth (black sooty mold), which constrains photosynthesis and causes a decrease in plant growth and adaptability to stock (Bailey, 2007).

13.7.2.8 Indirect Damage (Virus Transmission)

Indirect damage of aphids occurs by spreading plant viruses. Spotted alfalfa aphids spread viruses between plants by feeding and searching as they move between plants and paddocks. Imported viruses transmitted by spotted alfalfa aphids include alfalfa mosaic virus (AMV) in medic and lucerne pastures and bean yellow mosaic virus in clover (Bailey, 2007).

13.7.2.9 Monitor

The use of resistant varieties and border harvesting or strip cutting can be important for preserving and encouraging natural enemies to control spotted alfalfa aphids. The aphids' distribution is very patchy; hence at least five sampling points of 20 plants each should be inspected over the paddock. Aphids generally colonize the underside of leaves, so this area should be visually searched. Aphids are dislodged after rain, so monitoring after overhead irrigation and rain is not advised (Bailey, 2007).

13.7.2.10 Biological

Natural enemies such as predators, parasites, and fungi are important for controlling the aphids in alfalfa. As predatory insects, ladybird beetles and nabid bugs, *Nabis alternatus*, play a significant role in aphid control. Many other insects also prey on aphids, including the green lacewing, *Chrysopa oculata*; the minute pirate bug, *Orius tristicolor*; the big-eyed bug, *Geocoris bullatus*; and larvae of the syrphid flies, *Scaeva pyrastri* and *Eupeodes volucris*. An *Entomophthora* fungus and the parasite *Trioxys complanatus* are also important in controlling the aphid in the United States but are not found in Canada (Bailey, 2007). The aphid population will be affected by fungal disease (*Pandora neoaphidis* and *Conidiobolyus obscurus*), which is naturally occurring (Bailey, 2007).

13.7.2.11 Cultural

Cultivars which are less susceptible to spotted alfalfa aphids should be selected. Varieties of clovers and medics are available which are tolerant to spotted alfalfa aphids. Heavy grazing or cutting stands temporarily destroys aphid populations and may decrease the need for a spray. Allowing for a few uncut strips will help the survival of beneficial insects (Bailey, 2007).

13.7.2.12 Chemical

Several insecticides registered for controlling aphids in lucerne and pastures are available. Primicarb, compared with other broad-spectrum chemicals, has little impact on beneficial insects and has effective control on spotted alfalfa aphid. When aphids begin to move into crops, border spray in late summer and autumn will provide sufficient control without spraying the entire paddock. Insecticide recommendations will change every year.

According to Bailey (2007), when damage is apparent, different factors to consider include the degree of aphid tolerance of the cultivar, availability of moisture, and incidence of predators and parasitoids before deciding whether chemical control is necessary. In general, chemical control is possibly only cost effective for seed crops. In the presence of spotted alfalfa aphids in drought conditions, using resistant varieties is the only way to establish a stand.

13.7.2.13 Economic Threshold

The time of applying insecticide and plant heights is significant; for example, on seedling, an insecticide should be applied when two to three aphids appear per seedling, or plants with 2 to 3 inches height will tolerate four to five aphids per plant. However, alfalfa with 10 inches height will not need treatment until there are 50 aphids per stem, and on 20-inch tall alfalfa, twice as many aphids per stem would be needed to justify treatment (Bailey, 2007).

Spotted alfalfa aphid treatment thresholds:

Spring months: 40 aphids per stem
Summer months: 20 aphids per stem
After last cutting in the fall: 50 to 70 aphids per stem
Newly seeded alfalfa in lower desert: 20 aphids per stem

13.7.3 Lucerne Leafhopper

13.7.3.1 *Austroasca alfalfa*

Cicadellidae

13.7.3.1.1 Introduction

The Lucerne leafhopper is a moderate pest. Although it has widespread abundance, the vegetable leafhopper (jassid) is only a minor pest (Bailey and Goodyer, 2007).

13.7.3.1.2 Description

Adult leafhoppers are approximately 2–3mm in length and vary in color depending on species and life stage. They may be yellow, green, brown, or dark gray and typically elongated and wedge shaped. Wings are held roof-like over the body at rest. They are very active and jump readily or move sideways when disturbed. Nymphs are usually smaller in size and wingless (Berg et al., 2014; Bellati et al., 2012).

13.7.3.1.3 Life Cycle

Adults are wedge-shaped, yellow or yellowish-green, and about 3 mm long. Nymphs resemble adults but are wingless and smaller. Do not confuse these with the green or bright green vegetable leafhopper, which is about 3 mm long and seldom warrants concern. It produces a white stipple pattern on the leaves when it feeds (Bailey and Goodyer, 2007).

13.7.3.1.4 Host Plants

Leafhoppers attack various crops, including lucerne, legumes, and cereals (Bailey and Goodyer, 2007).

13.7.3.1.5 Damage

Leafhoppers suck sap, resulting in damage of fine pale dots in a patterned 'wriggly' or 'zigzag' line (Bailey and Goodyer, 2007).

- Both adults and nymphs are sapsuckers.
- Lucerne leafhoppers are phloem feeders who inject toxins, causing leaves to turn yellow and burn off (die) from the tips. These symptoms are called 'hopper burn.'
- Vegetable leafhoppers are xylem feeders, and their feeding causes lots of small white spots (from dead cells) on leaves. The resultant feeding patterns are known as 'stippling.'
- Severe infestations of the lucerne jassid can stunt plant growth and reduce yield, but even high vegetable jassid populations have negligible if any effect on yield.
- Damage is worse when plants are stressed.

13.7.3.1.6 Monitor

Collect samples weekly at vegetative, flowering, and pegging stages. Sample five leaves halfway up plants at six locations over the lucerne paddock (Bailey and Goodyer, 2007).

13.7.3.1.7 Biological

Predatory bugs and spiders will attack leafhoppers. Unnecessary sprays for leafhoppers will adversely affect these and other beneficial insects and trigger other pests such as *Heliothis* (*Helicoverpa*) (Bailey and Goodyer, 2007).

13.7.3.1.8 Cultural

Well-watered, vigorously growing crops can tolerate damage.

13.7.3.1.9 Chemical

As a guide, do spray if there are 20 or more adults and nymphs per sweep of a 35-cm-diameter net. Early cutting or grazing may be an acceptable alternative to spraying if hay stand is approaching maturity but could induce movement into nearby crops (Bailey and Goodyer, 2007).

13.7.3.1.10 Economic Threshold

Mainly a sporadic pest of grazing and hay stands in summer and autumn. Infestation promoted by hot, dry conditions and moisture-stressed, unirrigated stands are usually most at risk (Bailey and Goodyer, 2007).

13.7.4 Lucerne Leafhopper

13.7.4.1 *Austroasca alfalfa*

Cicadellidae: Hemiptera

13.7.4.1.1 Introduction

Austroasca alfalfa has a widespread abundance, and it is known as an actual pest, while vegetable leafhopper is a minor pest (Bailey and Goodyer, 2007).

13.7.4.1.2 Description

The leafhopper adults are around 2–3 mm in length with various colors depending on species and life cycle stage. The color varies from yellow, green, and brown to dark gray, and they are typically elongated and wedge shaped. During rest, the wings are held roof-like over the body. Leafhoppers are very active and jump eagerly or move sideways when bothered. Nymphs are smaller in size and usually wingless (Berg et al., 2014; Bellati et al., 2012).

13.7.4.1.3 Life Cycle

The adult is 3 mm long, wedge shaped, and yellow or yellowish-green in color. Nymphs are wingless and smaller but resemble adults. It can be confused with vegetable leafhopper, which has a green or bright green color and is about 3 mm long but seldom warrants concern and produces a white stipple pattern on the leaves when it feeds (Bailey and Goodyer, 2007).

13.7.4.1.4 Host Plants

Many crops will be damaged by attacks or leafhopper such as lucerne, legumes, and cereals (Bailey and Goodyer, 2007).

13.7.4.1.5 Damage

Leafhoppers cause damage by sucking the sap, and the symptom appears fine pale dots in a patterned 'wriggly' or 'zigzag' line, which has been observed within the affected crops (Bailey and Goodyer, 2007).

- Both the adults and nymphs cause damage by sucking the plants' sap.
- Lucerne leafhopper is a phloem feeder; besides sucking sap, the leafhopper injects a toxin, causing a yellowish color, and the plant finally burns off from the tips, known as 'hopper burn.'
- In contrast to lucerne leafhopper, vegetable leafhoppers are xylem feeders that cause lots of white spots from dead cells on leaves and the feeding patterns called 'stippling.'
- In case of a severe infestation of the lucerne jassid, plant growth will be stunted and reduce yield, while a high vegetable jassid population has little if any effect on yield.
- Stressed plants are more susceptible, and damage caused will be worse.

13.7.4.1.6 Monitor

Collect samples at different plant growth stages such as the vegetative, flowering, and pegging stage weekly. Lucerne sampling should be five leaves from the middle up from plants at six locations over the lucerne paddock (Bailey and Goodyer, 2007).

13.7.4.1.7 Biological

Predatory bugs and spiders will prey on leafhoppers, but nonselective sprays for leafhopper that adversely affect beneficial insects should be avoided because it may trigger other pests, *Heliothis* (*Helicoverpa*) (Bailey and Goodyer, 2007).

13.7.4.1.8 Cultural

Wells, management of irrigation, and vigorously growing crops can tolerate damage.

13.7.4.1.9 Chemical

Chemical spray will be necessary when there are 20 or more adults and nymphs per sweep of a 35-cm-diameter net. Early cutting or grazing may be an acceptable alternative to spraying if the hay stand is approaching maturity but could induce movement into nearby crops (Bailey and Goodyer, 2007).

13.7.4.1.10 Economic Threshold

As mentioned in a guide, lucerne leafhoppers' economic threshold is 20 or more lucerne leafhoppers per sweep of a 35-cm-diameter sweep net. Vegetable leafhoppers over 100 per sweep may require treatment (Bailey and Goodyer, 2007).

13.7.5 Lucerne Leaf Roller

13.7.5.1 (*Merophyas divulsana*)

(Tortricidae: Lepidoptera)

13.7.5.1.1 Introduction

Merophyas divulsana is a moth species, belongs to the family of Tortricidae, is known as the lucerne leaf roller, and is found throughout Australia and stippled in New Zealand (Bailey and Goodyer, 2007).

13.7.5.1.2 Description

Mature larvae are up to 15 mm long with yellowish-green to green color, and young larvae are pale with dark heads. Disturbed larvae will drop from plants on silken threads. Lucerne leaf roller moths are 8 mm long and yellow to light brown, with dark markings on the forewings (Bailey and Goodyer, 2007). The color of the caterpillar is yellowish-green to green with distinctively darker colored heads, up to 15 mm long. Antennae have three segments, filiform and hair-like. Lucerne leaf roller legs are broad overlapping scales. Tarsi have five segments. The head is smooth, with broad overlapping scales. The compound eyes are rounded and prominent. The thorax has three segments of broad overlapping scales. The front segment is considerably smaller. The hairy abdomen is broad overlapping scales with 7–11 segments. Spiracles are located on segments 1–7. The eggs are flat and oval, with pale yellow forewings (Bailey and Goodyer, 2007). The larvae bind leaves together at the growing tips of plant stems. To make shelter, several leaves are bent over and joined with silk.

13.7.5.1.3 Distribution

Leaf roller caterpillars usually feed on lucerne leaves and can completely defoliate a stand in a month. Leaves and tips are rolled over and joined with silk to form a shelter for the caterpillar. The larvae attack various crops and hence are considered a pest on various crops and herbaceous garden plants, including *Medicago sativa*, *Daucus carota*, *Lactuca sativa*, *Lonicera japonica*, *Sclerolaena muricata*, *Mentha spicata*, and *Rumex* species (Bailey and Goodyer, 2007).

13.7.5.1.4 Life Cycle

The females lay eggs on the upper side of leaves overall the year. They cause damage before flowering. The leaf roller has many generations per year (Bailey and Goodyer, 2007).

13.7.5.1.5 Host Plants

Leaf roller mainly causes damage on lucerne and is known as a significant pest of lucerne. It also causes damage to a range of other legumes, including clover and soybeans. The leaf roller's activity or impact is mostly restricted to the summer and autumn periods (Bailey and Goodyer, 2007).

13.7.5.1.6 Damage

Larvae web and roll leaves at the shoot tip and skeletonize the leaves from within the roll. Leaf rollers cause severe damage to the plant at the near flowering stage. Crops that are cut or grazed in spring, summer, and autumn are also at risk. The leaf roller caterpillar feeds on the terminal leaves, and flowering stems and heavily infested plants become stunted, significantly reducing the yield and quality of the hay. Crops in moisture-stressed and dryland conditions are most at risk (Bailey and Goodyer, 2007).

13.7.5.1.7 Monitor

Estimations of the number of stems on which regrowth terminals are rolled are necessary for controlling lucerne leaf rollers. Spraying is advisable in grazing or hay stand when about 30% of the terminals are rolled in the first half of regrowth. However, in general, it is not necessary to spray for lucerne leaf roller unless seed production is endangered. Early cutting or grazing reduces the economic damage caused by lucerne leaf rollers (Bailey and Goodyer, 2007).

13.7.5.1.8 Biological

Natural enemies such as the fungal pathogen *Zoophthora radicans* (*Erynia radicans*) and a nuclear polyhedrosis virus, plus the parasitoids *Apanteles tasmanica* and *Voriella uniseta*, are reported as effective in delaying or preventing economic damage (Bailey and Goodyer, 2007).

13.7.5.1.9 Cultural

Early cutting or grazing can be an acceptable alternative while the stand is approaching maturity. Cutting or grazing at four to five week intervals can keep leaf roller numbers low when eggs or tiny larvae are present. Considering the optimum soil moisture level is important (Bailey and Goodyer, 2007).

13.7.5.1.10 Chemical

When leaf roller numbers build up and seed production is endangered, or 30% of lucerne terminals are rolled, the standard chemical application is necessary.

13.7.5.1.11 Economic Threshold

The lucerne leaf roller is a sporadic pest of grazing or hay crops in spring, summer, and autumn. Heavily infested plants become stunted, but the yield and quality of hay are seriously affected and harshly reduce the pod set of seed crops. Damage to moisture-stressed unirrigated stands is often significant (Bailey and Goodyer, 2007).

13.7.6 Small Lucerne Weevil

13.7.6.1 *Atrichonotus taeniatulus*

(Coleoptera: Curculionidae)

13.7.6.1.1 Introduction

Besides the beetle, weevils are known as a common pest of grain crops in Australia. The characteristic appearance of weevils will remove confusion with other beetles, but distinguishing between the many species of weevil is challenging.

13.7.6.1.2 Description

Larvae are not similar to adults; adult weevils appear very unlike larvae. Adults have a hardened body, six elongated and prominent legs, and a downward curved head forming a 'snout.' Small lucerne weevil larvae are legless, have a maggot-like shape, and can be confused with fly larvae. However, weevil larvae possess a small, hardened head capsule. The adult weevil color is gray with some brownish mottling, and it is up to 10 mm long; larvae are creamy-white, up to 8 mm long, with small pointed brown jaws, and they live in the soil (Bailey, 2007).

13.7.6.1.3 Distribution

Small lucerne weevils are similar to flies, and they have to walk or be carried to spread; that is why infestation of small lucerne weevils is slow.

13.7.6.1.4 Life Cycle

Small lucerne weevils are 10 mm long; the color is gray with some brownish mottling. Larvae are 8 mm long, creamy white, legless grubs. During the day, small lucerne weevils typically hide in the soil around the base of plants; however, some will be found resting together in a group of three to four on a single leaf (Howes, 1990). Adults emerge in mid-February–March and lay eggs at the base of plants. In winter, these eggs hatch, and the larvae burrow into the soil and begin feeding on plant roots (Howes, 1990).

13.7.6.1.5 Host Plants

The small lucerne weevil causes damage mainly on canola, lucerne, and pasture legumes but is not considered a pest of cereals. The small lucerne weevil is a minor, irregular pest in New South Wales (Howes, 1990).

13.7.6.1.6 Damage

The small lucerne weevil attacks vegetative parts of crop plants such as roots, stems, shoots, buds, and leaves. As the small lucerne weevil cannot fly, infestations spread slowly; they have to walk or be carried to spread. Adults and larvae can cause damage to plants, but it also depends on the species, crop type, and time of year. Weevil feeding damage appears as scallop-shaped holes along the edge of leaves their distribution within paddocks (Howes, 1990). The adults feed on the cotyledon and leaves of plants, seedlings can be chewed entirely off at ground level, and in lucerne, damage is also caused by larvae when they burrow into or chew furrows in the taproot, which causes plant death and bare patches (Bailey, 2007). They usually feed at night and hide in the soil during the day (Howes, 1990).

13.7.6.1.7 Monitor

Monitoring the small lucerne weevil adult should be done after dark; the adult emerges during late spring and early summer.

13.7.6.1.8 Biological

No known natural enemies.

13.7.6.1.9 Cultural

Minimum tillage and stubble retention are usually favored for small lucerne weevils. We can reduce the population of weevils by cultivation, burning, and reducing the amount of stubble habitat for weevils. Rotations will be adequate to break the weevil life cycle if all lucerne and tap-rooted plants are killed. Use break crops such as cereals (with dryland lucerne) or horticultural crops re-sown. Be careful to avoid carrying adult weevils in the hay, farm machinery, and vehicles from infested areas to clean lucerne. All it takes to start an infestation is a single female (Howes, 1990).

13.7.6.1.10 Chemical

As the tiny lucerne weevil larvae are protected in the soil, insecticide cannot kill them. Spray will be effective when the adult emerges in February–March if applied twice. However, the first spray should be applied one week after the first weevil appears and followed up two to three weeks later. To prevent egg laying, timing is critical; sampling at night through February is the only way to find the first weevils. Chemical control of small lucerne weevils is difficult due to their secretive habits. Numerous species of weevil are patchy in their distribution within paddocks. For some species, seed treatments and foliar insecticides can provide a level of control (Howes, 1990).

13.7.6.1.11 Economic Threshold

Chemical control of small lucerne weevils will be cost effective when the adult numbers are high in February–March and at the time of germination.

REFERENCES

Anthony F (2005) *An Introductory Guide to Sources of Traditional Fodder and Forage and Usage*. Available at www.acbar.org/upload/1493193872857.pdf (Accessed 22 July 2020).

Arthur AL, Hoffmann AA, Umina PA (2014) Challenges in devising economic spray thresholds for a major pest of Australian canola, the redlegged earth mite (*Halotydeus destructor*). *Pest Management Science* 71: 1462–1470.

Bag J (2012) Preventing mouldy hay using propionate preservatives. *Field Crop News.* Available at http://fieldcropnews.com/2012/06/preventing-mouldy-hay-using-propionic-acid (Accessed 19 July 2020).

Bailey PT (2007) *Pests of Field Crops and Pastures: Identification and Control.* Australia: CSIRO Publishing.

Bailey PT and Goodyer G (2007) Lucerne leafhopper. In: Bailey PT (ed) *Pests of Field Crops and Pastures.* Australia: CSIRO Publishing, pp. 429–430.

Bell NL and Willoughby BE (2010) A review of the role of predatory mites in the biological control of Lucerne flea, *Sminthurus viridis* (L.) (Collembola: Sminthuridae) and their potential use in New Zealand. *Agricultural Research* 46: 141–146.

Bellati J, Mangano P, Umina P and Henry K (2012) *Insects of Southern Australian Broadacre Farming Systems Identification Manual and Education Resource.* Department of Agriculture and Food Western Australia.

Berg G, Faithfull IG, Powell KS, et al. (2014) Biology and management of the redheaded pasture cockchafer *Adoryphorus couloni* (Burmeister) (Scarabaeidae: Dynastinae) in Australia: A review of current knowledge. *Austral Entomology* 53: 144–158.

Brier (2007) Pulses—summer (Including peanuts). In: Bailey PT (ed) *Pests of Field Crops and Pastures.* Australia: CSIRO Publishing, pp. 69–175.

Capinera JL (2001) *Handbook of Vegetable Pests. Gulf Professional Publishing*, pp. 317–320. ISBN 978-0-12-158861-8.

Chaudhry HR and Mian MKS (2017) *Fodder Production.* Univ. coll. of Vet. & Ani. Sci. The Islamia Univ. of Bahawalpur.

Chen G, Vickery VR and Kevan DK (1967) A morphological comparison of antipodean *Teleogryllus* species. *Canadian Journal of Zoology* 45: 1215–1224.

Child JC (1991) *The Insects of Australia.* CSIRO. *Melbourne University Press*, pp. 384–385.

Dick Z (2012) Global Survey: Feed Production Reaches Record of 873 Million Tons. Available at https://web.archive.org/web/20141013170912/www.allaboutfeed.net/Process-Management/Management/2012/3/Global-Survey-Feed-production-reaches-record-of-873million-tonnes-AAF012981W/ (Accessed 7 July 2020).

Dikshit TSS, Kumar SN, Raizada RB and Srivastava MK (1989) *Toxicology* 43: 691–696. Available at www.microbiologyresearch.org/docserver/fulltext/micro/135/10/mic-135102601.pdf?expires=1595656864&id=id&accname=guest&checksum=D33128B340058699278C692A3C1C0AC5 (Accessed 18 July 2020).

Elder RJ (2007) Pasture—summer rainfall. In: Bailey PT (ed) *Pests of Field Crops and Pastures.* Australia: CSIRO Publishing, pp. 355–356.

FAO (2015) Alternative Fodder Production for Vulnerable Herders in the West Bank. Available at www.fao.org/3/a-i4759e.pdf (Accessed 22 July 2020).

Gashaw M (2016) Review on mycotoxins in feeds: Implications to livestock and human health. *Journal of Agricultural Research* 5: 0137–0144.

Giachino PM (2005) Results of the zoological missions to Australia of the Regional Museum of Natural Science of Turin, Italy. *Monographs Museo Regionale di Scienze Naturali Torino* 42: 239–268.

Howes KMW (1990) *Sitona Weevil (Sitona discoideus). Insect and Allied Pests of Extensive Farming.* Department of Agriculture, Western Australia. Bulletin No. 4185.

John PJ, Bakore N and Bhatnagar P (2001) Assessment of organochlorine pesticide residue levels in dairy milk and buffalo milk from Jaipur City, Rajasthan, India. *Environment International* 26: 231–236.

Kibedi J (2016) *An Investigation on the Ecological Significance of the Terrestrial Context in Predator-Prey Interactions between Echolocating Bats and the Australian Field Cricket.* School of Biomedical Science, pp. 1–207.

Koli P and Bhardwaj NR (2018) Status and use of pesticides in forage crops in India. *Pesticide Science* 43: 225–232. Available at https://europepmc.org/backend/ptpmcrender.fcgi?accid=PMC6240772&blobtype=pdf (Accessed 18 July 2020).

Lean IJ, Bruss ML, Troutt HF, et al. (1994) Bovine ketosis and somatotrophin: Risk factors for ketosis and effects of ketosis on health and production. *Research in Veterinary Science* 57: 200–209.

Liu AT, Ridsdill-Smith J and Tanveer K (1996) Pulse vary in their susceptibility to redlegged earth mite. *Australian Grain.*

Martens H and Schweigel M (2000) Grass tetany and other hypomagnesaemias. In: Herd T (ed) *Veterinary Clinics of North America: Food Animal Practice: Metabolic Disorders of Ruminants.* Philadelphia: Saunders Company, pp. 339–368.

Micic S, Henry K and Horne P (2007) *Identification and Control of Pest Slugs and Snails for Broadacre Crops in Western Australia.* Department of Agriculture and Food Western Australia, Perth Bulletin 4713.

Miles MM, Baker GJ and Hawthorne W (2007) Pulses—winter. In: Bailey PT (ed) *Pests of Field Crops and Pastures.* Australia: CSIRO Publishing, pp. 259–260.

Muck RE and Kung Jr L (1997) Effects of silage additives on ensiling. In: *Silage: Field to Feedbunk.* Ithaca NY: Northeast Regional Agricultural Engineering Service, pp. 187–199.

Oetzel GR and Goff JP (2008) Milk fever (parturient paresis) in cows, ewes, and doe goats. In: Anderson DE and Rings DM (eds) *Current Veterinary Therapy 5: Food Animal Practice*, 5th ed. Philadelphia: Saunders Company, pp. 130–134.

Pavri C (2007) Legume pastures. In: Bailey PT (ed) *Pests of Field Crops and Pastures.* Australia: CSIRO Publishing, p 413.

Robinson MT and Hoffmann AA (2001) The pest status and distribution of three cryptic blue oat mite species (*Penthaleus* spp.) and redlegged earth mite (*Halotydeus destructor*) in south-eastern Australia. *Experimental and Applied Acarology* 9: 699–716.

Savoie P and Jofriet JC (2003) Silage storage. In: Buxton DR, Muck RE and Harrison JH (eds) *Silage Science and Technology.* USA: Agronomy Publication, pp. 405–467.

Sharma HR, Kaushik A and Kaushik CP (2012) Organochlorine pesticide residues in drinking water in the rural areas of Haryana. *Environmental Monitoring Assessment* 95: 69–81. Available at https://sci-hub.st/https://link.springer.com/article/10.1007/s10661-011-1950-9 (Accessed 22 July 2020).

Taverner PD, Hopkins DC and Henry KR (1996) A method for sampling Lucerne flea, *Sminthurus viridis* L. (Collembola: Sminthuridae), in annual medic pastures. *AJE* 35: 197–199.

Tripathi HP, Singh AP, Upadhyay VS, et al. (1995) In: Kiran S and Schiere JB (eds) *Handbook for Straw Feeding Systems.* New Delhi. India: ICAR.

Van Barneveld RJ (1999) Understanding the nutritional chemistry of lupin (*Lupinus* spp.) seed to improve livestock production efficiency. *Nutrition Research Reviews* 12: 203. Available at https://sci-hub.st/10.1079/095442299108728938 (Accessed 22 July 2020).

Weinberg ZG (2008) Preservation of forage crops by solid-state lactic acid fermentation-ensiling. In: Pandey AS and Larroche CRC (eds) *Current Developments in Solid-state Fermentation.* New York: Springer.

14 Toxins of Preserved Fodders

A Threat to Livestock

Asad Manzoor, Misbah Ijaz, Muhammad Tahir Mohyuddin, and Faiza Hassan

CONTENTS

DOI: 10.1201/9781003055365-14

14.1 FODDER PRESERVATION AND STORAGE

Depending on available resources and weather conditions, fodder can be preserved mainly in two forms, hay and silage. Hay is termed dried fodder, whereas silage is termed wet fodder. These fodder forms are fed to animals to fulfill demand in developing countries' fodder shortage periods.

14.2 REASONS FOR FODDER PRESERVATION

As a developing country, Pakistan has limited resources. Being an agriculture-based country, agriculture is a significant source of GDP, and in the year 2018–19, about 20.9% of the country's GDP was shared by the agriculture sector. If we further split this, the livestock sector contributed around 60.5% of this share. This livestock sector tries to fulfill the daily requirement of milk and meat of human beings and contributes to wool and hides. In return, the only need livestock has is the supply of good-quality ample fodder and water daily. However, the limited resources available and the cultivation of cash crops instead of fodder crops are the major factors in shortcoming this requirement.

Moreover, there are harsh weather conditions (December–January being too cold and May–June being too hot), which constitute a significant obstacle in fodder production. These periods are categorized as lean periods due to the shortage of fodder during these periods. Owing to owners' poor financial condition, feeding wheat straw or dried and matured sorghum stalks is practiced during these periods (Mehmood et al., 2020). Both these alternatives of fodder are low quality in terms of nutritional value and palatability. Hence, these factors are significant contributors to a decrease in animal production and reproduction around the year and ultimately decrease the number of small-scale farmers or household/backyard farming of livestock.

On the other hand, during early spring or during the country's monsoon season, leguminous plants grow very fast, and multiple cuts of these crops can be achieved during this period. These surplus fodders can then be preserved in hay or silage and then fed to animals during lean periods. It will fulfill animals' daily requirements and bridge the gap, and a considerable loss in terms of decreased milk production due to a shortage of fodder can be prevented by adopting this technique. Moreover, there is a general belief that preserved fodders' nutritional values and palatability, especially silage, are enriched compared to raw forage crops.

14.3 TYPES OF PRESERVED FODDERS

As mentioned earlier, there are two major types of preserved fodders:

1. Hay or dry fodder
2. Silage or wet fodder

14.3.1 Hay

Drying of the whole plant to biologically inactivate its microbiological decay and enzymatic activity is called haymaking, while the product is called hay. Mostly legumes, cereals, and grasses are used to make hay. The primary objective is to reduce water contents (up to approximately 15%) of the fodder, so harvesting at the proper stage and seasonal factors are considered significant during haymaking. There are three main types of hay:

1. Leguminous
2. Non-leguminous
3. Mixed

Based on haymaking technique, hay can be classified into

1. **Long hay**
   It is the most traditional type of hay, in which an entire plant with a comparatively long stem is harvested, dried, and stored.
2. **Chopped hay**
   Where mechanization is available, this technique suits well depending on drying conditions. Its handling is easy, as it is less bulky compared to long hay.
3. **Hand trussed hay**
   It is the most commonly used technique in manual haymaking. It usually decreases shattering.
4. **Wafered and pelleted hay**

Wafered hay is reffered to as chopped 3-inches diameter hay, whereas pelleted hay is termed when hay is ground with a hammer mill and pressed into a half inch pellet and about quarter inch long.

14.3.1.1 Techniques for Haymaking

There are generally two types of techniques used for haymaking, which are as follows.

14.3.1.1.1 Field Method It is the technique in which fodder crops are dried post-harvesting on the ground in an open area. It is further divided into two sub-techniques.

14.3.1.1.1.1 Swath The area of the entire field is used. The harvested crop simply lies in the field in as thin a layer as possible.

14.3.1.1.1.2 Windrows This technique uses almost one third of the area of the land. It is made by staking together one or more swaths.

Drying of the fodder is faster in the windrow technique compared to its counterpart. To achieve good-quality hay using field technique, harvesting of the crop is carried out after evaporation of dew from the crop. After harvesting, the crop is allowed to dry in the field by turning it every four to five hours. By adopting this technique, moisture contents are reduced to 40–75% by the end of the day if there is proper sunshine with low humidity. On the second day, only one to two turnings are required, and moisture contents fall by up to 25%. At this stage, hay is ready to be stored in the form of bales or on stands. This technique is not suitable during rainy seasons.

14.3.1.1.2 Mechanical Method

It is further divided into two techniques.

14.3.1.1.2.1 Fence Method In this technique, wire fences along with angled iron posts are used to dry the fodder crop. This technique is suitable to make hay using lucerne, berseem, legume fodders, and groundnut haulms. The advantage of this technique is that it reduces protein losses 2–3%.

14.3.1.1.2.2 Forced Air Batch Hot air is blown through the forage to dry it in a barn. This technique can be used during the rainy season or conditions with high humidity levels. Depending upon capabilities and available resources, approximately 1 ton of hay can be produced per day with an additional cost of approximately 60 rs/ton. Nevertheless, this technique has many advantages over other techniques in nutritional value and prevention of losses.

14.3.2 Silage

Preservation of green succulent roughage generating a controlled and strict anaerobic fermentation process by pressing chopped green fodder in water and airtight containers is called ensiling, and the product achieved is termed silage. The whole process of ensiling is based on the principle of conversion of sugars present in fodder being ensiled into lactic acid, hence leading to reduction of pH to 4.0 or lower. It ultimately leads to the inhibition of biological activities that decay or spoil fodder. The most basic or perhaps the only requirement to make good silage is creating and maintaining an anaerobic (oxygen-free) microenvironment as early as possible.

The anaerobic acids (lactic acid) produced during the ensiling usually halt bacterial and mold growth and the inactivation of putrefying organisms, acting as a preservative of the fodder. Due to these reasons, nutrient losses of the fodder are reduced. Ensiling is considered a better option for fodder preservation compared to haymaking.

Various factors on which one will decide that a crop is suitable for silage making include that dry matter of the crop should be 30–45%, along with the proportion of water-soluble sugars from 8–10%. Moreover, one will also look for a ratio between buffer capacity and water-soluble carbohydrates of the crop and the ratio of sugars to crude proteins. All these factors are essential, as they will decide the production of lactic acid.

While practicing ensiling, it should be evident that the fodder being used has a dry matter of 30–45%. Moreover, wrapping the material on all sides with polyethylene adds to the improvement in silage quality. Silo filling should be in uniform layers, with a layer of 20–30 cm thickness at a time, and it should preferably be carried out on a clear, bright, sunny day. Perfect compaction must be achieved, and for the removal of air pockets, trampling can be helpful. It is better to keep the top in a convex or dome shape while ensiling.

14.3.2.1 Process of Ensiling

14.3.2.1.1 Harvesting of the Crop

Good- to medium-quality silage is gained by using good-quality fodders and grasses with a dry matter content ranging from 15–35%. This dry matter range can be found in various crops, like at the dough stage in corn. This DM concentration is present at the emergence of the ear, flowering stage in sorghum, and most grasses in pearl millet. Similarly, harvesting oats at the milking stage generates this range of dry matter. Sometimes partial wilting of legumes is required to reduce water content up to 70%.

14.3.2.1.2 Preparation of the Silo

The silo should be thoroughly clean, and if the floor and side walls are not paved, then a layer of about 10 cm thickness of straw or waste fodder should be first laid down on the ground. A better alternative is to place the polythene layer with or without this layer mentioned previously. This polythene layer is also strongly recommended in paved or cemented floor settings to improve the silage quality. The chopped fodder filling should be carried out layer by layer quickly, where the thickness of the layer must not be greater than 30 cm. Proper compaction is necessary to evacuate entrapped air while filling the silos. For increasing the fermentation process in fodders with lower sugar contents (legumes, etc.), a sprinkling of dried or liquid molasses is recommended layer by layer. The entire silo should be filled like this.

14.3.2.1.3 Closing the Silo

After the proper filling and compaction of the silo, with a convex or dome shape at the top, the silo should be covered immediately with a layer of straw and or waste fodder, followed by a polythene layer. The polythene or plastic sheet should be of an appropriate thickness (250–275 microns) to avoid oxygen and water entry into the silage. Afterward, to keep these plastic sheets intact/in place, sufficient weight should be put on them. Putting a layer of mud on the plastic sheet can help to achieve good-quality silage. During this practice, cracks and folds should be properly removed to prevent air and water entry into the silo. The time taken to complete the fermentation process is four to five weeks, generally.

14.3.2.1.4 Opening the Silo

Removal of the cover in a proper way is an essential step in the maintenance of well-preserved silage. The plastic sheet should be removed from a small designated area of the silo by ensuring that the minimum possible silage area is exposed to the environment. Before taking the silage for feeding, the upper layer of silage should be discarded if moldy in appearance. Care should be taken during the feed-out phase that the remaining silage not be exposed to the external environment to avoid aerobic deterioration of the silage at the storage phase.

14.3.2.2 Properties of Good Silage

- It should be free from the growth of molds.
- It should have a pleasing aroma, that is, a pleasantly fruity odor.
- Well-preserved silage is a yellowish-green color.

- The texture of the silage should be non-sticky and free flowering.
- It should have an increased (3–4%) palatability compared to raw fodder.
- It must have an enhanced nutritive value.
- It must have a pH value ranging from 4.0 to 4.5.
- The proportion of lactic acid should be higher compared to other acids present.
- There should be an increased value of ammoniacal-nitrogen and decreased nitrate-nitrogen. However, the proportion of former must not be higher than 15% of the silage's total nitrogen contents.

14.3.2.3 Advantages of Ensiling

- Increased palatability compared to raw fodder and hay.
- Increased nutritive value compared to raw fodder and hay.
- More nutrients per area unit are available for animal feeding if the fodder is harvested at the proper stage, that is, pre-flowering.
- Losses like shattering, leaching, and bleaching are experienced during haymaking and can be prevented.
- It is less affected by environmental factors like rainfall if proper covering or filling of silos was carried out.

14.3.2.4 Disadvantages of Ensiling

- More labor is required for the filling of silos.
- The initial investment is required, such as for the construction of silos.
- Due to lower dry matter content, transportation and handling are more laborious as compared to hay.
- Slight sloppiness can lead to huge losses due to aerobic deterioration of the silage.

14.4 TOXINS OF PRESERVED FODDERS

As far as haymaking is concerned, this method has been gradually replaced by silage making due to more benefits and advantages of silage than hay. The major toxins found in hay are plant toxins, which are discussed in detail in the next section.

14.5 TOXINS COMMONLY FOUND IN SILAGE

Many factors influence the quality of silage, but the major ones are the forage composition and suitable silage-making technique. Maintenance of anoxic conditions and attaining a swift low pH by lactic acid fermentation are considered cornerstones in silage preservation (Pahlow et al., 2003). Biological and chemical silage additives can achieve these. There is a crucial role of lactic acid bacteria in the ensiling process. These bacteria can convert fermentable carbohydrates found in fodder into lactic acid majorly and acetic acid to a lesser degree. Well-preserved and managed silage is thought to be excellent feed for animals with less or no harm to animals or human beings. Instead, bacteria present in such silage are considered to have a prebiotic effect on animals (Driehuis et al., 2018).

A problem arises when there is some breach in the making or management of silage, which leads to faulty production of silage or detrimental effects on silage quality. Many potential threats can arise from such silage and can have harmful effects on silage quality. When fed to animals, this low-quality silage will result in deleterious effects on animal health and production and threaten human beings' health, that is, end-users. We can broadly categorize these threats into;

1. Microbial hazards
2. Chemical hazards
3. Plant hazards

14.5.1 Microbial Hazards

As described earlier, low pH and anoxic conditions are essential principles of ensiling, but insufficient reduction in silage pH or oxygen availability leads to the development of unwanted bacteria in the silage. These unwanted bacteria reduce silage quality by decreasing its nutritional value (e.g., yeast and butyric acid bacteria) and pose a threat to animal and human health (e.g., *Bacillus cereus, Escherichia coli, Clostridium botulinum, Listeria monocytogenes*, molds, etc.). The potential hazards may be these microbes or metabolites produced by these like various endotoxins, exotoxins, and mycotoxins.

14.5.1.1 *Bacillus cereus* (*B. cereus*)

Bacilli and paenibacilli are aerobic or facultative anaerobic spore-forming bacteria. *B. licheniformis*, *B. pumilus*, *B. coagulans*, *B. sphaericus*, and *B. cereus* are primarily associated with silage. *B. cereus*, being a food-borne pathogen in human beings, is of particular concern. The spores of *B. cereus* may come into milk through the soil, silage, and other feeds and bedding materials. In various studies, soil and sawdust bedding for animals kept indoors are reported as a significant source of *B. cereus* spores as contaminants of milk (Christiansson et al., 1999; Magnusson et al., 2007). Depending on soil type, the season of sampling, and sampling site, the concentration of *B. cereus* ranges from 10^1 to 10^6 spores/g (Driehuis, 2013). Due to contamination of silage by soil, *B. cereus* spores in silage are considered inevitable. In a study in the Netherlands during 2007, Vissers and coworkers reported an average concentration of 3×10^2 spores/g in mixed silage of grass and corn fed to dairy cows. They concluded silage was a significant source of raw milk contamination with *B. cereus* spores. Surface layers and moldy silage areas are the sites where high concentrations (10^8 to 10^9 spores/g) of aerobic spore-forming bacteria can be found (Pahlow et al., 2003). In a well-managed and adequately acidified silage, growth of *B. cereus* is not expected, as pH below 4.6 is supposed to inhibit bacterium growth (Browne and Dowds, 2002; Biesta-Peters et al., 2010). However, in areas of aerobic deterioration of silage where pH is elevated above 4.6, growth of *B. cereus* can occur.

The spores of *B. cereus* present in silage and other feeds, when ingested by animals, pass unchanged from the gastrointestinal tract and are shed through fecal material. These spores could be transferred to raw milk during milking operations if proper hygienic measures are not adopted. *B. cereus* is categorized as a significant organism spoiling pasteurized milk and its products when stored in refrigeration settings (Heyndrickx and Scheldeman, 2002). At temperatures as low as 5°C, there are chances that spores of psychrotrophic *B. cereus* may germinate into bacterium even in pasteurized milk and milk products. These products' shelf life is limited by spores of psychotropic *B. cereus*, as higher concentrations result in curdling and off-flavors. *B. cereus* spores can even be present in powdered dairy products, including nonfat milk powder, sweet whey powder, and whey protein concentrate (Miller et al., 2015).

14.5.1.2 *Clostridium botulinum* (*C. botulinum*)

Clostridium species are a class of bacteria that are Gram-positive, obligately anaerobic, and spore-forming in nature. The requirements for the growth of clostridia include a high (more than 4.5) pH value, high (0.952–0.971) water activity, and high (70%) forage moisture contents. Therefore, if the pH of silage is reduced to a level below 4 in three days through rapid acidification, clostridial growth can be inhibited in silage (Muck et al., 2003). There is variation in critical pH for bacterial growth inhibition depending on the crop's moisture content. Hence, clostridial growth can be favored by the factors that predispose the plant to a higher moisture content at the time of harvest or that delay decrease in pH of silage (Driehuis et al., 2018). There are two types of clostridial bacteria present in silage, non-pathogenic and pathogenic. In the non-pathogenic class, major ones are *C. beijerinckii, C. tyrobutyricum, C. sporogenes,* and *C. butyricum*, while the pathogenic class includes *C. difficile, C. perfringens, C. botulinum*, and *C. tetani* (Driehuis, 2013; Doyle et al., 2015). Among these pathogenic clostridia, *C. botulinum* is usually associated with silage. *C. botulinum* is a heterogeneous group of bacteria that can produce a neurotoxin termed botulinum. Botulinum is believed to be the most potent toxin and protein toxin in nature. There are four discrete groups of *C. botulinum*, Groups I, II, III, and IV. *C. botulinum* Groups I and II usually cause botulism in human beings, whereas Group III is thought to be responsible for botulism in animals and birds (Carter and Peck, 2015). There are seven different types of botulinum, identified so far as A, B, C, D, E, F, and G. Botulism in animals is usually caused by botulinum type C and D and occasionally by type B, whereas the remaining types and type B are responsible for botulism in human beings (Carter and Peck, 2015; Table 14.1 adapted from Driehuis et al., 2018). There is also a significant difference in phenotypic characters of different groups of *C. botulinum*. These phenotypic characters include minimum and optimum temperature for growth, carbohydrate metabolism, proteolytic activity, and spore heat resistance.

A higher fatality rate (5–10%) in humans and more than 10% in animals make botulism a severe disease in both (Galey et al., 2000; Sobel et al., 2004; Payne et al., 2011). Ingestion of botulinum-contaminated feed or water results in intoxication in cattle, termed botulism. Several studies have reported silage as the sole reason for botulinum toxin intoxication in cattle (Myllykoski et al., 2009; Payne et al., 2011;

TABLE 14.1
Classification and Properties of Different Groups of *Clostridium botulinum*

Item	Group I	II	III	IV
Botulinum toxin type	A, B, F, H[1]	B, E, F	C, D	G
Associated with botulism in	Humans	Humans	Animals	–
Proteolysis	+	–	–	+
Optimum growth temperature	37°C	30°C	40°C	37°C
Minimum growth temperature	12°C	3°C	15°C	
Spore heat resistance[2]	D121°C = 0.21 min	D82.2°C = 2.4 min	D104°C = 0.1–0.9 min	D104°C = 0.8–1.1 min

1 More than one toxin type may be formed.
2 D-value (decimal reduction time) at a specified temperature (i.e., time to a 10-fold reduction in viable spores).

Kummel et al., 2012). The clinical signs exhibited by affected animals are muscular weakness, collapse, paralysis, downed cows, uncoordinated stumbling walk with knuckling, constipation, and flaccid tail paralysis. Also, the animal cannot eat and or drink due to paralysis of the face, tongue, jaw, shallow and abdominal breathing, drooling, drooping eyelids due to paralysis of eyelids, and many other associated signs.

It is very well established that there is a shallow concentration (1 spore/L) of *C. botulinum* spores in unaffected cows' raw milk. Even though the routine pasteurization process (72–75°C for 15 seconds) does not inactivate the *C. botulinum* spores present in milk, it is not very common to see the multiplication of *C. botulinum* in such products (Anderson et al., 2011). In a study in 2010, Lindstrom and coworkers cautioned that infant dairy products (milk powder, dried whey proteins) need special attention because there is a potential risk of infant botulism caused by spores of *C. botulinum* (Lindstrom et al., 2010). It was endorsed by a previous report in the UK during 2001, where *C. botulinum* spores (1–4 spores/kg) found in a dried infant formula resulted in infant botulism (Brett et al., 2005).

The spores of *C. botulinum* can be transmitted to milk during milking operations similar to other spore-forming bacteria. As it is evident that teats and other parts of the animals' skin are soiled by spores of these potential health hazards already present in the soil, feed, or bedding, proper hygienic measures during milking are required. Several studies report that these methods are inefficient in removing 100% of contaminants, as only a fraction of these spores is scrubbed/washed off (Magnusson et al., 2006; Vissers et al., 2007).

14.5.1.3 Shiga Toxin-Producing *Escherichia coli*

One of the crucial etiologies of food-borne diseases is shiga toxin-producing *Escherichia coli* (STEC). There might be a mild infection to a more complicated one resulting in hemolytic uremic syndrome, hemorrhagic colitis, or end-stage renal disease (Majowicz et al., 2014). *E. coli* is a member of the Enterobacteriaceae family, a typical family-like Gram-negative and facultative anaerobic one. Most *E. coli* are non-pathogenic and are constituent of the normal intestinal microbiota of animals and humans. They are also part of the normal epiphytic microflora of forage crops.

The major serotype, O157:H7, causes STEC infection in human beings, followed by O103, O26, O145, and O111 (Gould et al., 2013). The main natural reservoir of STEC is cattle. Despite shedding pathogens in their feces, cattle are not affected by nor do they show signs of STEC infection (Callaway et al., 2006).

There is a difference in opinion on the effect of ensiling on STEC. Some studies report that natural or experimental contamination of raw fodder with STEC, specially O157:H7, when tested after ensiling did not yield growth of the pathogen (Byrne et al., 2002; Bach et al., 2002; Pedroso et al., 2010; Duniere et al., 2011; Ogunade et al., 2016; Ogunade et al., 2017). Contrary to this, some other studies report that STEC or non-STEC *E. coli* survive and multiply after ensiling (Fenlon and Wilson, 2000; Chen et al., 2014; Ogunade et al., 2016). This survival is supposed to be due to poorly fermented, poorly processed silage, where a slower decline in pH was achieved. Another possible reason for the survival of STEC in silage is a relatively high (more than 5) pH (Pedroso et al., 2010; Duniere et al., 2011; Ogunade et al., 2016; Ogunade et al., 2017). It can be concluded from these reports that STEC contamination of silage is not out of the question. It can occur during phases where chances of aerobic deterioration are higher, such as during feed-out. During these phases, air entry results in significant STEC growth, contributing to organisms' cycling in farm premises. Contamination of raw milk is assumed to be mainly through feces, but alternatively, it can also be shed in milk when the animal experiences *E. coli* mastitis.

14.5.1.4 *Mycobacterium bovis (M. bovis)*

Per FAO, 2015, due to its adverse effects on animal production and human health, bovine tuberculosis has gained a place in problems of global concern. The reason for this is its zoonotic nature and a rapidly increasing incidence rate. It has been proved that *M. bovis* can survive in the form of cyst or dormant stage during ensiling (Phillips, 2005). A pH as low as 5 inhibits pathogen growth, but a linear increase in growth rate is observed at pH 7.0 or above. It is clear from this report that aerobic deterioration of silage resulting in a higher pH encourages *M. bovis* (Rao et al., 2001).

Diseased wildlife like badgers, *Meles meles*, and so on shed *M. bovis*, which can contaminate forage crops like grass, grain-bearing maize, and whole cereal crops in the field before harvesting or at the feed troughs and silo feed-out face (Garnett et al., 2002). Hence, it is necessary to take practical steps or adopt biosecurity measures to prevent wildlife entry to feeding troughs and exposed silo feed-out faces to prevent the spread of this pathogen of serious concern.

In affected cattle, no signs are seen in the early stages of the disease, while later on, the animal shows emaciation, weakness, lethargy, anorexia, low-grade fever, pneumonia, moist and chronic cough, and swelling of the palatable lymph nodes.

14.5.1.5 *Listeria monocytogenese* (*L. monocytogenese*)

Humans and animals can suffer from a severe systemic disease called listeriosis caused by *L. monocytogenese*. A higher mortality rate, the severity of the disease, and increased numbers of cases categorize *L. monocytogenese* as a pathogen of serious concern globally (EFSA and ECDC, 2016).

There are two forms of listeriosis in humans, a non-invasive illness affecting the gastrointestinal tract or invasive disease. Typical signs of mild gastroenteritis like vomiting, diarrhea, and fever are observed in non-invasive forms, usually in healthy individuals. Neonates and adults suffering from underlying immunodepression in old age, pregnancy, and immunosuppression are mostly affected by invasive disease forms. There are very few outbreaks of invasive listeriosis, but these outbreaks significantly impact the high (20–30%) case fatality rate (Lomonaco et al., 2015). *L. monocytogenese* is ranked as the food-borne pathogen causing the highest mortality per year in developed countries (Scallan et al., 2011).

Listeriosis is a major food-borne illness in ruminants where silage is fed as the primary feed source. The result of these infections can be encephalitis and uterine infections. Furthermore, abortion in the late term of pregnancy can occur due to this uterine infection. Moreover, direct contact with silage may lead to an infection caused by *L. monocytogenese*, termed 'silage eye' (Erdogan, 2010).

L. monocytogenese is a Gram-positive, facultative anaerobe widely present in the environment. Various sources can yield its presence, such as surface water, soil, vegetative material, animals, and human beings' feces. It can adapt to a wide range of environmental conditions and grow in temperatures ranging from 0–45°C, pH ranging from 4.3 to 9.6, and salt concentrations up to 12% (Gandhi and Chikindas, 2007; Van der Veen et al., 2008). The bacterium can survive for an extended period in milieus where it cannot grow, such as in well-preserved silage. The degree of anaerobiosis and variation in pH are significant factors for *L. monocytogenese* in silage. At the time of ensiling, poor compaction or entry of air into the silage during fermentation contributes to pathogen survival. Similarly, aerobic deterioration of the silage is also linked with the survival of the pathogen. Vilar and coworkers (2007) reported that a rise in pH results in increased *L. monocytogenese* in the silage.

There is a causal relationship between feeding poor-quality silage to cattle, sheep, and goats and listeriosis (Ho et al., 2007). *L. monocytogenese* quickly passes through the animal's gastrointestinal tract by surviving therein and is shed in the feces not only by diseased animals but also by asymptomatic ones (Unnerstad et al., 2000; Nightingale et al., 2004; Vilar et al., 2007).

There is a direct relationship between the occurrence of *L. monocytogenese* at higher levels in silage and contamination of raw milk of dairy animals (Tasci et al., 2010). As with other bacteria, *L. monocytogenese* is also transferred to raw milk during milking operations. Occasionally, it comes directly into the milk animal if the animal is suffering from *L. monocytogenese* mastitis. It is mostly found in an insufficient number in raw milk and almost negative in processed milk, as due to its heat-susceptible nature, *L. monocytogenese* is supposed to be killed/inactivated during pasteurization (Van Kessel et al., 2011; Marshall et al., 2016).

14.5.1.6 Molds

Two types of molds can contaminate silage, molds resulting in crop infestation during the pre-harvesting or growth phase and molds which grow in poorly managed silage. Examples of the first category include *Alternaria* and *Fusarium* species, endophytic symbionts of grasses, and cereals like *Neotyphodium* and *Claviceps* species. *Penicillium paneum* and *Penicillium roqueforti*, *Byssochlamus nivea*, *Aspergillus fumigatus*, *Rhizopus nigricans Chrysonilia sitophila*, and *Monascus ruber* are major molds of the second category (CAST, 2003; Driehuis, 2013). Aerobic deterioration of silage during the storage or feed-out phase usually results in the development of silage molds mentioned previously. All these listed molds can produce mycotoxins except *Chrysonilia sitophila* and *Rhizopus nigricans* (CAST, 2003; Driehuis, 2013).

Mycotoxins can be present anywhere worldwide, as proved by a global survey conducted by Schatzmayr and Streit (2013). They found 72% of samples taken from feed and feed commodities had detectable levels of one or more than one mycotoxin, including deoxynivalenol, aflatoxins, ochratoxins A zearalenone, and fumonisins. There are two major concerns regarding the presence of these mycotoxins in feed. The first is their potential hazard to animal health and production, leading to significant economic losses, while the second is related to human health, as these are potentially hazardous to human health. Aflatoxin is of significant concern for human health among the mycotoxins present in silage.

Aspergillus fumigatus is ranked as a significant health hazard issue due to two main reasons. First, its mycotoxins produced in silage pose severe threats for health and production of animals, and second, its spores are a severe threat not only for animals but also to human beings, as inhalation of these spores results in severe respiratory problem termed aspergillosis (Driehuis et al., 2018; Ogunade et al., 2018). *Aspergillus fumigatus*–contaminated silage results in bovine aspergillosis. In human beings suffering from asthma or cystic fibrosis of the lungs leading to altered lung function, a hypersensitive response is produced by *Aspergillus fumigatus*, and this disease is termed 'farmer's lung' (Driehuis et al., 2018). *Aspergillus fumigatus* has also been listed as a risk factor for hemorrhagic bowel syndrome in cattle by Puntenney et al., 2002.

14.5.1.6.1 Mycotoxins Present in Silage

The frequently found mycotoxins in silage are fumonisins, trichothecenes, ZEA, aflatoxins, roquefortine C and mycophenolic acids (Driehuis et al., 2008a; Schmidt et al., 2015).

14.5.1.6.1.1 Fumonisins

Several *Fusarium* species produce fumonisins (Bennett and Klich, 2003). It is now well established that more than 28 forms of fumonisins are designated as A, B, C, and P-series (Ogunade et al., 2018). Among these, fumonisin B and B-series, notably B_1, are categorized as most toxic with particular reference to contamination of feed (Schmaile and Munkvold, 2009). *Fusarium proliferatum* and *Fusarium verticilliodes* are the significant sources of B_1 fumonisin. These are among the significant reasons for a prevalent and vital disease of corn termed '*Fusarium* corn rot' (Whitlow and Hagler, 2005; Gxasheka et al., 2015). The main predisposing factor for the secretion of fumonisins from *Fusarium* includes dry and hot periods followed by humid environment episodes and insects (de Wolf et al., 2005). Per reports of Rodrigues and Naehrer, 2012, with a prevalence rate of 64% and concentration averaging at 1965 ug/kg in a study conducted in Asia, Europe, and the Americas, fumonisins are perhaps the most frequent health hazard to livestock.

As fumonisins are structurally similar to sphingosine (which is a significant sphingolipid present in high amounts in nervine tissue; Michaelson et al., 2016), these toxins lead to leucoencephalomalacia by interrupting biosynthesis of sphingolipids, and this condition in horses is termed 'moldy corn poisoning.' When fed to calves, fumonisins lead to nephrotoxicity at a level ranging up to 1000 ug/kg body weight. These also result in a decrease in milk production in dairy animals. Although the secretion of fumonisin B_1 in milk is negligible, it still poses a severe public health threat, as it is potentially carcinogenic for human beings (Loi et al., 2017).

14.5.1.6.1.2 Trichothecenes

There are many trichothecene toxins, like T-2 toxins, HT-2 toxins, T-2 triol, T-2 tetraol, seripentril, deoxynivalenol (DON), nivalenol, and acetated DON, and so on produced by several mold spp. (Marin et al., 2013).

14.5.1.6.1.2.1 Deoxynivalenol

Fusarium nivale, Fusarium graminearum, Fusarium culmorum, Fusarium roseum, Fusarium poae, and *Fusarium tricinetum* generally produce DON. Its production is generally favored by wet and cold periods followed by dry shorter periods (De Wolf et al., 2005).

The previously reported study of Rodrigues and Naehrer (2012) found its prevalence around 59% and its level ranging up to 1104 ug/kg. Various studies report its worldwide presence in ensiled fodder at a much higher level (Driehuis et al., 2008b; Storm et al., 2008; Gallo et al., 2015; Kosicki et al., 2016).

It generally causes vomiting, diarrhea, reproductive problems, and death in simple-stomach animals. It is generally considered less hazardous to ruminants, as some ruminal microflora actively degrade and cause the DON into de-epoxy DON, which is considered less toxic or nontoxic (Marczuk et al., 2012). It generally does not affect milk production but can result in reduced reproductive performance of the animal.

14.5.1.6.1.2.2 Zearalenone

Several *Fusarium* species like *F. roseum, F. croockwellenese, F. graminearum*, and *F. culmorum* produce an estrogenic metabolite called Zearalenone (ZEA; Saeger et al., 2003). Corn crops with high moisture contents, moldy hay, and pelleted feeds are natural hosts of *F. graminearum*. In grain crops like barley, corn, and sorghum, occurrence of ZEA can be detected in low concentration along with DON. The factors favoring its growth include alternating lower to moderate temperature (11–14°C and then 27°C, respectively) and higher humidity (De Wolf et al., 2005).

ZEA is structurally similar to estrogen, enabling the toxin to mimic estrogen (Saeger et al., 2003). Several reproductive problems, including hyper-estroginism, mammary gland enlargement, and vaginitis, can result (Marczuk et al., 2012). Ruminants are considered less susceptible to ZEA toxicity, as ruminal flora can convert ZEA into its alpha and beta metabolites. Despite the higher receptor affinity of alpha metabolites, its conversion to beta form in the liver and low absorption rate makes it less toxic (Malekinejad et al., 2006; Fink-Gremmels, 2008). It can cause a reduction in feed and water intake, decreased milk production, diarrhea, and reproductive failure in dairy animals.

14.5.1.6.1.3 Aspergillus Toxins

14.5.1.6.1.3.1 Aflatoxins

These are a group of mutagenic and carcinogenic toxic substances produced by several Aspergillus species, including *A. nominus, A. flavus*, and *A. parasiticus* (Bennett and Klich, 2003). The ability of Aspergillus to grow over a vast range of humidity and temperature results in the occurrence of natural and metabolite forms of aflatoxins in food and feed ingredients (Tulayakul et al., 2005). The most common natural forms of aflatoxins are B_1 and $G_{1,}$ while B_2 and G_2 are dihydro-derivatives. The main toxin categorized from this group is $AFB_{1,}$ which is ranked as the most potent carcinogenic. Due to its harmful effects and stressed campaign against the toxin recently, the toxins' level and incidence are relatively low in well-preserved fodder (Keller et al., 2013; Kosicki et al., 2016). Various comparative studies indicate a higher incidence of aflatoxins in trench silos instead of silo bags. It might be that trench silos are more exposed to different environmental conditions, being made on bare ground or a concrete base (Gonzalez-Pereyra et al., 2011; Alonso et al., 2013).

In lactating animals, chronic exposure through diets results in several adverse effects, including decreased health and production of animals, suppressed immunity, decreased liver functionality, and more susceptibility to diseases even if they are vaccinated (Ogunade et al., 2018). On the other hand, acute aflatoxin toxicity results in symptoms like ataxia, inappetence, enlarged liver, and rough coat (Whitlow and Hagler, 2005; Ogunade et al., 2018). Due to non-specific symptoms of the condition, accurate diagnosis is generally tricky. Some other studies have also reported decreased milk quality (loss of milk proteins) along with milk quantity (Queiroz et al., 2012).

Ruminal microflora can degrade a portion of AFB_1 into aflatoxin (Upadhaya et al., 2010). Cytochrome P-450 enzyme converts this toxin into aflatoxin-8, 9-epoxide. This form of the toxin causes damage to the chromosomes by binding with forming adducts on DNA and hence is more toxic, mutagenic, and carcinogenic (Jouany et al., 2009). Other metabolic forms of the toxin found in the liver include AFP_1, $AFM_{1,}$ and AFQ_1 (Wu et al., 2009; Jafarian-Dehkordi and Pourradi, 2013).

This toxin is excreted in the form AFB_1 and AFM_1 in the milk of dairy animals. This contaminated milk and its products pose an essential threat to public health, as both AFB_1 and AFM_1 are group I human carcinogens (IARC, 2002).

14.5.1.6.1.3.2 Ochratoxin

These are secondary metabolites of *A. ochraceus* (*P. verrucosum*), *A. carnonarius*, and *A. niger* (Keller et al., 2013). Ochratoxin's (OTA) nature is chloro-isocumarin, categorized as the group's highest toxic chemical compound with immunotoxic and carcinogenic properties and inhibition of glucose metabolism (Yiannikouris and Jouany, 2002). Due to the low tolerance level of fungi producing OTA to elevated acetic acid and Co_2, OTA's concentration is relatively lower than other significant mycotoxins, such as ZEA, DON, and fuminosins (Gallo et al., 2015; Kosicki et al., 2016; Biomin, 2016).

Rumen microflora rapidly and extensively degrade OTA into a less toxic metabolite, OTA-alpha (Fink-Gremmels, 2008). It might be a reason for the tolerance of ruminants to OTA, but sometimes its conversion to OTA-C may result in toxicity similar to OTA. Other favoring factors of toxicity include high grain feed, which leads to decreased (5.5–5.8) rumen pH (Pantaya et al., 2016), or high feed intake may exceed the levels by suppressing detoxifying capacity of the rumen. The outcomes exhibited by OTA toxicity include reduced milk production and excretion of OTA in milk.

14.5.2 Chemical Hazards

14.5.2.1 Butyric Acid

The presence of butyric acid can be attributed to low fermentation of silage where fermentation of water-soluble carbohydrates lactic acid produced butyric acid. This process is generally facilitated by clostridial bacteria already present on fodder. Development of these obligate anaerobic spore-farming bacteria can occur in poorly farmed silage during storage or feed-out, which results in not only the formation of butyric acid but also a range of other toxic substances like NH_3, amines, carbon dioxide, hydrogen, and acetic acid (Pahlow et al., 2003). This development of *Clostridia* in silage results in severe consequences related to animal health issues. These effects are exerted either in the way reduction in silage uptake and risk of clinical ketosis increases manyfold due to absorption of butyric acid from the rumen into blood or pathogenic *C. botulinum* develops in the silage. The risk factors for butyric acid silage ketosis in dairy cows are presented in Table 14.2 (adapted from Driehuis et al., 2018)

It is recommended that if it is confirmed that there are elevated levels of butyric acid in the silage, it should never be fed to close-up dry cows or freshly calved

TABLE 14.2
Risk Factors of Silage Ketosis Caused by Butyric Acid in Dairy Cows

Item	Risk Factor for Butyric Acid Silage Ketosis in Dairy Cows
Crop	Alfalfa, clover, ryegrass, hay crop
Silage	
DM	<250 g/kg of fresh weight
pH	>4.4
Butyric acid	>10 g/kg of DM
Daily intake of silage	>5 kg of DM/cow
Stage of lactation	<50 DIM
Daily intake of butyric acid from silage	>50 g/cow

animals but can be given to other categories of animals. Even higher concentrations should not be fed to any animal but can be used as fertilizer.

14.5.2.2 Biogenic Amines and Ammonia

Amines with biological activity (biogenic amines) are generally present in silage, with hazardous feed consumption and animal health outcomes. Hydrolysis of peptide bonds leads to the formation of these amines. Enzymatic decarboxylation of a free amino acid (AA) by the activity of bacterial enzymes, plant peptidases, and proteases is a significant source of this hydrolysis (Driehuis et al., 2018). Tyramine, histamine, cadaverine, spermine, spermidine, and putrescine are amines of concern, resulting in nausea, headaches, vomiting, diarrhea, and hypertension in human beings (Gardini et al., 2016). Accumulation of these amines can occur in various dairy products, especially cheese (Linares et al., 2011).

Slow acidification after ensiling is a significant risk factor accumulation of these amines. Most probably, decarboxylation of AA by enterobacteria during the early aerobic stage leads to the formation of these amines in the initial phase of ensiling. On the other hand, subsequent growth of proteolytic *Clostridia* like *C. sporogenes* and *C. bifermentans* is responsible for producing these amines in later phases of ensiling (Rooke and Hatfield, 2003). Low fermentation of silage leads to increased amines and ammonia concentration in silage, contributing to reduced silage intake and animal utilization. These elevated levels of amines and increased concentration of ammonia lead to reduced silage intake. However, it further exacerbates the situation, as this high level of ammonia goes directly into the liver without going into general circulation and is converted into urea in the liver, leading to loss of energy (approximately 23 KJ/g of nitrogen). Hence, the animal may go into a negative energy balance. Furthermore, impaired fertility in cyclic animals is also associated with increased urea concentration in blood circulation (Driehuis et al., 2018).

14.5.2.3 Nitrate, Nitrite, and Oxide Gases of Nitrogen

As a signal for plant tissue growth and role as a macronutrient, Nitrate (NO_3) is an essential component of crops. Crops harvested in early spring may have higher concentrations of nitrates. This elevation is generally considered due to the uptake of nitrogen from fertilizer by the plant and soil nitrogen following soil mineralization, which results from a decreased rate of synthesis of plant proteins due to lowered environmental temperature. Drought is categorized as one of the significant reasons for elevated nitrate levels in crops, resulting from reduced tissue nitrate reductase activity leading to reduced photosynthesis.

Although nitrate itself is not toxic, nitrate toxicity can result from fodder intake with higher nitrate concentrations. Usually, nitrate is converted into ammonia (NH_3) in the rumen, where nitrite (NO_2^-) is an intermediate product of this pathway. This means high nitrate consumption leads to the accumulation of nitrite in the rumen, absorbed in the blood. In the blood, methemoglobin is produced by the reaction of nitrite and hemoglobin. This methemoglobin is unable to transport oxygen to the tissue from the lungs. This condition is termed methemoglobinemia, with a range of complications like decreased feed intake, respiratory distress, abortion, coma, and sudden death (Binta and Mushi, 2012; Lee and Beauchemin, 2014).

About the effect of ensiling on nitrates, usually, a concentration of nitrates decreases during ensiling compared to raw crops due to the fermentation process. Lactic acid bacteria and enterobacteria degrade nitrate in the initial phases of ensiling. Ammonia and nitrous oxide (N_2O) gases are end products of this degradation, with nitrogen dioxide (NO_2) and nitric oxide (NO) gases being intermediate products. Prevention of clostridial growth is taken as a positive outcome of reduction of nitrate to nitrogen dioxide. However, if all the nitrogen dioxide is converted to ammonia, a relatively high silage pH will be achieved. This higher pH is considered insufficient to prevent clostridial growth during the storage phase.

'Silo-fillers disease' is a severe respiratory hazard to both animals and human beings, resulting from the production of nitrous oxide and nitrogen dioxide gases in silage. These gases react with the water in environmental air, resulting in nitrous oxide and nitric oxide gases. Damage to lung tissue and respiratory distress leading to asphyxiation result from inhalation of these gases, leading to this disease's signs and symptoms.

14.5.3 Plant Hazards

These include various substances contained by plants at the time of ensiling or originate during a flawed ensiling process. This list includes many hazards, but the major ones are pyrrolizidine alkaloids, phytoestrogens, mimosine, prussic acid, tropane alkaloids, and ergot alkaloids.

14.5.3.1 Pyrrolizidine Alkaloids

Pyrrolizidine alkaloids (PAs) are mostly found in various plants, especially in families Compositae, Leguminosae, and Boraginaceae (Prakash et al., 1999). These result in liver cirrhosis in human beings. Other affected species are sheep, horses, and cattle. Cattle and horses are considered more prone to toxicity than small herbivores like sheep and goats. It is believed that ensiling does not affect the presence of PA in plants, while some researchers state that the process of ensiling may degrade some of PA depending on the part of the plant used, plant maturity, and levels of PA in plants (Berendonk et al., 2010; Becerra-Jiminez et al., 2013; Chizzola et al., 2015).

14.5.3.2 Phytoestrogens

These are compounds contained by the plants, but they are analogous to animal estrogen in nature and effects (Pace et al., 2006). The most common examples of these compounds are ligans, coumestans, and isoflavones. Coumestrol, an example of coumestans, is found in many plants like in alfalfa or red clover. These compounds' significant effect on animal health is believed to be reduced fertility, decreased lambing percentage, temporary or permanent infertility, low development of udder, and rectal/vaginal prolapse. Susceptible animals include sheep, goats, and cattle (Driehuis et al., 2018). There are mixed reports by considering the effects of ensiling on phytoestrogens as some studies indicate that phytoestrogens' levels increased after silage making, mainly when red clover was used. Contrary to this, other studies revealed a decreased level of isoflavones after ensiling. These differences can be due to various reasons like the difference in cultivars, the maturity of the crop, degree of wilting, levels of phytoestrogens before ensiling, environmental temperatures, nature of silage additives and duration of ensiling, and so on (Sarelli et al., 2003; Moravcova et al., 2003; Daems et al., 2016; Driehuis et al., 2018).

14.5.3.3 Mimosine

As a non-protein amino acid, it has a structure similar to tyrosine. Its toxicity is due to its nature as an antagonist to tyrosine, thus inhibiting protein synthesis in the body, ultimately reducing animal production (Angthong et al., 2007). A tropical legume (*Leucaena*) was termed a 'miracle tree' due to its high nutritional value and long life. It is used as livestock feed was limited due to high mimosine contents (Driehuis et al., 2018). Susceptible animals include cattle, especially steers; in later stages, it causes impaired thyroid function and reduced feed intake at levels 10 g/kg of dry matter. It is now established that proper ensiling may reduce mimosine levels in silage compared to raw feed. It is believed that there is a negative correlation between the rate of degradation and pH and a positive correlation between the rate of degradation and lactic acid concentration (Chen et al., 2014; Phesatcha and Wanapat, 2015).

14.5.3.4 Prussic Acid

Prussic acid (hydrocyanic acid) is a product of decomposition of nontoxic cyanide and glucosides in cyanogenic plants, is highly toxic. Examples of these plants are sorghum and Sudan grass (Toaima et al., 2014). Susceptible animals include ruminants like sheep, goats, cattle, buffalo, and so on. The affected animal shows elevated respiratory rate, irregular pulse, staggering and oozing of froth from the mouth, respiratory paralysis, and ultimately death (Barnhart and Dewell, 2011). It is well established that ensiling may decrease this fatal compound in fodder but cannot eliminate it from raw fodder (Kallah et al., 1997; Driehuis et al., 2018).

14.5.3.5 Ergot Alkaloids

Grass endophytic fungi like *Claviceps* spp., *Neotyphodium* spp., and many others produce mycotoxins, which are ergot alkaloids in nature. These are mostly found in forages, including ryegrass, sorghum, and tall fescue (*Festuca arundinacea*). *Claviceps africana* infection in sorghum results in the production of dihydroergosine (Molloy et al., 2003); infection of tall fescue and ryegrass with endophyte *Neotyphodium* spp. results in the production of ergovaline as a major ergot alkaloid (Crews, 2015). Susceptible animals include cattle, buffalo, sheep, and goats. The animals' significant signs after ingestion of ergot alkaloids include reduced milk production, decreased body weight, a decline in fertility, hyperthermia, mastitis, and feed refusal (Schneider et al., 1996; Lean, 2001; Naude et al., 2005). There are different schools of thought concerning the effect of ensiling on the silage concentration of ergot alkaloids. Some studies have reported no change in ergot alkaloids in levels after ensiling (Roberts et al., 2002).

Contrary to this, some other reports indicate even up to a 50% fall in levels of dihydroergosine or ergovaline. However, this, in turn, increases the overall level of ergot alkaloids in the preserved fodder (Mikołajczak et al., 2005; Blaney et al., 2010; Roberts et al., 2011; Roberts et al., 2014). This overall increase in ergot alkaloids levels is attributed to the by-products of degradation of the ergot mentioned previously:

alkaloids, like lysergic acid produced after the degradation of ergovaline. Hence, it can be concluded that ensiling is not an appropriate technique to decrease ergot alkaloid levels in fodder, as it fails to decrease ergot to safe levels; instead, it increases the overall level of these potentially toxic substances in the fodder.

14.5.3.6 Tropane Alkaloids

Jimson weed (*Datura stramonium*) contains tropane alkaloids naturally, and this weed usually contaminates hay and silage (Cortinovis and Caloni, 2013). Although the highest amount of tropane alkaloids is contained in ripe seeds of the weed with strong anticholinergic properties, all the weed parts are toxic (Chang et al., 1999; DeFrates et al., 2005; Soler-Rodriguez et al., 2006). Hyoscyamine, scopolamine, and atropine are major toxic tropane alkaloids in jimson weed (DeFrates et al., 2005). The signs of jimson weed toxicity are dry mouth, tachycardia, mydriasis, incoordination, convulsions, and coma, shown by almost all livestock spp. (Anadon et al., 2012). Although very little information is available on the effect of ensiling on tropane alkaloid levels, some studies have reported 100% mortality by consuming tropane alkaloids present in ensiled corn contaminated with jimson weed (Binev et al., 2006). This gives rise to the assumption that ensiling does not lower tropane alkaloid levels in fodder.

14.6 IMPACT OF TOXINS ON PUBLIC HEALTH

One can easily understand by going through the previous sections that many toxins of preserved fodder are contaminants of milk in one way or another. These toxins deteriorate milk quality by either hindering the pasteurization process and decreasing the shelf life of pasteurized milk or its products and pose a severe health issue for the end consumers: human beings. These contaminants also result in economic losses by discarding the preserved fodder, contaminated milk, and milk products and treatment losses rendered on animal and human infections/ailments.

If we talk of bacterial toxins/contaminations of preserved fodder, *B. cereus* is one of the significant bacteria found in preserved fodder, and it causes some serious health issues in animals. This microbe is categorized as one of the major pathogens for human beings, too, which was previously considered to be associated with food poisoning. Recent reports also associate it with a very severe and fatal non-gastrointestinal tract infection along with some eye infection. Similarly, *C. botulinum* produces neurotoxins, which lead to food-borne botulism in humans (Table 14.1). Early signs of the condition include weakness, fatigue, blurred vision, diarrhea, vomiting, abdominal swelling, and ultimately flaccid paralysis. This condition can be fatal in 5% to 10% of cases. Infection with STEC results in stomach pain, nausea, vomiting, diarrhea, and fever. Infection or ingestion of toxins by other bacteria causes health hazards to human beings.

When transmitted to human beings, mycotoxins present in milk or meat pose a severe threat of fatal conditions (Table 14.3). As mentioned earlier, aflatoxin M_1 is categorized as a group 2B human carcinogen by the International Agency for Research on Cancer (IARC). Likewise, ochratoxin A is renally toxic, leading to kidney failure and damaging other vital organs of the body. Mycotoxins are considered so harmful to human

TABLE 14.3
Permissible Levels of Some Mycotoxins in Human and Animal Feeds

		Intended Use	
Mycotoxin	**Agency**	**Ruminant Feed**	**Human Consumption**
Aflatoxins	US Food and Drug Administration (FDA)	Dairy animals = 20 μg/kg Breeding cattle = 100 μg/kg Beef cattle = 300 μg/kg	All human foods (20 μg/kg) Milk AFM_1 = 0.50 μg/kg
	European Commission (EC)	Dairy animals = 5 μg/kg All other ruminants (20 μg/kg)	4 μg/kg in all cereals and cereal products with the exception of corn and rice subjected to physical treatment (10 μg/kg) 0.1 μg/kg in foods for infants and young children 0.05 μg/kg in milk (aflatoxin M_1)
Total fumonisins	FDA	Breeding ruminants = 15 mg/kg Beef ruminant (30 mg/kg)	Degermed dry milled corn products = 2 mg/kg Popcorn = 3 mg/kg Whole or partially degermed dry milled corn products and dry milled corn bran (4 mg/kg)
	EC	50 mg/kg for adult ruminants (>4 mo old)	1 mg/kg in corn intended for direct human consumption 0.2 mg/kg in foods for infants and young children
Ochratoxins	FDA	Currently not stipulated	Currently not stipulated
	EC	Currently not stipulated	0.003 mg/kg in cereals intended for direct human consumption 0.0005 mg/kg in foods for infants and young children
Zearalenone	FDA	Currently not stipulated	Currently not stipulated
	EC	Ruminants = 0.5 mg/kg	0.075 mg/kg in cereals intended for direct human consumption
Deoxynivalenol	FDA	Beef and feedlot cattle (older than 4 months) = 10 mg/kg) Dairy cattle (older than 4 months) = 5 mg/kg	1 mg/kg in wheat products intended for human consumption
	EC	Ruminants = 5 mg/kg	0.75 mg/kg in cereals intended for direct human consumption 0.2 mg/kg in foods for infants and young children
T-2 and HT-2	FDA	Currently not stipulated	Currently not stipulated
	EC	Currently not stipulated	0.05 mg/kg in foods intended for direct human consumption except for corn (0.1 mg/kg) and oats (0.2 mg/kg) 0.015 mg/kg in foods for infants and young children

health that specific legislative conditions exist for minimum and maximum limits of mycotoxins in animal and human feed in many developed countries. The regulatory levels of some common mycotoxins for ruminant feed and human food are given in Table 14.3 (adopted from Ogunade et al., 2018).

REFERENCES

Alonso VA, Pereya CM, Keller LAM, Dalcero AM, Rosa CAR, Chiacchiera SM and Cavaglieri LR (2013) Fungi and mycotoxins in silage: An overview. *Journal of Applied Microbiology* 115(3): 637–643.

Anadon A, Martinez-Larranaga MR and Castellano V (2012) Poisonous plants of Europe. In: Gupta RC (ed) *Veterinary Toxicology: Basic and Clinical Principles*, 2nd ed. Cambridge, MA: Academic Press, pp. 1080–1094.

Anderson NM, Larkin JW, Cole MB, Skinner GE, Whiting RC, Gorris LG, Rodriguez A, Buchanan R, Stewart CM, Hanlin JH, Keener L and Hall PA (2011) Food safety objective approach for controlling *Clostridium botulinum* growth and toxin production in commercially sterile foods. *Journal of Food Protection* 74(11): 1956–1989.

Angthong W, Cheva-Isarakul B, Promma S and Cheva-Isarkul B (2007) Beta-carotene, mimosine and quality of *leucaena* silage kept at different duration. *Kasetsart Journal—Natural Science* 41(2): 282–287.

Bach SJ, McAllister TA, Baah J, Yanke LJ, Veira DM, Gannon VPJ and Holley RA (2002) Persistence of *Escherichia coli* O157:H7 in barley silage: Effect of a bacterial inoculant. *Journal of Applied Microbiology* 93(2): 288–294.

Barnhart SK and Dewell GA (2011) Prussic acid poisoning potential in frosted forages. *Integrated Crop Management News*. Paper 304.

Becerra-Jiminez J, Kuschak M, Roeder E and Wiedenfeld H (2013) Toxic pyrrolizidine alkaloids as undesired contaminants in food and feed: Degradation of the PAs from *Senecio jacobaea* in silage. *Pharmazie* 68(7): 636–639.

Bennett JW and Klich M (2003) Mycotoxins. *Clinical Microbiology Reviews* 16(3): 497–516.

Berendonk C, Cerff D, Hunting K, Wiedenfeld H, Becerra J and Kuschak M (2010) Pyrrolizidine alkaloid level in *Senecio jacobaea* and *Senecio erraticus*—the effect of plant organ and forage conservation. In: Proc. 23rd Gen. Meet Eur. Grassld Fed. Grassland in a Changing World, Kiel, Germany, vol. 15, pp. 669–671.

Biesta-Peters EG, Reij MW, Joosten H, Gorris LGM and Zwietering MH (2010) Comparison of two optical-density-based methods and a plate count method for estimation of growth parameters of *Bacillus cereus*. *Applied and Environmental Microbiology* 76(5): 1399–1405.

Binev R, Valchev I and Nikolov J (2006) Clinical and pathological studies on intoxication in horses from freshly cut jimson weed (*Datura stramonium*)-contaminated maize intended for ensiling: Clinical communication. *Journal of South African Veterinary Association* 77(4): 215–219.

Binta MG and Mushi EZ (2012) Environmental factors associated with nitrate poisoning in livestock. *Journal of Petroleum and Environmental Biotechnology* 3: 6. https://doi.org/10.4172/2157-7463.1000131.

Biomin (2016) *Mycotoxin Survey 2016 Third Quarter* (July to Sept 2016) Accessed 15 January 2017. https://nutricionanimal.info/wp-content/uploads/2016/11/Mycotoxin-Survey-Presentation-Q3-2016-1.pdf.

Blaney BJ, Ryley MJ and Boucher BD (2010) Early harvest and ensilage of forage sorghum infected with ergot (*Claviceps africana*) reduces the risk of livestock poisoning. *Australian Veterinary Journal* 88(8): 311–312.

Brett MM, McLauchlin J, Harris A, O'Brien S, Black N, Forsyth RJ, Roberts D and Bolton FJ (2005) A case of infant botulism with a possible link to infant formula milk powder: Evidence for the presence of more than one strain of *Clostridium botulinum* in clinical specimens and food. *Journal of Medical Microbiology* 54(8): 769–776.

Browne N and Dowds BC (2002) Acid stress in the food pathogen *Bacillus cereus*. *Journal of Applied Microbiology* 92(3): 404–414.

Byrne CM, O'kiely P, Bolton DJ, Sheridan JJ, McDowell DA and Blair IS (2002) Fate of *Escherichia coli* O157:H7 during silage fermentation. *Journal of Food Protection* 65(12): 1854–1860.

Callaway TR, Edrington TS, Brabban AD, Keen JE, Anderson RC, Rossman ML, Engler MJ, Genovese KJ, Gwartney BL, Reagan JO, Poole TL, Harvey RB, Kutter EM and Nisbet DJ (2006) Fecal prevalence of *Escherichia coli* O157, *Salmonella, Listeria*, and bacteriophage infecting *E. coli* O157:H7 in feedlot cattle in the southern plains region of the United States. *Food-borne Pathogens and Disease* 3(3): 234–244.

Carter AT and Peck MW (2015) Genomes, neurotoxins and biology of *Clostridium botulinum* Group I and Group II. *Research in Microbiolology* 166(4): 303–317.

CAST (2003) *Mycotoxins: Risks in Plant, Animal, and Human Systems*. Task Force Report No. 139. Council for Agricultural Science and Technology, Ames, IA.

Chang SS, Wu ML, Deng JF, Lee CC, Chin TF and Liao SJ (1999) Poisoning by datura leaves used as edible wild vegetables. *Veterinary and Human Toxicology* 41(4): 242–245.

Chen XZ, Feng F, Liu QH and Zhang JG (2014) Degrading mimosine and tannins of *Leucaena leucocephala* by ensiling. *Applied Mechanics and Materials* 618: 349–353.

Chizzola R, Bassler G, Winter S, Zebeli Q and Kriechbaum M (2015) Persistence of alkaloids of typical poisonous plants autumn crocus and marsh ragwort in grass silage. *Veterinary Medicine Austria* 102(12): 285–292.

Christiansson A, Bertilsson J and Svensson B (1999) *Bacillus cereus* spores in raw milk: Factors affecting the contamination of milk during the grazing period. *Journal of Dairy Science* 82(2): 305–314.

Cortinovis C and Caloni F (2013) Epidemiology of intoxication of domestic animals by plants in Europe. *Veterinary Journal* 197(2): 163–168.

Crews C (2015) Analysis of ergot alkaloids. *Toxins (Basel)* 7(6): 2024–2050.

Daems F, Decruyenaere V, Agneessens R, Lognay G, Romnee JM and Froidmont E (2016) Changes in the isoflavone concentration in red clover (*Trifolium pratense* L.) during ensiling and storage in laboratory-scale silos. *Animal Feed Science and Technology* 217: 36–44.

De Wolf E, Kuldau G, Lipps P, Munkvold G, Vincelli P, Woloshuk C and Mills D (2005) *Subject: Mouldy Grains, Mycotoxins and Feeding Problems*. Accessed 5 November 2016. http://www.oardc.ohiostate.edu/ohiofieldcropdisease/Mycotoxins/myco-pagedefault.htm.

DeFrates LJ, Hoehns JD, Sakornbut EL, Glascock DG and Tew AR (2005) Antimuscarinic intoxication resulting from the ingestion of moonflower seeds. *Annals of Pharmacotherapy* 39 (1): 173–176.

Doyle CJ, Gleeson D, Jordan K, Beresford TP, Ross RP, Fitzgerald GF and Cotter PD (2015) Anaerobic sporeformers and their significance with respect to milk and dairy products. *International Journal of Food Microbiology* 197: 77–87.

Driehuis F (2013) Silage and the safety and quality of dairy foods: A review. *Agriculture and Food Science* 22(1): 16–34.

Driehuis F, Spanjer MC, Scholten JM and Te Giffel MC (2008a) Occurrence of mycotoxins in feedstuffs of dairy cows and estimation of total dietary intakes. *Journal of Dairy Science* 91(11): 4261–4271.

Driehuis F, Spanjer MC, Scholten JM and Te Giffel MC (2008b) Occurrence of mycotoxins in maize, grass and wheat silage for dairy cattle in the Netherlands. *Food Additives and Contaminants Part B Surveillance* 1(1): 41–50.

Driehuis F, Wilkinson JM, Jiang Y, Ogunade IM and Adesogan AT (2018) Silage review: Animal and human health risks from silage. *Journal of Dairy Science* 101(5): 4093–4110.

Duniere L, Gleizal A, Chaucheyras–Durand F, Chevallier I and Thevenot-Sergentet D (2011) Fate of *Escherichia coli* O26 in corn silage experimentally contaminated at ensiling, at opening or after aerobic exposure and protective effect of various bacterial inoculants. *Applied and Environmental Microbiology* 77(24): 8696–8704.

EFSA (European Food Safety Authority) and ECDC (European Centre for Disease Prevention and Control) (2016) The European Union summary report on trends and sources of zoonoses, zoonotic agents and food-borne outbreaks in 2015. *EFSA J.* 14: 4634–4865.

Erdogan HM (2010) Listerial keratoconjunctivitis and uveitis (silage eye). *Veterinary Clinics of North America and Food Animal Practice* 26(3): 505–510.

FAO (2015) Strengthening control of bovine tuberculosis at the animal human-ecosystem interface. *Food and Agriculture Organization of the United Nations.* Accessed 17 May 2020. http://www.fao.org/ag/againfo/programmes/en/empres/news240315.html.

Fenlon DR and Wilson J (2000) Growth of *Escherichia coli* O157 in poorly fermented laboratory silage: A possible environmental dimension in the epidemiology of *E. coli* O157. *Letters in Applied Microbiology* 30(2): 118–121.

Fink-Gremmels J (2008) Mycotoxins in cattle feeds and carry-over to dairy milk: A review. *Food Additives and Contaminants Part A Chemistry Analysis Control Exposure and Risk Assessment* 25(2): 172–180.

Galey FD, Terra R, Walker R, Adaska J, Etchebarne MA, Puschner B, Fisher E, Whitlock RH, Rocke T, Willoughby D and Tor E (2000) Type C botulism in dairy cattle from feed contaminated with a dead cat. *Journal of Veterinary Diagnostic Investigations* 12(3): 204–209.

Gallo A, Giuberti G, Frisvad JC, Bertuzzi T and Nielsen KF (2015) Review on mycotoxin issues in ruminants: Occurrence in forages, effects of mycotoxin ingestion on health status and animal performance and practical strategies to counteract their negative effects. *Toxins (Basel)* 7(8): 3057–3111.

Gandhi M and Chikindas ML (2007) *Listeria*: A food-borne pathogen that knows how to survive. *International Journal of Food Microbiology* 113(1): 1–15.

Gardini F, Ozoqui Y, Suzzi G, Tabanelli G and Ozogui F (2016) Technological factors affecting biogenic amine content in food: A review. *Frontiers in Microbiology* 7: 1218. https://doi.org/10.3389/fmicb.2016.01218.

Garnett BT, Delahay RJ and Roper TJ (2002) Use of cattle farm resources by badgers (*Meles meles*) and risk of bovine tuberculosis (*Mycobacterium bovis*) transmission. Proceedings of Biological Sciences 269 (1499): 1487–1491.

Gonzalez Pereyra ML, Chiacchiera SM, Rosa CAR, Sager R, Dalceroa AM and Cavaglieri L (2011) Comparative analysis of the mycobiota and mycotoxins contaminating corn trench silos and silo bags. *Journal of the Science of Food and Agriculture* 91(8): 1474–1481.

Gould LH, Mody RK, Ong KL, Clogher P, Cronquist AB, Garman KN, Lathrop S, Medus C, Spina NL, Webb TH, White PL, Wymore K, Gierke RE, Mahon BE and Griffin PM (2013) Emerging Infections Program Foodnet Working Group. Increased recognition of non-O157 Shiga toxin-producing *Escherichia coli* infections in the United States during 2000–2010: Epidemiologic features and comparison with *E. coli* O157 infections. *Food-borne Pathogens and Disease* 10(5): 453–460.

Gxasheka M, Wang J, Tyasi TL and Gao J (2015) Scientific understanding and effects on ear rot diseases in maize production: A review. *International Journal of Soil and Crop Sciences* 3(4): 77–84.

Heyndrickx M and Scheldeman P (2002) Bacilli associated with spoilage in dairy and other food products. In: Berkeley R, Heyndrickx M, Logan N and De Vos P (eds) *Applications and Systematics of Bacillus and Relatives.* Oxford, UK: Blackwell Science, pp. 64–82.

Ho AJ, Ivanek R, Grohn YT, Nightingale KK and Wiedmann M (2007) *Listeria monocytogenes* fecal shedding in dairy cattle shows high levels of day-to-day variation and includes outbreaks and sporadic cases of shedding of specific *L. monocytogenes* subtypes. *Preventive Veterinary Medicine* 80(4): 287–305.

IARC (2002) Some traditional herbal medicines, some mycotoxins, naphthalene and styrene. Pages 9–13 in *IARC Monographs on the Evaluation of Carcinogenic Risks to Humans.* Vol. 82. Lyon: IARC Scientific Publication.

Jafarian-Dehkordi A and Pourradi N (2013) Aflatoxin M1 contamination of human breast milk in Isfahan, Iran. *Advanced Biomedical Research* 2: 86. doi:10.4103/2277-9175.122503. eCollection 2013.

Jouany JP, Yiannikouris A and Bertin G (2009) Risk assessment of mycotoxins in ruminants and ruminant products. *Options Mediterraneennes A* 85: 205–224.

Kallah MS, Baba M, Alawa JP, Muhammad IR and Tanko RJ (1997) Ensiling quality of columbus grass (*Sorghum almum*) grown in northern Nigeria. *Animal Feed Science and Technology* 68(2): 153–163.

Keller LAM, Gonzalez Pereyra ML, Keller KM, Alonso VA, Oliveira AA, Almeida TX, Barbosa TS, Nunes LMT, Cavaglieri LR and Rosa CAR (2013) Fungal and mycotoxins contamination in corn silage: Monitoring risk before and after fermentation. *Journal of Stored Products Research* 52: 42–47.

Kosicki R, Błajet-Kosicka A, Grajewski J and Twaruzek M (2016) Multiannual mycotoxin survey in feed materials and feedingstuffs. *Animal Feed Science and Technology* 215: 165–180.

Kummel J, Krametter-Froetscher R, Six G, Brunthaler R, Baumgartner W and Altenbrunner-Martinek B (2012) Descriptive study of botulism in an Austrian dairy herd: A case report. *Veterinární Medicína* 57(3): 143–149.

Lean IJ (2001) Association between feeding perennial ryegrass (*Lolium perenne* cultivar grasslands impact) containing high concentrations of ergovaline, and health and productivity in a herd of lactating dairy cows. *Australian Veterinary Journal* 79(4): 262–264.

Lee C and Beauchemin KA (2014) A review of feeding supplementary nitrate to ruminant animals: Nitrate toxicity, methane emission and production performance. *Canadian Journal of Animal Science* 94(4): 5457–5570.

Linares DM, Martin MC, Ladero V, Alvarez MA and Fernandez M (2011) Biogenic amines in dairy products. *Critical Reviews in Food Science and Nutrition* 51(7): 691–703.

Lindstrom M, Myllykoski J, Sivela S and Korkeala H (2010) *Clostridium botulinum* in cattle and dairy products. *Critical Reviews in Food Science and Nutrition* 50(4): 281–304.

Loi M, Fanelli F, Liuzzi VC, Logrieco AF and Mule G (2017) Mycotoxin biotransformation by native and commercial enzymes: Present and future perspectives. *Toxins (Basel)* 9(4): 111. https://doi.org/10.3390/toxins9040111.

Lomonaco S, Nucera D and Filipello V (2015) The evolution and epidemiology of *Listeria monocytogenes* in Europe and the United States. *Infection Genetics and Evolution* 35: 172–183.

Magnusson M, Christiansson A and Svensson B (2007) *Bacillus cereus* spores during housing of dairy cows: Factors affecting contamination of raw milk. *Journal of Dairy Science* 90(6): 2745–2754.

Magnusson M, Christiansson A, Svensson B and Kolstrup C (2006) Effect of different premilking manual teat-cleaning methods on bacterial spores in milk. *Journal of Dairy Science* 89(10): 3866–3875.

Majowicz SE, Scallan E, Jones-Bitton A, Sargeant JM, Stapleton J, Angulo FJ, Yeung DH and Kirk MD (2014) Global incidence of human shiga toxin-producing *Escherichia coli* infections and deaths: A systematic review and knowledge synthesis. *Food-borne Pathogens and Disease* 11(6): 447–455.

Malekinejad H, Maas-Bakker R and Fink-Gremmels J (2006) Species differences in the hepatic biotransformation of zearalenone. *Veterinary Journal* 172(1): 96–102.

Marczuk J, Obremski K, Lutnicki K, Gajecka M and Gajecki M (2012) Zearalenone and deoxynivalenol mycotoxicosis in dairy cattle herds. *Polish Journal of Veterinary Science* 15(2): 365–372.

Marin S, Ramos AJ, Cano-Sancho G and Sanchis V (2013) Mycotoxins: Occurrence, toxicology, and exposure assessment. *Food and Chemical Toxicology* 60: 218–237.

Marshall JC, Soboleva TK, Jamieson P and French NP (2016) Estimating bacterial pathogen levels in New Zealand bulk tank milk. *Journal of Food Protection* 79(5): 771–780.

Mehmood T, Haq ZU, Mahmood S, Nawaz MK, Asam HM and Shafi MK (2020) Forage preservation technology for sustainable livestock industry in rainfed areas of Pakistan: A review. *Pure and Applied Biology* 9(3): 1849–1855.

Michaelson LV, Napier JA, Molino D and Faure JD (2016) Plant sphingolipids: Their importance in cellular organization and adaption. *Biochimica et Biophysica Acta (BBA)—Molecular and Cell Biology of Lipids* 1861(9): 1329–1335.

Mikołajczak J, Podkowka L, Podkowka Z and Staszak E (2005) Effects of endophyte infection of grasses on the chemical composition, quality and stability of silage. *Folia Biologica* 53(4): 67–72.

Miller RA, Kent DJ, Watterson MJ, Boor KJ, Martin NH and Wiedmann M (2015) Spore populations among bulk tank raw milk and dairy powders are significantly different. *Journal of Dairy Science* 98(12): 8492–8504.

Molloy JB, Moore CJ, Bruyeres AG, Murray SA and Blaney BJ (2003) Determination of dihydroergosine in sorghum ergot using an immunoassay. *Journal of Agricultural and Food Chemistry* 51(14): 3916–3919.

Moravcova J, Kleinova T, Loucka R, Tyrolova I, Kvasnicka F, Dusek M and Cerovsky M (2003) Effect of additives on coumestrol content in laboratory alfalfa silages. *Czech Journal of Animal Science* 48(10): 425–431.

Muck RE, Moser LE and Pitt RE (2003) Postharvest factors affecting ensiling. In: Buxton DR, Muck RE and Harrison JH (eds) *Silage Science and Technology*. Madison, Wisconsin, USA: American Society of Agronomy, pp. 251–304.

Myllykoski J, Lindstrom M, Keto-Timonen R, Soderholm H, Jakala J, Kallio H, Sukura A and Korkeala H (2009) Type C bovine botulism outbreak due to carcass contaminated non-acidified silage. Epidemiology and Infection 137(2): 284–293.

Naude TW, Botha CJ, Vorster JH, Roux C, Van der Linde EJ, Van der Walt SI, Rottinghaus GE, Van Jaarsveld L and Lawrence AN (2005) *Claviceps cyperi*, a new cause of severe ergotism in dairy cattle consuming maize silage and teff hay contaminated with ergotised *Cyperus esculentus* (nut sedge) on the Highveld of South Africa. *Onderstepoort Journal of Veterinary Research* 72(1): 23–37.

Nightingale KK, Schukken YH, Nightingale CR, Fortes ED, Ho AJ, Her Z, Grohn YT, McDonough PL and Wiedmann M (2004) Ecology and transmission of *Listeria monocytogenes* infecting ruminants and in the farm environment. *Applied and Environmental Microbiology* 70(8): 4458–4467.

Ogunade IM, Jiang Y, Tuppia CM, Queiroz OCM, Drouin P and Adesogan AT (2018) Silage review: Mycotoxins in silage: Occurrence, effects, prevention, and mitigation. *Journal of Dairy Science* 101(5): 4034–4059.

Ogunade IM, Jiang Y, Kim DH, Pech Cervantes AA, Arriola KG, Vyas D, Weinberg ZG, Jeong KC and Adesogan AT (2017) Fate of *Escherichia coli* O157:H7 and bacterial diversity in corn silage contaminated with the pathogen and treated with chemical or microbial additives. *Journal of Dairy Science* 100(3): 1780–1794.

Ogunade IM, Kim DH, Jiang Y, Weinberg ZG, Jeong KC and Adesogan AT (2016) Control of *Escherichia coli* O157:H7 in contaminated alfalfa silage: Effects of silage additives. *Journal of Dairy Science* 99(6): 4427–4436.

Pace V, Carbone K, Spirito F, Iacurto M, Terzano MG, Verna M, Vincenti F and Settineri D (2006) The effects of subterranean clover phytoestrogens on sheep growth, reproduction and carcass characteristics. *Meat Science* 74(4): 616–622.

Pahlow G, Muck RE, Driehuis F, Oude-Elfereink SJWH and Spoelstra SF (2003) Microbiology of ensiling. In: Buxton DR, Muck RE and Harrison JH (eds) *Silage Science and Technology*. Madison, Wisconsin, USA: American Society of Agronomy, pp. 31–93.

Pantaya D, Morgavi DP, Silberberg M, Chaucheyras-Durand F, Martin C, Wiryawan KG and Boudra H (2016) Bioavailability of aflatoxin B1 and ochratoxin A, but not fumonisin B1 or deoxynivalenol, is increased in starch-induced low ruminal pH in nonlactating dairy cows. *Journal of Dairy Science* 99(12): 9759–9767.

Payne JH, Hogg RA, Otter A, Roest HI and Livesey CT (2011) Emergence of suspected type D botulism in ruminants in England and Wales (2001 to 2009), associated with exposure to broiler litter. *Veterinary Record* 168(24): 640. doi: 10.1136/vr.d1846.

Pedroso AF, Adesogan AT, Queiroz OCM and Williams SK (2010) Control of *Escherichia coli* O157:H7 in corn silage with or without various inoculants: Efficacy and mode of action. *Journal of Dairy Science* 93(3): 1098–1104.

Phesatcha K and Wanapat M (2015) Improvement of *Leucaena* silage nutritive value and *in vitro* ruminal fermentation by molasses and urea supplementation. Asian-Australas. *Journal of Animal Science* 29(8): 1136–1144.

Phillips CJC (2005) The epidemiology and control of bovine tuberculosis. In: Smithe LT (ed) *Focus on Tuberculosis Research*. New York, NY: Nova Biomedical Books, pp. 203–247.

Prakash AS, Pereira TN, Reilly PE and Seawright AA (1999) Pyrrolizidine alkaloids in human diet. *Mutation Research/Genetic Toxicology and Environmental Mutagenesis* 443(1–2): 53–67.

Puntenney SB, Wang Y and Forsberg NE (2002) Keeping them out of the rough: Practical insights into hemorrhagic bowel syndrome. *Journal of the American Veterinary Medical Association* 331: 686–689.

Queiroz OCM, Han JH, Staples CR and Adesogan AT (2012) Effect of adding a mycotoxin-sequestering agent on milk aflatoxin M1 concentration and the performance and immune response of dairy cattle fed an aflatoxin B1-contaminated diet. *Journal of Dairy Science* 95(10): 5901–5908.

Rao M, Struer TL, Aldwell FE and Cook GM (2001) Intracellular pH regulation by *Mycobacterium smegmatis* and *Mycobacterium bovis* BCG. *Microbiology* 147(4): 1017–1024.

Roberts C, Kallenbach R and Hill N (2002) Harvest and storage method affects ergot alkaloid concentration in tall fescue. *Crop Management* 1(1): 1–3.

Roberts CA, Davis DK, Looper ML, Kallenbach RL, Rottinghaus G and Hill NS (2014) Ergot alkaloid concentrations in high-and low-moisture tall fescue silage. *Crop Science* 54: 1887–1892.

Roberts CA, Kallenbach RL, Rottinghaus GE and Hill NS (2011) Ergovaline and ergot alkaloid concentrations change in conserved tall fescue. *Forage and Grazinglands* 9(1): 1–9.

Rodrigues I and Naehrer K (2012) A three-year survey on the worldwide occurrence of mycotoxins in feedstuffs and feed. *Toxins (Basel)* 4(9): 663–675.

Rooke JA and Hatfield RD (2003) Biochemistry of ensiling. In: Buxton DR, Muck RE and Harrison JH (eds) *Silage Science and Technology*. Madison, Wisconsin, USA: American Society of Agronomy, pp. 95–139.

Saeger SD, Sibanda L and Peteghem CV (2003) Analysis of zearalenone and alpha-zearalenol in animal feed using high-performance liquid chromatography. *Analytica Chimica Acta* 487(2): 137–143.

Sarelli L, Tuori M, Saastamoinen I, Syrjala-qvist L and Saloniemi H (2003) Phytoestrogen content of birdsfoot trefoil and red clover: Effects of growth stage and ensiling method. *Acta Agriculturae Scandinavica, Section A-Animal Science* 53(1): 58–63.

Scallan E, Hoekstra RM, Angulo FJ, Tauxe RV, Widdowson MA, Roy SL, Jones JL and Griffin PM (2011) Food-borne illness acquired in the United States—Major pathogens. *Emerging Infectious Diseases* 17(1): 7–15.

Schatzmayr G and Streit E (2013) Global occurrence of mycotoxins in the food and feed chain: Facts and figures. *World Mycotoxin Journal* 6(3): 213–222.

Schmaile DG and Munkvold GP (2009) Mycotoxins in crops: A threat to human and domestic animal health. *The Plant Health Instructor*. https://doi.org/10.1094/PHI-I-2009–0715–01.

Schmidt P, Novins KICO, Junges D, Almeida R and de Souza CM (2015) Concentration of mycotoxins and chemical composition of corn silage: A farm survey using infrared thermography. *Journal of Dairy Science* 98(9): 6609–6619.

Schneider DJ, Miles CO, Garthwaite I, Halderen AV, Wessels JC and Lategan HJ (1996) First report of field outbreaks of ergot-alkaloid toxicity in South Africa. *Onderstepoort Journal of Veterinary Research* 63(2): 97–108.

Sobel J, Tucker N, Sulka A, McLaughlin J and Maslanka S (2004) Food-borne botulism in the United States, 1990–2000. *Emerging Infectious Diseases* 10(9): 1606–1611.

Soler-Rodriguez F, Martin A, Garcia-Cambero JP, Oropesa AL and Perez-Lopez M (2006) Datura stramonium poisoning in horses: A risk factor for colic. *Veterinary Record* 158(4): 132–133.

Storm IM, Sorensen JL, Rasmussen RR, Nielsen KF and Thrane U (2008) Mycotoxins in silage. *Stewart Postharvest Review* 4(6): 1–12.

Tasci F, Turutoglu H and Ogutcu H (2010) Investigations of *Listeria* species in milk and silage produced in Burdur province. *Kafkas Üniversitesi Veteriner Fakültesi Dergisi* 16(Suppl. A): S93—S97.

Toaima SA, Lamlom MM, Abdel-Wahab TI and Abdel-Wahab SI (2014) Allelopathic effects of sorghum and Sudan grass on some following winter field crops. *International Journal of Plant and Soil Science* 3: 599–622.

Tulayakul P, Sakuda S, Dong KS and Kumagai S (2005) Comparative activities of glutathione-S-transferase and dialdehyde reductase toward aflatoxin B1 in livers of experimental and farm animals. *Toxicon* 46(2): 204–209.

Unnerstad H, Romell A, Ericsson H, Danielsson-Tham ML and Tham W (2000) *Listeria monocytogenes* in faeces from clinically healthy dairy cows in Sweden. *Acta Veterinaria Scandinavica* 41(2): 167–171.

Upadhaya SD, Park MA and Ha JK (2010) Mycotoxins and their biotransformation in the rumen: A review. *Asian-Australasian Journal of Animal Sciences* 23(9): 1250–1260.

Van der Veen S, Moezelaar R, Abee T and Wells-Bennik MHJ (2008) The growth limits of a large number of *Listeria monocytogenes* strains at combinations of stresses show serotype- and niche specific traits. *Journal of Applied Microbiology* 105(5): 1246–1258.

Van Kessel JAS, Karns JS, Lombard JE and Kopral CA (2011) Prevalence of *Salmonella enterica, Listeria monocytogenes*, and *Escherichia coli* virulence factors in bulk tank milk and in-line filters from U.S. dairies. *Journal of Food Protection* 74(5): 759–768.

Vilar MJ, Yus E, Sanjuan ML, Dieguez JL and Rodriguez- Otero FJ (2007) Prevalence of and risk factors for *Listeria* species on dairy farms. *Journal of Dairy Science* 90(11): 5083–5088.

Vissers MMM, Driehuis F, Te Giffel MC, De Jong P and Lankveld JMG (2007) Minimizing the level of *Bacillus cereus* spores in farm tank milk. *Journal of Dairy Science* 90(7): 3286–3293.

Whitlow LW and Hagler WM (2005) Mycotoxins in dairy cattle: Occurrence, toxicity, prevention and treatment. *Proc. Southwest Nutr. Conf.* 124–138.

Wu Q, Jezkova A, Yuan Z, Pavlikova L, Dohnal V and Kuca K (2009) Biological degradation of aflatoxins. *Drug Metabolism Reviews* 41(1): 1–7.

Yiannikouris A and Jouany JP (2002) Mycotoxins in feeds and their fate in animals: A review. *Animal Research* 51(2): 81–99.

15 Quality Seed Production of Winter Fodder Crops

Imran Khan, Farhana Bibi, Faisal Mahmood, Muqarab Ali, M. Shahid Ibni Zamir, M. Umer Chattha, Muhammad Shakeel Hanif, Mohsin Nawaz, Momina Iqbal, Tahira Amjad, Sajid Hussain, M. Talha Aslam, Umair Ashraf, and Sajid Usman

DOI: 10.1201/9781003055365-15

15.1 INTRODUCTION

Fodder crops are usually grown for livestock feed. The farmer who owns a dairy farm or goat farm grows fodder crops for feed purposes and reduces his or her financial burden. Mixed fodder crops are recommended for better production, and these are the primary source of nutrients. There are four types of fodder crops: cereal fodder, grasses, legume fodder, and tree fodder (Singh et al., 2018; Bama and Babu, 2016).

Legumes and grasses also have purposes other than animal feeding, so these come under forage crops (Boller et al., 2010; Pirhofer-Walzl et al., 2012). Legumes are rich in protein and minerals, and grasses contain crude protein, crude fibers, and other minerals (Solati et al., 2018; Temba et al., 2016). Pakistan has 176.4 million domestic animals and is the third-largest livestock rearing country (Fazal Abbas et al., 2019). Since there is a negligible amount of livestock being farmed on a modern basis, most of the national livestock is being reared by small animal holders. The areas in the Indian subcontinent face severe fodder scarcity during two lean periods, June–July and December–January (Iqbal et al., 2015). The Indian subcontinent faces fodder scarcity due to a lack of quality seed production.

There are two main reasons for the lack of fodder seed availability:

- It is challenging to estimate crop-wise or variety-wise seed requirements due to the absence of reliable data on the crop-wise area under different fodder crops.
- Fodder seed production is highly unorganized because the production of food crop seeds is the main focus of large public-sector seed companies, and private-sector seed companies focus on high-value crops like vegetables, hybrids, and genetically modified (GM) crops. Moreover, only a few organized private companies are involved in the production of sorghum and Sudan grass hybrid.

During seed production, the atmosphere has significant effects on the performance of the seeds produced (Zinsmeister et al., 2020; MacGregor et al., 2015). In several plant species, factors such as the age and location of the mother plant may affect seed properties in the fruit, inflorescence, or foliage, often pursued by the dimorphic form of seeds or the fruits in which they are produced (Foroughi et al., 2014; Jha et al., 2010). Temperature is the dominant factor regulating seed properties at the parent plant (MacGregor et al., 2015).

This chapter will discuss how to increase the quality seed production of fodder crops like berseem, alfalfa, mustard, Persian clover, and Indian clover. We should adopt useful techniques to gain quality seed production of fodder crops and fill the gap.

15.2 QUALITY SEED PRODUCTION

Seed plays a vital role in the growth of any crop and enhancing and sustaining food production. Pure, genetically healthy, and dynamic seed is required for commercial seed production. Seed standards of germination, uniformity, genetic purity, freedom from seed-borne pathogens, and other prescribed parameters are established by good-quality seed (Rao et al., 2017). Fodder requires extra care and attention for its seed production because of its cross-pollinated nature. Prescribed standards, achieved by seed certification, are necessary to maintain and make seed available to the public on time. Depending on the stage of multiplication, the Association of Official Seed Certifying Agencies (AOSCA) divides the process of seed production into different classes:

15.2.1 Breeder Seed

Breeder seed is directly supervised by the originating or sponsoring plant-breeding institution or organization. The seed production center has to provide a certificate of good kind of seed in the case of breeder seed, and tags should be provided in case of other classes of seed.

15.2.2 Foundation Seed

It is progeny of breeder seed controlled to sustain specific genetic purity and varietal difference.

15.2.3 Registered Seed

Registered seed is the progeny of breeder seed or foundation seed controlled to maintain adequate genetic purity and varietal identity.

15.2.4 Certified Seed

The progeny of breeder, foundation, and registered seed handled to sustain satisfactory genetic purity and varietal difference.

15.3 CONSTRAINTS IN FORAGE SEED PRODUCTION

Feed shortages and the low quality of available feed are the significant constraints to increased livestock productivity throughout the globe. Many problems or constraints have been seen in forage seed production, such as little area availability because of traditional crop production, irrigation problems, and little or no rainfall.

Most cultivated fodder crops, range grasses, and pasture legumes have low seed production ability because of certain factors such as genetic, physiological,

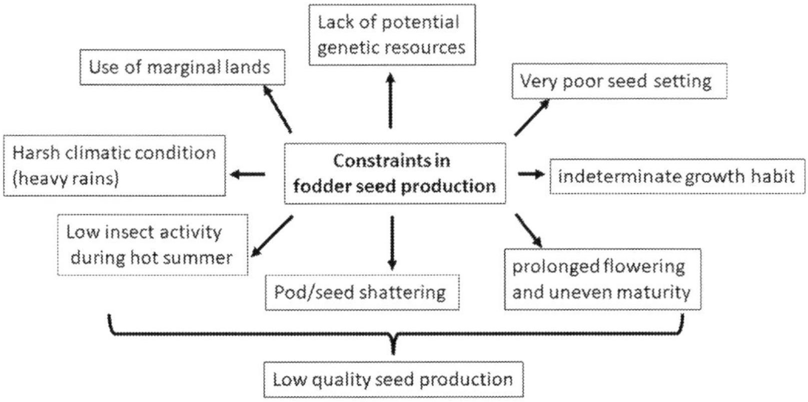

FIGURE 15.1 Reasons for low-quality seed production.

environmental, and organizational factors (Kharb and Dahiya, 2000; P. Singh and Singh, 2011). The constraints and problems associated with the seed production of fodder crops are shown in Figure 15.1.

15.3.1 Low Seed Production Potential

There are substantial differences in seed-producing ability among forages and grasses. The yield of seed varies from region to region and year to year; for example, the yield of sorghum forage ranged between 4 and 19 q/ha concerning locations and years in the coordination of seed multiplication trials.

15.3.2 The Case of Cultivated Fodder Crops

- Low seed setting and low seed production ability of most varieties because of continued emphasis on breeding for vegetative growth potential led them to become low seed producers.
- Many forage species are of indeterminate growth habit. The vegetative phases go simultaneously, resulting in a low seed setting.
- Inadequate movement of photosynthates from vegetative to reproductive structures.
- Non-synchronization of flowering, prolonged flower drops, and uneven maturity.
- Pod/seed shattering (e.g., teosinte) and lodging of the seed crop.
- Improper management of monetary and non-monetary inputs, especially in multi-cut forage crops.
- Low insect activity during hot summer months results in low seed productivity in entomophilous allogamous legumes species (alfalfa, berseem).
- Preferential use of low capability marginal lands and rainfed conditions for raising most forage crops.

15.3.3 The Case of Range Grasses and Pasture Legumes

Organized seed production in range grasses and legumes is somewhat more problematic than cultivated fodder crops due to the following reasons:

- They tend to be non-synchronized within populations (cowpea, sitaro, stylos) and within a panicle (*Cenchrus* spp.). Defoliation, adjusting irrigation, and fertilizer applications are essential to achieve synchronized flowering.
- Slow pod formation and maturity in pasture legumes.
- Prolonged head emergence and uneven maturity in range grasses.
- Many of the forage grasses are higher polyploids coupled with apomixes, for example, Cenchrus, Panicum, and Poa. Both obligate and facultative apomixis exist in grasses.
- Pod dehiscence, seed shedding, and shattering (siratro, stylos) cause heavy losses during harvesting and result in low-quality seeds.
- Poor sink formation. The realization of sink is only 10–20% in grasses as against 75% in cereals.
- A more significant proportion of blank seeds (about 30%) is also a problem of grasses that adversely affects the seed yield and reduces the germination by 20–30%.
- Lodging of seed crops adversely affects seed yield as well as seed quality.
- High cost involved in the collection of seeds. There is a common seed collection problem in most range grasses because of non-synchronized flowering, maturity, and severe seed shedding.
- Lack of information on management practices and seed quality.
- Protocols of seed certification and seed quality for the majority of native grasses have also so far not been worked out.

15.3.4 Less Availability of Extension Machinery

No single government agency takes the responsibility to improve the production technology of forage among the farming community.

15.3.5 Less Interest in Fodder Seed Production

The majority of farmers give the lowest priority to forage crops due to their overdependence on cash crops such as paddy, wheat, and cotton. They are not interested in fodder seed production; instead, they harvest crops before the seed set to feed the animals. This is because:

- The problem of low and uncertain market demand and unorganized marketing system of forage seeds.
- Seed setting is also erratic and low due to prevailing adverse weather conditions during flowering, seed feeling, and maturity.

- The improved varieties have not been reached due to non-availability of seed materials.
- The multiplication ratio for forage crops is very narrow, and sometimes it is not beneficial to producing agencies; therefore, they avoid producing seed.

15.4 SEED PRODUCTION OF MAJOR FODDER CROPS

15.4.1 Berseem

15.4.1.1 Introduction

Berseem is considered an important fodder and forage legume. It originated in Syria and was then introduced in Egypt during the 6th century (Hannaway et al., 2005). It was introduced in Sindh in 1904, where it was well adapted to the climate of irrigated tracts of the sub-continent and rapidly spread throughout India. In 1924, it was cultivated in the Peshawar region; then, it moved toward Punjab's irrigated tracts. Berseem is now a widely spread clover in South African irrigated planes (Oushy, 2008). It is a significant fodder crop, as it can supply nutrient-enriched green fodder in abundant quantity compared with Indian and Persian clover. It is a multi-cut fodder crop and gives green fodder for a long time (November to April). Berseem has high digestibility and palatability, which help to increase milk production.

15.4.1.2 Climatic and Soil Requirements

A cool climate but not below 20°C is good for its optimum growth. It is mostly cultivated in canal-irrigated areas. Berseem grows best on well-drained medium to heavy loam soils and requires frequent irrigation. It can tolerate salinity to some extent and can be used for reclamation of saline soils.

15.4.1.3 Selection of Seed Plot

The seed production of berseem crops requires leveled and approachable land free from weeds and volunteer plants. Soil should be well drained and must have adequate irrigation facilities. Seed should not be grown on the same land on which the same crop variety was grown in the preceding year unless the same variety is confirmed by seed certification standards.

15.4.1.4 Source of Seed

Seed plays a vital role in enhancing and sustain food production. Genetically pure and true type seed should be used for seed production purposes. A healthy, dynamic, and genetically pure seed is required for commercial seed production. It is better to get seed from government institutes along with the relevant documentation.

15.4.1.5 Land Preparation

Sowing is done on adequately leveled land, which must be free from weeds. The seedbed is prepared with three plowings, each followed by planking. Early or late sowing results in inadequate seed production.

15.4.1.6 Time of Sowing

The berseem crop is sown from the last week of September to the first week of October for fodder and seed production. It can also be sown at the end of November or January for seed production. It can provide two to three cuttings of green fodder before leaving the crop for seed production.

15.4.1.7 Seed Rate and Method of Sowing

A seed rate of 25 kg/ha is recommended for berseem fodder, and the seed should be broadcast in standing water. Seed should be inoculated with *Rhizobium trifoli* when it is sown on soil under rice cultivation and berseem has not been sown for the last three years. Seed should be mixed with an equal quantity of soil to ensure uniform distribution. To get a kasni (*Cichorium intybus*)-free seed, the fodder seed should be socked in a 10% solution of common salt (NaCl), then sieve out the floating kasni (*Cichorium intybus*) seed. One 50 kg DAP is enough to meet its fertilizer requirement; because it is a leguminous crop, it can fix atmospheric N and improve soil fertility.

15.4.1.8 Isolation

Isolation is essential for quality seed production to prevent contamination by cross-pollination with other varieties and to mix seed during harvesting. It is essential to maintain a proper distance from other varieties of the same crop to get genetically pure seed.

15.4.1.9 Fertilizer Application

Along with 50 kg/ha of phosphorus, 15 ton/ha farmyard manure is applied at sowing time. Nitrogen and phosphorus at the rate of 28 kg/ha and 74 kg/ha, respectively, are applied, where farmyard manure is not applied.

15.4.1.10 Weed Control

Trifolium resupinatum and *Cichorium intybus* weeds should be removed entirely from the seed crop. To avoid other weeds, delayed sowing (in the second week of October) is preferred because a fall in temperature reduces weeds drastically. Fluchloralin at 0.45 kg/ha or oxyfluorfen at 0.1 kg/ha followed by imazethapyr at 0.075 kg/ha appeared more useful for effective weed control in berseem.

15.4.1.11 Flowering and Pollination for Seed Production

The berseem seed yield depends on the last cut for green fodder and then leaving it for seed purpose. It varies according to the type of variety, soil, and climate. To get higher yield (starting from flowering initiation) two sprays of 2% potassium nitrate are applied at weekly intervals. Two sprays of salicylic acid 7.5 g in 100 L of water per acre can also be applied.

15.4.1.12 Monitoring and Inspection

At least two inspections for genetic seed purity should be done from crop approaches to flowering to harvesting. The first inspection should be done at the time of flowering to remove off-types based on leaf shape, size, and color. The second inspection should be after flowering nearing harvest in order to remove all weeds.

15.4.1.13 Harvesting

The first cutting of fodder is done after 50 days of sowing, and subsequent cutting can be done in winter at 40 days and 30 days in the spring season. Thus, it can give four to six cuttings, and the last cut is to be taken on or before 15th March, and the crop is then left for seed production. Crop for seed purposes is matured the last week of May and the start of June.

15.4.1.14 Storage

Seed should be store in damp-proof stores to avoid any type of deterioration or contamination. The walls of the store should be whitewashed. Seed should be of true type seed, genetically pure, and viable and should give the farmer maximum yield to enhance his production and, ultimately, economic status. Seed production should be done by following all the seed production standards.

15.4.1.15 Essential Tips for the Seed Production of Forage Crops

BERSEEM

- For seed production, berseem can be sown at the end of November after the harvesting of basmati rice. It can provide three cuttings of green fodder and then be left for seed production.
- Fields must be isolated from contaminants (other varieties and from the same variety).
- Land should free from volunteer plants.
- After three cuttings of green fodder, leave the crop for seed production.
- Apply restricted irrigation, but irrigation at the flowering and seed setting stage is critical.
- Shaftal, kashni, and other weeds should be altogether removed from the seed crop. There should be no objectionable weed plant like chicory (kasni) *Chicorium intybus* L. in the case of foundation seed; they can be permitted up to 0.05% in the case of certified seed.

15.4.2 Alfalfa

15.4.2.1 Introduction

Alfalfa (lucerne) is a perennial, multi-cut, and nitrogen-fixing crop that can be grown in dry environments. It is also called the 'queen' of fodders. Alfalfa is considered a vital forage crop worldwide due to its highly nutritious, digestible, palatable, and nitrogen-fixing nature. It also serves as a source of proteins and vitamins in livestock feed. Alfalfa is grown mainly for fodder and forage purposes. Seed yield importance is considered secondary. Farmers usually face a good-quality lucerne seed shortage due to the unavailability of quality seed production. The main reason behind low-quality seed production is the farmers' interest in getting more cutting for fodder production. They leave the crop in April to produce seed, which results in poor quality and low seed production. Farmers are advised to adopt best management practices to get the best seed quality of lucerne.

15.4.2.2 Climatic and Soil Conditions

The regions have sunny and clear summer days with little or no rainfall at the flowering stage, much favoring lucerne seed yield. It performs best in areas with rainfall between 30 and 40 inches and altitude up to 9800 feet. These climatic conditions play a vital role in the pollinating activity of bees. For quality seed production, well-drained and loamy soils are most suitable. Alfalfa seed varieties can also be grown on sandy or clay loam soils.

15.4.2.3 Selection of Seed Plot

Alfalfa requires well-leveled and weed-free land for quality seed production. Alfalfa cannot tolerate waterlogging, so the plot should have good drainage ability to give the best results. There should no lucerne volunteer plants grown previously.

15.4.2.4 Source of Seed

For better seed production, seed should be genetically pure, vigorous, dynamic, and true type. It's better to get seed from any government institute, along with the relevant documents.

15.4.2.5 Land Preparation

For the ideal seedbed preparation, three to four shallow plowings followed by planking should be done. Seed should not be placed too deep in the soil for proper germination. Seed should be sown in lines.

15.4.2.6 Sowing Time

October–November is the best sowing time for good-quality seed production. Early or delayed sowing results in low quality and less quantity of seed.

15.4.2.7 Seed Rate

The seed rate varies according to the area. In New Zeeland, 6–12 kg seed per hectare is being practiced. Experiments were conducted to check the seed rate, and 1 kg per hectare is recommended as the best seed rate for higher seed yield potential. However, in Pakistan, 10–12 kg seed per hectare is also being practiced for this purpose. Before sowing, the seed should be treated with fungicide to avoid diseases affecting plant density and uniformity. Seed purity and seed germination tests should be conducted to get optimum plant density. The seed should be treated with strains of *Rhizobium meliloti* of rhizobium culture 200 g/10 kg for best nitrogen use. Rhizobium bacteria die in excess heat, so seeds should be sown in the evening.

15.4.2.8 Isolation

The fields where seed varieties are grown should be separated from those grown for fodder purposes. Field isolation is also recommended for varieties that do not confirm to purity requirements: at least 400 m for foundation and 100 m for certified seed. This field isolation is necessary for good-quality seed production of alfalfa.

15.4.2.9 Fertilizer Requirement

Twenty to 25 kg N, 50–60 kg P_2O_5 and 40–45 kg K_2O per hectare are recommended for seed production. All these fertilizers should be applied at the time of seed sowing.

15.4.2.10 Irrigation

For the whole cropping period, irrigation should be applied when needed, but irrigation must be applied at the pod filling stage.

15.4.2.11 Harvesting

Harvest the crop after 70–75 days of sowing. For fodder, purpose cuttings should be taken at an interval of two to three weeks. For seed purposes, the last cutting should be done in the last week of February to the middle of March. Almost 300–375 kg seed per hectare can be obtained by taking the last cutting in March. However, most farmers take the last cuttings in April, which has adverse effects on seed yield.

15.5 MINOR FODDER CROPS

15.5.1 Mustard

15.5.1.1 Introduction

Mustard belongs to the Brassicaceae family. Mustard was commonly grown by the Indus civilization of 2500–1700 BCE. In west Asia and Europe, wild forms of mustard are found (Zohary et al., 2012). Oriental mustard originated in the Himalayas, and now it is commonly grown in India, Canada, the United States, United Kingdom, and Denmark. In Argentina, the United States, and Chile, black mustard is grown. In 2010, 57% of the world's production was from Nepal and Canada (FAOSTAT, 2012). In Europe, white mustard is used as a cover crop. About 700 million lbs of mustard are consumed annually.

Mustard is extensively used as fodder, green manure, and cover crops in temperate agriculture (Vaughan and Hemingway, 1959). The protein contents of mustard in total dry matter range from 12–16%, and total digestible content is 55–60% at the harvesting stage. Crude proteins and energy levels are higher at the early stages.

15.5.1.2 Climate and Soil

For vegetative and reproductive growth, mustard requires cool temperatures, and it is well adapted in temperate regions. Up to flowering, it grows best under relatively cool temperatures. More heat can cause a reduction in yield, seed size, and oil contents, but after flowering, it can tolerate high temperatures. The short growth cycle is 70 days, and the long growth is almost 380 days. Germination requires a low soil temperature of about 40°F. Waterlogged soils can also hamper the growth of mustard. The optimum pH of soil for mustard is 6.0–8.0. It performs best at neutral pH (Hannaway et al., 2005).

15.5.1.3 Seedbed Preparation

Mustard requires a firm seedbed with adequate moisture content because this type of soil allows rapid and uniform seed germination. The seedbed should be reasonably level and free of weeds and crop residues. Shallow tillage deep enough to kill weeds should keep soil moist.

15.5.1.4 Plot Isolation

The plot where crops will be grown for quality seed production should be isolated from other fields. This field should not contain weed plants or other volunteer plants from the same or different varieties.

15.5.1.5 Sowing Time and Seed Rate

It is a winter crop grown from mid-September to October–November. Neither early nor too late sowing is recommended where the crop is grown for quality seed production. For yellow mustard, 20 to 30 lbs/ha seed are required, and the white mustard seed rate is 12 to 17 lbs/ha. A higher seed rate is used in heavy fertile soils where seedling emergence is difficult. Seed should be shallowly planted at 0.5- to 1-inch depth. In dry soil conditions, the depth should be increased.

15.5.1.6 Row to Row Spacing

To produce the best quality seed, the row to row spacing in case of line sowing should be 30–45 cm, and plant to plant should be 3–4 cm (Pande et al., 1998).

15.5.1.7 Fertilizer Requirement

FYM should be applied (75–120 qtl) to the field at land preparation time. The N:P ratio for mustard should be 62.5:20 kg per hectare, which can be obtained by applying urea and superphosphate. Commonly recommended NPK in Pakistani soils is 225:150:125 kg per hectare.

15.5.1.8 Irrigation

Two irrigations (30–35 days after sowing and again 55 to 60 days after sowing) are needed. The water requirement is 300–400 mm per year.

15.5.1.9 Weed and Disease Control

Weeds can lower the crop yield. It is challenging to remove weed seeds, so they can lower market grades. Shallow tillage can also reduce the weed seeds. Roundup (glyphosate) is applied to control Canada thistle. For grasses and broadleaf weeds, Trifluralin is recommended. Mustard is insensitive to 4-chloro-2-methly phenoxy acetic acid, so these are not recommended and should be avoided. Mustard is vulnerable to several diseases like downy mildew, mosaic virus, white rust, and leaf spots. These diseases lead to 10–75% crop losses (Madhusoodanan and Bogunovic, 2004). Cultural practices such as crop rotation and seed treatment are most useful to control diseases.

15.5.1.10 Harvesting

Mustard matures early and has high yield potential. Harvesting should be done when all seeds turn black and moisture content becomes less than 15%. These parameters lead to better seed quality and seed potential. The plant should be cut beneath the height of the lowest seed pot. A wheat-mustard thresher has better spike tooth drum and blower, reducing the visible and mechanical damage (Shahbazi, 2011).

15.5.1.11 Drying and Storage

At 10% moisture content, seeds can be stored safely. Drying equipment used for corn and mustard requires little modification when used for mustard drying. To prevent the loss of smaller seeds, a fine screen is needed. Air temperature for seed drying should not exceed 150°F, and seed temperature should stay below 120°F.

15.5.2 Persian Clover

15.5.2.1 Introduction

Persian clover (*Trifolium resupinatum*) is also known as shaftal clover and belongs to the Leguminosae family. It is known as an annual, semi-erect branched legume and is similar to berseem but shorter. It is highly palatable and nutritive fodder. Persian clover is considered an areal winter annual crop. The germination of the seed occurs in fall, and the plant grows throughout the winter. The flower stems develop in spring, and seeds are produced in late spring or early summer, after which plants die.

15.5.2.2 Climatic and Soil Conditions

Persian clover (*T. resupinatum* L.) can germinate in various types of soils and climatic conditions, but well-drained and loamy soils are considered best. Light to heavy loam soils with irrigation facilities are best. Cool and moist weather is suitable for better growth

15.5.2.3 Seedbed Preparation

Three to four cultivations are required, each followed by planking to get a fine seedbed for good germination.

15.5.2.4 Method of Sowing and Seed Rate

The recommended seed rate for the broadcasting method in standing water is 12–16 kg/ha.

15.5.2.5 Sowing Time

The best planting time suggested for Persian clover is the whole month of October to mid-November.

15.5.2.6 Fertilizer

This crop does not require fertilizer, but two and half bags of di-ammonium phosphate (DAP) per hectare increase seed yield.

15.5.2.7 Irrigation

The first irrigation should be applied within eight to ten days of planting, and the remaining irrigations should be applied at an interval of 15–20 days.

15.5.2.8 Interculture and Weeding

This crop is usually sown by the broadcasting method, and due to this sowing method, there is no need to have an interculture practice. Kasni (*Cichorium intybus*) can reduce fodder yield and even destroy the quality of seed; therefore, it should rogue out from the field earlier or at planting time.

15.5.2.9 Pests and Diseases

This crop usually remains free from pests and diseases.

15.5.2.10 Harvesting

The best time to harvest Persian clover is when many capsules appear light brown. A mower with lifter guards is used for cuttings; in the case of lodging, use a heavy short weed bar without guards. Curing in windrows is suggested, and the rolling of heads in the windrow will reduce shattering. A grain combine can thresh the crop with a pick-up attachment.

15.5.2.11 Seed Production

Persian clover is suitable for seed production. In April–May, the seeds ripen under ideal conditions and give sufficient seed yield from 750 to 1000 kg per hectare. Its flowers are self-fertile and self-pollinating, which helps in seed setting even under unfavorable weather conditions; honeybees help for nectar and pollen, which can help in increasing seed production. Heavy rains at seed maturity are the reason for seed loss because the seed capsules break off easily from the heads.

15.5.3 Indian Clover

15.5.3.1 Introduction

It belongs to the family Leguminosae and is locally known as senji (Xiong et al., 2020). There are two types of Indian clover/senji, as local senji and desi senji. Desi senji originated in Indo-Pak. It is an annual legume forage crop. Local senji originated in Bukhara. It is considered a crop of Asia and Europe's temperate zone, from where it moved toward Argentina and southern Australia. Its cultivation was started in the United States, giving the highest green fodder return of all clovers. It is not widely grown in Pakistan because animals do not like it due to the presence of coumarin, which is a bitter substance.

It can be sown in moist areas of Peshawar and dry areas of Sindh and Baluchistan due to its adaptability. It can be sown as a relay crop in standing cotton and maize crops, and it can be a source of food for animals in a lean period. It can also be grown in regions suffering from water shortage because it requires less water than berseem and alfalfa. It contains crude protein 16.57%, and other digestible nutrients present in it are 60%.

15.5.3.2 Climatic and Soil Conditions

It has a wide range of climatic and soil adaptability. It is usually grown in subcontinental Punjab areas, but it can also be cultivated in cooler rainfed areas. It requires loam to heavy soils for its propagation in dry and cool climates. On the other hand, it is also fodder of hilly areas, plains, sandy, and average soil types.

15.5.3.3 Seedbed Preparation

One to two plowings followed by planking is enough. It does not require too much preparation before sowing. Seed should be treated with tetramethylthiuram disulfide @ 2g/kg seed.

15.5.3.4 Sowing Time

September to November is the optimum sowing time to get more good-quality seed. The best time for the planting of clover as a relay crop in standing cotton is the second to third week of September, and in November, it can be sown as a catch crop.

15.5.3.5 Isolation

Isolation is necessary to prevent contamination by cross-pollination with other varieties and mix of seeds during harvesting. It is important to get pure seed (free from impurities).

15.5.3.6 Fertilizer

Fifty kg DAP (containing 46% phosphorus and 18% nitrogen) is enough to meet its fertilizer requirements. Apply NPK with 30:40:20 kg/ha as base dose and 30 kg N per hectare as a top dressing after 30 days of sowing. There is no severe problem of weed propagation during the initial phase of crop growth.

15.5.3.7 Pests and Their Control

Weevil is an essential pest of Indian clover, and dimethyl dithiophosphate 50 EC@400 ml/100–150 L of water per acre can help to control it.

15.5.3.8 Harvesting

Do not take cuttings from crops grown for seed purposes to get maximum seed yield

15.5.4 Sweet Clover

Two known sweet clover species are *Melilotus albus* (sweet white clover) and *Melilotus officinalis* (sweet yellow clover). *Melilotus* spp. can be annual or biennial in growth habit and belongs to the family Fabaceae. It is challenging to differentiate yellow and white sweet clover in their vegetative state (Jepson and Hickman, 1993); they are only distinguished by flower color. *Melilotus albus* is originated in Eurasia and was then introduced into America, South Africa, New Zealand, and Australia. It was introduced as a forage crop in North America in 1664 and brought to Alaska in 1913 as a nitrogen-fixing and forage crop (Conn et al., 2011). The Hawaiian Sugar Growers Association planted white clover before 1920 (Wagner et al., 1999). *Melilotus officinalis* is native to Europe and Asia and was introduced as a forage crop in North America. Both species are common in the upper Midwest and plains regions of the United States. In the United States, it gives the highest green fodder return of all clovers. In Pakistan, it is not extensively planted because of its bitter taste and coumarin, which is not liked by animals.

15.5.4.1 Climatic and Soil Condition

Sweet clover is an aggressive crop; due to this reason, it can tolerate a wide range of climatic and soil conditions, but moisture is essential for seedling establishment. It can be successfully grown on alkaline or saline soils.

15.5.4.2 Land Preparation and Fertilizers

Sweet clover is sensitive to different herbicides used to overcome the broad-leaved weeds. Therefore, cultural practices before sowing can be helpful to minimize weed competition. Its nitrogen requirements can be fulfilled by nitrogen fixation because it is a legume crop, and basal application of DAP is enough to fulfill its remaining nutrient requirements. Soil testing is required to check the pH of soil because sweet clover is sensitive to acidic pH. Soil pH has tremendous importance in fertilization, and it should be brought to 6.5 by the liming process.

15.5.4.3 Sowing Time and Seed Rate

The average seed rate is 15 kg per hectare for closely spaced stands. However, the best seed production is obtained from the thin stands. Sweet clover should be sown early in the spring to obtained favorable moisture content for seedling establishment.

15.5.4.4 Isolation

The flower structure of sweet clover stimulates cross-pollination by insects, and different bees are common sweet clover pollinators. One to two colonies of bees per acre are considered appropriate for adequate seed production. When different insects sit on the flower petals of flowers, stigma and anthers bend and contact insects (Bare, 1979). As a cross-pollinated crop, the sweet clover seed crop requires proper isolation like other cross-pollinated crops. An interval of two to three years from other seed crops and five years from sweet clover are required.

15.5.4.5 Irrigation and Rouging

It requires relatively less water as compared to berseem and alfalfa. Therefore, three to four irrigations are enough for its whole growing period. The first irrigation is recommended after ten days of sowing and the next ones at intervals of 20–25 days. Rouging is required to get more seed purity because not more than one off-type plant is allowed in 10 cm^2. The period for this practice should be between the flowering and harvesting stages.

15.5.4.6 Seed Production

Yellow sweet clover generally produces less seed than white clover (Turkington et al., 1978). Seed production of sweet clover is reduced if plants are damaged or planted on infertile soils and due to unfavorable weather conditions that disturb insect visitation.

15.5.4.7 Harvesting

Sweet clover has a somewhat indeterminate growth habit, and pods have a loose attachment to the stem; this results in the loss of ripe pods. Windrowing of the crop when 50–60% of pods have turned brown helps get a high yield of good-quality seed. After few days of curing, the windrow should be collected and threshed carefully to avoid broken seeds. Harvesting should be done early in the morning to reduce the risk of shedding.

15.5.4.8 Insect Pest Management

Weevils and aphids damage the crop, and to control this problem, spraying the seed crop with Carbaryl at a rate of ½ kg per acre in 300 liters of water is recommended. Weevils damage the leaves by chewing, and this damage is more severe when growth is slow.

15.5.4.9 Drying and Seed Treatment

If harvesting is done by windrowing, it usually contains green vegetative material and must be dried and cleaned. Scarification is needed to remove the hulls in combined harvesting because it leaves a high quantity of unhulled seed. It is also helpful to reduce the challenging seed content.

15.6 STORAGE

Up to seed production, we have completed one step of quality seed production. The second step is processing, which includes drying, cleaning, grading, and storage. Storage is essential to preserve seed of the highest quality from one planting season to another. Different operations should done before storage, such as the following.

Seed drying refers to reducing the moisture to safer levels for sustaining seed vigor and viability by minimizing the attack of different pests and pathogens. High moisture leads to mold development, and seeds deteriorate. The moisture content of seed is important during storage because it influences the longevity of seed. High moisture content leads to respiration, biological heating, and mold growth. Therefore, it is crucial to reduce seed moisture content to a safe level for storage. For safer storage, the moisture level should be less than 12%.

After drying, the next step is cleaning and grading of the seed. After cleaning and grading, the seed should be treated with appropriate fungicide to avoid any seed or soil-borne disease during storage and also during the next growing season. The seed should also be packed in appropriate bags or containers. The choice of packing material depends upon the local weather conditions to avoid any damage during storage and also during transportation and marketing. The bags or packing material (bags, containers, etc.) is also labeled with essential information like crop, variety, class of seed, and name and address of the producer, and the tags according to the class of the seed should also contain information like physical purity, genetic purity, and moisture and germination of percentage of the seed.

In seed stores, fumigation is often applied as part of management to avoid the attack of any pest during storage time. It is essential to use correct chemicals on the particular seed lot at the correct dosage. Different fumigants are used during storage such as ethylene dibromide and aluminum phosphide. The essential steps before, during, and after storage are given in the following.

CONSIDERATIONS FOR STOREROOM

- Clean the storeroom before seed storage.
- The walls of the stores should be whitewashed.
- Spray the stores and their vicinity with Cypermethrin.
- After spraying, stores should be closed for four to six hours.
- After staking of seed bags in stores, fumigation should done with appropriate fumigant (aluminum phosphide or ethylene dibromide).
- Ensure the store is airtight after fumigation.
- Protective measures like gloves, clothing, and face masks should be used during fumigation, and after treatment all handling material should be disposed of properly.
- Repeat the fumigation process depending upon the time of storage and also the local weather conditions.

REFERENCES

Bama KS and Babu C (2016) Perennial forages as a tool for sequestering atmospheric carbon by best management practices for better soil quality and environmental safety. *Forage Research* 42(3): 149–157.

Bare JE (1979) *Wildflowers and Weeds of Kansas.*

Boller B, Posselt UK, Veronesi F, Boller B, Posselt UK and Veronesi F (2010) *Fodder Crops and Amenity Grasses.* Springer.

Conn JS, Werdin-Pfisterer NR, Beattie KL and Densmore RV (2011) Ecology of invasive *Melilotus albus* on Alaskan glacial river floodplains. *Arctic, Antarctic, and Alpine Research* 43(3): 343–354.

FAOSTAT Countries by Commodity (2012) *UN Food and Agriculture Organization.* Retrieved 08–May–2012. www.fao.org/faostat/en/#data.

Fazal Abbas MI, Akhtar Z, Mehmood K, Hyder MZ and Iqbal U (2019) A study on the correlation of serum electrolytes and trace elements along with associated risk factors in diarrheic buffalo and cattle calves. *Pakistan Journal of Zoo*logy 51(3): 1–4.

Foroughi A, Gherekhloo J and Ghaderi-Far F (2014) Effect of plant density and seed position on mother plant on physiological characteristic of cocklebur (*Xanthium strumarium*) seeds. *Planta Daninha* 32(1): 61–68.

Hannaway DB, Daly C, Cao W, Luo W, Wei Y-R, Zhang W, . . . Li X-L (2005) Forage species suitability mapping for China using topographic, climatic and soils spatial data and quantitative plant tolerances. *Agricultural Sciences in China* 4(9): 660.

Iqbal MA, Ahamd B, Shah MH and Ali K (2015) A study on forage sorghum (*Sorghum bicolor* L.) production in perspectives of white revolution in Punjab, Pakistan: Issues and future options. *American-Eurasian Journal* of *Agriculture and Environmental Science* 15(4): 640–647.

Jepson WL and Hickman JC (1993) *The Jepson Manual: Higher Plants of California.* Univ of California Press.

Jha P, Norsworthy JK, Riley MB and Bridges W (2010) Shade and plant location effects on germination and hormone content of Palmer amaranth (*Amaranthus palmeri*) seed. *Weed Science* 58(1): 16–21.

Kharb RPS and Dahiya BS (2000) Influence of natural ageing of seeds on field performance in pigeonpea (*Cajanus cajan* L. Millsp.). *Seed Research* 28(2): 149–152.

MacGregor DR, Kendall SL, Florance H, Fedi F, Moore K, Paszkiewicz K, . . . Penfield S (2015) Seed production temperature regulation of primary dormancy occurs through control of seed coat phenylpropanoid metabolism. *New Phytologist* 205(2): 642–652.

Madhusoodanan S and Bogunovic OJ (2004) Safety of benzodiazepines in the geriatric population. *Expert Opinion on Drug Safety* 3(5): 485–493.

Oushy H (2008) Fact Sheet: Egyptian Clover (4): 1–19.

Pande S, Bakr MA and Johansen C (1998) *Recent Advances in Research and Management of Botrytis Gray Mold of Chickpea.* International Crops Research Institute for the Semi-Arid Tropics.

Pirhofer-Walzl K, Rasmussen J, Høgh-Jensen H, Eriksen J, Søegaard K and Rasmussen J (2012) Nitrogen transfer from forage legumes to nine neighbouring plants in a multi-species grassland. *Plant and Soil* 350(1): 71–84.

Rao NK, Dulloo ME and Engels JMM (2017) A review of factors that influence the production of quality seed for long-term conservation in genebanks. *Genetic Resources and Crop Evolution* 64(5): 1061–1074.

Shahbazi F (2011) Impact damage to chickpea seeds as affected by moisture content and impact velocity. *Applied Engineering in Agriculture* 27(5): 771–775.

Singh P and Singh K (2011) Analysis of association among different morphological traits in fodder barley. *Range Management and Agroforestry* 32(2): 92–95.

Singh T, Ramakrishnan S, Mahanta SK, Tyagi VC and Roy AK (2018) Tropical forage legumes in India: Status and scope for sustaining livestock production. *Forage Groups.*

Solati Z, Manevski K, Jørgensen U, Labouriau R, Shahbazi S and Lærke PE (2018) Crude protein yield and theoretical extractable true protein of potential biorefinery feedstocks. *Industrial Crops and Products* 115: 214–226.

Temba MC, Njobeh PB, Adebo OA, Olugbile AO and Kayitesi E (2016) The role of compositing cereals with legumes to alleviate protein energy malnutrition in Africa. *International Journal of Food Science & Technology* 51(3): 543–554.

Turkington RA, Cavers PB and Rempel E (1978) The Biology of Canadian Weeds.: 29. Melilotus alba Desr. and M. officinalis (L.) Lam. *Canadian Journal of Plant Science* 58(2): 523–537.

Vaughan JG and Hemingway JS (1959) The utilization of mustards. *Economic Botany* 13(3): 196–204.

Wagner WL, Bruegmann M, Herbst DR and Lau JQ (1999) Hawaiian vascular plants at risk: 1999. *Bishop Museum Occasional Papers.*

Xiong Y, Xiong Y, He J, Yu Q, Zhao J, Lei X, . . . Zhang X (2020) The complete chloroplast genome of two important annual clover species, *Trifolium alexandrinum* and T. resupinatum: Genome structure, comparative analyses and phylogenetic relationships with relatives in Leguminosae. *Plants* 9(4): 478.

Zinsmeister J, Leprince O and Buitink J (2020) Molecular and environmental factors regulating seed longevity. *Biochemical Journal* 477(2): 305–323.

Zohary D, Hopf M and Weiss E (2012) *Domestication of Plants in the Old World: The Origin and Spread of Domesticated Plants in Southwest Asia, Europe, and the Mediterranean Basin.* Oxford University Press on Demand.

Index

Note: numbers in **bold** indicate a table. Numbers in *italics* indicate a figure on the corresponding page.

Printed in the United States
by Baker & Taylor Publisher Services